VOLUME NINETY SEVEN

SEMICONDUCTORS AND SEMIMETALS

Semiconductors for Photocatalysis

SERIES EDITOR

CHENNUPATI JAGADISH

Distinguished Professor
Department of Electronic Materials Engineering
Research School of Physics and Engineering
Australian National University
Canberra, ACT2601, Australia

VOLUME NINETY SEVEN

SEMICONDUCTORS AND SEMIMETALS

Semiconductors for Photocatalysis

Edited by

ZETIAN MI
University of Michigan,
Ann Arbor, MI, United States

LIANZHOU WANG
Nanomaterials Centre,
School of Chemical Engineering and Australian Institute for Bioengineering and Nanotechnology,
The University of Queensland,
Brisbane, QLD, Australia

CHENNUPATI JAGADISH
Research School of Physics and Engineering,
Australian National University,
Canberra, ACT, Australia

ELSEVIER

ACADEMIC PRESS
An imprint of Elsevier

Academic Press is an imprint of Elsevier
50 Hampshire Street, 5th Floor, Cambridge, MA 02139, United States
525 B Street, Suite 1800, San Diego, CA 92101-4495, United States
The Boulevard, Langford Lane, Kidlington, Oxford OX5 1GB, United Kingdom
125 London Wall, London, EC2Y 5AS, United Kingdom

First edition 2017

ISBN: 978-0-12-811727-9
ISSN: 0080-8784

Publisher: Zoe Kruze
Acquisition Editor: Poppy Garraway
Editorial Project Manager: Shellie Bryant
Production Project Manager: Vignesh Tamil
Cover Designer: Greg Harris

Typeset by SPi Global, India

CONTENTS

CONTRIBUTORS

Xiaoxia Chang
Key Laboratory for Green Chemical Technology of Ministry of Education, School of Chemical Engineering and Technology, Tianjin, University; Collaborative Innovation Center of Chemical Science and Engineering, Tianjin, China. (ch12)

Xiaobo Chen
University of Missouri, Kansas City, MO, United States. (ch11)

Sheng Chu
McGill University, Montreal, QC, Canada. (ch6)

James R. Durrant
Imperial College London, London, United Kingdom. (ch1)

Miharu Eguchi
Eelectronic Functional Materials Group, Polymer Materials Unit, National Institute for Materials Science, Tsukuba, Japan. (ch7)

Thomas Fischer
Institute of Inorganic Chemistry, University of Cologne, Cologne, Germany. (ch5)

Katsushi Fujii
Institute of Environmental Science and Technology, The University of Kitakyushu, Kitakyushu, Fukuoka; RIKEN Center for Advanced Photonics, Wako, Saitama, Japan. (ch4)

Robert Godin
Imperial College London, London, United Kingdom. (ch1)

Jinlong Gong
Key Laboratory for Green Chemical Technology of Ministry of Education, School of Chemical Engineering and Technology, Tianjin, University; Collaborative Innovation Center of Chemical Science and Engineering, Tianjin, China. (ch12)

Yakup Gönüllü
Institute of Inorganic Chemistry, University of Cologne, Cologne, Germany. (ch5)

Hong Guo
McGill University, Montreal, QC, Canada. (ch6)

Shu Hu
School of Engineering and Applied Sciences, Yale University, New Haven, CT, United States. (ch3)

Chennupati Jagadish
Department of Electronic Materials Engineering, Research School of Physics and Engineering, The Australian National University, Canberra, ACT, Australia. (ch8)

Andreas Kafizas
Imperial College London, London, United Kingdom. (ch1)

Siva Krishna Karuturi
Department of Electronic Materials Engineering, Research School of Physics and Engineering, The Australian National University, Canberra, ACT, Australia. (ch8)

Xianghua Kong
McGill University, Montreal, QC, Canada. (ch6)

Jennifer Leduc
Institute of Inorganic Chemistry, University of Cologne, Cologne, Germany. (ch5)

Gang Liu
Shenyang National Laboratory for Materials Science, Institute of Metal Research, Chinese Academy of Sciences, Shenyang; School of Materials Science and Engineering, University of Science and Technology of China, Hefei, China. (ch10)

Yan Liu
University of Missouri, Kansas City, MO, United States; College of Environment, Sichuan Agricultural University, Chengdu, China. (ch11)

Kazuhiko Maeda
School of Science, Tokyo Institute of Technology, Tokyo, Japan. (ch7)

Sanjay Mathur
Institute of Inorganic Chemistry, University of Cologne, Cologne, Germany. (ch5)

Zetian Mi
McGill University, Montreal, QC, Canada; University of Michigan, Ann Arbor, MI, United States. (ch6)

Takayoshi Oshima
School of Science, Tokyo Institute of Technology, Tokyo, Japan. (ch7)

Jian Pan
Curtin University, Perth, WA, Australia. (ch10)

Zhenhua Pan
School of Engineering and Applied Sciences, Yale University, New Haven, CT, United States. (ch3)

Aida Raauf
Institute of Inorganic Chemistry, University of Cologne, Cologne, Germany. (ch5)

Georges Siddiqi
School of Engineering and Applied Sciences, Yale University, New Haven, CT, United States. (ch3)

Hark Hoe Tan
Department of Electronic Materials Engineering, Research School of Physics and Engineering, The Australian National University, Canberra, ACT, Australia. (ch8)

Srinivas Vanka
McGill University, Montreal, QC, Canada. (ch6)

Dunwei Wang
Boston College, Merkert Chemistry Center, Chestnut Hill, MA, United States. (ch2)

Lianzhou Wang
Nanomaterials Centre, School of Chemical Engineering and Australian Institute for Bioengineering and Nanotechnology, The University of Queensland, Brisbane, QLD, Australia. (ch9)

Songcan Wang
Nanomaterials Centre, School of Chemical Engineering and Australian Institute for Bioengineering and Nanotechnology, The University of Queensland, Brisbane, QLD, Australia. (ch9)

Tuo Wang
Key Laboratory for Green Chemical Technology of Ministry of Education, School of Chemical Engineering and Technology, Tianjin, University; Collaborative Innovation Center of Chemical Science and Engineering, Tianjin, China. (ch12)

Xiaogang Yang
Key Laboratory of Micro-Nano Materials for Energy Storage and Conversion of Henan, Xuchang University, Henan, China. (ch2)

Rowena Yew
Department of Electronic Materials Engineering, Research School of Physics and Engineering, The Australian National University, Canberra, ACT, Australia. (ch8)

Jung-Ho Yun
Nanomaterials Centre, School of Chemical Engineering and Australian Institute for Bioengineering and Nanotechnology, The University of Queensland, Brisbane, QLD, Australia. (ch9)

PREFACE

Artificial photosynthesis, which mimics the nature photosynthesis process to store solar energy into energy-rich chemical fuels, is considered as a promising method for providing a carbon-neutral, renewable, and scalable source of energy. Since the first demonstration of photoelectrochemical water splitting by Fujishima and Honda in 1972, a multitude of materials and designs have been explored over nearly five decades. While the earlier work mostly focused on the simple water-splitting reaction, recent efforts start to tackle the more difficult, but also more important reactions such as CO_2 reduction. The basic idea of artificial photosynthesis may sound deceptively simple: light absorbers harvest the energy delivered by photons; the energy is then used to excite charges; negative charges enable reduction reactions; and positive charges fuel oxidation reactions. However, to realize a viable artificial photosynthetic system for large-scale practical application, we need to carry out photosynthesis at efficiencies far higher than what is possible by the natural processes, which, to date, has been one of the most daunting challenges faced by the scientific community. It has been recognized that solutions to this challenge will likely rely on the combined knowledge of solid-state physics, electrochemistry, catalysis, photochemistry, spectroscopy, materials science, and device engineering.

Our goal for this book is to provide an overview of the latest breakthrough research and development in semiconductor photocatalysis, solar fuel production, and artificial photosynthesis. It includes a broad range of topics that cover a wide variety of materials and many important aspects of solar fuels and in-depth discussions on materials design, growth/synthesis, engineering, characterization, and photoelectrochemical studies. This book includes two parts. Part I is concerned about the fundamentals of semiconductor photoelectrodes and photoelectrochemical water splitting, and Part II discusses the emerging nanostructured photocatalysts and photoelectrodes for solar water splitting, pollutant degradation, and CO_2 reduction.

Part I consists of five chapters. Chapter 1 is focused on charge carrier dynamics in metal oxide photoelectrodes that occur during water oxidation, which has been identified as the bottleneck in limiting the overall water-splitting efficiency. Chapter 2 discusses the fundamental principles of photophysics and photochemistry that govern the process of artificial photosynthesis by semiconductors. Chapter 3 presents the latest development of III–V semiconductor photoelectrodes for light-driven chemical conversion and the fundamental understanding of the physical chemistry of the III–V semiconductor/liquid electrolyte interface. Chapter 4 focuses on the recent progress made on III-nitride planar semiconductor photoelectrodes, including their photoelectrochemical properties and stability analysis. Chapter 5 presents the use of rare earth-containing materials for photoelectrochemical water-splitting application.

Part II comprises Chapters 6–12. Chapter 6 reviews the state-of-the-art research activities made on III-nitride nanowire arrays for artificial photosynthesis, including water splitting and CO_2 reduction via photocatalytic and photoelectrochemical approaches. Chapter 7 presents the recent progress of layered metal oxide and their nanosheets for application in photocatalytic water splitting. Chapter 8 discusses the template-assisted fabrication method to produce nanostructured materials of controlled morphology and optoelectronic properties and their application in photocatalytic solar energy conversion. Chapter 9 provides the insight for the underlying fundamentals of bifunctional photocatalytic and photoelectrochemical systems for solar fuel production, organic pollutant degradation, and value-added chemical/electricity generation. Chapter 10 describes the facet engineering of photocatalysts for solar water splitting. Chapter 11 reviews the recent progress on the synthesis, properties, and photocatalysis of black TiO_2 nanomaterials for photocatalytic H_2 generation, degradation of organic pollutants, and reduction of CO_2. Chapter 12 describes the fundamental understanding of CO_2 photoreduction on the surface of heterogeneous catalysts as well as recent advances in the CO_2 reduction on visible light photocatalysts. The challenges and perspectives of CO_2 photoreduction for future development are also presented.

This book is well suited for students and researchers working in the fields of photocatalysis, solar fuels, electrochemistry, photoelectrochemistry, nanotechnology, material science, and renewable energy. The in-depth discussions on the design, growth/synthesis, and characterization of a broad range of nanostructured materials will benefit the readers for innovative semiconductors and device design and development towards more efficient utilization of abundant solar energy.

Zetian Mi,
University of Michigan, Ann Arbor
Lianzhou Wang,
University of Queensland, Brisbane
Chennupati Jagadish,
Australian National University, Canberra
Editors

PART I

Fundamentals of Semiconductor Photoelectrodes

CHAPTER ONE

Charge Carrier Dynamics in Metal Oxide Photoelectrodes for Water Oxidation

Andreas Kafizas[1], Robert Godin, James R. Durrant
Imperial College London, London, United Kingdom
[1]Corresponding author: e-mail address: a.kafizas@imperial.ac.uk

Contents

Semiconductors and Semimetals, Volume 97
ISSN 0080-8784
http://dx.doi.org/10.1016/bs.semsem.2017.02.002

1. BACKGROUND

Since the first demonstration of photoelectrochemical water splitting using titanium dioxide (TiO_2) (Fujishima and Honda, 1972), this route to renewable hydrogen fuel has been pursued with great endeavor (Osterloh, 2008). The need for a renewable fuel has never been so pressing, where excessive use of fossil fuels has resulted in pervasive and lasting damage to the earth's climate and ecosystems (Broecker, 1975; Gruber, 2011). This has resulted in an agreement by the world's leading economies to both reduce their CO_2 emissions and restrict global warming to 2°C above preindustrial era temperatures (Fawcett et al., 2015). In order to meet these targets, and maintain our standard of living, carbon neutral and renewable energy technologies need to be improved, which includes photoelectrochemical water splitting devices. Renewable energy research has primarily focused on technologies that harness sunlight, as it is our largest renewable energy source—the amount reaching the Earth every hour being twice the total energy we consume each year (Morton, 2006). Currently, the fastest growing renewable energy technology is the photovoltaic cell, which converts sunlight into electricity, but is limited by the need to use this energy at the time of generation, as there is currently no integrated, scalable means of storing it (Beaudin et al., 2010).

Natural photosynthesis is an inspiring example of how sunlight can renewably produce chemical fuels, which can be stored and used only when needed. This has been exploited by countries such as Brazil, where plant sugars are used as a feedstock for producing biofuels (such as sugarcane to ethanol). However, low solar to biomass conversion efficiency is primarily constraining biofuel production to economies with excess arable land (Hall et al., 2011). Alternatively, artificial photosynthetic strategies are being explored that use sunlight to drive the synthesis of chemical fuels, with inorganic semiconducting materials being most widely studied (Tachibana et al., 2012). In terms of efficiency, the most promising artificial route is water splitting; this produces the carbon-free fuel, hydrogen, from the most environmentally benign and freely available feedstock, water (Gust et al., 2011). Water splitting has several notable advantages over the use of photovoltaic cells: (i) hydrogen can be stored, (ii) hydrogen is compatible with fuel cell technology and can be converted into electrical energy with greater efficiency than its combustion (Busby, 2005), (iii) hydrogen can be injected into existing natural gas supplies (Melaina et al., 2013), and (iv) hydrogen can be converted into conventional liquid fuels (Dry, 2002) and used in areas of transportation that cannot be powered by battery technology (e.g., shipping and aviation). However, solar-driven hydrogen generation from water is not yet cost competitive with fossil fuel-derived hydrogen, and so has seen little practical implementation to date.

During water splitting, hydrogen and oxygen gas is formed and is best understood from the two redox half reactions:

$$4H^{+} + 4e^{-} \rightarrow 2H_2 \quad E^{\circ} = 0\,V \tag{1a}$$

$$2H_2O + 4h^{+} \rightarrow O_2 + 4H^{+} \quad E^{\circ} = 1.23\,V \tag{1b}$$

For a single semiconducting material to demonstrate overall solar-driven water splitting, it must possess a conduction band more negative than the proton reduction reaction (Eq. 1a) and a valence band more positive than the water oxidation reaction (Eq. 1b). The oxidation of water is the kinetic "bottleneck" in the overall reaction (Fillol et al., 2011; Zhong and Gamelin, 2010); thus, there is a pressing need to develop more active photocatalysts that can drive this reaction.

Current approaches to integrated solar-driven water splitting can broadly be divided into two approaches: (i) nanoparticle suspensions and (ii) photoelectrodes. Nanoparticle suspensions possess the benefits of high surface area and a minimized distance charge needs to travel to reach the surface and react. More than 200 inorganic semiconductor materials have been examined (Martin, 2015), where quantum efficiencies as high as $\approx 6\%$ ($\lambda = 420$ nm) have been observed with particles of GaN:ZnO solid solution (Maeda et al., 2008). However, the major drawback of nanoparticle suspensions is the presence of both oxidation and reduction sites on the same material, which inadvertently facilitate short-circuit reactions (i.e., the parasitic reduction of oxygen to superoxide instead of protons to hydrogen fuel). There is also the added concern of forming a potentially explosive mixture of H_2 and O_2 within the same medium. Photoelectrodes can circumvent these problems as H_2 and O_2 are formed on separate sites (H_2 at the cathode and O_2 at the anode) and can be kept separate through the use of appropriate membranes, although with an increase in system complexity and therefore, most probably cost. Using semiconductors with complimentary bandgaps, a synergetic improvement in solar-to-hydrogen efficiency can be achieved in a tandem system, where theoretical efficiencies have been estimated to reach

~27% vs ~17% for a single material (Bolton et al., 1985), with greater flexibility in material energetics.

Some of the most promising materials for water oxidation are metal oxide semiconductors. Metal oxides are often water stable and show a high level of resistance to photocorrosion. Moreover, most metal oxides show low toxicity and can be grown using low cost methods that are often up-scalable. Although higher water splitting efficiencies have been found using nonoxide-based materials, such as c-Si, GaP, InP, and GaAs (Walter et al., 2010), such materials, often associated with photovoltaic devices, readily corrode in water and cannot be grown cheaply.

The most popular metal oxide photoanodes for water oxidation are TiO_2, α-Fe_2O_3, $BiVO_4$, and WO_3. Fig. 1 summarizes the solar absorption properties and band potentials of these four metal oxides (Chen and Wang, 2012; Xu and Schoonen, 2000). Of these four materials, only TiO_2 possesses the appropriate conduction and valence band energies for overall water splitting (both water reduction and oxidation). However, its major drawback is its wide bandgap (≤410 nm, 3.0 eV), which limits its function to only a small fraction of the solar spectrum (≤5% of total photons). Of the four

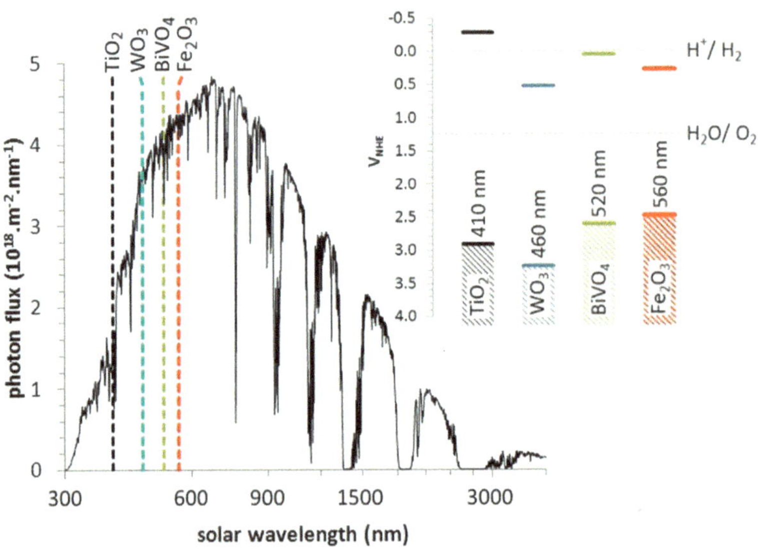

Fig. 1 The solar absorption properties of TiO_2 (~5%), WO_3 (~8%), $BiVO_4$ (~12%), and α-Fe_2O_3 (~17%), band potentials, and bandgap energies.

materials, α-Fe_2O_3 possesses the narrowest bandgap ($\leq$560 nm, 2.2 eV) and can absorb ~17% of the solar spectrum (Sivula et al., 2011). However, as a photoanode, it possesses a late-onset potential for water oxidation, and typically does not show any photocatalytic activity until ~0.9 V_{RHE} is applied. Compared with α-Fe_2O_3, $BiVO_4$ ($\leq$520 nm, 2.4 eV) and WO_3 ($\leq$460 nm, 2.7 eV) have successively wider bandgaps, but earlier onset potentials ($BiVO_4$~0.7 V_{RHE}, WO_3~0.5 V_{RHE}). TiO_2 is stable at a wide range of pH and applied potentials (Wahl et al., 1995), whereas α-Fe_2O_3 is stable at positive potentials (V_{RHE}) in nonacidic pH, WO_3 is stable at positive potentials in acidic pH, and $BiVO_4$ is relatively stable at positive potentials in neutral pH. Of the four materials, WO_3 has the most oxidizing valence band position and shows variable selectivity for water oxidation (Hill and Choi, 2012).

Although α-Fe_2O_3 can, in theory, absorb nearly $1/5^{th}$ of the solar spectrum, the best performing photoanodes based on α-Fe_2O_3 (~4.3 mA cm^{-2} at 1.23 V_{RHE}) have shown photocatalytic currents far shy of the theoretical maximum (~12.6 mA cm^{-2}) (Kim et al., 2013a). Similar limitations are observed in $BiVO_4$, which in theory can achieve a maximum photocurrent of ~7.6 mA cm^{-2} but experimentally has shown values closer to ~4.4 mA cm^{-2} at 1.23 V_{RHE} (Kim and Choi, 2014). These shortcomings have been attributed to a number of fundamental physical properties. In the case of α-Fe_2O_3, this is due to: (i) the conduction band potential having insufficient energy to reduce water, (ii) a large requisite overpotential for water oxidation, (iii) a relatively low absorption coefficient, (iv) poor electron conductivity, and (v) a short hole diffusion length (Sivula et al., 2011). Hole diffusion lengths are typically on the nm scale but vary considerably: ~2–4 nm for α-Fe_2O_3, ~10 nm for TiO_2 (Salvador, 1984), and is less of a limitation at ~100 nm for $BiVO_4$ (Rettie et al., 2013), and ~150 nm for WO_3 (Wang et al., 2014). $BiVO_4$ shares the undesirable characteristics (i)–(iv). The main limitations in WO_3 are due to (i) and (iii) (Wang et al., 2014), While TiO_2 is mainly impeded by (iii) and (v). Importantly, the common problem shared by all these oxides is their low absorption coefficients, resulting in absorption lengths greater than the charge carrier diffusion lengths. This has made nanostructuring these materials paramount to improving their activity and is one of the most commonly pursued strategies (Cowan and Durrant, 2012). Other strategies include: (a) impurity doping either to improve electron conductivity (Parmar et al., 2012) or narrow the bandgap (Tang et al., 2011); (b) surface treatments, such as the addition of surface phosphate groups (Jing et al., 2012a) or acid treatment

(Yang et al., 2016); (c) surface passivation, where ultrathin layers of a wide bandgap semiconductor is grown on the material surface, e.g., Al_2O_3 (Forster et al., 2015); and (d) the use of surface cocatalysts; such as Co_3O_4 (Feckl et al., 2015) and cobalt phosphate (CoPi) (Barroso et al., 2011). Moreover, recent research has shown that a very promising method, perhaps vital for producing photoanodes with efficiencies reaching the theoretical limits, is to form a heterojunction between two active semiconductor materials. This can improve charge–charge separation and synergistically enhance activity, and has been demonstrated in a number systems, e.g., $WO_3/BiVO_4$ (Grigioni et al., 2015), rutile TiO_2/anatase TiO_2 (Liu et al., 2011), etc. It is clear that there are numerous useful strategies that can be explored to improve the activity of photoanodes for water splitting; however, it is less clear how these changes in material architecture impact on the charge carrier dynamics and kinetics of water oxidation. As such, there is an increasing need to develop techniques that probe charge carrier behavior. By increasing our understanding of the processes that limit water oxidation efficiency, such as the kinetics of charge carrier recombination or slow water oxidation kinetics, we can develop strategies to counteract these problems through rational design.

This review focuses on the use of TiO_2, α-Fe_2O_3, $BiVO_4$, and WO_3 as photoanodes for water oxidation and studies that probe charge carrier dynamics in these materials. This review will discuss each material in turn and interpret the results of these charge carrier dynamic studies to build an inclusive and complimentary picture of the processes that occur during water oxidation. There will be a general view on how changes in material structure impact on charge carrier dynamics and the reaction kinetics of water oxidation.

The charge carrier dynamics of metal oxide photoanodes have been studied using a number of techniques, including: (i) time-resolved microwave conductivity (TRMC) and terahertz (THz) spectroscopy, (ii) time-correlated single photon counting (TCSPC), (iii) transient absorption spectroscopy (TAS), (iv) photoinduced absorption spectroscopy (PIAS), and (v) electrochemical impedance spectroscopy (EIS). The basis of each technique will be discussed in turn.

1.1 TRMC and THz Spectroscopy

TRMC and THz spectroscopy are pump–probe techniques. The pump is an ultra-bandgap laser pulse used to excite the semiconductor material, and the probe is either microwaves (~0.3–300 GHz, TRMC) or far-infrared radiation (~0.3–3 THz, THz spectroscopy). Both methods track time-resolved changes in photoinduced conductivity, often with nanosecond resolution. Conventional conductivity techniques are hampered by the need to apply electrical contacts; however, TRMC and THz spectroscopy can wirelessly measure transient changes in photoinduced conductivity (Kunst and Beck, 1986). In single-crystalline semiconductors (e.g., materials such as c-Si that possess few lattice defects), the decay of photoinduced conductivity is directly related to charge carrier recombination (De Haas and Warman, 1982). However, in materials with numerous lattice defects (a common attribute of metal oxides), the decay of photoinduced conductivity is also due to charge carrier trapping processes and complicates the interpretation of data (Saeki et al., 2014).

In most cases, TRMC and THz spectroscopy have been used to study the kinetics of charge carrier trapping and recombination processes in metal oxide nanopowders. However, to our knowledge, it has not been used to study the charge carrier dynamics of metal oxide-based-photoanodes. Nevertheless, TRMC and THz spectroscopy can provide useful insight into how physical changes in a material such as crystal phase (Colbeau-Justin et al., 2003) or impurity doping (Abdi et al., 2013) impact upon charge carrier mobility, trapping, and recombination.

1.2 TCSPC

TCSPC measures the time-resolved light emission of semiconductor materials due to photoluminescence, often with nanosecond resolution. An ultra-bandgap laser pulse is used to excite the semiconductor material. When these charge carriers recombine, energy is released in the form of light (radiative recombination) or heat (nonradiative recombination). TCSPC measures the transient emission of light due to radiative recombination. In metal oxides, radiative recombination mostly falls within the visible region of the electromagnetic spectrum. Of note, radiative recombination is the less dominant decay pathway in metal oxides.

TCSPC has mostly been used to study the kinetics of charge carrier recombination in metal oxide nanopowders and thin films, although such studies are complicated by the relatively low photoluminescence yields of most oxides. To our knowledge, TCSPC has not been used to study the reaction kinetics of charge carriers in metal oxide-based photoanodes during water oxidation. However, an alternative time-resolved photoluminescence method has been used to study ZnO (Appavoo et al., 2015). Steady-state photoluminescence measurements have been used to probe the water oxidation reaction at various applied potentials (Rex et al., 2016) and also to identify the intermediates formed during water oxidation in the absence of an applied potential (Imanishi et al., 2007).

1.3 TAS

TAS is a pump–probe technique. The pump is an ultra-bandgap laser pulse, used to excite the semiconductor material, and the probe is typically visible or near-infrared subbandgap light. It has been shown that transient changes in absorption arise from the generation of charge carriers that absorb in this region of the electromagnetic spectrum. A typical TAS apparatus is shown in Fig. 2.

TAS can be used to measure transient changes in photoinduced absorption on either the ultrafast (femto- to nanosecond) or slower (microsecond–second) timescales. On the ultrafast timescale, processes such as geminate recombination can be observed (recombination that occurs before the charge carriers separate) (Sachs et al., 2016), typically on the femto- to picosecond timescale. Nongeminate processes, such as bimolecular recombination, are observed starting from the picosecond timescale (Pendlebury et al., 2014). Moreover, the spectral signatures of electrons and holes can be determined with the help of chemical scavengers (Wang et al., 2015). As such, TAS can be used to selectively monitor holes or electrons, and can be used to determine the kinetics of photocatalytic processes such as alcohol oxidation (Tamaki et al., 2006), dioxygen reduction (Xiao-e et al., 2004), and water oxidation (Zhang et al., 2012). TAS can also be used to monitor charge transfer processes, such as charge separation across a heterojunction (Kafizas et al., 2016), or charge transfer to a surface cocatalyst (Reynal et al., 2013). As such, TAS can monitor the generation, recombination, trapping, charge transfer, and reaction of photogenerated charges in semiconductors. TAS has been used to monitor the charge carrier dynamics during photoelectrochemical water oxidation on TiO_2, α-Fe_2O_3, and $BiVO_4$ photoanodes. Moreover, there have been numerous studies on how changes in the material structure (such as nanostructuring, doping, surface passivation, the use of cocatalysts, etc.) impact on the charge carrier dynamics and water oxidation kinetics, which we will cover in more detail during this review.

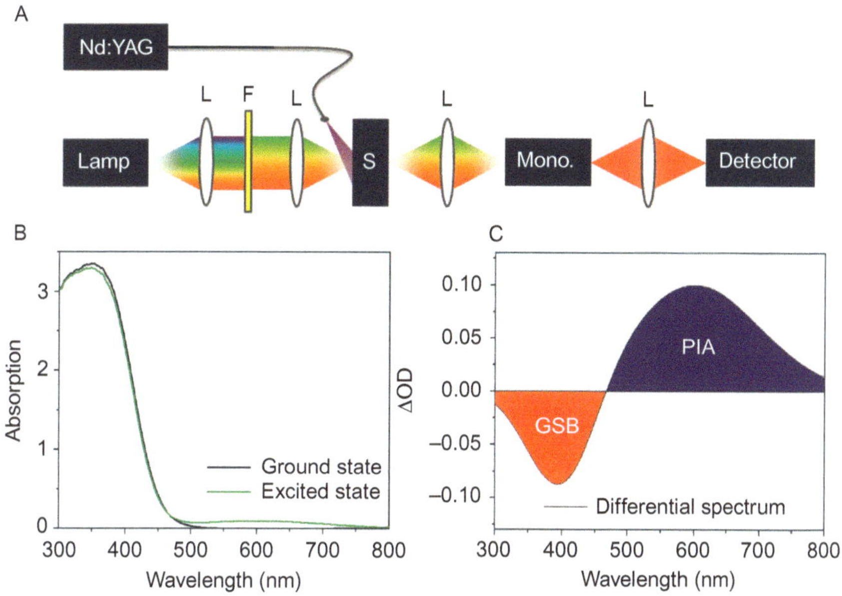

Fig. 2 (A) Typical arrangement of a TAS apparatus. A neodymium:yttrium aluminium garnet (Nd:YAG) laser is used to excite the sample (S). Light from a lamp is used to probe transient changes in absorption. This light is passed through a series of lenses (L) so that it can be focussed on the sample (S), and subsequently the monochromator (Mono.) and then the detector. A long-pass filter (F) is sometimes used when this probe light can cause a photoexcitation in the sample. (B) A typical transient absorption spectrum for a semiconducting metal oxide before (ground state) and after (excited state) a laser excitation pulse. (c) The differential spectrum between the ground and excited state spectra. A negative absorption, due to a ground-state bleach (GSB), is observed *blue* of the band edge and a positive absorption, due to a photoinduced absorption (PIA), is observed *red* of the band edge.

1.4 PIAS

PIAS is a slow timescale (millisecond–second) technique that has only recently been used to study the water oxidation reaction on metal oxide photoanodes (Le Formal et al., 2015). PIAS is analogous to TAS but instead utilizes a continuous-wave light source to photoexcite the material over a period of seconds, so probing quasi-steady-state irradiation conditions. As in TAS, the change in absorption due to photogenerated charge is measured using a visible light source. When a photoanode is held above the onset potential for water oxidation, photoexcitation over a period of several seconds results in the accumulation of hole carriers at the material surface. By simultaneously measuring the extracted photocurrent, one can model the reaction order and kinetics of the water oxidation reaction with respect to changes in hole carrier concentration, which we will cover in more detail later.

1.5 EIS Spectroscopy

EIS measures changes in impedance in an electrochemical system after a small sinusoidal potential (or current) of fixed frequency is applied (Retter and Lohse, 2005). The impedance is simply a measure of the ability of a circuit to resist the flow of electrical current. In a typical EIS measurement, one measures the change in impedance at a range of frequencies. For photoelectrochemical EIS measurements of the water oxidation reaction, this is typically measured between 10,000 and 0.01 Hz under constant illumination using a continuous-wave light source. EIS data are modeled as a network of passive electrical circuit elements, called an equivalent circuit. Each element represents a barrier to charge transfer and typically include elements for the double-layer capacitance, electron transfer resistance, and electrolyte resistance (Baram and Ein-eli, 2010). EIS has been used to monitor the water oxidation reaction on TiO_2, α-Fe_2O_3, $BiVO_4$, and WO_3 photoanodes, and in some cases, the influence of changes in the material structure (such as surface passivation and the use of cocatalysts), which we will cover in more detail later (Klahr et al., 2012a,b).

2. TiO_2

The first example of semiconductor photocatalysis can be traced back more than 90 years to observations of the chalking of TiO_2-based paints (Renz, 1921). Of all the photocatalytic materials that have since been studied, TiO_2 is still the most popular due to its high activity, durability, and low cost (Mills and Le Hunte, 1997). TiO_2 is a versatile photocatalyst and has been shown to degrade a wide range of organic pollutants, bacteria, cancer cells, and viruses and has applications in air and water remediation, and self-sterilizing surfaces (Hashimoto et al., 2005). For photocatalytic applications, anatase ($E_{bg} \sim 3.2$ eV) and rutile ($E_{bg} \sim 3.0$ eV) are the two most studied polymorphs (Fujishima et al., 2008). Despite possessing the same chemical composition, the different arrangements of TiO_6 octahedra in anatase (four-edge sharing partly distorted octahedra) and rutile (two-edge sharing non-distorted octahedra) result in different physicochemical properties

(Wang et al., 2015). Moreover, TiO_2 was the first photoanode material to demonstrate photoelectrochemical water splitting (Fujishima and Honda, 1972). The development of TiO_2 photoanodes has resulted in solar-to-hydrogen efficiencies above 1% and photocurrent densities of 2.6 mA cm^{-2} at 1 sun irradiation (Liu et al., 2011). Given the robust nature of TiO_2, it is increasingly being used as a protection layer for materials that photocorrode in water, but harness more of the solar spectrum, such as Cu_2O (Paracchino et al., 2011), c-Si, and III–V semiconductors (Hu et al., 2014).

2.1 TRMC and THz Spectroscopy

Although TRMC and THz spectroscopy has not been used to study TiO_2 photoanodes under operation, it has been employed to study the effect of materials preparation and reaction conditions on time-resolved charge trapping and recombination processes.

Carneiro et al. found that increasing the crystallinity of anatase powders resulted in prolonged photoinduced conductivity ($t_{50\%} > 5$ μs; $t_{50\%}$ represents the time where 50% of the initial intensity has decayed) and a greater photocatalytic activity in oxidizing methylene blue dye (Carneiro et al., 2010). They also found that composites containing both the anatase and rutile phases also showed prolonged photoinduced conductivity, which was attributed to an improved separation of photogenerated charge (Carneiro et al., 2011). Colbeau-Justin et al. found that charge carrier trapping was significantly faster in rutile ($t_{50\%} \sim 15$ ns) compared with anatase ($t_{50\%} \sim 1.2$ μs) (Colbeau-Justin et al., 2003). Colbeau-Justin et al. also studied the change in TRMC in commercially available TiO_2 powders when coated with a Pt cocatalyst and found evidence of charge transfer from TiO_2 to Pt on the sub-ns timescale (Emilio et al., 2006). Using TRMC they also found that plasmonic Au nanoparticles, in the presence of visible light, could inject electrons into TiO_2 (Mendez-Medrano et al., 2016). Saeki et al. used frequency modulated TRMC and THz spectroscopy to determine the change in trap depth and relative populations of "trapped" and free photogenerated electrons in anatase (Saeki et al., 2014). They found that the trap depth changed from ~110 to ~75 meV from 5 ns to 5 μs and that the respective population of "trapped" electrons increased from ~80% to ~98% over this period. However, they were unable to discern if these "trapped" electrons were merely a relaxation of a free electron to a trap state or its recombination with a hole.

2.2 TCSPC

TCSPC has not been used to study TiO_2 photoanodes but has been used to study the kinetics of radiative recombination processes in anatase and rutile, and their composites.

Ohno et al. found that rutile particles suspended in water showed rapid radiative recombination ($t_{50\%} < 1$ ns) (Fujihara et al., 2000). Yi et al. found that N-doping anatase nanorods decreased the rate of radiative recombination from $t_{50\%} \sim 2.5$ ns to $t_{50\%} \sim 5$ ns (Lee et al., 2014). Lai et al. found that radiative recombination increased in anatase nanotube arrays ($t_{50\%} \sim 70$ ns) when loaded with Ag nanoparticles ($t_{50\%} \sim 30$–68 ns) (Ge et al., 2016). However, Zhu et al. found that radiative recombination decreased in H-doped anatase nanotube arrays ($t_{50\%} \sim 1$ ns) (Xu et al., 2013). Park et al. showed that radiative recombination in P25 powders and "black" rutile TiO_2 varied between $t_{50\%} \sim 0.4$ and 0.8 ns depending on their processing conditions (Zhang et al., 2016). Thus far, these TCSPC measurements were focused on the nanosecond timescale. Li et al. studied the kinetics of radiative recombination on much slower timescales from the micro- to millisecond (Ma et al., 2014a). They found that radiative recombination was inhibited in anatase:rutile composites, compared with pure phase materials, on both the microsecond ($t_{50\%} \sim 2$ μs) and millisecond ($t_{50\%} \sim 25$ μs) timescales, which was attributed to an enhanced separation of photogenerated charge at the anatase:rutile junction.

2.3 TAS

Much of the pioneering work on the TAS of TiO_2 was conducted on powders, as opposed to photoanodes. Through the use of chemical scavengers, Tamaki et al. showed that the relatively broad transient absorption spectrum of anatase TiO_2 was composed of Gaussian-shaped electron and hole signals, centered near 800 and 500 nm, respectively (Yoshihara et al., 2004). Tamaki et al. also studied the pH dependence of these spectral signals and found that trapped hole states were consistently observed in the blue region between ~400 and 550 nm (Yoshihara et al., 2007). Durrant et al. showed that the electron (~800 nm) and hole (~550 nm) signals in rutile TiO_2 were centered at similar wavelengths to anatase (Kafizas et al., 2016). Computational studies of anatase and rutile have shown that trapped electrons and holes give rise to states that absorb visible light in the red and blue-green regions, respectively (Cheng et al., 2014; Nunzi et al., 2016; Zawadzki, 2013).

Durrant et al. studied the charge carrier dynamics of anatase and rutile films over ultrafast timescales (femto- to nanosecond) (Sachs et al., 2016). In both polymorphs, charge trapping occurred within several picoseconds. In anatase, the recombination of these charges varied more prominently with laser intensity and ranged from $t_{50\%} \sim 1100$ to 100 ps with laser intensities from ~50 to 700 μJ cm^{-2} (Fig. 3A). In rutile, recombination was substantially faster than anatase over these timescales and ranged from $t_{50\%} \sim 200$ to 100 ps over similar laser intensities (Fig. 3B). In addition, Durrant et al. found that the recombination dynamics in dense films were analogous to mesoporous films, which indicated bulk rather than surface recombination is the primary determinant of charge carrier lifetime in the absence of space charge layer formation.

Durrant et al. also studied the recombination dynamics of anatase and rutile over slower timescales and found a stark difference in behavior (Wang et al., 2015). Over these slower timescales (from 10 μs), recombination in rutile ($t_{50\%} \sim 1$ ms) was slower than anatase ($t_{50\%} \sim 80$ μs) and showed dynamics indicative of tunneling-mediated recombination. In contrast, anatase showed decay dynamics indicative of trapping/detrapping limited bimolecular recombination. Assessing the decay dynamics over the ultrafast and slower timescales, early picosecond–nanosecond timescale recombination was faster in rutile than in anatase; however, a portion of charge carriers that survived into the microsecond timescale were longer lived in rutile.

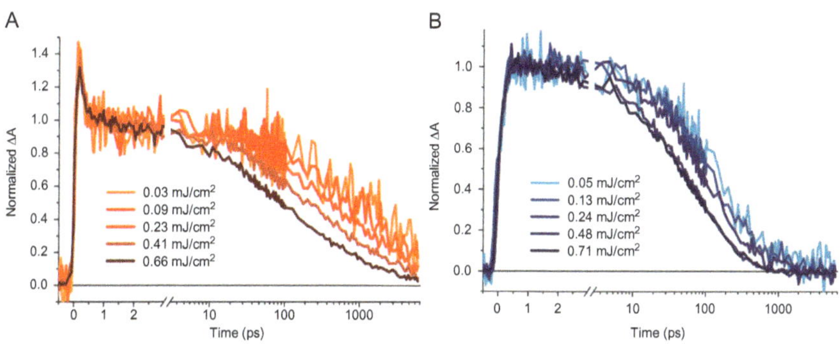

Fig. 3 Ultrafast TAS of mesoporous films of (A) anatase and (B) rutile measured in argon. The transient absorption decay kinetics at 1200 nm are shown (normalized from 1 ps) following 355 nm excitation (various excitation densities). *Reprinted with permission from Sachs, M., Pastor, E., Kafizas, A., Durrant, J.R., 2016. Evaluation of surface state mediated charge recombination in anatase and rutile TiO_2. J. Phys. Chem. Lett. 3742–3746. Copyright (2016) American Chemical Society.*

The reaction kinetics of TiO_2 powders have been investigated for a number of photocatalytic processes, including alcohol oxidation (reaction initiates within picoseconds) (Tamaki et al., 2006) and water oxidation (reaction initiates within microseconds) (Yamakata et al., 2001). Charge transfers from TiO_2 onto water reduction (Patrocinio et al., 2015) and oxidation (Meekins and Kamat, 2011) catalysts have also been investigated as well as the effects of impurity doping (Tang et al., 2011).

There have been numerous transient absorption studies of the charge carrier dynamics in TiO_2 photoanodes during water oxidation. To our knowledge, there have been no ultrafast TAS studies (fs–ns) under these conditions, with all such studies focusing on the slower timescales (μs–s) where water oxidation takes place (typically ms–s). Tang et al. were the first to identify the requirement for long-lived photogenerated holes to oxidize water (Tang et al., 2008). Cowan et al. investigated the water oxidation reaction on mesoporous anatase photoanodes (Cowan et al., 2010). They found that under a positive applied potential, long-lived holes were formed on anatase that oxidized water on the millisecond timescale. These long-lived holes were observed from the photocurrent onset (0.05 V_{RHE}). More long-lived holes were observed at higher applied potentials, and plateaued at 0.45 V_{RHE}, coinciding with the plateau of the photocurrent. Water was oxidized with ~100% Faradaic efficiency; however, the quantum yield (~8%) under laser-pulsed illumination (355 nm) was low, where electron–hole recombination was the dominant loss pathway. From transient photocurrent measurements, Cowan et al. also measured the electron diffusion length in mesoporous anatase photoanodes (~8.5–12.5 μm, between 0.2 and 0.4 V_{RHE}) (Leng et al., 2010). Also, by comparing their transient photocurrents with their long-lived hole signals, they were able to estimate the extinction coefficient of a hole (~3000 M^{-1} cm^{-1} at 460 nm) (Cowan et al., 2013).

Cowan et al. also investigated the charge carrier dynamics of the water oxidation reaction on rutile photoanodes (Fig. 4) (Pesci et al., 2013). Hydrogen-treated samples (black rutile) showed both enhanced photocatalytic and light-harvesting properties. Their TAS studies showed that electron–hole recombination was suppressed in hydrogen-treated rutile. Moreover, from moderate positive applied potentials (≥0.2 V_{RHE}) long-lived holes were observed (Fig. 4A) and showed comparatively similar water oxidation kinetics to anatase photoanodes (Cowan et al., 2010). Black rutile did not show any improvement in water oxidation kinetics vs standard rutile

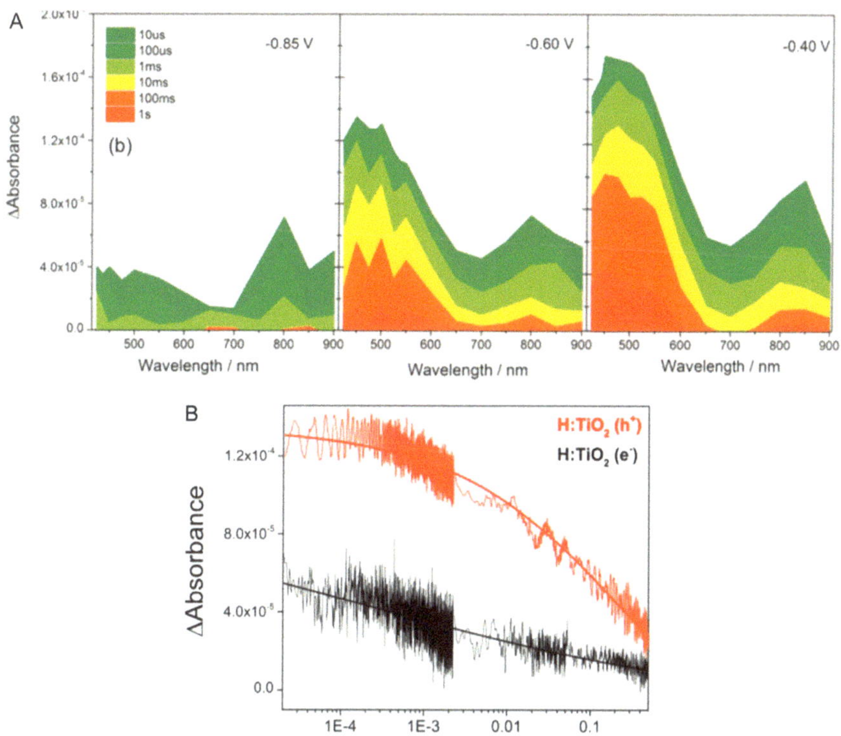

Fig. 4 TAS of a hydrogen-treated rutile nanowire photoanode, measured during water oxidation in 1 M NaOH (pH 13.6). (A) The change in transient absorption spectrum at several timescales (10 μs–1 s) for three applied potentials −0.85, −0.60, and −0.40 $V_{Ag/AgCl}$ (~0.15, 0.40, and 0.60 V_{RHE}, respectively). (B) Decay kinetics of holes at 500 nm and electrons at 800 nm at −0.60 $V_{Ag/AgCl}$ (~0.40 V_{RHE}). TAS was recorded using a 355-nm laser pulse (70 μJ cm^{-2}, 0.33 Hz repetition rate). *Reprinted with permission from Pesci, F.M., Wang, G., Klug, D.R., Li, Y., Cowan, A.J., 2013. Efficient suppression of electron–hole recombination in oxygen-deficient hydrogen-treated TiO_2 nanowires for photoelectrochemical water splitting. J. Phys. Chem. C 117, 25837–25844. Copyright (2013) American Chemical Society.*

and showed that the suppression of electron–hole recombination was the primary cause of the improved activity in this material. Despite the visible light absorption of black rutile, there were no long-lived photogenerated states formed under visible light excitation. This was in agreement with the low quantum efficiencies observed at $\lambda > 400$ nm and showed that black rutile was not able to form the long-lived holes required to oxidize water under visible excitation. Moreover, Cowan et al. showed that there was no change in the reaction kinetics of holes with water when the temperature was increased from ~25°C to 55°C, which showed that there is no distinct thermal barrier to hole transfer (Cowan et al., 2011).

Jing et al. studied the effect of adding surface phosphate groups onto anatase TiO_2 photoanodes (Jing et al., 2012b). They found that the addition of phosphate doubled the photocurrent density in mesoporous films. TAS studies showed that the lifetime of photogenerated holes was substantially improved by the addition of phosphate (Jing et al., 2012a) and that these phosphate groups did not catalyze reactions between photogenerated holes and water.

The charge carrier dynamics of several TiO_2 composite materials, with applications in water splitting, have also been studied. Kafizas et al. studied anatase:rutile TiO_2 composites and found that hole transfer from rutile to anatase occurred on the sub-microsecond timescale, substantially increasing the lifetime and yield of photogenerated charge (Kafizas et al., 2016). Morais et al. studied anatase TiO_2:reduced graphene oxide composites and found that the injection of photogenerated electrons from anatase into reduced graphene oxide increases the lifetime and yield of photogenerated holes in anatase, almost doubling the activity compared with pure anatase (Morais et al., 2016). Barreca et al. studied the charge carrier dynamics in an α-Fe_2O_3/TiO_2 heterojunction and found that hole transfer from α-Fe_2O_3 into the TiO_2 surface layer increased both the water oxidation kinetics and photocatalytic efficiency (Barreca et al., 2015).

2.4 EIS

To our knowledge, EIS has rarely been used to investigate the dynamics of water oxidation processes on TiO_2 photoanodes. Cao et al. used EIS to investigate the photoelectrochemical oxidation of salicylic acid on anatase TiO_2 photoanodes (Leng et al., 2005). Sutter et al. used EIS to study the water oxidation reaction on photoanodes of anatase nanotube arrays (Cachet and Sutter, 2015). As the applied potential was increased, they found that the recombination rate constant decreased more readily in base (from 0.8 V_{RHE}, pH 13.3) compared with near neutral conditions (from 1.2 V_{RHE}, pH 6.0). This was attributed to a lower photoelectrochemical stability of the nanotube array in neutral solution. Fig. 5 shows the change in rate constants for charge transfer (the reaction of holes with water) and recombination with applied potential. The rate constant for charge transfer was approximately constant above -0.2 $V_{Ag/AgCl}$ (~1.2 V_{RHE}) and ranged between ~100 and 300 s^{-1}, whereas the recombination constant decreased

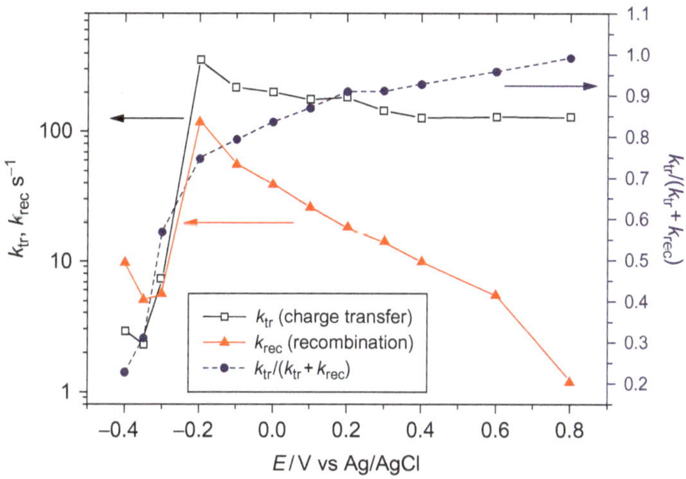

Fig. 5 Evolution of the rate constants for charge transfer (the reaction of holes with water) and recombination as a function of applied potential—determined from EIS of an anatase TiO_2 photoanode during photoelectrochemical water oxidation. *Reprinted with permission from Cachet, H., Sutter, E.M.M., 2015. Kinetics of water oxidation at TiO_2 nanotube arrays at different pH domains investigated by electrochemical and light-modulated impedance spectroscopy. J. Phys. Chem. C 119, 25548–25558. Copyright (2015) American Chemical Society.*

steadily from ~100 s^{-1} at −0.2 $V_{Ag/AgCl}$ to ~1 s^{-1} at 0.8 $V_{Ag/AgCl}$. This indicated that the reaction kinetics of the water oxidation reaction did not change substantially with applied potential, but rather, the competing process of charge carrier recombination was suppressed.

3. α-Fe_2O_3

Hematite (α-Fe_2O_3) is a promising photoanode material due to its significant light absorption, chemical stability, and abundance (Sivula et al., 2011). It is the most thermodynamically stable form of iron oxide under ambient conditions and forms a corundum structure where the O^{2-} anions are arranged in a hexagonal closed-packed lattice along the [001] direction and the Fe^{3+} cations occupy two-thirds of the octahedral interstices in the (001) basal planes. The bandgap falls between ~2.0 and 2.2 eV, corresponding to an absorption edge near 560 nm. Hematite's strong absorption of yellow to ultraviolet light, and transmission of orange to infrared light, gives it a characteristic red color. The first demonstration of an α-Fe_2O_3

photoanode for water oxidation was in 1976 (Hardee and Bard, 1976). Subsequent studies have shown that the performance of α-Fe_2O_3 as a photoanode is limited by its poor optoelectronic properties, which result in low light-harvesting efficiencies, and the requirement of a large overpotential for photocatalytic water oxidation to occur. Nanostructuring strategies have decoupled the mismatch between light absorption depth and hole collection length, resulting in more promising activities (Lin et al., 2011). However, the biggest unresolved problem in α-Fe_2O_3 photoanodes is poor charge separation (Zandi and Hamann, 2015). To our knowledge, the highest photocurrent achieved on an α-Fe_2O_3-based photoanode is 4.3 mA cm^{-2} (at 1.23 V_{RHE} and 1 sun irradiance, based on Pt-doped α-Fe_2O_3 with a worm-like nanostructure coated with a CoPi cocatalyst) (Kim et al., 2013a).

3.1 TRMC Measurements

Lin et al. examined the TRMC of both planar and nanonet-structured α-Fe_2O_3 (Lin et al., 2011). They found that photoinduced conductivity was lost more rapidly in thicker films of α-Fe_2O_3, taking 2.5 ms in a 25-nm thick film and 1.3 ms in a 100-nm thick film. These corresponded to $t_{50\%}$'s of ~2 and 1 μs, respectively, and indicated bulk recombination was faster than surface-mediated recombination.

3.2 TCSPC

Like TRMC studies, TCSPC has not been conducted on α-Fe_2O_3 photoanodes during water oxidation. However, Kwon et al. did show that SnO_2 sandwich layers formed between FTO and "worm-like" mesoporous nanostructures of α-Fe_2O_3 possess increased photoinduced luminescence lifetimes (Liu et al., 2013). Tao et al. showed that Sn-doped α-Fe_2O_3 has a marginally increased photoinduced luminescence lifetime compared with α-Fe_2O_3 (Qin et al., 2015).

3.3 TAS

There have been numerous transient absorption studies on the charge carrier dynamics of α-Fe_2O_3 photoanodes during water oxidation. Pendlebury et al. investigated the charge carrier dynamics on both the ultrafast (femto- to nanosecond) and slower (microsecond–second) timescales (Pendlebury et al., 2014). They showed that an increase in applied potential from flat band (0.5 V_{RHE}) to high anodic bias (1.4 V_{RHE}) results in substantial changes in the lifetime of photogenerated charge (Fig. 6). Similar transient absorption spectra and kinetics were observed by Milan et al. who investigated the ultrafast transient absorption behavior of α-Fe_2O_3 at 1.23 V_{RHE} on the fs–ns timescale (Milan et al., 2016). Back to the work of Pendlebury et al., above 0.8 V_{RHE}, an initially positive absorption signal observed at 580 nm ($t < 50$ ps) flips to a negative absorption (i.e., a bleach, $t >$ ns) due to the trapping of photogenerated electrons by localized states located close to the conduction band edge (Fig. 6B). At positive potentials these states are unoccupied (i.e., oxidized) within the space charge region, enabling ground-state absorption, and the observation of a bleach signal after electron trapping (Fig. 6D). Similar behavior was observed by Kaunisto et al. in their study of α-Fe_2O_3:TiO_2 composites, where the bleach at 580 nm was observed from 1.23 V_{RHE} (Ruoko et al., 2015). Below 0.6 V_{RHE}, photogenerated holes are short-lived and quantitative electron–hole recombination occurs before the millisecond (Barroso et al., 2013). However, from 1.1 V_{RHE} and above, this signal becomes biphasic and long-lived

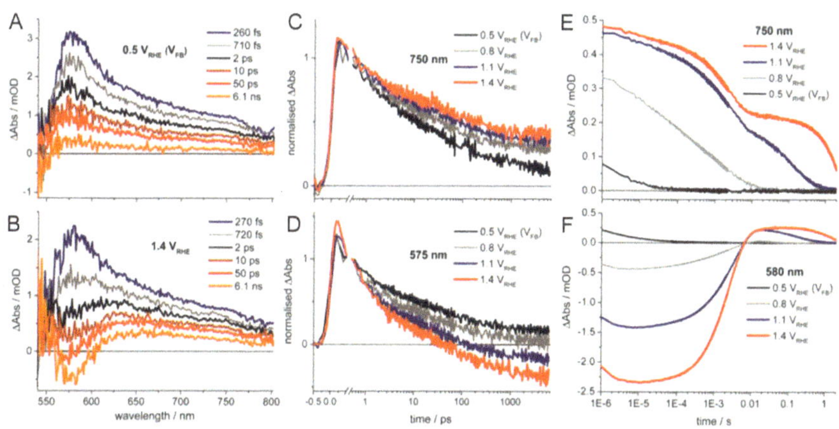

Fig. 6 TAS of a Si-doped hematite photoanode, measured during water oxidation in 0.1 M NaOH (pH 12.8) at various applied potentials. The femto- to nanosecond transient absorption spectra at (A) 0.5 V_{RHE} and (B) 1.4 V_{RHE} and kinetics (normalized at 0.5 ps) at various applied potentials probed at (C) 750 nm and (D) 575 nm, following a 355-nm laser pulse (500 Hz, ~110 μJ cm^{-2} $pulse^{-1}$). The corresponding microsecond–second transient absorption kinetics at (E) 750 nm and (F) 580 nm, following a 355-nm laser pulse (0.33 Hz, ~200 μJ cm^{-2} $pulse^{-1}$). *Reprinted with permission from Pendlebury, S.R., Wang, X., Formal, F. Le, Cornuz, M., Kafizas, A., Tilley, S.D.,* Grätzel, M., *Durrant, J.R., 2014. Ultrafast charge carrier recombination and trapping in hematite photoanodes under applied bias. JACS 136, 9854. Copyright (2014) American Chemical Society.*

(~2–4 s) (Pendlebury et al., 2011) (Fig. 6E and F). This biphasic decay is due to two competing processes: (i) electron–hole recombination on the μs–ms timescale and (ii) the reaction of surface holes with water on the ms–s timescale. Florian et al. showed that this electron–hole recombination on the μs–ms timescale is due to a process called "back electron–hole recombination" (Formal et al., 2014). This occurs when photogenerated electrons, initially extracted from the material and electrode layer, come back into the material and recombine with surface holes. Preventing back electron–hole recombination in α-Fe_2O_3 requires the application of a strong anodic bias (≥1.4 V_{RHE}) and is needed to yield holes with lifetimes comparable to the water oxidation rate, making it the primary cause for the onset potential of water oxidation being ~500 mV anodic of flat band.

Further transient absorption studies on the charge carrier dynamics of α-Fe_2O_3 photoanodes have focused on slower timescales, over microseconds–seconds. Pendlebury et al. found that under these pulsed excitation conditions, the kinetics of water oxidation was independent of the concentration of photogenerated holes, indicating the mechanism of water oxidation on α-Fe_2O_3 occurs via sequential single-hole transfer steps (Pendlebury et al., 2012). Cowan et al. monitored the reaction kinetics of photogenerated holes with water at various temperatures (~25–55°C) and found that the reaction kinetics became faster with increasing temperature (Cowan et al., 2011). An Arrhenius plot of the temperature dependence showed that the activation energy of hole transfer was 0.45 ± 0.04 eV. This activation energy was independent of the applied potential. Therefore, the increase in activity found in α-Fe_2O_3 at more positive potentials is attributed to the inhibition of electron–hole recombination, providing more long-lived holes that are able to oxidize water, and not to a major change in the rate of the hole transfer step.

Barroso et al. studied the effect of modifying the surface of α-Fe_2O_3 photoanodes by adding either an ultrathin passivation layer of Ga_2O_3 or a CoPi surface "catalyst" (Barroso et al., 2012). Both surface modifications caused the onset potential for water oxidation to cathodically shift by ~200 mV. However, their transient absorption studies showed that the addition of CoPi did not increase the reaction kinetics of water oxidation (i.e., CoPi did not formally act as a catalyst). Rather, the CoPi "catalyst" increased the yield of long-lived holes in α-Fe_2O_3 that could oxidize water, which was attributed to an enhancement in the magnitude of the space charge region at the α-Fe_2O_3/CoPi interface. Surface modifications with either Ga_2O_3 or CoPi passivate surface states and thus enhance the width of the space charge layer at a given voltage bias. This was more obvious in CoPi-modified α-Fe_2O_3 (given its redox active nature), yielding a three orders of magnitude increase in the lifetime of photogenerated holes in the absence of applied potential (Barroso et al., 2011). A similar effect was observed by Bein et al. in their transient absorption studies of Co_3O_4 on α-Fe_2O_3 (Feckl et al., 2015). The enhancement in activity was attributed to the suppression of electron–hole recombination and not to an increase in catalytic activity

Cowan et al. found that acid treatment can substantially inhibit electron–hole recombination losses in α-Fe_2O_3 photoanodes (Yang et al., 2016). They also studied the role of oxygen vacancies in α-Fe_2O_3 by heating anodes in oxygen deficient atmospheres and found that oxygen vacancies retard back electron–hole recombination (Forster et al., 2015).

3.4 PIAS

In their pioneering work, Le Formal et al. were the first to use PIAS to study the reaction kinetics of photogenerated holes on α-Fe_2O_3 photoanodes (Le Formal et al., 2015). Exciting the material with continuous wave illumination (as opposed to the pulsed laser excitation used in TAS), PIAS can track the optical absorption of photogenerated charge over seconds, allowing one to study the behavior of charge carriers at conditions typical to an applied device (i.e., continuous wave solar light). Le Formal et al. compared the accumulation of surface holes on α-Fe_2O_3 (Fig. 7A), with the photocurrent extracted at several light intensities (Fig. 7B). When the light was turned on, the hole concentration increased until it reached a steady-state plateau (~2–5 s). When the light was turned off, the hole concentration decreased (~5–10 s), assigned to their dissipative reaction with water. Changes in photocurrent were sharper than changes in hole concentration, reaching a steady state almost instantly as the excitation light source was turned on and off. Strikingly, they found that the steady-state hole absorption increased nonlinearly with light intensity, in contrast to the photocurrent which showed a near-linear dependence (Fig. 7C). Using the extinction coefficient of a hole in α-Fe_2O_3 (~640 M^{-1} cm^{-1} at 650 nm), Le Formal et al. converted their optical signals into a surface hole density. The differences in behavior at steady state were assessed using a simple kinetic model, where the change in surface hole concentration $[h^+]$ is dependant on the flux of photogenerated holes I_{ph} (i.e., the photocurrent), leading to the formation of holes, and the rate of water oxidation, leading to the consumption of holes:

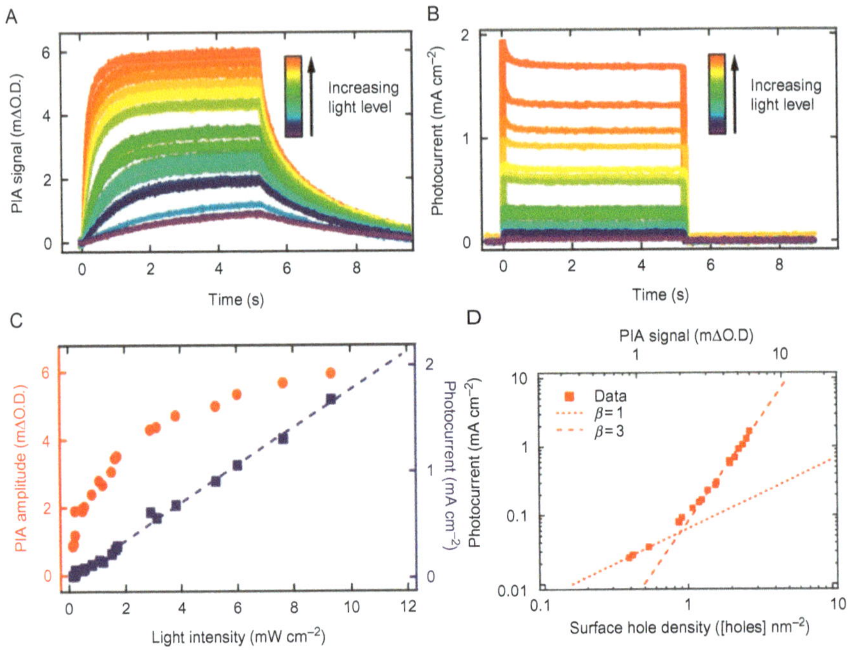

Fig. 7 PIAS of a Si-doped hematite photoanode, measured during water oxidation in 1 M NaOH (pH 13.6) at 1.5 V_{RHE}. The behavior was assessed at various light intensities using a continuous wave 365 nm LED light source (~0.1–10 mW cm^{-2}), which was switched on for 5 s and then off for 5 s, where panel (A) is the photoinduced absorption of holes at 650 nm, panel (B) is the simultaneously measured photocurrent, panel (C) is the steady-state absorption and photocurrent, and panel (D) is a logarithmic plot of the surface hole density and photocurrent at steady state. *Reprinted with permission from Le Formal, F., Pastor, E., Tilley, S.D., Mesa, C.A., Pendlebury, S.R., Grätzel, M., Durrant, J.R., 2015. Rate law analysis of water oxidation on a hematite surface. J. Am. Chem. Soc. 137, 6629–6637 http://pubs.acs.org/doi/full/10.1021/jacs.5b02576. Copyright (2015) American Chemical Society.*

$$\frac{d[h^+]}{dt} = I_{ph} - k_{WO}.[h^+]^{\beta} \tag{2a}$$

where k_{WO} is the rate constant and β is the reaction order. At the steady-state conditions in which the surface hole concentration does not change with time, Eq. 2a can be simplified to:

$$I_{ph} = k_{WO} \cdot [h^+]^{\beta} \tag{2b}$$

and the terms can be separated using logarithms:

$$\log_{10}(I_{ph}) = \log_{10}k_{WO} + \beta\log_{10}[h^+] \tag{2c}$$

Thus, a logarithmic plot of the steady-state photocurrent against hole concentration (Fig. 6D) yields a gradient of β (the reaction order) and intercept k_{WO} (the rate constant). This revealed two distinct regions: one at low hole densities ($<1\ h^+ nm^{-2}$), where a first-order reaction was observed, and one at high hole densities ($>1\ h^+ nm^{-2}$) where a third-order reaction was observed. The first-order reaction observed at low hole densities ($<1\ h^+ nm^{-2}$) was consistent with transient absorption studies (Cowan et al., 2011), where water oxidation proceeds via single-hole oxidation steps. The observation of third-order behavior at higher hole densities was attributed to a change in water oxidation mechanism, where multiple holes (or oxidized equivalents) participate in the rate-determining step. If we consider surface holes to be located on Fe sites, the transition to a third-order reaction occurred when approximately each oxidized Fe atom had, on average, one nearest neighbor oxidized. This suggested surface hole diffusion was unimportant, and multihole water oxidation occurs when nearest neighbor Fe atoms are oxidized. Moreover, the transition from a first- to third-order reaction results in dramatically accelerated reaction kinetics for higher hole accumulation (e.g., at a hole density of $\sim 3\ h^+ nm^{-2}$, a photocurrent of ~ 2 mA cm^{-2} is observed in the third-order region, and by extrapolating the first-order regime, we predict a photocurrent of just ~ 0.2 mA cm^{-2} at a hole density of $\sim 3\ h^+ nm^{-2}$). This study showed that a multihole oxidation of water can occur on α-Fe_2O_3, similar to the manganese cluster within photosystem II (Dasgupta et al., 2008).

3.5 EIS

Hamann et al. used EIS to study the water oxidation reaction on α-Fe_2O_3 photoanodes grown by atomic layer deposition (Klahr et al., 2012a). They compared the water oxidation reaction with the oxidation a hole scavenger, ferrocyanide. Mott–Schottky plots of the ferrocyanide oxidation reaction were identical in the dark and under illumination, which indicated surface state holes do not have a role in oxidizing ferrocyanide, but rather, valence band holes are involved (analogous to methanol oxidation; Klahr et al., 2015). Although water oxidation did not occur at low applied potentials, Hamann et al. found that a fraction of holes were trapped at surface states. It was postulated that these surface states (possibly surface hydroxide species) were intermediates in the water oxidation reaction, as these states are only observed in the presence of water and not in the presence of ferrocyanide. However, a critical number of these surface intermediates need to be generated in order for water oxidation to proceed. Hamann et al. found that the lifetime of these surface state holes increases by an order of magnitude from ~0.05 s at 1.0 V_{RHE} (where no photocurrent was observed) to ~1 s at 1.5 V_{RHE} (where significant photocurrent was observed).

Le Formal et al. studied the water oxidation reaction on α-Fe_2O_3 photoanodes using both TAS and EIS (Le Formal et al., 2014). Back electron–hole recombination was observed using both techniques, with a rate constant that ranged from ~100 to 1 s^{-1} from 0.8 to 1.3 V_{RHE}. In α-Fe_2O_3, this recombination processes occurs from 0.8 to 1.3 V_{RHE} as the space charge field can separate the photogenerated charge but is not large enough to block the back flow of electrons induced by accumulated holes at the interface. However, above 1.3 V_{RHE} (~0.8 V_{fb}), the space charge width and field are large enough to repulse electrons and inhibit this recombination path. This implies that the space charge layer has two distinct roles in inhibiting electron–hole recombination in α-Fe_2O_3 and operates on two very different time scales. First, the space charge layer needs to be large enough to retard fast recombination (<10 ms), leading to spatial separation of electrons and holes and the accumulation of long-lived holes at the surface. Second, the space charge layer (and the associated electric field) needs to be increased significantly more to stop the back electron flow (10 ms–10 s) induced by the accumulated holes. Both recombination processes involve the recombination of electrons and holes; they are however distinct as the latter process either requires a thermal activation or tunnels across the space charge layer and therefore occurs on much slower time scales with a different dependence on applied potential.

Hamann et al. investigated the effect of doping α-Fe_2O_3 photoanodes with Ti and found that the incorporation of Ti atoms dramatically enhances the water oxidation performance (optimum ~3 at.% Ti) (Zandi et al., 2013). Ti doping inhibited the formation of the "dead layer," an imperfect crystal layer that forms at the α-Fe_2O_3:FTO interface due to lattice mismatch, which resulted in the combination of improved bulk properties (hole collection length) and surface properties (water oxidation efficiency). In contrast to undoped α-Fe_2O_3, EIS studies of Ti-doped α-Fe_2O_3 showed that the charge transfer resistance from surface state holes to solution decreased (i.e., the reaction kinetics of holes with water increased) and surface state capacitance increased (i.e., the number of surface state holes increased). This showed that Ti doping increased both the activity and number of surface state holes. Hamann et al. also investigated the differences between α-Fe_2O_3 photoanodes grown by atomic layer deposition and electrodeposition (Zandi et al., 2016). EIS studies indicated electrodeposited materials were more catalytically active for water oxidation. This was due the higher degree of preferred growth in the [110] diffraction plane, which results in more surface terminated Fe–OH groups and a more highly active surface toward water oxidation.

Bahnemann et al. investigated the effect of ultrathin (<2 nm thick) TiO_2 passivation layers grown on the surface of α-Fe_2O_3 photoanodes (Ahmed et al., 2015). The passivation layers were grown using a simple dip-coating method, and cathodically shifted the water oxidation onset potential by nearly 200 mV. EIS studies showed that the TiO_2 passivation layer both suppressed electron–hole recombination and increased catalytic activity, where the rate constant for water oxidation at 1.2 V_{RHE} increased from $\sim$20 s^{-1} on bare α-Fe_2O_3 to $\sim$55 s^{-1} on TiO_2 coated α-Fe_2O_3. Hamann et al. also investigated the effect of CoPi and found that the CoPi layer can accept and store photogenerated holes from α-Fe_2O_3 (Klahr et al., 2012b). This increased charge separation and reduced surface state recombination. Hamann et al. also studied the addition of the surface cocatalyst Co_3O_4 (Riha et al., 2013). The optimal coating was less than 1 monolayer thick and cathodically shifted the water oxidation onset potential by 100–200 mV. EIS indicated that the addition of the cocatalyst increased the catalytic activity of the surface to water oxidation by an order of magnitude.

4. $BiVO_4$

$BiVO_4$ has emerged as one of the most promising photoanode materials (Park et al., 2013). It is composed of inexpensive elements and has an optical bandgap of $\sim$2.4 eV, with commercial use as a yellow pigment. Kudo et al. were the first to demonstrate the material's ability to photocatalytically oxidize water (Kudo et al., 1998). It is generally found that the monoclinic scheelite structure of $BiVO_4$ is the most photocatalytically active phase. In the scheelite structure, each V^{5+} cation is coordinated by four O^{2-} anions in a tetrahedral site, and each Bi^{3+} cation is coordinated by eight O^{2-} anions from eight neighboring $VO_4{}^{3-}$ tetrahedral units. Recent studies of $BiVO_4$ have focused on their use as photoanodes to oxidize water. Undoped $BiVO_4$ possesses poor electrical conductivity, where studies have shown that doping with either W or Mo often results in a more conductive and active photoanode (Parmar et al., 2012). To further improve the efficiency of $BiVO_4$ photoanodes, strategies such as nanostructuring, forming heterojunctions, and the addition of cocatalysts have also been explored (Liang and Messinger, 2014). Of note, $BiVO_4$ is more susceptible to photocorrosion than TiO_2, α-Fe_2O_3, and WO_3 (Chen and Wang, 2012). To our knowledge, the highest photocurrent achieved on a $BiVO_4$-based photoanode is ~4.4 mA cm^{-2} (at 1.23 V_{RHE} and 1 sun irradiance), based on a nanoporous $BiVO_4$ structure with a surface Ni:FeOOH cocatalyst (Kim and Choi, 2014).

4.1 TRMC and THz Spectroscopy

The TRMC of $BiVO_4$ was studied by van de Krol et al. (Abdi et al., 2013). They found that carrier mobility ($cm^2\ V^{-1}\ s^{-1}$) in undoped $BiVO_4$ (0.04) was substantially lower than α-Fe_2O_3 (~5), WO_3 (~10), and TiO_2 (~3–26) (Kafizas et al., 2013). Carrier lifetimes of $t_{50\%}\sim 40$ ns and diffusion lengths of ~70 nm were observed. W-doping decreased carrier mobility and introduced carrier traps. However, the increase in carrier density and conductivity improved the activity of the photoanode, which demonstrated that bulk electronic conductivity is one of the main performance bottlenecks of $BiVO_4$.

Using THz spectroscopy, van de Krol et al. also examined ultrafast changes in photoinduced conductivity (Ziwritsch et al., 2016). From 1 ps to 1 ns, the localization of charge carriers caused carrier mobility to decrease by an order of magnitude. This was due to both hole polaron formation and charge carrier trapping. Self-trapped carriers possessed a mobility of 0.02 $cm^2\ V^{-1}\ s^{-1}$ and a thermal hopping activation energy of ~90 meV.

4.2 TCSPC

Ng et al. examined the effect of altering the ratio of exposed [010] and [110] crystal faces on monoclinic $BiVO_4$ nanocrystals (Tan et al., 2016). They found that the [010] face was dominant in plate-like nanocrystals and the [110] face was dominant in truncated bipyramidal nanocrystals. TCSPC showed that electron–hole recombination was slower in plate-like nanocrystals ($t_{50\%}\sim 1$ ns) compared with truncated bipyramidal nanocrystals ($t_{50\%}\sim 0.8$ ns).

4.3 TAS

The ultrafast charge carrier behavior in $BiVO_4$ photoanodes was investigated by van de Krol et al. using TAS (Ravensbergen et al., 2014). Hole carriers were observed in the visible region between ~475 and 700 nm, with the majority being trapped within 5 ps. They attributed a shift in the ground-state bleach at ~450 nm to electron trapping processes, which relaxed with a time constant of 40 ps, followed by deeper trapping on the 2.5 ns timescale. On time scales longer than 10 ns, trap-limited recombination was observed and followed a power law decay, spanning timescales up to the microsecond. Unlike α-Fe_2O_3 (Pendlebury et al., 2014), no spectral or kinetic differences were observed on the application of an applied potential, which indicated early timescale recombination processes in $BiVO_4$ are not inhibited by the action of an applied voltage.

Huang et al. investigated the effect of W-doping (Pattengale et al., 2016). They found that the number of intrinsic hole traps decreased, inhibiting electron–hole recombination and increasing the electron lifetime. X-ray absorption fine structure measurements showed that the local structure at Bi centers was less distorted in W-doped $BiVO_4$ and potentially altered the nature of these hole trap states.

Kamat et al. investigated the ultrafast carrier dynamics in $WO_3/BiVO_4$ heterojunction photoanodes (Grigioni et al., 2015). Through hole scavenging studies with sodium sulfite, they confirmed the transient absorption of photogenerated holes in the visible region (~475–750 nm). Again, a bleach due to electron trapping in $BiVO_4$ was observed at ~450 nm. When in contact with WO_3, this bleach signal decayed more rapidly and was attributed to the injection of electrons from $BiVO_4$ into WO_3 (with a time constant of ~18 ps). Buonsanti et al. observed the same effect in their study of the $WO_3/BiVO_4$ system (Loiudice et al., 2015). In addition, they found that the hole signals were far more long-lived in the heterojunction (up to ~1 s) compared with $BiVO_4$ alone (up to ~10 ms).

The transient kinetics of water oxidation on $BiVO_4$ photoanodes was studied by Ma et al. over the microsecond–second timescale, where water oxidation takes place (Ma et al., 2014b). Using chemical scavengers, they found that photogenerated holes most strongly absorbed at ~550 nm. The hole carrier dynamics were studied at a range of applied potentials (Fig. 8A). At potentials greater than ~0.6 V_{RHE}, a long-lived component was observed, which coincided with the photocurrent onset. This was assigned to the long-lived hole states that oxidized water. Each transient decay in hole absorption was composed of three competing processes: (i) bimolecular electron–hole recombination, which followed a power law decay and (ii) the slower processes of either back electron–hole recombination or (iii) water oxidation by holes, which followed exponential decays (Fig. 8B). Transient photocurrent studies showed that back electron–hole

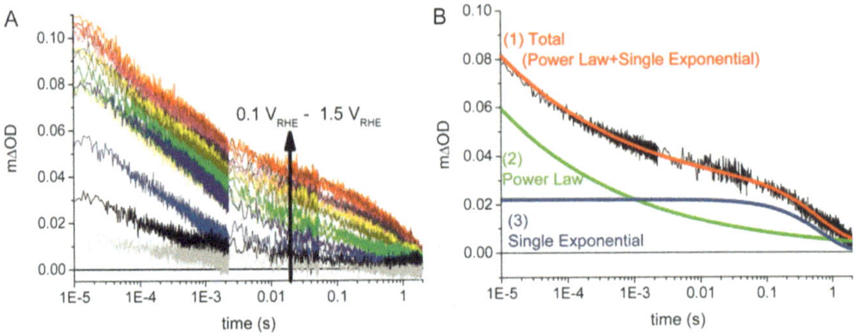

Fig. 8 TAS of a $BiVO_4$ photoanode, measured during water oxidation in 0.05 M K_2HPO_4 + 0.05 M KH_2PO_4 (pH 6.7) at various applied potentials. (A) Transient decay kinetics of photogenerated holes at 550 nm, following a 355-nm laser pulse (~0.8 mJ cm^{-2}). (B) The fitting (*red*) to a hole decay measured at 1.2 V_{RHE}, composed of a power law (*green*) and single exponential (*blue*) function. *Ma, Y., Pendlebury, S.R., Reynal, A., Formal, F. Le, Durrant, J.R., 2014. Dynamics of photogenerated holes in undoped $BiVO_4$ photoanodes for solar water oxidation. Chem. Sci. 5, 2964–2973—Published by The Royal Society of Chemistry.*

recombination was turned off at potentials above ~1.4 V_{RHE}. At similar applied potentials, bimolecular recombination was shown to decrease substantially with a decrease in laser power. At strong anodic bias, the water oxidation kinetics of holes were independent of excitation intensity, and therefore, hole density ($t_{50\%}$ ~ 500 ms). This indicated that under the experimental conditions employed, water oxidation did not involve a concerted, multihole reaction, but rather, sequential single-hole water oxidation steps.

Ma et al. also studied the effect of modifying the surface of $BiVO_4$ with CoPi (Ma et al., 2015). The addition of CoPi caused an ~200 mV cathodic shift in the photocurrent onset, and substantially retarded back electron–hole recombination. However, the rate constant for water oxidation did not substantially increase (~1 s^{-1} in $BiVO_4$ vs ~0.7 s^{-1} in CoPi-modified $BiVO_4$). This led them to conclude that the primary function of CoPi on $BiVO_4$ was to retard back electron–hole recombination across the space charge layer, with no evidence of catalytic water oxidation via CoPi being observed.

4.4 PIAS

Similar to Le Formal et al.'s PIAS studies of the water oxidation reaction on α-Fe_2O_3 (Le Formal et al., 2015), Ma et al. studied the same reaction on $BiVO_4$ (Ma et al., 2016b). Analogous to α-Fe_2O_3, the reaction order changed with hole density from first order for $[h^+] < 1$ nm^{-2} to third order for

$[h^+] > 1\ nm^{-2}$. The first-order behavior at low surface hole densities indicated that the rate-limiting step in water oxidation involved single-hole transfers, where hydroxyl radical formation has been suggested as the key intermediate for the single-hole oxidation of water. The third-order reaction kinetics under high surface hole densities is consistent with, for example, a multihole reaction mechanism involving the formation of an O–O bridging bond between two reaction sites. The effective turnover frequency (TOF) of each surface hole increase from $0.3\ s^{-1}$ at $0.1\ h^+ nm^{-2}$ to $14\ s^{-1}$ at $5\ h^+ nm^{-2}$ (corresponding to conditions of ~1 sun irradiance). TOFs on $BiVO_4$ were faster than those found on α-Fe_2O_3 ($5\ s^{-1}$ at $\sim 5\ h^+ nm^{-2}$), which may result from its deeper valence band energy or other factors such as lateral surface hole mobility. Of note, these water oxidation kinetics on $BiVO_4$ are faster in terms of TOF per surface hole/catalytic site compared to those reported for CoPi, which range between 0.01 and $0.001\ s^{-1}$ (Klingan et al., 2014). The faster kinetics on $BiVO_4$ can be attributed to its deep valence band energy providing a large overpotential for water oxidation. The effect of CoPi-modified $BiVO_4$ was studied in more detail by Ma et al., where they found that $\leq 5\%$ of photocatalytic water oxidation proceeds from CoPi sites (Ma et al., 2016a). This was due to the combination of slow hole transfer kinetics from $BiVO_4$ to CoPi and slower water oxidation kinetics on CoPi.

4.5 EIS

Lee et al. studied a wide range of dopants in $BiVO_4$ and used EIS to understand the effect of each dopant (Parmar et al., 2012). They found that W- or Mo-doped $BiVO_4$ photoanodes showed markedly improved charge transfer at the photoanode:electrolyte interface and suppressed electron–hole recombination. Lee et al. also studied the effect of phosphate oxoanion doping (Jo et al., 2012). Phosphate doping substantially improved water oxidation activity, where EIS measurements showed that this was due to a decrease in charge transfer resistance at the semiconductor:electrolyte interface.

Lee et al. also used EIS to study the charge carrier dynamics in WO_3/$BiVO_4$ heterojunction photoanodes (Hong et al., 2011). They found that the charge transfer resistance in $WO_3/BiVO_4$ heterojunctions was similar to pure WO_3 and substantially lower than pure $BiVO_4$. Grimes et al. studied both planar and nanorod structured $WO_3/BiVO_4$ heterojunctions (Su et al., 2011). Using EIS, they determined the average transit time for photogenerated electrons to reach the back contact to be ~3 ms in planar structures under front illumination. In nanorod structures the average transit time was substantially higher (~35 ms), which was attributed to a difference in crystal phase (planar = monoclinic, nanorod = hexagonal). Gimenez et al. also studied the $WO_3/BiVO_4$ heterojunction using EIS and found that the enhancement activity was primarily due to photogenerated electron injection from $BiVO_4$ to WO_3, which reduced bulk electron–hole recombination (Shi et al., 2016).

5. WO_3

The simplest crystal structure of WO_3 is an "ideal" cubic ReO_3-type structure and forms a three-dimensional network of corner-sharing WO_6 octahedra. However, cubic WO_3 is seldom observed experimentally. At room temperature, monoclinic γ-WO_3 is the most thermodynamically stable crystal phase. WO_3 is an indirect bandgap semiconductor (E_{bg} ~ 2.6–2.8 eV) and can utilize ~8%–12% of the solar spectrum. Manassen et al. were the first to demonstrate the use of WO_3 as a photoanode for water oxidation (Hodes et al., 1976), and efforts to improve the activity of this material have since focused on the exploration of various nanostructures, such as nanowires, nanotubes, nanobelts, and nanoplatelets (Liu et al., 2012). To our knowledge, the highest photocurrent achieved on a WO_3-based photoanode is ~2.4 mA cm^{-2} (at 1.23 V_{RHE} and 1 sun irradiance), based on a monoclinic nanoneedle WO_3 structure (Fabrega et al., 2016). Relatively few studies have focused on the charge carrier dynamics of WO_3, and much remains to be understood.

5.1 THz Spectroscopy

Schmuttenmaer et al. studied the transient change in photoinduced conductivity of commercial WO_3 powders over the ultrafast timescale (Regan et al., 2016). Charge carriers were longer lived in nanopowders, as opposed to granular powders. Moreover, in the presence of an electrolyte, surface states were pacified more greatly in nanopowders resulting in longer-lived photoinduced conductivity ($t_{50\%}$ increase from ~15 to ~45 ps).

5.2 TAS

Klug et al. investigated the transient absorption signals in mesoporous WO_3 using chemical scavengers (Pesci et al., 2011). In the presence of a methanol hole scavenger, long-lived electrons were observed in the red (centered at ~950 nm), and in the presence of a silver nitrate electron scavenger, long-lived holes were observed in the blue (centered at ~500 nm). A nonlinear dependence was observed between the kinetics of charge carrier recombination and the intensity of laser excitation, which was indicative of trap/detrap-limited bimolecular recombination. In the presence of silver nitrate, long-lived holes were produced with a lifetime that extended into the millisecond–second timescale that facilitated water oxidation. Choi et al. examined the effect of ultrathin Al_2O_3 passivation layers on WO_3, grown by atomic layer deposition (Kim et al., 2013b). At strong anodic bias, they found that the number of photogenerated holes increased substantially in materials coated with Al_2O_3.

5.3 EIS

Bisquert et al. studied the difference between anodically grown and colloidal WO_3 (Tacca et al., 2012). EIS of anodically grown WO_3 showed well-defined space charge effects. In contrast, EIS of colloidal WO_3 did not show any evidence of space charge effects, with behavior closer to chemical capacitance. Electron diffusion lengths were determined from transient photocurrent measurements and ranged between ~0.8 and 2.2 μm.

6. COMPARATIVE SUMMARY OF METAL OXIDE CHARGE CARRIER DYNAMICS

The charge carrier dynamics of metal oxide photoelectrodes have been investigated using various techniques. TRMC, THz spectroscopy, and TCSPC have been used to study charge carrier relaxation and recombination processes in TiO_2, α-Fe_2O_3, $BiVO_4$, and WO_3, but have not been used to study the action of these charge carriers under an applied potential during photoelectrochemical water oxidation. For TRMC and THz spectroscopy studies this may be due to the experimental difficulties of measuring transient changes in microwave absorption in water, as water is a strong absorber of microwave radiation. In addition, interpreting TRMC and THz spectroscopy signals can be difficult, as both electron–hole recombination and electron trapping can be responsible for a decrease in microwave conductivity (Schneider et al., 2014). TCSPC methods also have not been used to study the charge carrier dynamics of metal oxide photoanodes during water oxidation, which may be attributed to nonradiative recombination being the dominant pathway in metal oxides.

TAS and EIS are the most commonly used methods to probe charge carrier dynamics in metal oxides and have been used to study the reaction kinetics of photogenerated holes with water as well as the competing process of electron–hole recombination. Perhaps the most versatile technique is TAS, as it allows one to investigate the charge carrier dynamics from femtoseconds, where charge carriers are generated, up to seconds, where the kinetically slow process of water oxidation occurs. TAS and transient photocurrent methods have identified three key loss processes in metal oxide photoanodes that occur on different timescales: (i) geminate recombination (~ps), (ii) bulk electron–hole recombination (typically ps–μs), and (iii) back electron–hole recombination (typically μs–ms). In some materials, these processes are inherently more difficult to inhibit or nullify.

Of note, there are gaps in our knowledge of some materials, making it difficult to compare the charge carrier dynamics across all timescales. For instance, the charge carrier dynamics in TiO_2 photoanodes have not been studied on ultrafast timescales under the action of applied potential. Therefore, to better compare the charge carrier dynamics in metal oxide photoanodes, there should be studies that encompass a range of materials. Also, comparing EIS and TAS studies of similar materials, some contradictory conclusions have been reached, for instance, on the role of surface cocatalysts such as CoPi. In order to eradicate such contradictions, more complimentary studies that use both TAS and EIS methods need to be undertaken (Formal et al., 2014).

Table 1 summarizes some key information about the metal oxides studied herein. Of particular importance is the ratio between the maximum photocurrent achieved in practice for a given material and what is theoretically possible. Looking at the best performing TiO_2-based photoanode, we find that there is little room for improvement (only ~13% losses). However, in α-Fe_2O_3 we find that there are substantial losses (~66% losses). The high quantum efficiencies achieved in TiO_2 can be attributed to a number of properties, which include its comparatively faster water oxidation kinetics and the modest applied potentials at which back electron–hole recombination turns off. Making further comparisons between TiO_2 and α-Fe_2O_3, ultrafast TAS studies have shown that electron–hole recombination is two orders of magnitude slower in TiO_2 (Pendlebury et al., 2014). Although TiO_2 shows the least losses of any metal oxide, its wide bandgap limits its performance under sunlight. This limitation has made it imperative for other promising materials to be studied and improved. Numerous strategies have emerged, which include coupling TiO_2 with narrow bandgap materials. In

Table 1 A Summary of the Best Performing Metal Oxide Photoanodes for Water Oxidation, and Their Typical Reaction Kinetics With Water

	E_{bg}	Max J_{ph}	Theo J_{ph}	Max J_{ph}:theo J_{ph}	BER	Water oxidation kinetics (s^{-1})		
	eV	mA cm^{-2}		%	V_{RHE}	TAS	PIAS	EIS
TiO_2	3.0–3.2	2.6[a]	3.0	87	0.8	~5	—	~200
WO_3	2.6	2.4[b]	5.3	45	—	—	—	—
$BiVO_4$	2.4	4.4[c]	7.6	58	1.4	~1	~14	—
Fe_2O_3	2.0	4.3[d]	12.6	34	1.3	~0.7	~5	~20

Notes: E_{bg} is the typically reported optical bandgap (TiO_2 varies from 3.0 to 3.2 eV depending on crystal phase); max J_{ph} is the highest reported photocurrent at 1.23 V_{RHE} and 1 sun irradiance, to our knowledge, of the following materials: *a* = anatase: rutile TiO_2 nanorods (Liu et al., 2011), *b* = monoclinic WO_3 nanoneedles (Fabrega et al., 2016), *c* = nanoporous monoclinic $BiVO_4$ coated with a Ni:FeOOH (Kim and Choi, 2014), and *d* = Pt-doped α-Fe_2O_3 with a worm-like nanostructure coated with a CoPi co-catalyst (Kim et al., 2013a); theo J_{ph} is the maximum theoretical photocurrent achievable at 1 sun irradiance given the optical bandgap; max J_{ph}:theo J_{ph} is ratio between what has been achieved and what is possible; BER is back electron–hole recombination and applied potential at which it turns off; typical rate constants of water oxidation derived independently from three techniques: TAS, PIAS, and EIS.

fact, the Cu_2O/TiO_2 heterojunction forms the most active metal oxide photocathode to date, which utilizes the robust properties of TiO_2 to serve as a protection layer and inhibit photocorrosion in the Cu_2O layer (Paracchino et al., 2011). The coupling of materials within a heterojunction is certainly one of the most promising strategies for improving the activity of metal oxide-based photoanodes, where perhaps the most studied system is the $WO_3/BiVO_4$ junction (Hong et al., 2011). Studies of the charge carrier dynamics in these heterojunction systems have started to provide some unique insight into how activity is improved. Ultrathin passivation layers and surface cocatalysts can also improve activity, and their impact on the charge carrier dynamics and reaction kinetics has also been studied. It is still a matter of debate if any improvement in catalytic activity is realized when CoPi or Co_3O_4 cocatalysts are applied, and thus highlights the need to study alternative cocatalysts of promise.

7. FUTURE OUTLOOK

Studies of the charge carrier dynamics that occur in metal oxide photoanodes during water oxidation have provided important insight for this half-reaction, acknowledged as the kinetic bottleneck of water splitting. Now that it has been shown that multihole reactions with water do indeed occur at metal oxide surfaces (Le Formal et al., 2015; Ma et al., 2016b), there is a growing need to understand the mechanism by which holes oxidize water to assist in the development of more catalytically active surfaces. Techniques that elucidate the chemical nature of the surface intermediates formed during water oxidation will help us develop this understanding. Using time-resolved FTIR spectroscopy, Nakato et al. showed that oxyhydroxide (Ti–O–OH) and bridging peroxo bonds (Ti–O–O–Ti) were formed at the surface of TiO_2 during photocatalytic water oxidation (Nakamura and Nakato, 2004). Using the same technique, Hamann et al. showed that $Fe^{IV}{=}O$ intermediates were formed on α-Fe_2O_3 photoanodes and provided the first direct evidence of high valency iron–oxo intermediates (Zandi and Hamann, 2016). Similarly, Frei et al. observed superoxide formation during electrocatalytic water oxidation on Co_3O_4 (Zhang et al., 2014).

High energy X-ray techniques have made it possible to measure time-resolved structural changes during photocatalysis. This has been used to determine the structural changes that occur in the catalytic site of PS II during water oxidation, with femtosecond resolution (Kern et al., 2014; Kupitz et al., 2014). X-ray photoelectron spectroscopy has been used to study water oxidation on TiO_2 photoanodes at various applied potentials and monitor changes in surface-binding energetics (Lichterman et al., 2015). X-ray absorption spectroscopy has also been used to examine structural changes in proton reduction catalysts such as MoS_2 (Lassalle-kaiser et al., 2015) and water oxidation catalysts such as MnO_x (Gorlin et al., 2013) and CoPi (Kanan et al., 2010). Interestingly, high energy X-ray techniques have also made it possible to probe the local structural changes that occur where charge carriers are trapped in metal oxide photocatalysts (Neppl et al., 2016; Santomauro et al., 2015). There has also been an increase in the number of computational studies of this reaction (Ertem et al., 2015).

The development of the aforementioned techniques will be crucial if we are to improve our understanding of the water oxidation reaction. When coupled with methods that probe charge carrier dynamics, this will result in a more complete picture, spanning early timescale charge carrier generation to the formation and reaction of the surface intermediates that mediate water oxidation. Only by improving our knowledge of the kinetic and mechanistic aspects of this reaction, will we be able to develop more active materials through rational design.

REFERENCES

Abdi, F.F., Savenije, T.J., May, M.M., Dam, B., Van De Krol, R., 2013. The origin of slow carrier transport in $BiVO_4$ thin film photoanodes: a time-resolved microwave conductivity study. J. Phys. Chem. Lett. 4, 2752–2757.

Ahmed, M.G., Kretschmer, I.E., Kandiel, T.A., Ahmed, A.Y., Rashwan, F.A., Bahnemann, D.W., 2015. A facile surface passivation of hematite photoanodes with TiO_2 overlayers for efficient solar water splitting. ACS Appl. Mater. Interfaces 7, 24053–24062.

Appavoo, K., Liu, M., Black, C.T., Sfeir, M.Y., 2015. Quantifying bulk and surface recombination processes in nanostructured water splitting photocatalysts via in situ ultrafast spectroscopy. Nano Lett. 15, 1076–1082.

Baram, N., Ein-eli, Y., 2010. Electrochemical impedance spectroscopy of porous TiO_2 for photocatalytic applications. J. Phys. Chem. C 2, 9781–9790.

Barreca, D., Carraro, G., Gasparotto, A., Maccato, C., Warwick, M.E.A., Kaunisto, K., Sada, C., Turner, S., Gönüllü, Y., Ruoko, T.P., Borgese, L., Bontempi, E., Van Tendeloo, G., Lemmetyinen, H., Mathur, S., 2015. Fe_2O_3-TiO_2 nano-heterostructure photoanodes for highly efficient solar water oxidation. Adv. Mater. Interfaces 2, 1–11.

Barroso, M., Cowan, A.J., Pendlebury, S.R., Gratzel, M., Klug, D.R., Durrant, J.R., 2011. The role of cobalt phosphate in enhancing the photocatalytic activity of α-Fe_2O_3 toward water oxidation. J. Am. Chem. Soc. 133, 14868.

Barroso, M., Mesa, C.A., Pendlebury, S.R., Cowan, A.J., Hisatomi, T., Sivula, K., Grätzel, M., Klug, D.R., Durrant, J.R., 2012. Dynamics of photogenerated holes in surface modified α-Fe_2O_3 photoanodes for solar water splitting. Proc. Natl. Acad. Sci. U.S.A. 109, 15640.

Barroso, M., Pendlebury, S.R., Cowan, A.J., Durrant, J.R., 2013. Charge carrier trapping, recombination and transfer in hematite (α-Fe_2O_3) water splitting photoanodes. Chem. Sci. 4, 2724.

Beaudin, M., Zareipour, H., Schellenberglabe, A., Rosehart, W., 2010. Energy storage for mitigating the variability of renewable electricity sources: an updated review. Energy Sustain. Dev. 14, 302–314.

Bolton, J.R., Strickler, S.J., Connolly, J.S., 1985. Limiting and realizable efficiencies of solar photolysis of water. Nature 316, 495–500.

Broecker, W.S., 1975. Climatic change: are we on the brink of a pronounced global warming? Science 189, 460–463.

Busby, R.L., 2005. Hydrogen and Fuel Cells: A Comprehensive Guide. PennWell Books, Oklahoma, USA.

Cachet, H., Sutter, E.M.M., 2015. Kinetics of water oxidation at TiO_2 nanotube arrays at different pH domains investigated by electrochemical and light-modulated impedance spectroscopy. J. Phys. Chem. C 119, 25548–25558.

Carneiro, J.T., Savenije, T.J., Moulijn, J.A., Mul, G., 2010. Toward a physically sound structure-activity relationship of TiO_2-based photocatalysts. J. Phys. Chem. C 114, 327–332.

Carneiro, J.T., Savenije, T.J., Moulijn, J.A., Mul, G., 2011. How phase composition influences optoelectronic and photocatalytic properties of TiO_2. J. Phys. Chem. C 115, 2211.

Chen, S., Wang, L.-W., 2012. Thermodynamic oxidation and reduction potentials of photocatalytic semiconductors in aqueous solution. Chem. Mater. 24, 3659.

Cheng, J., Vandevondele, J., Sprik, M., 2014. Identifying trapped electronic holes at the aqueous TiO_2 interface. J. Phys. Chem. C 118, 5437.

Colbeau-Justin, C., Kunst, M., Huguenin, D., 2003. Structural influence on charge-carrier lifetimes in TiO_2 powders studied by microwave absorption. J. Mater. Sci. 38, 2429–2437.

Cowan, A.J., Durrant, J.R., 2012. Long-lived charge separated states in nanostructured semiconductor photoelectrodes for the production of solar fuels. Chem. Soc. Rev. 42, 2281–2293.

Cowan, A.J., Tang, J., Leng, W., Durrant, J.R., Klug, D.R., 2010. Water splitting by nanocrystalline TiO_2 limited by charge recombination in a complete photoelectrochemical cell exhibits efficiencies. J. Phys. Chem. C 114, 4208–4214.

Cowan, A.J., Barnett, C.J., Pendlebury, S.R., Barroso, M., Sivula, K., Grätzel, M., Durrant, J.R., Klug, D.R., 2011. Activation energies for the rate-limiting step in water photooxidation by nanostructured α-Fe_2O_3 and TiO_2. J. Am. Chem. Soc. 133, 10134.

Cowan, A.J., Leng, W., Barnes, P.R.F., Klug, D.R., Durrant, J.R., 2013. Charge carrier separation in nanostructured TiO_2 photoelectrodes for water splitting. Phys. Chem. Chem. Phys. 15, 8772.

Dasgupta, J., Ananyev, G.M., Dismukes, G.C., 2008. Photoassembly of the water-oxidizing complex in photosystem II. Coord. Chem. Rev. 252, 347–360.

De Haas, M.P., Warman, J.M., 1982. Photon-induced molecular charge separation studied by nanosecond time-resolved microwave conductivity. Chem. Phys. 73, 35–53.

Dry, M.E., 2002. High quality diesel via the Fischer-Tropsch process—a review. J. Chem. Technol. Biotechnol. 77, 43–50.

Emilio, C.A., Litter, M.I., Kunst, M., Bouchard, M., Colbeau-Justin, C., 2006. Phenol photodegradation on platinized-TiO_2 photocatalysts related to charge-carrier dynamics. Langmuir 22, 3606–3613.

Ertem, M.Z., Kharche, N., Batista, V.S., Hybertsen, M.S., Tully, J.C., Muckerman, J.T., 2015. Photoinduced water oxidation at the aqueous GaN (10-10) interface: deprotonation kinetics of the first proton-coupled electron-transfer step. ACS Catal. 5, 2317–2323.

Fabrega, C., Murcia-Lopez, S., Monllor-Satoca, D., Prades, J.D., Hernandez-Alonso, M.D., Penelas, G., Morante, J.R., Andreu, T., 2016. Efficient WO_3 photoanodes fabricated by pulsed laser deposition for photoelectrochemical water splitting with high faradaic efficiency. Appl. Catal. B Environ. 189, 133–140.

Fawcett, A.A., Iyer, G.C., Clarke, L.E., Edmonds, J.A., Hultman, N.E., McJeon, H.C., Rogelj, J., Schuler, R., Alsalam, J., Asrar, G.R., Creason, J., Jeong, M., McFarland, J., Mundra, A., Shi, W., 2015. Can Paris pledges avert severe climate change? Science 350, 1168–1169.

Feckl, J.M., Dunn, H.K., Zehetmaier, P.M., Muller, A., Pendlebury, S.R., Zeller, P., Fominykh, K., Kondofersky, I., Doblinger, M., Durrant, J.R., Scheu, C., Peter, L., Fattakhova-Rohlfing, D., Bein, T., 2015. Ultrasmall Co_3O_4 nanocrystals strongly enhance solar water splitting on mesoporous hematite. Adv. Mater. Interfaces 2, 1–8.

Fillol, J.L., Codolà, Z., Garcia-Bosch, I., Gómez, L., Pla, J.J., Costas, M., 2011. Efficient water oxidation catalysts based on readily available iron coordination complexes. Nat. Chem. 3, 807–813.

Forster, M., Potter, R.J., Ling, Y., Yang, Y., Klug, D.R., Li, Y., Cowan, A.J., 2015. Oxygen deficient α-Fe_2O_3 photoelectrodes: a balance between enhanced electrical properties and trap-mediated losses. Chem. Sci. 6, 4009–4016.

Fujihara, K., Izumi, S., Ohno, T., Matsumura, M., 2000. Time-resolved photoluminescence of particulate TiO_2 photocatalysts suspended in aqueous solutions. J. Photochem. Photobiol. A Chem. 132, 99–104.

Fujishima, A., Honda, K., 1972. Electrochemical photolysis of water at a semiconductor electrode. Nature 238, 37–38.

Fujishima, A., Zhang, X., Tryk, D., 2008. TiO_2 photocatalysis and related surface phenomena. Surf. Sci. Rep. 63, 515–582.

Ge, M.-Z., Cao, C.-Y., Li, S.-H., Tang, Y.-X., Wang, L.-N., Qi, N., Huang, J.-Y., Zhang, K.-Q., Al-Deyab, S.S., Lai, Y.-K., 2016. In situ plasmonic Ag nanoparticle anchored TiO_2 nanotube arrays as visible-light-driven photocatalysts for enhanced water splitting. Nanoscale 8, 5226–5234.

Gorlin, Y., Lassalle-Kaiser, B., Benck, J.D., Gul, S., Webb, S.M., Yachandra, V.K., Yano, J., Jaramillo, T.F., 2013. In situ X-ray absorption spectroscopy investigation of a bifunctional manganese oxide catalyst with high activity for electrochemical water oxidation and oxygen reduction. J. Am. Chem. Soc. 135, 8525–8534.

Grigioni, I., Stamplecoskie, K.G., Selli, E., Kamat, P.V., 2015. Dynamics of photogenerated charge carriers in WO_3/$BiVO_4$ heterojunction photoanodes. J. Phys. Chem. C 119, 20792.

Gruber, N., 2011. Warming up, turning sour, losing breath: ocean biogeochemistry under global change. Philos. Trans. A Math. Phys. Eng. Sci. 369, 1980–1996.

Gust, D., Kramer, D., Moore, A., Moore, T.A., Vermaas, W., 2011. Engineered and artificial photosynthesis: human ingenuity enters the game. MRS Bull. 33, 383–387.

Hall, C.A.S., Dale, B.E., Pimentel, D., 2011. Seeking to understand the reasons for different energy return on investment (EROI) estimates for biofuels. Sustainability 3, 2413–2432.

Hardee, K.L., Bard, A.J., 1976. Semiconductor electrodes. J. Electrochem. Soc. 123, 1024.

Hashimoto, K., Irie, H., Fujishima, A., 2005. TiO_2 photocatalysis: a historical overview and future prospects. Jpn. J. Appl. Phys. 44, 8269.

Hill, J.C., Choi, K., 2012. Effect of electrolytes on the selectivity and stability of n-type WO_3 photoelectrodes for use in solar water oxidation. J. Phys. Chem. C 116, 7612–7620.

Hodes, G., Cahen, D., Manassen, J., 1976. Tungsten trioxide as a photoanode for a photoelectrochemical cell (PEC). Nature 260, 312–313.

Hong, S.J., Lee, S., Jang, J.S., Lee, J.S., 2011. Heterojunction $BiVO_4$/WO_3 electrodes for enhanced photoactivity of water oxidation. Energy Environ. Sci. 4, 1781.

Hu, S., Shaner, M.R., Beardslee, J.A., Lichterman, M., Brunschwig, B.S., Lewis, N.S., 2014. Amorphous TiO_2 coatings stabilize Si, GaAs, and GaP photoanodes for efficient water oxidation. Science 344, 1005–1009.

Imanishi, A., Okamura, T., Ohashi, N., Nakamura, R., Nakato, Y., 2007. Mechanism of water photooxidation reaction at atomically flat TiO_2 (rutile) (110) and (100) surfaces: dependence on solution pH. J. Am. Chem. Soc. 129, 11569–11578.

Jing, L., Cao, Y., Cui, H., Durrant, J.R., Tang, J., Liu, D., Fu, H., 2012a. Acceleration effects of phosphate modification on the decay dynamics of photo-generated electrons of TiO_2 and its photocatalytic activity. Chem. Commun. 48, 10775–10777.

Jing, L., Zhou, J., Durrant, J.R., Tang, J., Liu, D., Fu, H., 2012b. Dynamics of photogenerated charges in the phosphate modified TiO_2 and the enhanced activity for photoelectrochemical water splitting. Energy Environ. Sci. 5, 6552.

Jo, W.J., Jang, J.W., Kong, K.J., Kang, H.J., Kim, J.Y., Jun, H., Parmar, K.P.S., Lee, J.S., 2012. Phosphate doping into monoclinic $BiVO_4$ for enhanced photoelectrochemical water oxidation activity. Angew. Chem. Int. Ed. Engl. 51, 3147–3151.

Kafizas, A., Noor, N., Carmalt, C.J., Parkin, I.P., 2013. TiO_2-based transparent conducting oxides; the search for optimum electrical conductivity using a combinatorial approach. J. Mater. Chem. C 1, 6335.

Kafizas, A., Wang, X., Pendlebury, S.R., Barnes, P.R.F., Ling, M., Sotelo-Vazquez, C., Quesada-Cabrera, R., Li, C., Parkin, I.P., Durrant, J.R., 2016. Where do photogenerated holes go in anatase: rutile TiO_2? A transient absorption spectroscopy study of charge transfer and lifetime. J. Phys. Chem. A 120, 715–723.

Kanan, M.W., Yano, J., Surendranath, Y., Dincă, M., Yachandra, V.K., Nocera, D.G., 2010. Structure and valency of a cobalt-phosphate water oxidation catalyst determined by in situ X-ray spectroscopy. J. Am. Chem. Soc. 132, 13692–13701.

Kern, J., Tran, R., Alonso-Mori, R., Koroidov, S., Echols, N., Hattne, J., Ibrahim, M., Gul, S., Laksmono, H., Sierra, R.G., Gildea, R.J., Han, G., Hellmich, J., Lassalle-Kaiser, B., Chatterjee, R., Brewster, A.S., Stan, C.A., Glöckner, C., Lampe, A., DiFiore, D.,

Milathianaki, D., Fry, A.R., Seibert, M.M., Koglin, J.E., Gallo, E., Uhlig, J., Sokaras, D., Weng, T.-C., Zwart, P.H., Skinner, D.E., Bogan, M.J., Messerschmidt, M., Glatzel, P., Williams, G.J., Boutet, S., Adams, P.D., Zouni, A., Messinger, J., Sauter, N.K., Bergmann, U., Yano, J., Yachandra, V.K., 2014. Taking snapshots of photosynthetic water oxidation using femtosecond X-ray diffraction and spectroscopy. Nat. Commun. 5, 4371.

Kim, T.W., Choi, K.-S., 2014. Nanoporous $BiVO_4$ photoanodes with dual-layer oxygen evolution catalysts for solar water splitting. Science 343, 990–994.

Kim, J.Y., Magesh, G., Youn, D.H., Jang, J.-W., Kubota, J., Domen, K., Lee, J.S., 2013a. Single-crystalline, wormlike hematite photoanodes for efficient solar water splitting. Sci. Rep. 3, 37–38.

Kim, W., Tachikawa, T., Monllor-Satoca, D., Kim, H., Majima, T., Choi, W., 2013b. Promoting water photooxidation on transparent WO3 thin films using an alumina overlayer. Energy Environ. Sci. 6, 3732.

Klahr, B., Gimenez, S., Fabregat-Santiago, F., Bisquert, J., Hamann, T.W., 2012a. Electrochemical and photoelectrochemical investigation of water oxidation with hematite electrodes. Energy Environ. Sci. 5, 7626.

Klahr, B., Gimenez, S., Fabregat-Santiago, F., Bisquert, J., Hamann, T.W., 2012b. Photoelectrochemical and impedance spectroscopic investigation of water oxidation with "Co-Pi"-coated hematite electrodes. J. Am. Chem. Soc. 134, 16693–16700.

Klahr, B.M., Gimenez, S., Zandi, O., Fabregat-Santiago, F., Hamann, T.W., 2015. Competitive photoelectrochemical methanol and water oxidation with hematite electrodes. ACS Appl. Mater. Interfaces 7, 7653–7660.

Klingan, K., Ringleb, F., Zaharieva, I., Heidkamp, J., Chernev, P., Gonzalez-Flores, D., Risch, M., Fischer, A., Dau, H., 2014. Water oxidation by amorphous cobalt-based oxides: volume activity and proton transfer to electrolyte bases. ChemSusChem 7, 1301–1310.

Kudo, A., Ueda, K., Kato, H., Mikami, I., 1998. Photocatalytic O_2 evolution under visible light irradiation on $BiVO_4$ in aqueous $AgNO_3$ solution. Catal. Lett. 53, 229–230.

Kunst, M., Beck, G., 1986. The study of charge carrier kinetics in semiconductors by microwave conductivity measurements. J. Appl. Phys. 60, 3558–3566.

Kupitz, C., Basu, S., Grotjohann, I., Fromme, R., Zatsepin, N.A., Rendek, K.N., Hunter, M.S., Shoeman, R.L., White, T.A., Wang, D., James, D., Yang, J.-H., Cobb, D.E., Reeder, B., Sierra, R.G., Liu, H., Barty, A., Aquila, A.L., Deponte, D., Kirian, R.A., Bari, S., Bergkamp, J.J., Beyerlein, K.R., Bogan, M.J., Caleman, C., Chao, T.-C., Conrad, C.E., Davis, K.M., Fleckenstein, H., Galli, L., Hau-Riege, S.P., Kassemeyer, S., Laksmono, H., Liang, M., Lomb, L., Marchesini, S., Martin, A.V., Messerschmidt, M., Milathianaki, D., Nass, K., Ros, A., Roy-Chowdhury, S., Schmidt, K., Seibert, M., Steinbrener, J., Stellato, F., Yan, L., Yoon, C., Moore, T.A., Moore, A.L., Pushkar, Y., Williams, G.J., Boutet, S., Doak, R.B., Weierstall, U., Frank, M., Chapman, H.N., Spence, J.C.H., Fromme, P., 2014. Serial time-resolved crystallography of photosystem II using a femtosecond X-ray laser. Nature 513, 261–265.

Lassalle-kaiser, B., Merki, D., Vrubel, H., Gul, S., Yachandra, V.K., Hu, X., Yano, J., Merisiers, L.O., 2015. Evidence from in situ X-ray absorption spectroscopy for the involvement of terminal disulfide in the reduction of protons by an amorphous molybdenum sulfide electrocatalyst. J. Am. Chem. Soc. 137, 314.

Le Formal, F., Pendlebury, S.R., Cornuz, M., Tilley, S.D., Graetzel, M., Durrant, J.R., 2014. Back electron–hole recombination in hematite photoanodes for water splitting. J. Am. Chem. Soc. 136, 2564.

Le Formal, F., Pastor, E., Tilley, S.D., Mesa, C.A., Pendlebury, S.R., Grätzel, M., Durrant, J.R., 2015. Rate law analysis of water oxidation on a hematite surface. J. Am. Chem. Soc. 137, 6629–6637.

Lee, M., Yun, H.J., Yu, S., Yi, J., 2014. Enhancement in photocatalytic oxygen evolution via water oxidation under visible light on nitrogen-doped TiO_2 nanorods with dominant reactive {102} facets. Catal. Commun. 43, 11–15.

Leng, W.H., Zhang, Z., Zhang, J.Q., Cao, C.N., 2005. Investigation of the kinetics of a TiO_2 photoelectrocatalytic reaction involving charge transfer and recombination through surface states by electrochemical impedance spectroscopy. J. Phys. Chem. B 109, 15008–15023.

Leng, W.H., Barnes, P.R.F., Juozapavicius, M., O'Regan, B.C., Durrant, J.R., 2010. Electron diffusion length in mesoporous nanocrystalline TiO_2 photoelectrodes during water oxidation. J. Phys. Chem. Lett. 1, 967.

Liang, Y., Messinger, J., 2014. Improving $BiVO_4$ photoanodes for solar water splitting through surface passivation. Phys. Chem. Chem. Phys. 16, 12014–12020.

Lichterman, M.F., Hu, S., Richter, M.H., Crumlin, E.J., Axnanda, S., Favaro, M., Drisdell, W., Hussain, Z., Mayer, T., Brunschwig, B.S., Lewis, N.S., Liu, Z., Lewerenz, H.-J., Gerischer, H., Schottky, W., Schottky, W., Schottky, W., Gerischer, H., Gerischer, H., Heller, A., Heller, A., Schwanitz, K., Mankel, E., Hunger, R., Mayer, T., Jaegermann, W., Sander, M., Jaegermann, W., Lewerenz, H.J., Mayer, T., Lebedev, M.V., Hunger, R., Jaegermann, W., Mayer, T., Lebedev, M., Hunger, R., Jaegermann, W., Bernède, J.C., Houari, S., Nguyen, D., Jouan, P.Y., Khelil, A., Mokrani, A., Cattin, L., Predeep, P., Heller, A., Chang, K.C., Miller, B., Kobayashi, K., Aikawa, Y., Sukigara, M., Heller, A., Chang, K.C., Miller, B., Lewis, N.S., Bard, A.J., Bocarsly, A.B., Fan, F.R.F., Walton, E.G., Wrighton, M.S., Bard, A.J., Allongue, P., Cachet, H., Nozik, A.J., Memming, R., Miller, B., Heller, A., Menezes, S., Lewerenz, H.J., Lattimer, J.R.C., Brunschwig, B.S., Lewis, N.S., Gray, H.B., Walter, M.G., Warren, E.L., McKone, J.R., Boettcher, S.W., Mi, Q., Santori, E.A., Lewis, N.S., Seah, M.P., Dench, W.A., Trasatti, S., Salmeron, M., Schlögl, R., Starr, D.E., Liu, Z., Hävecker, M., Knop-Gericke, A., Bluhm, H., Crumlin, E.J., Bluhm, H., Liu, Z., Klyushin, A.Y., Rocha, T.C.R., Hävecker, M., Knop-Gericke, A., Schlögl, R., Kaichev, V.V., Prosvirin, I.P., Bukhtiyarov, V.I., Casalongue, H.S., Kaya, S., Viswanathan, V., Miller, D.J., Friebel, D., Hansen, H.A., Nørskov, J.K., Nilsson, A., Ogasawara, H., Zemlyanov, D., Aszalos-Kiss, B., Kleimenov, E., Teschner, D., Zafeiratos, S., Hävecker, M., Knop-Gericke, A., Schlögl, R., Gabasch, H., Unterberger, W., Hayek, K., Klötzer, B., Lu, Y.-C., Crumlin, E.J., Veith, G.M., Harding, J.R., Mutoro, E., Baggetto, L., Dudney, N.J., Liu, Z., Shao-Horn, Y., Axnanda, S., Crumlin, E.J., Mao, B., Rani, S., Chang, R., Karlsson, P.G., Edwards, M.O.M., Lundqvist, M., Moberg, R., Ross, P., Hussain, Z., Liu, Z., Hu, S., Shaner, M.R., Beardslee, J.A., Lichterman, M.F., Brunschwig, B.S., Lewis, N.S., Tersoff, J., Kamat, P.V., Tvrdy, K., Baker, D.R., Radich, J.G., Emfietzoglou, D., Nikjoo, H., Radecka, M., Rekas, M., Trenczek-Zajac, A., Zakrzewska, K., Schwanitz, K., Weiler, U., Hunger, R., Mayer, A.T., Jaegermann, W., Fabregat-Santiago, F., Garcia-Belmonte, G., Bisquert, J., Bogdanoff, P., Zaban, A., Kobayashi, H., Mori, T., Namba, K., Nakato, Y., Kobayashi, H., Yamashita, Y., Mori, T., Nakato, Y., Komeda, T., Nishioka, Y., 2015. Direct observation of the energetics at a semiconductor/liquid junction by operando X-ray photoelectron spectroscopy. Energy Environ. Sci. 8, 2409–2416.

Lin, Y., Yuan, G., Sheehan, S., Zhou, S., Wang, D., 2011. Hematite-based solar water splitting: challenges and opportunities. Energy Environ. Sci. 4, 4862.

Liu, M., de Leon Snapp, N., Park, H., 2011. Water photolysis with a cross-linked titanium dioxide nanowire anode. Chem. Sci. 2, 80–87.

Liu, X., Wang, F., Wang, Q., 2012. Nanostructure-based WO_3 photoanodes for photoelectrochemical water splitting. Phys. Chem. Chem. Phys. 14, 7894–7911.

Liu, J., Shahid, M., Ko, Y.-S., Kim, E., Ahn, T.K., Park, J.H., Kwon, Y.-U., 2013. Investigation of porosity and heterojunction effects of a mesoporous hematite electrode on photoelectrochemical water splitting. Phys. Chem. Chem. Phys. 15, 9775–9782.

Loiudice, A., Cooper, J.K., Hess, L.H., Mattox, T.M., Sharp, I.D., Buonsanti, R., 2015. Assembly and photocarrier dynamics of heterostructured nanocomposite photoanodes from multicomponent colloidal nanocrystals. Nano Lett. 15, 7347–7354.

Ma, L., Jia, I., Guo, X., Xiang, L., 2014a. Time-resolved photoluminescence of anatase/rutile TiO_2 phase junction revealing charge separation dynamics. Chin. J. Catal. 35, 108–119.

Ma, Y., Pendlebury, S.R., Reynal, A., Formal, F. Le, Durrant, J.R., 2014b. Dynamics of photogenerated holes in undoped $BiVO_4$ photoanodes for solar water oxidation. Chem. Sci. 5, 2964–2973.

Ma, Y., Le Formal, F., Kafizas, A., Pendlebury, S.R., Durrant, J.R., 2015. Efficient suppression of back electron/hole recombination in cobalt phosphate surface-modified undoped bismuth vanadate photoanodes. J. Mater. Chem. A 3, 20649–20657.

Ma, Y., Kafizas, A., Pendlebury, S.R., Le Formal, F., Durrant, J.R., 2016a. Photoinduced absorption spectroscopy of CoPi on $BiVO_4$: the function of CoPi during water oxidation. Adv. Funct. Mater. 26, 4951–4960.

Ma, Y., Mesa, C.A., Pastor, E., Kafizas, A., Francàs, L., Le Formal, F., Pendlebury, S.R., Durrant, J.R., 2016b. Rate Law analysis of water oxidation and hole scavenging on a $BiVO_4$ photoanode. ACS Energy Lett. 1, 618–623.

Maeda, K., Teramura, K., Domen, K., 2008. Effect of post-calcination on photocatalytic activity of $(Ga_{1-x}Zn_x)(N_{1-x}O_x)$ solid solution for overall water splitting under visible light. J. Catal. 254, 198–204.

Martin, D.J., 2015. Investigation Into High Efficiency Visible Light Photocatalysts for Water Reduction and Oxidation. Springer. http://www.springer.com/gp/book/9783319184876.

Meekins, B.H., Kamat, P.V., 2011. Role of water oxidation catalyst IrO_2 in shuttling photogenerated holes across TiO_2 interface. J. Phys. Chem. Lett. 2, 2304–2310.

Melaina, M.W., Antonia, O., Penev, M., 2013. Blending Hydrogen into Natural Gas Pipeline Networks: A Review of Key Issues. U.S. Department of Energy. https://energy.gov/eere/fuelcells/downloads/blending-hydrogen-natural-gas-pipeline-networks-review-key-issues.

Mendez-Medrano, M.G., Kowalska, E., Lehoux, A., Herissan, A., Ohtani, B., Rau, S., Colbeau-Justin, C., Rodriguez-Lopez, J.L., Remita, H., 2016. Surface modification of TiO_2 with Au nanoclusters for efficient water treatment and hydrogen generation under visible light. J. Phys. Chem. C 120, 25010–25022.

Milan, R., Cattarin, S., Comisso, N., Baratto, C., Kaunisto, K., Tkachenko, N.V., Concina, I., 2016. Compact hematite buffer layer as a promoter of nanorod photoanode performances. Sci. Rep. 6, 35049.

Mills, A., Le Hunte, S., 1997. An overview of semiconductor photocatalysis. J. Photochem. Photobiol. A Chem. 108, 1–35.

Morais, A., Longo, C., Araujo, J.R., Barroso, M., Durrant, J.R., Nogueira, A.F., 2016. Nanocrystalline anatase TiO_2/reduced graphene oxide composite films as photoanodes for photoelectrochemical water splitting studies: the role of reduced graphene oxide. Phys. Chem. Chem. Phys. 18, 2608–2616.

Morton, O., 2006. Solar energy: a new day dawning? Silicon Valley sunrise. Nature 443, 19–22.

Nakamura, R., Nakato, Y., 2004. Primary intermediates of oxygen photoevolution reaction on TiO_2 (rutile) particles, revealed by in situ FTIR absorption and photoluminescence measurements. J. Am. Chem. Soc. 126, 1290–1298.

Neppl, S., Mahl, J., Tremsin, A., Rude, B., Qiao, R., Yang, W., Guo, J., Gessner, O., 2016. Towards efficient time-resolved X-ray absorption studies of electron dynamics at photocatalytic interfaces. Faraday Discuss. 194, 659–682.

Nunzi, F., Angelis, F. De, Selloni, A., 2016. Ab initio simulation of the absorption spectra of photoexcited carriers in TiO_2 nanoparticles. J. Phys. Chem. Lett. 7, 3597–3602.
Osterloh, F.E., 2008. Inorganic materials as catalysts for photochemical splitting of water. Chem. Mater. 20, 35–54.
Paracchino, A., Laporte, V., Sivula, K., Grätzel, M., Thimsen, E., 2011. Highly active oxide photocathode for photoelectrochemical water reduction. Nat. Mater. 10, 456.
Park, Y., McDonald, K.J., Choi, K.-S., 2013. Progress in bismuth vanadate photoanodes for use in solar water oxidation. Chem. Soc. Rev. 42, 2321–2337.
Parmar, K.P.S., Kang, H.J., Bist, A., Dua, P., Jang, J.S., Lee, J.S., 2012. Photocatalytic and photoelectrochemical water oxidation over metal-doped monoclinic $BiVO_4$ photoanodes. ChemSusChem 5, 1926–1934.
Patrocinio, A.O.T., Schneider, J., França, M.D., Santos, L.M., Caixeta, B.P., Machado, A.E.H., Bahnemann, D.W., 2015. Charge carrier dynamics and photocatalytic behavior of TiO_2 nanopowders submitted to hydrothermal or conventional heat treatment. RSC Adv. 5, 70536–70545.
Pattengale, B., Ludwig, J., Huang, J., 2016. Atomic insight into the W-doping effect on carrier dynamics and photoelectrochemical properties of $BiVO_4$ photoanodes. J. Phys. Chem. C 120, 1421–1427.
Pendlebury, S.R., Barroso, M., Cowan, A.J., Sivula, K., Tang, J., Grätzel, M., Klug, D., Durrant, J.R., 2011. Dynamics of photogenerated holes in nanocrystalline α-Fe_2O_3 electrodes for water oxidation probed by transient absorption spectroscopy. Chem. Commun. (Camb.) 47, 716–718.
Pendlebury, S.R., Cowan, A.J., Barroso, M., Sivula, K., Ye, J., Grätzel, M., Klug, D.R., Tang, J., Durrant, J.R., 2012. Correlating long-lived photogenerated hole populations with photocurrent densities in hematite water oxidation photoanodes. Energy Environ. Sci. 5, 6304.
Pendlebury, S.R., Wang, X., Formal, F. Le, Cornuz, M., Kafizas, A., Tilley, S.D., Grätzel, M., Durrant, J.R., 2014. Ultrafast charge carrier recombination and trapping in hematite photoanodes under applied bias. J. Am. Chem. Soc. 136, 9854.
Pesci, F.M., Cowan, A.J., Alexander, B.D., Durrant, J.R., Klug, D.R., 2011. Charge carrier dynamics on mesoporous WO_3 during water splitting. J. Phys. Chem. Lett. 2, 1900.
Pesci, F.M., Wang, G., Klug, D.R., Li, Y., Cowan, A.J., 2013. Efficient suppression of electron–hole recombination in oxygen-deficient hydrogen-treated TiO_2 nanowires for photoelectrochemical water splitting. J. Phys. Chem. C 117, 25837–25844.
Qin, D.-D., Li, Y.-L., Wang, T., Li, Y., Lu, X.-Q., Gu, J., Zhao, Y.-X., Song, Y.-M., Tao, C.-L., 2015. Sn-doped hematite films as photoanodes for efficient photoelectrochemical water oxidation. J. Mater. Chem. A 3, 6751–6755.
Ravensbergen, J., Abdi, F.F., van Santen, J.H., Frese, R.N., Dam, B., van de Krol, R., Kennis, J.T.M., 2014. Unraveling the carrier dynamics of $BiVO_4$: a femtosecond to microsecond transient absorption study. J. Phys. Chem. C 118, 27793–27800.
Regan, K.P., Koenigsmann, C., Sheehan, S.W., Konezny, S.J., Schmuttenmaer, C.A., 2016. Size-dependent ultrafast charge carrier dynamics of WO_3 for photoelectrochemical cells. J. Phys. Chem. C 120, 14926–14933. http://dx.doi.org/10.1021/acs.jpcc.6b04390.
Renz, C., 1921. Lichtreaktionen der Oxyde des Titans, Cers und der Erdsäuren. Helv. Chim. Acta 4, 961–968.
Retter, U., Lohse, H., 2005. Electrochemical impedance spectroscopy. In: Scholz, F. (Ed.), Electroanalytical Methods. Springer Berlin Heidelberg, Berlin, Heidelberg, pp. 149–166.
Rettie, A.J.E., Lee, H.C., Marshall, L.G., Lin, J.F., Capan, C., Lindemuth, J., McCloy, J.S., Zhou, J., Bard, A.J., Mullins, C.B., 2013. Combined charge carrier transport and photoelectrochemical characterization of $BiVO_4$ single crystals: intrinsic behavior of a complex metal oxide. J. Am. Chem. Soc. 135, 11389–11396.

Rex, R.E., Yang, Y., Knorr, F.J., Zhang, J.Z., Li, Y., McHale, J.L., 2016. Spectroelectrochemical photoluminescence of trap states in H-treated rutile TiO_2 nanowires: implications for photooxidation of water. J. Phys. Chem. C 120, 3530–3541.

Reynal, A., Lakadamyali, F., Gross, M.A., Reisner, E., Durrant, J.R., 2013. Parameters affecting electron transfer dynamics from semiconductors to molecular catalysts for the photochemical reduction of protons. Energy Environ. Sci. 6, 3291.

Riha, S.C., Klahr, B.M., Tyo, E.C., Vajda, S., Hamann, T.W., Martinson, A.B.F., 2013. Atomic layer deposition of a submonolayer catalyst for the enhanced photoelectrochemical performance of water oxidation with hematite. ACS Nano 7, 2396–2405.

Ruoko, T.P., Kaunisto, K., Bärtsch, M., Pohjola, J., Hiltunen, A., Niederberger, M., Tkachenko, N.V., Lemmetyinen, H., 2015. Subpicosecond to second time-scale charge carrier kinetics in hematite-titania nanocomposite photoanodes. J. Phys. Chem. Lett. 6, 2859–2864.

Sachs, M., Pastor, E., Kafizas, A., Durrant, J.R., 2016. Evaluation of surface state mediated charge recombination in anatase and rutile TiO_2. J. Phys. Chem. Lett. 7, 3742–3746.

Saeki, A., Yasutani, Y., Oga, H., Seki, S., 2014. Frequency-modulated gigahertz complex conductivity of TiO_2 nanoparticles: interplay of free and shallowly trapped electrons. J. Phys. Chem. C 118, 22561–22572.

Salvador, P., 1984. Hole diffusion length in n-TiO_2 single crystals and sintered electrodes: photoelectrochemical determination and comparative analysis. J. Appl. Phys. 55, 2977–2985.

Santomauro, F.G., Lübcke, A., Rittmann, J., Baldini, E., Ferrer, A., Silatani, M., Zimmermann, P., Grübel, S., Johnson, J.A., Mariager, S.O., Beaud, P., Grolimund, D., Borca, C., Ingold, G., Johnson, S.L., Chergui, M., 2015. Femtosecond X-ray absorption study of electron localization in photoexcited anatase TiO2. Sci. Rep. 5, 14834.

Schneider, J., Matsuoka, M., Takeuchi, M., Zhang, J., Horiuchi, Y., Anpo, M., Bahnemann, D.W., 2014. Understanding TiO_2 photocatalysis: mechanisms and materials. Chem. Rev. 114, 9919–9986.

Shi, X., Herraiz-Cardona, I., Bertoluzzi, L., Lopez-Varo, P., Bisquert, J., Park, J.H., Gimenez, S., 2016. Understanding the synergistic effect of WO_3–$BiVO_4$ heterostructures by impedance spectroscopy. Phys. Chem. Chem. Phys. 18, 9255–9261.

Sivula, K., Le Formal, F., Gratzel, M., 2011. Solar water splitting: progress using hematite (a-Fe_2O_3) photoelectrodes. ChemSusChem 4, 432–449.

Su, J., Guo, L., Bao, N., Grimes, C.A., 2011. Nanostructured WO_3/$BiVO_4$ heterojunction films for efficient photoelectrochemical water splitting. Nano Lett. 11, 1928–1933.

Tacca, A., Meda, L., Marra, G., Savoini, A., Caramori, S., Cristino, V., Bignozzi, C.A., Pedro, V.G., Boix, P.P., Gimenez, S., Bisquert, J., 2012. Photoanodes based on nanostructured WO_3 for water splitting. ChemPhysChem 13, 3025–3034.

Tachibana, Y., Vayssieres, L., Durrant, J.R., 2012. Artificial photosynthesis for solar watersplitting. Nat. Photonics 6, 511–518.

Tamaki, Y., Furube, A., Murai, M., Hara, K., Katoh, R., Tachiya, M., 2006. Direct observation of reactive trapped holes in TiO_2 undergoing photocatalytic oxidation of adsorbed alcohols: evaluation of the reaction rates and yields. J. Am. Chem. Soc. 128, 416.

Tan, H.L., Wen, X., Amal, R., Ng, Y.H., 2016. $BiVO_4$ {010} and {110} relative exposure extent: governing factor of surface charge population and photocatalytic activity. J. Phys. Chem. Lett. 7, 1400–1405.

Tang, J., Durrant, J.R., Klug, D.R., 2008. Mechanism of photocatalytic water splitting in TiO_2. Reaction of water with photoholes, importance of charge carrier dynamics, and evidence for four-hole chemistry. J. Am. Chem. Soc. 130, 13885.

Tang, J., Cowan, A.J., Durrant, J.R., Klug, D.R., 2011. Mechanism of O_2 production from water splitting: nature of charge carriers in nitrogen doped nanocrystalline TiO_2 films and factors limiting O_2 production. J. Phys. Chem. C 115, 3143.

Wahl, A., Ulmann, M., Carroy, A., Jermann, B., Dolata, M., Kedzierzawski, P., Chatelain, C., Monnier, A., Augustynski, J., 1995. Photoelectrochemical studies pertaining to the activity of TiO_2 towards photodegradation of organic compounds. J. Electroanal. Chem. 396, 41–51.

Walter, M.G., Warren, E.L., Mckone, J.R., Boettcher, S.W., Mi, Q., Santori, E.A., Lewis, N.S., 2010. Solar water splitting cells Chem. Rev. 110, 6446.

Wang, G., Lu, X., Li, Y., 2014. Low-cost nanomaterials for photoelectrochemical water splitting. In: Lin, Z., Wang, J. (Eds.), Low-Cost Nanomaterials—Toward Greener and More Efficient Energy Applications. Springer-Verlag, London, pp. 267–295.

Wang, X., Kafizas, A., Li, X., Moniz, S.J.A., Reardon, P.J.T., Tang, J., Parkin, I.P., Durrant, J.R., 2015. Transient absorption spectroscopy of anatase and rutile: the impact of morphology and phase on photocatalytic activity. J. Phys. Chem. C 119, 10439–10447.

Xiao-e, L., Green, A.N.M., Haque, S.A., Mills, A., Durrant, J.R., 2004. Light-driven oxygen scavenging by titania/polymer nanocomposite films. J. Photochem. Photobiol. A Chem. 162, 253–259.

Xu, Y., Schoonen, A.A., 2000. The absolute energy positions of conduction and valence bands of selected semiconducting minerals. Am. Mineral. 85, 543.

Xu, C., Song, Y., Lu, L., Cheng, C., Liu, D., Fang, X., Chen, X., Zhu, X., Li, D., 2013. Electrochemically hydrogenated TiO_2 nanotubes with improved photoelectrochemical water splitting performance. Nanoscale Res. Lett. 8, 391.

Yamakata, A., Ishibashi, T.-A., Onishi, H., 2001. Water- and oxygen-induced decay kinetics of photogenerated electrons in TiO_2 and Pt/TiO_2: a time-resolved infrared absorption study. J. Phys. Chem. B 105, 7258.

Yang, Y., Forster, M., Ling, Y., Wang, G., Zhai, T., Tong, Y., Cowan, A.J., Li, Y., 2016. Acid treatment enables suppression of electron-hole recombination in hematite for photoelectrochemical water splitting. Angew. Chemie. Int. Ed. 55, 3403–3407.

Yoshihara, T., Katoh, R., Furube, A., Tamaki, Y., Murai, M., Hara, K., Murata, S., Arakawa, H., Tachiya, M., 2004. Identification of reactive species in photoexcited nanocrystalline TiO_2 films by wide-wavelength-range (400 – 2500 nm) transient absorption spectroscopy. J. Phys. Chem. B 108, 3817.

Yoshihara, T., Tamaki, Y., Furube, A., Murai, M., Hara, K., Katoh, R., 2007. Effect of pH on absorption spectra of photogenerated holes in nanocrystalline TiO_2 films. Chem. Phys. Lett. 438, 268–273.

Zandi, O., Hamann, T.W., 2015. The potential versus current state of water splitting with hematite. Phys. Chem. Chem. Phys. 17, 22485–22503.

Zandi, O., Hamann, T.W., 2016. Determination of photoelectrochemical water oxidation intermediates on haematite electrode surfaces using operando infrared spectroscopy. Nat. Chem. 8, 778–783.

Zandi, O., Klahr, B.M., Hamann, T.W., 2013. Highly photoactive Ti-doped [small alpha]-Fe_2O_3 thin film electrodes: resurrection of the dead layer. Energy Environ. Sci. 6, 634–642.

Zandi, O., Schon, A.R., Hajibabaei, H., Hamann, T.W., 2016. Enhanced charge separation and collection in high-performance electrodeposited hematite films. Chem. Mater. 28, 765–771.

Zawadzki, P., 2013. Absorption spectra of trapped holes in anatase TiO_2. J. Phys. Chem. C 117, 8647–8651.

Zhang, L., Mohamed, H.H., Dillert, R., Bahnemann, D., 2012. Kinetics and mechanisms of charge transfer processes in photocatalytic systems: a review. J. Photochem. Photobiol. C Photochem. Rev. 13, 263–276.

Zhang, M., Respinis, M. De, Frei, H., 2014. Time-resolved observations of water oxidation intermediates on a cobalt oxide nanoparticle catalyst. Nat. Chem. 6, 362–367.

Zhang, K., Wang, L., Kim, J.K., Ma, M., Veerappan, G., Lee, C.-L., Kong, K., Lee, H., Park, J.H., Kim, K., Ma, M., Veerappan, G., Lee, C.-L., Kong, K., Hyoyoung, L., Park, J.H., 2016. An order/disorder/water junction system for highly efficient co-catalyst-free photocatalytic hydrogen generation. Energy Environ. Sci. 9, 499–503.

Zhong, D.K., Gamelin, D.R., 2010. Photoelectrochemical water oxidation by cobalt catalyst ("Co – Pi")/α-Fe_2O_3 composite photoanodes: oxygen evolution and resolution of a kinetic bottleneck. J. Am. Chem. Soc. 132, 4202–4207.

Ziwritsch, M., Müller, S., Hempel, H., Unold, T., Abdi, F.F., van de Krol, R., Friedrich, D., Eichberger, R., 2016. Direct time-resolved observation of carrier trapping and polaron conductivity in $BiVO_4$. ACS Energy Lett. 1, 888–894.

CHAPTER TWO

Photophysics and Photochemistry at the Semiconductor/Electrolyte Interface for Solar Water Splitting

Xiaogang Yang*, Dunwei Wang[†,1]
*Key Laboratory of Micro-Nano Materials for Energy Storage and Conversion of Henan, Xuchang University, Henan, China
[†]Boston College, Merkert Chemistry Center, Chestnut Hill, MA, United States
[1]Corresponding author: e-mail address: dunwei.wang@bc.edu

Contents

1. INTRODUCTION

One of the most important challenges faced by modern humanities is how to utilize sufficient energy to enjoy the convenience enabled by civilization without devastating the environment (B.P., 2014). A careful

Semiconductors and Semimetals, Volume 97
ISSN 0080-8784
http://dx.doi.org/10.1016/bs.semsem.2017.03.001

examination of the thermodynamics of the Earth may help us find the keys to this challenge. As a pseudo-isolated thermodynamic system in the University, the Earth exchanges energy with the surroundings through two main sources, sunlight as the incoming energy and reflection (of the sunlight) and IR irradiation as the outgoing one. Any disruption to this dynamic balance has the potential to lead to dramatic changes to the biosphere, thereby threatening the very existence of our species. The rapid rise of temperatures we witnessed since the industrial revolution is a good example of such disruptions (Demars et al., 2016; Grätzel, 2001). From this consideration, we see that the majority of our future energy supplies should come from sunlight; any other sources of energy (e.g., fossil fuels or nuclear) will generate waste heat that is not part of the existing dynamic balance on a global scale, potentially altering the delicate dynamic balance as described earlier (Stephens et al., 2012). In other words, we foresee that a sustainable future should be powered by solar energy, which brings us to a critical technical challenge associated with the diurnal and intermittent nature of solar energy. How to harvest and store solar energy on a large scale remains outstanding. While a multitude of technologies have been proposed and tested, at scales large and small, the most promising one may lie in what has been demonstrated by mother Nature—namely the photosynthesis (Boekema et al., 2006; Satoh et al., 2005). This is because the chemical bonds hold great promise for storing energies with high densities; the products also allow for convenient transportation, redistribution, and utilization of the stored energies. It is within this context that significant research has been undertaken to study the process now popularly known as *artificial photosynthesis* (Nocera, 2012). The idea is deceptively simple: Light absorbers harvest the energy delivered by photons; the energy is then used to excite charges; negative charges enable reduction reactions; and positive charges fuel oxidation reactions. When it comes to discussions on the feasibility of the process, we have a formidable prior success—the natural photosynthesis, which is the ultimate source of the fossil fuels that provide >80% of today's global energy supply. Yet, to match the rapid industrial developments, the ever-growing living standards, and a seemingly ever-larger population, we need to carry out photosynthesis at efficiencies far higher than what is possible by the natural processes. For a good part of the past 5 decades, scientists have been hard at work to meet this critical challenge. In line with the main topics of this book, one of the approaches would be to use semiconductors for the purpose of artificial photosynthesis (in Fig. 1). This line of research started with the pioneering work by P.J. Boddy in the late 1960s (Boddy, 1968). The interest

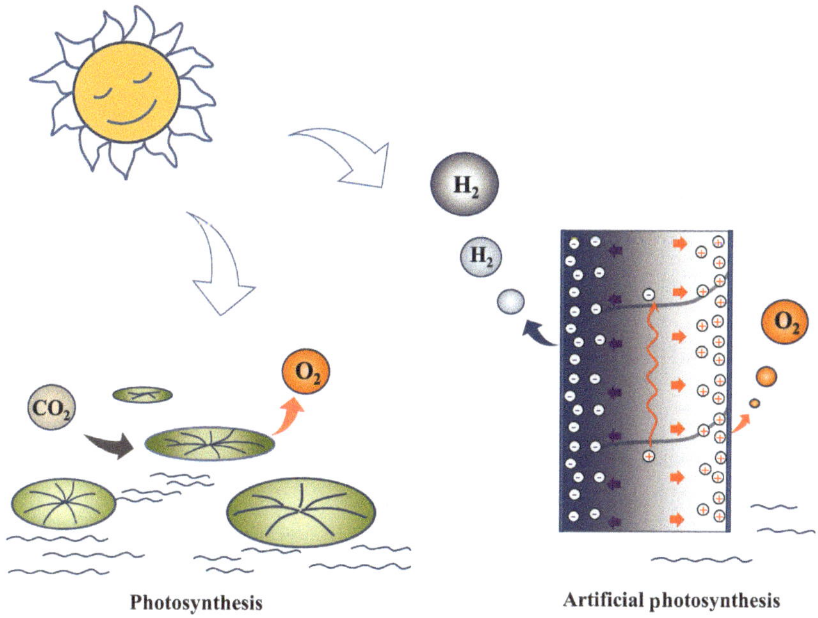

Fig. 1 Schematic illustration of natural (*left*) and artificial (*right*) photosynthesis.

then picked up pace following the seminal work by Honda et al. (Fujishima and Honda, 1972) and it was further fueled by the energy crisis in the 1970s. While there has been ups and downs during the past years (Bard and Fox, 1995; Grätzel, 2001), we have seen continued efforts toward to goal of using semiconductors to direct solar-to-chemical energy conversion (Morrison, 1980; Sato, 1998). While the earlier works mostly focused on the simple water splitting reactions, recent efforts start to tackle the more difficult, but also more important, reactions such as CO_2 reduction (Kumar et al., 2012). Our goal for this chapter is to provide an overview of the fundamental principles of photophysics and photochemistry that govern the process of artificial photosynthesis by semiconductors. The chapter is thus arranged briefly for the researchers.

To appreciate the fundamentally important processes involved in semiconductor-based photoelectrochemistry (PEC), let us start with a quick examination of water splitting, which is the first step of natural photosynthesis.

$$H_2O(l) \rightarrow H_2(g) + \frac{1}{2}O_2(g) \quad (1)$$

Thermodynamically, Eq. (1) is an uphill reaction ($\Delta G^{o} = 237.14 kJ/mol$) under standard conditions (Dean, 1999). *Based on electrochemical thermodynamics, the potential corresponds to −1.229 V, where the negative sign indicates the reverse reaction would be spontaneous.* The way Eq. (1) is balanced indicates that two electrons are involved. When carried out in an electrochemical cell, the electrolysis can proceed through two separate half reactions (Eqs. 2 and 3) (Ma et al., 2015).

$$2H^{+} + 2e^{-} \rightarrow H_2(g) \quad E^{o}_{HER} = 0V \text{ vs SHE} \tag{2}$$

$$O_2(g) + 4H^{+} + 4e^{-} \rightarrow 2H_2O(l) \quad E^{o}_{OER} = 1.229V \text{ vs SHE} \tag{3}$$

A key component of the PEC water splitting cell is the light absorber, whose function can be fulfilled by semiconductors. Upon illumination, the semiconductor is photoexcited, producing energetic electrons in the conduction band and leaving holes in the valence band. Under appropriate band bending conditions near the surface, these electrons and holes can be separated to drive the reduction and oxidation, respectively, generating H_2 and O_2. At the most rudimental level, we see that a PEC system first and foremost features one (or more) diode, whose functions are governed by the photophysics developed the studies of photovoltaics. Furthermore, we see that a PEC system also features at least two electrochemical catalytic components for reduction and oxidation, the detailed operations of which can be described by electrochemical principles that govern charge transfer across an electrode/electrolyte interface. With this understanding in mind, we conceived this chapter to study these critical components of a PEC cell. Given that numerous books (Memming, 2015; Morrison, 1980; Peter, 2013; Sato, 1998; van de Krol, 2012) have been previously published on similar topics, we aim to provide what was previously missing, namely the behaviors of the semiconductor/water interface (Liu et al., 2014; Thorne et al., 2015). As such, our discussions of the photophysics within the semiconductor, and the electrochemistry of the catalyst/water interface will be limited (Roger et al., 2017). Instead, we concentrate our attention on the semiconductor/catalyst/water interface. Our primary goal for this chapter is to articulate how the addition of a catalyst may influence the behaviors of the semiconductor, and whether the catalysts indeed function as a catalyst in a PEC system. It is further noted that for convenience we base our discussions on n-type semiconductors. Similar discussions, however, can be readily developed for p-type ones.

2. THERMODYNAMIC DRIVING FORCE FOR WATER SPLITTING: THE FUNDAMENTALS OF SEMICONDUCTOR PHYSICS

The first topic we wish to touch on is the energetics of the photogenerated charges within the semiconductor. This topic is important because the energetics determines the driving force of the desired PEC reactions. In studying this topic, we must be aware that the energy scale has been previously studied by two distinct groups of scientists, solid-state physicists, and electrochemists (Trasatti, 1986). Due to historical reasons, these two communities used very different terminologies. As will be seen in Fig. 2, the languages used by these communities are so different that signs of the values taken to measure the energy are opposite. Consider the commonly used scales by solid-state physicists as an example. To measure the free energy of electrons within a solid, physicists often employ the concept of *work function* ($\chi = q\phi$; where ϕ is the electric potential below the vacuum level and q is the elementary charge of an electron), which represents the energy needed to liberate a bound electron to the level of vacuum ($\phi_{vac} = 0$). It is shown in Fig. 2 that most materials feature a negative work function because electrons in a solid are typically more stable than when they are *free* (in vacuum). By contrast, electrochemists used the concept of *electrochemical potential* to measure the free energy of charges (Bard and Faulkner, 2000). For instance, a "reduction potential" would measure the energy needed

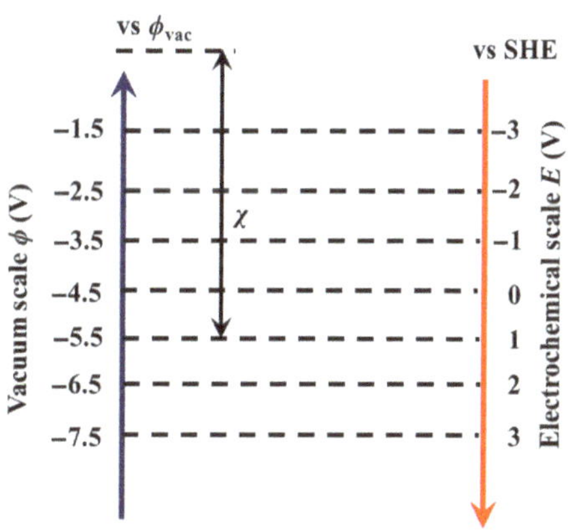

Fig. 2 Relationship of the two different ways to measure energy by solid-state scientists (*left*) and electrochemists (*right*).

to donate an electron in an electrochemical reaction. Due to empirical reasons, most electrochemical potentials are measured and presented relative to a reference potential (such as the standard hydrogen electrode, or SHE; NHE is also commonly used that represents normal hydrogen electrode). Now let us consider two representative systems, A and B. Suppose that electrons are more tightly bound in System A than in System B. When measured by *work function*, a more negative value would be obtained for System A. However, because System A should be more easily reduced due to the stronger binding energies of electrons, it would feature a more *positive* electrochemical potential. Already, we see that the differences in the conventions can be a major source of confusion for novice beginners. We corroborate different potentials of work function vs electrochemical potential in Fig. 2, where SHE potential is calibrated at −4.5 V (−4.44 V by IUPAC) vs vacuum (Trasatti, 1986). It is seen that the two scales can be readily unified by Eq. (4), where electrochemical scale is shown as E (vs SHE) and ϕ represents the measure against vacuum (Eq. 4) (Morrison, 1980):

$$\phi = -4.5\mathrm{V} - E \tag{4}$$

2.1 Semiconductor Basics

The hallmark of a semiconductor is the existence of a forbidden gap between the energy bands (Chattopadhyay and Rakshit, 2006; Seeger, 2004). As depicted in Fig. 3, the highest filled band is known as valence band and the lowest empty band is the conduction band. Note the "filled" and "empty" state would only be observed at absolute 0 K. At any finite temperature (e.g., room temperature), thermal ionization would populate the conduction band to a given extent, giving rise to mobile electrons in the conduction band and mobile holes in the valence band. A common strategy to increase mobile charges is to introduce impurities known as dopants. When dopants that are more electron deficient (relative to the host materials) are present, they *accept* electrons from the valence band (hence known as

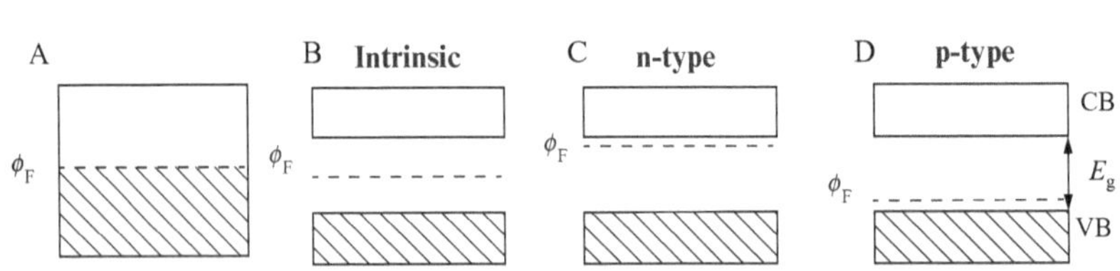

Fig. 3 Band structures for metals (A) and semiconductors with different doping types (B–D).

acceptors; a good example is B in Si), producing more mobile holes in the materials than mobile electrons. The semiconductor would be known as p-type. Conversely, more electron-rich dopants are n-type dopants that increase the relative concentration of mobile electrons. The concentrations of mobile electrons and holes in a semiconductor are correlated to the Fermi level following Eqs. (5) and (6):

$$n \approx N_{\mathrm{CB}} \exp\left[-\frac{q(\phi_{\mathrm{CB}}-\phi_{\mathrm{F}})}{kT}\right] \quad (5)$$

$$p \approx N_{\mathrm{VB}} \exp\left[-\frac{q(\phi_{\mathrm{F}}-\phi_{\mathrm{VB}})}{kT}\right] \quad (6)$$

where the N_{CB} and N_{VB} are the inherent charge carrier concentrations, the ϕ_{CB} and ϕ_{VB} are the conduction band minimum and valence band maximum, respectively. k is the Boltzmann's constant, T is the absolute temperature.

Charge carrier concentration relationship with the Fermi level as described earlier represents what is expected under equilibrium conditions. When a semiconductor is illuminated, the system is moved away from equilibrium because the energy delivered by absorbed photons will excite additional electrons from the valence band to the conduction band. As is true in photovoltaic devices, if the bands are flat (e.g., Fig. 3), the photogenerated charges are distributed within the semiconductor randomly without order. Heterogeneity is critical for the collection of photogenerated charges. Next, we use a simple model based on semiconductor/metal junctions (i.e., a Schottky-type diode) (Rhoderick and Williams, 1988) in Fig. 4 to develop descriptions for a semiconductor/electrolyte interface.

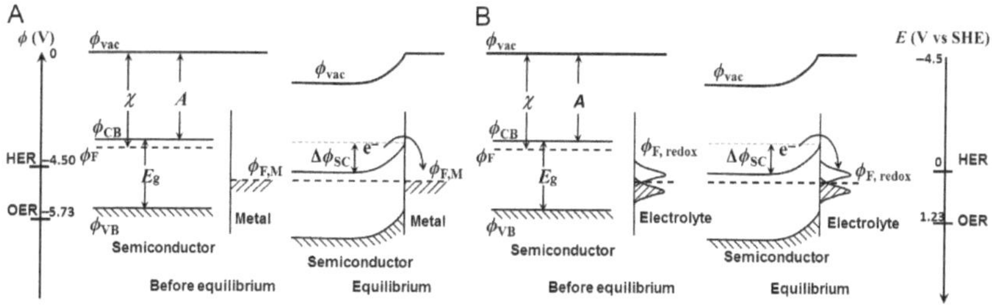

Fig. 4 Band diagrams of semiconductor/metal (A) and semiconductor/electrolyte (B) interfaces. The relative energy levels to vacuum (*left*) and electrochemical reference potentials (*right*) are correlated for easy reading. The drawings are based on an n-type semiconductor. χ and A are the semiconductor work function and electron affinity, respectively. The band bending is $\Delta\phi_{\mathrm{SC}}=\phi_{\mathrm{F,SC}}-\phi_{\mathrm{F,redox}}$ (or $\phi_{\mathrm{F,M}}$).

As shown in Fig. 4A, before contact, the metal may be described by its Fermi level $\phi_{F,m}$ or work function ($\chi_M = q\phi_{F,M}$), and the semiconductor is characterized by its electron affinity A ($A = -q\phi_{CB}$) and work function χ ($\chi_{SC} = q\phi_{F,SC}$). Upon contact, if the Fermi level of metal ($\phi_{F,M}$) is lower than semiconductors ($\phi_{F,SC}$) ($\chi_M < \chi_{SC}$), electrons would flow from the semiconductor to the metal. The net result is that positive charges accumulate within the semiconductor near the surface, moving the Fermi level away from the conduction band edge and toward the valence band edge. It is qualitatively shown in Fig. 4A that the effect (depletion of electrons near the surface) is most pronounced at the surface. At a distance sufficiently far away from the surface, the effect diminishes. This distance is known as the depletion width (W_{SC}) and can be calculated based on the dielectric constant $\varepsilon_0\varepsilon_{SC}$ of the semiconductor (van de Krol, 2012), the difference of the Fermi level difference (in vacuum) and the carrier concentration N_D:

$$W_{SC} = \sqrt{\frac{2\varepsilon_0\varepsilon_{SC}}{qN_D}\left(\Delta\phi_{SC} - \frac{kT}{q}\right)} \tag{7}$$

where $\Delta\phi_{SC} = \phi_{F,SC} - \phi_{F,M}$ corresponds to the band bending, determined by the Fermi level difference between the semiconductor and the metal.

Under equilibrium, we see that the Fermi level would be flat throughout the system. The relative movement of the Fermi level to the band edges would then be manifested by the bending of the band edge positions across the depletion region. Intuitively, we see that the bending of the band produces a barrier (or a built-in field) to counteract the spontaneous flow of electrons from the semiconductor to the metal. This barrier is known as the *Schottky* barrier.

In a simplistic fashion, the redox pairs in the electrolyte can be regarded as electron acceptors (or donors), providing the function nearly identical to a metal contact. Consequently, the electrochemical potential E as described by the *Nernst equation* of an electrolyte may be treated as the "Fermi level" ($\phi_{F,redox}$) of the system, much similar to that of a metal contact. This understanding is depicted in Fig. 4B. If the Fermi level $\phi_{F,SC}$ of n-type semiconductor is higher (less negative) than the electrolyte ($\phi_{F,SC} > \phi_{F,redox}$) before contact, a similar band bending within the semiconductor near the surface would develop in much the same fashion as in a semiconductor/metal case, forming a Schottky-type junction (Krishnan, 2007).

Depending on the relative positions of the Fermi level of the semiconductor and the electrochemical potential of the electrolyte, at least four types of band bending could form on the surface of the semiconductor (Grätzel, 2001). Again, using n-type semiconductor as an example, we describe these possibilities in Fig. 5. First, when the Fermi level of the semiconductor equals to the electrochemical potential of the electrolyte ($\phi_F = \phi_{F,redox}$), no net electric charges would flow through the interface, and there is no excess charge on either side of the junction. This leads to a *flat band* condition

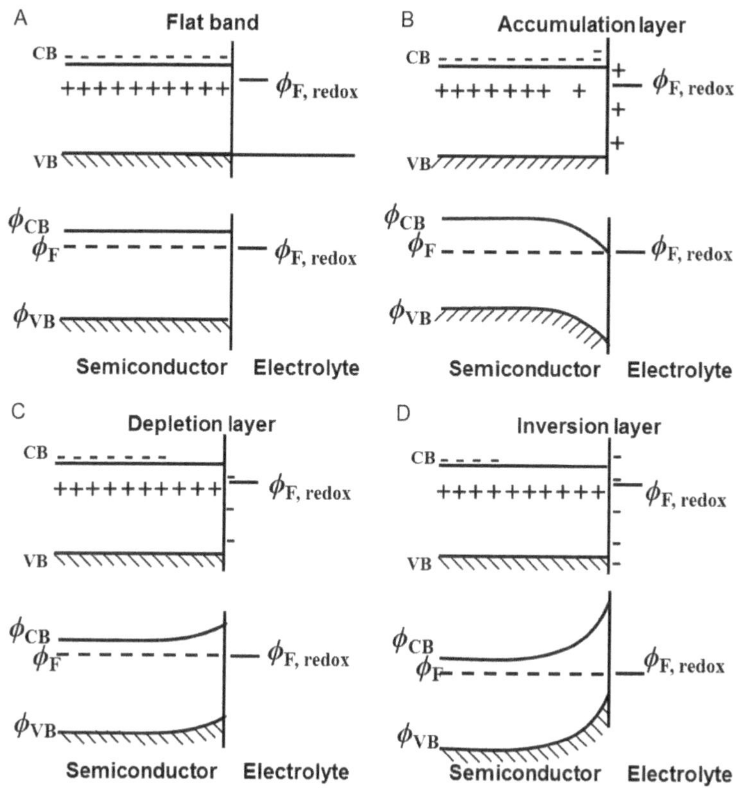

Fig. 5 Schematic illustrations of the electronic energy levels at the interface between an n-type semiconductor and an electrolyte containing a redox couple. The four cases indicated are: (A) flat band potential, where no space charge layer exists in the semiconductor; (B) accumulation layer, where excess electrons have been injected into the solid producing a downward bending of the conduction and valence band toward the interface; (C) depletion layer, where electrons have moved from the semiconductor to the electrolyte, producing an upward bending of the bands; and (D) inversion layer, where the electrons have been depleted below their intrinsic level, enhancing the upward band bending and rendering the semiconductor p-type at the surface. *Adapted from Grätzel, M., 2001. Photoelectrochemical cells. Nature, 414, 338–344, with permissions, Copyright © 2001, Rights Managed by Nature Publishing Group.*

of the semiconductor (Fig. 5A). Second, when the electrochemical potential of the electrolyte is higher than the Fermi level of the semiconductor ($\phi_{F,redox} > \phi_F$), electrons would accumulate on the semiconductor side, and one would obtain the condition of *accumulation* (Fig. 5B). Third, when the electrochemical potential of the electrolyte is lower than the Fermi level of the semiconductor ($\phi_{F,redox} < \phi_F$), the electrons would be depleted on the semiconductor side, giving rise to the *depletion* condition (Fig. 5C). Fourth, when the electrochemical potential of the electrolyte is much lower than the Fermi level of the semiconductor, electrons would be depleted on the surface of the semiconductor to such a point that holes become the majority carrier. That is, the Fermi level at the surface would surpass the mid-gap point and approach the vicinity of the valence band, creating the situation known as *inversion* (Fig. 5D). In practice, the formation of an inversion layer is difficult because the majority charge exchange rate with the electrolyte may not be fast enough to match the rate at which they are thermally generated, or because the minority charge carriers could exchange with the electrolyte at a non-negligible rate, under which condition we would achieve *deep depletion* situation.

2.2 Quantitative Analysis of the Depletion Region

The built-in field formed by the band bending is crucial to PEC, because it is the fundamental reason for charge separation at the semiconductor/electrolyte interface. In accordance with Schottky diodes formed by solid-solid junctions, the depletion region is also known variably as the depletion layer or the space charge region. Many of the discussions from semiconductor physics can be borrowed to further describe the details of this region in a quantitative fashion (van de Krol, 2012).

In Fig. 6, we set the band bending start point as zero ($x=0$) and use the width of the space charge region $x = W_{SC}$ at the surface as the boundaries. In the bulk region, the net charge density equals to zero, because the negative charge (electron) concentration is the same to the ionized dopant sites ($x<0$, $n_e = n_h = N_D$). At a certain position x in the space charge region, the charge density $\rho(x)$ can be related to the electric potential $\phi(x)$ as Eq. (8):

$$\rho(x) = qn_h - qn_e(x) = qN_D\left(1 - \exp\left[-\frac{q(\phi(x) - \phi_{CB})}{kT}\right]\right) \tag{8}$$

As has been well documented by other authors (Peter, 2013; van de Krol, 2012), the charge distribution, the electric field, and the electric potential in

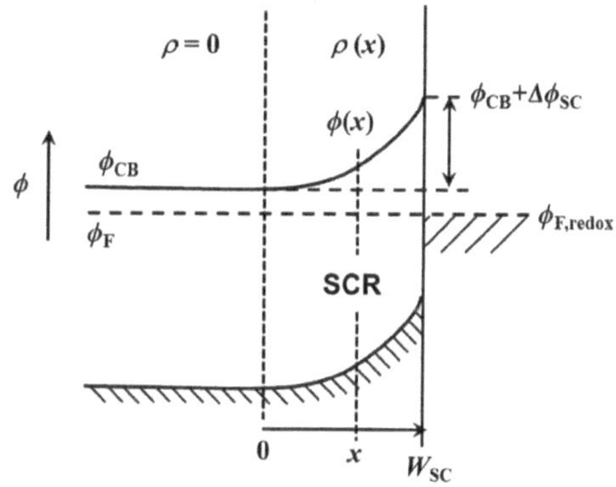

Fig. 6 A detailed view of band bending phenomenon and the space charge region. ρ is the net charge density.

the semiconductor can be calculated by solving the Poisson's equation. By integrating Eq. (8) from the boundaries ($x=0$–x), we can determine the number of charges as described in Eq. (9):

$$Q(x) \approx \sqrt{2q\varepsilon_0\varepsilon_{SC}N_D A^2\left(\phi(x)-\phi_{CB}-\frac{kT}{q}\right)} \quad (x \neq 0) \tag{9}$$

where A is the semiconductor area. The electric potential at a given position x can then be calculated as:

$$\phi(x) = \frac{qN_D}{2\varepsilon_0\varepsilon_{SC}}x^2 + \phi_{CB} \quad (x > 0) \tag{10}$$

For the typical semiconductors, such as Si and $BiVO_4$, the potential change ϕ and the width of the depletion layer can be simply estimated, respectively (in Fig. 7A and B).

Last before we wend this section, we note that the extent of band bending as defined by $\Delta\phi_{SC}$ depends on the applied potential following $\Delta\phi_{SC}=\phi_{F,SC}-\phi_{F,redox}$, where $\phi_{F,SC}$ equals to the applied potential (Fig. 8). Such a dependence suggests that the capacitance of the space charge region, which can be conveniently measured using alternating current (AC) techniques, varies with the applied potential quantitatively. This has been popularly used in electrochemical characterization known as Mott–Schottky method, the details of which will be discussed in the next section.

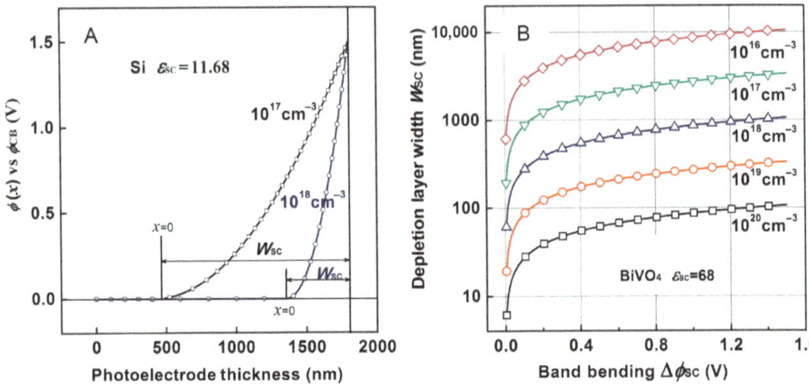

Fig. 7 Examples of quantitative analysis of: (A) band bending on Si (dopant level 10^{17}, 10^{18} cm^{-3}) and (B) the relationship of depletion width W_{SC} and band bending $\Delta\phi_{SC}$ on $BiVO_4$ (dopant levels between 10^{16}–10^{20} cm^{-3}).

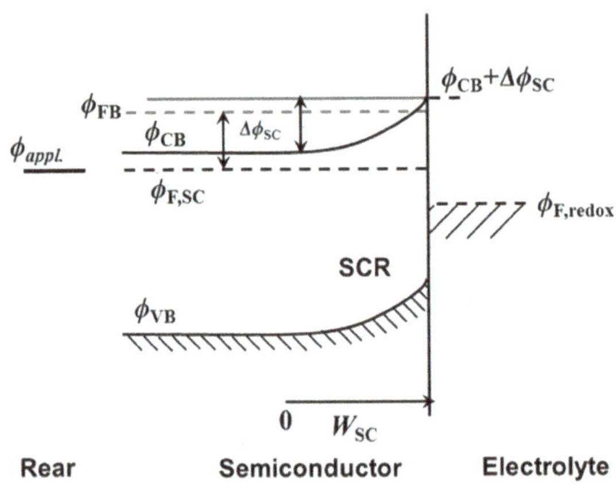

Fig. 8 Surface band structures of an n-type semiconductor under applied potential ϕ_{Appl}.

2.3 Flat Band Condition and Mott–Schottky Relationship

Following the discussions from last section, let us next examine how to best take advantage of the dependence of the depletion region on the applied potentials.

Experimentally, this measurement is often carried out by varying the applied electronic potential ($\phi_{Appl.}$) from the rear side of the electrode while probing the degree of band bending ($\Delta\phi_{SC}=\phi_{FB}-\phi_{Appl.}$) through the changes of the capacitance (Fig. 8). The space charge in the depletion layer

(Q_{SC}) under the applied potential can be obtained according to Eq. (9), with the boundary $\phi(x)$ ranged from ϕ_{CB} to $\phi_{CB} + \left(\phi_{FB} - \phi_{Appl.}\right)$:

$$Q_{SC} = \sqrt{2q\varepsilon_0\varepsilon_{SC}N_DA^2\left(\phi_{FB} - \phi_{Appl.} - \frac{kT}{q}\right)} \tag{11}$$

Therefore, the capacitance of the depletion region can be calculated according to the $C_{SC} = dQ_{SC}/d\Delta\phi_{SC}$. Considering that ($\Delta\phi_{SC} = \phi_{FB} - \phi_{Appl.}$), the exact solution is written as the Mott–Schottky equation (Eq. 12) for an n-type semiconductor (Morrison, 1980):

$$\frac{1}{C_{SC}^2} = \frac{1}{2\varepsilon_0\varepsilon_{SC}qN_DA^2}\left(\phi_{FB} - \phi_{Appl.} - \frac{kT}{q}\right) \tag{12}$$

We note that when the applied potential $\phi_{Appl.}$ equals to a certain value (ϕ_{FB}), the built-in voltage $\Delta\phi_{SC}$ becomes zero. This condition is known as the *flat band condition*. The applied potential then reports on the Fermi level of the photoelectrode relative to the electrochemical potential of the electrolyte. Such information is critical because the relative difference between the two energy levels determine the extent of the band bending, which in turn determines the charge separation capability of the system.

More specifically, the capacitance C_{SC} is normally measured as a function of the applied potential $\phi_{Appl.}$ For an n-type photoelectrode (Fig. 9), the

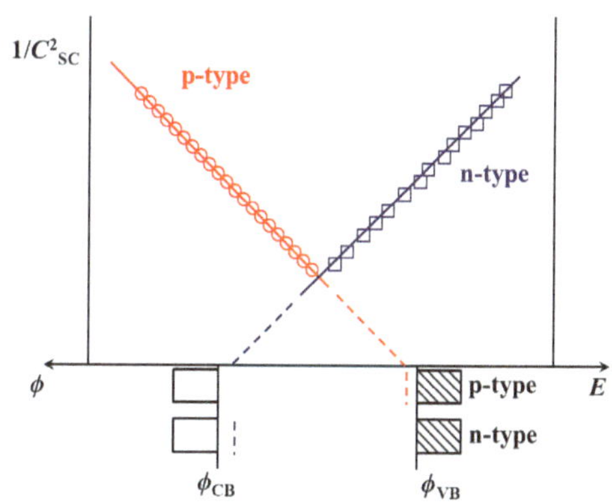

Fig. 9 Typical Mott–Schottky plots for n-type (*blue*) and p-type semiconductors (*red*). *Adapted from reference of Peter, L.M., 2016. Semiconductor electrochemistry. In: Giménez, S., Bisquert, J. (Eds.) Photoelectrochemical Solar Fuel Production: From Basic Principles to Advanced Devices. Cham: Springer International Publishing, with permissions, Copyright © 2016, Springer International Publishing Switzerland.*

obtained $C_{SC}{}^{-2}-f$ exhibits a linear relationship with a negative slope. (Similarly, a p-type semiconductor would feature a positive slope in the Mott–Schottky plot.) As can be seen from Eq. (12), an important parameter of the semiconductor, namely the doping level, can be extracted from this slope. Under normal conditions, the kT/q part is small enough to be negligible. As such, we can extrapolate the linear relationship of the Mott–Schottky curve to $1/C_{SC}{}^2=0$, obtaining the intersection at ϕ-axis as $\phi_{Appl}=\phi_{FB}$.

2.4 Quasi-Fermi Level and Photovoltage

By focusing on the semiconductor/electrolyte interface under equilibrium conditions (in dark), the discussions presented above do not offer a full picture of the working mechanisms of the system under light. We next deal with this topic. We know that illumination excites additional electrons from the valence band to the conduction band, moving the system away from equilibrium (Fig. 10A). This causes the redistribution of charges in the depletion region, resulting in the splitting of the quasi-Fermi levels of electrons and holes (Peter, 2007). Let us illustrate this understanding using an n-type semiconductor as an example. When light ($h\nu > E_g$) strikes the semiconductor, a great number of electrons and holes are "generated." The relative change of carrier concentrations is particularly dramatic for the minority carrier (holes in this example). The best way to describe the system is to use quasi-Fermi level of electrons ($\phi^*_{F,n}$) and holes ($\phi^*_{F,p}$), which can account for the concentration of photogenerated charges. They are:

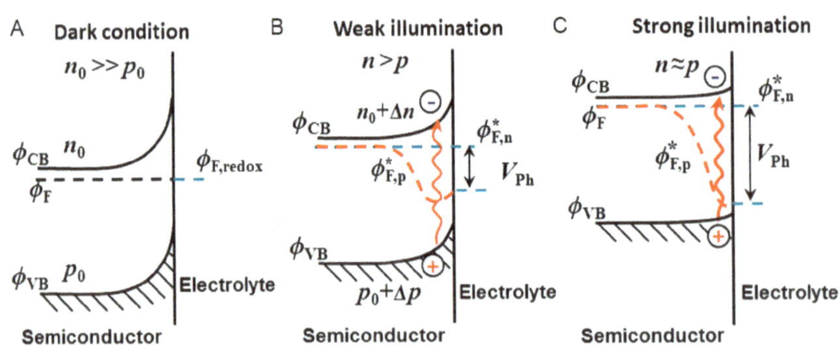

Fig. 10 Band diagrams of an n-type photoanode under different conditions. (A) At equilibrium, (B) quasi-equilibrium with relatively weak illumination, and (C) quasi-equilibrium with strong illumination.

$$n = n_0 + \Delta n = N_{CB} \exp\left[-q\left(\phi_{CB} - \phi^*_{F,n}\right)/kT\right] \quad (13)$$

$$p = p_0 + \Delta p = N_{VB} \exp\left[-q\left(\phi^*_{F,p} - \phi_{VB}\right)/kT\right] \quad (14)$$

where n_0 and p_0 correspond to the majority and minority charge carrier density under equilibrium in the dark, while the Δn and Δp correspond to the additional carriers generated by illumination. For an n-type semiconductor ($n_0 >> p_0$) and medium illumination intensity ($\Delta p = \Delta n \approx n_0$), we would observe relatively little difference between the equilibrium Fermi level and the quasi-Fermi level of electrons: $\phi^*_{F,n} \approx \phi_{F,n}$ The quasi-Fermi level of holes, on the other hand, is significantly different from the equilibrium Fermi level (Fig. 10B). The difference ($\phi^*_{F,n} - \phi^*_{F,p}$) between the quasi-Fermi level of electron and holes defines the upper limit of the achievable photovoltage (V_{ph}) of the semiconductor/electrolyte system (Qi and Wang, 2012). Under strong illuminations, the system would approach (but never fully reach) the flat band condition as shown in Fig. 10C, under which condition the upper limit of V_{ph} is still defined by the difference between the quasi-Fermi levels of electrons and holes.

3. THE SEMICONDUCTOR/ELECTROLYTE INTERFACE

An important assumption we made by introducing the preceding discussions is that an electrolyte would behave like an ideal metal, with the only difference being to use the electrochemical potential of the electrolyte to replace the work function of metal. However, in reality, the semiconductor/liquid interface presents complexities far greater than one expects from a semiconductor/metal one. For instance, the chemisorption on the semiconductor surface is unique to the semiconductor/liquid interface. The relatively weak screening effect of the electrolyte as compared to metal is also unique. We next discuss this aspect of the system in details.

3.1 Surface Hydroxylation and the Helmholtz Layer

When considering the semiconductor/liquid interface, it helps to recognize two major influences, namely the chemical adsorption on the surface and the influence by ions within the electrolyte near the surface. Because the context within which we wish to focus our discussions concerns H_2O splitting, it should be a good practice to start the discussion with aqueous systems. The dominant adsorptions are:

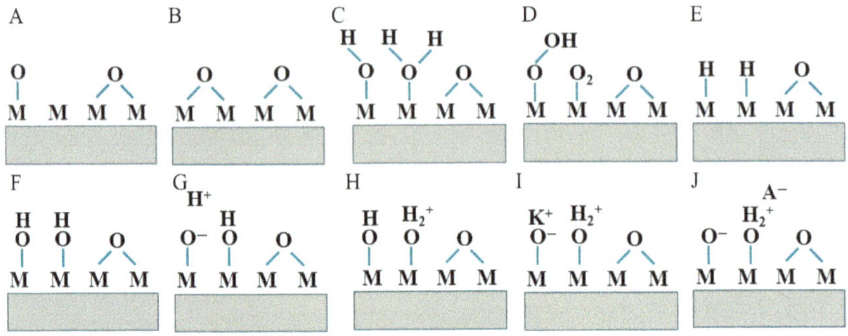

Fig. 11 Schematic representation of selected possible surface chemical nature for a metal oxide photoelectrode in aqueous electrolytes.

(1) O and H_2O adsorption on the unsaturated metal sites (Fig. 11B–D). Such adsorptions could take place in ambient atmosphere through bridge bonding or single bonding (Doyle and Lyons, 2016; Imanishi et al., 2007). These adsorptions may be regarded as a thin oxide or hydroxide layer. Under certain conditions, hydrogen can also be adsorbed on the metal sites (Fig. 11E).

(2) Partly due to the adsorptions as described earlier, abundant acid–base sites are expected on the surface functional groups on a semiconductor. When immersed in an electrolyte, these sites can undergo protonation and/or deprotonation (Fig. 11F–H). As such, the surface may be positively or negatively charged depending on the pH of the electrolyte relatively to the pK_a (or pK_b) of the surface. Only when the pH is at pK_a does the net total charge on the surface equal to zero (point of zero charge or PZC). At any other pH, the surface potential would have a profound influence on the electronic properties of the semiconductor.

(3) Beyond protons and hydroxyls, other ions are expected to interact with the semiconductor surface due to electrostatic forces (Fig. 11I and J). For instance, cations such as Na^+, K^+ can form a strong counterion shell to balance surface negative charges. Conversely, anions such as F^- may form strong bonding connect with Ge or Si surfaces (Grätzel, 2009).

Surface species due to electrostatic interactions may be better described by theories developed by pioneers such as Dybe-Huckel, Gouy-Chapman, Helmholtz, and Stern et al. In a simplistic fashion, we can borrow the theory by Helmholtz that treats the potential away from the surface of a solid into the electrolyte as a linear dependence on the distance $\psi = \psi_0\left(1 - \frac{x}{x_{HP}}\right)$, where ψ is the potential at position x away from the solid surface into the electrolyte and x_{HP} is the position beyond which the potential is zero. This model (Fig. 12) assumes a tight layer of counter ions adsorbed on the surface to balance the surface charges (the inner Helmholtz layer, or IHL), followed by solvated ions that are more loosely bound as the outer Helmholtz layer (OHL) (Srinivasan, 2006). In electrolytes with high ionic strength, the HL is compact. In electrolytes of low ionic strength, there may be insufficient ions available at the OHL plane to compensate all adsorbed charges at the IHL plane. The excess charges are then compensated in a region that extends much beyond the outer Helmholtz plane, forming the so-called Gouy Layer.

Next, we expand upon the simplified Helmholtz equation and describe the potential drop across the Helmholtz layer in more details. First, let us consider how the potential changes with pH. Assume M–OH as the surface species which forms equilibrium with H^+ through protonation and deprotonation. When the concentration of [M–O$^-$] and [M–OH$_2^+$] are equal, the net charge on the semiconductor is zero, namely PZC or isoelectronic point. Away from PZC, the absolute Helmholtz layer potential ϕ_{HL} changes with pH following (van de Krol, 2012):

$$\phi_{HL} = \phi_{HL}^{o} + \frac{2.3kT}{e}(pH - PZC) \tag{15}$$

where e is the elementary charge of an electron. Eq. (15) clearly indicates that the Helmholtz potential changes 59 mV per pH unit at 25°C.

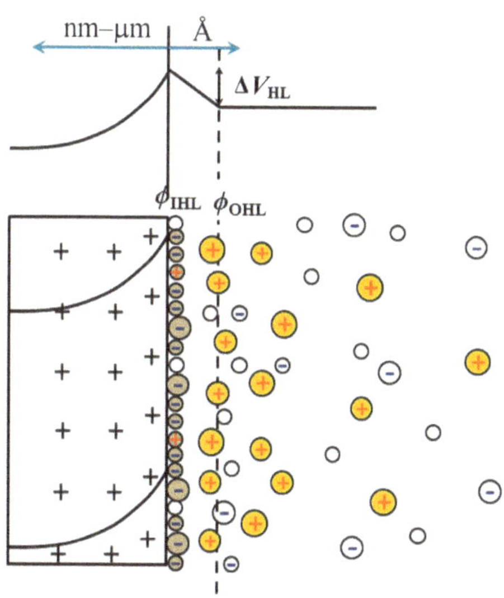

Fig. 12 Scheme of the Helmholtz layer at the semiconductor surface.

Second, it is reasonable to see that the charges within the depletion region of the semiconductor (e.g., minority charge accumulation) would have a profound influence on the species adsorbed on the surface of the semiconductor, simply because of the electrostatic effects.

Notwithstanding, a potential drop across the Helmholtz layer can be described:

$$\Delta V_{HL} = \frac{\varepsilon_{SC}}{\varepsilon_{HL}} \left(\frac{\Delta \phi_{SC} q N_D}{2 \varepsilon_{SC} \varepsilon_0} \right)^{1/2} W_{HL} \tag{16}$$

ε_{HL} is the relative permittivity and W_{HL} is the thickness of the Helmholtz layer, respectively. The potential drop across the Helmholtz layer is usually negligible compare with $\Delta \phi_{SC}$.

Note: when the doping level of the semiconductor is higher (10^{19} cm^{-3}) or the surface state cannot be negligible, the potential drop in Helmholtz layer can be very high ($\Delta V_{HL} \sim 0.1 - 0.5 V$).

3.2 Fermi-Level Pinning Effect

We see from the earlier discussions that the degree of band bending is highly sensitive to at least two important factors, the flat band potential of the semiconductor and the electrochemical potential of the electrolyte. Furthermore, we see that the potential drop within the Helmholtz layer is also critically important. Next, let us examine how this potential drop may be influenced by the electrolyte, the semiconductor, and their interactions at the interface. We will use the change of pH to demonstrate how the relative positions of the semiconductor band edges and the electrochemical potential of the electrolyte change and the implications. First, under ideal conditions, Eq. (15) suggests that the potential drop within the Helmholtz layer tracks pH following a 59 mV/decade relationship at room temperature. Given that the electrochemical potentials of water oxidation and reduction (van de Krol, 2012) also track the pH following the same trend (Eqs. 17 and 18), the relative positions of the band edge energies of the semiconductor and the electrochemical potential are fixed. We call such a condition band edge pinned, where the increase of the applied potentials will be fully utilized to increase the degree of band banding and, hence, the charge separation capabilities under lighting conditions (see, e.g., Fig. 8).

$$\begin{aligned}\phi_{F,HER} &= \phi^{o}_{F,HER} + \frac{RT}{4F}\ln\left(\frac{p^2_{H_2}}{[H^+]^4}\right)\\ &= \phi^{o}_{F,HER} + \frac{2.3kT}{e}\left(\log p^{1/2}_{H_2} + pH\right)\end{aligned} \tag{17}$$

$$\begin{aligned}\phi_{F,OER} &= \phi^{o}_{F,OER} - \frac{RT}{4F}\ln\left(p_{O_2}[H^+]^4\right)\\ &= \phi^{o}_{F,OER} + \frac{2.3kT}{e}\left(-\log p^{1/4}_{O_2} + pH\right)\end{aligned} \tag{18}$$

The ideal band edge pinning effect can be achieved when the surface adsorbed species are the same as the electrochemical reaction species, and no other chemical adsorptions take place. For most metal oxide semiconductors studied for water splitting reactions, the situation is a reasonable assumption. However, when other semiconductors (e.g., Si or III–V semiconductors) are used, the assumption rarely holds true. In addition to the deviation of the 59 mV/decade shift following the change of pH by the semiconductor band edge positions, the potential drop within the semiconductor may also fail to follow the same trend. Under extreme conditions, the relative positions of the band edge energies become completely decoupled from the electrochemical potential of the electrolyte, where we consider the system Fermi level pinned (Fig. 13). The variation of the electrolyte pH causes negligible band bending in semiconductor side but a potential shift in Helmholtz layer (Fig. 13A and B). Here, by Fermi level pinning (Bard et al., 1980), we suggest that the relative position of the Fermi level at the surface as compared to the band edge energies is independent of the applied potential. As a result, the increase of the applied potential does little to increase the degree of band bending (Fig. 13C), which is analogous to the illuminated condition (Fig. 13D). Other reasons responsible for the Fermi

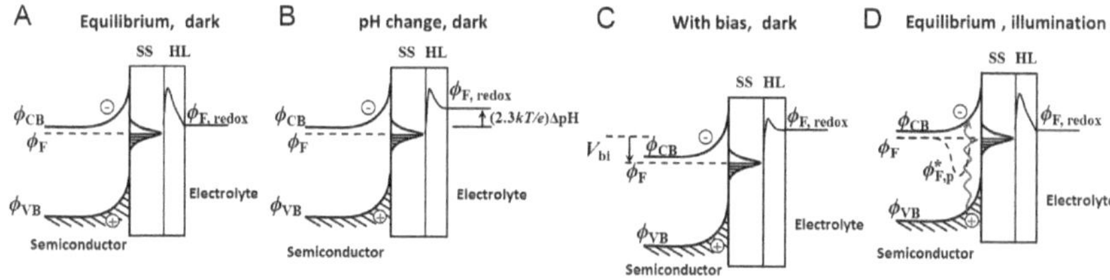

Fig. 13 Fermi level pinning effect due to surface state of an n-type semiconductor in contact with an electrolyte: (A) under dark; (B) dependence on pH variation; (C) effects of externally applied, where $V_{bi} = -\phi_{bi} = \phi_{Appl.} - \phi_{FB}$; and (D) under illumination with severe Fermi level pinning.

level pinning include large concentrations of surface states due to structural defects and/or chemisorbed species.

Literatures showed the interface may not be ideal model described as previously, either by the intrinsic surface exposing or by the surface modification. The surface states, the passivation or the electrocatalyst modification, can significantly influence on these parts.

Experimentally, Fermi level pining effect has a direct impact on the Mott–Schottky method to quantitatively study the space charge region. If we use the general $R_S(R_{SC}C_{SC})(R_{HL}C_{HL})$ equivalent circuit, the relationship between the applied potential $\phi_{Appl.}$ ($\Delta\phi_{SC} = \phi_{Appl.} - \phi_{FB}$ for n-semiconductor) and the space charge region $1/C_{SC}^2$ is no longer linear following Eq. (12) in Fig. 14 (Liu et al., 2013). At the energy level of surface states, the measured space charge capacitance C_{SC} changes little due to the change of applied potentials, creating a plateau in the Mott–Schottky plot (Klahr et al., 2012). In other words, one must exercise cautions when interpreting the AC electrochemical data for the quantification of, for example, the flat band potentials. Errors may occur due to the "charging" effect of the surface states. Conversely, one may take advantage of the charging effect to quasi-quantitatively characterize the surface states, the result of which can be borrowed for the understanding of the surface processes.

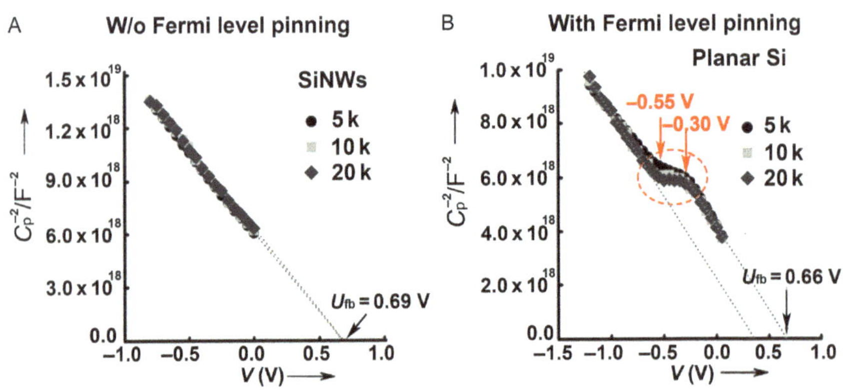

Fig. 14 Mott–Schottky plots of p-type Si: (A) without Fermi level pinning effect and (B) with Fermi level pinning condition, where the space charge region capacitance in *red circle* region is fixed due to the unchanged band bending. *Adapted from the reference of Liu, R., Stephani, C., Han, J.J., Tan, K.L., Wang, D., 2013. Silicon nanowires show improved performance as photocathode for catalyzed carbon dioxide photofixation. Angew. Chem. Int. Ed. 52, 4225–4228, with permission, Copyright © 2013 Wiley VCH Verlag GmbH & Co. KGaA, Weinheim.*

4. CHARGE TRANSFER ACROSS THE SEMICONDUCTOR/ ELECTROLYTE INTERFACE

All previous discussions primarily concern the energetics of the semiconductor/electrolyte interface. That is, the discussions help us understand how the interface defines the charge separation capabilities. As far as a chemical system is concerned, the nature of charge transfer at the interface is of equal importance. We aim to discuss this aspect next. In a simplistic fashion, charge transfer across this interface may be first treated as a simple electrochemical system, in which the potential of the photoelectrode comprise of two components, the applied potential and the photovoltage ($V = V_{app} + V_{ph}$). Beyond this point, the photoelectrode can be conveniently treated as one indifferent from a metal electrode, at least for the consideration of the charge transfer kinetics. According to the Butler–Volmer equation (van de Krol, 2012), we have:

$$j = j_0\left(\exp\left(\frac{(1-\alpha)e\eta}{kT}\right) - \exp\left(\frac{-\alpha e\eta}{kT}\right)\right) \tag{19}$$

where α is the charge transfer coefficient (normally 0.5 for metal electrodes). e is the elementary charge of an electron, η corresponds to the overpotential ($\eta = \left|\phi_{appl.} - \phi_{F,redox}\right|$). The Butler–Volmer equation describes the relationship between the electrical current and the electrode potential. Under dark conditions, the current–potential relationship can be described as:

$$j_n = j_n^0\left(\exp\left(\frac{-e\eta}{kT}\right) - 1\right) \tag{20}$$

where j_n^0 is the saturated exchange current of reverse bias on n-type semiconductor.

To quantitatively describe the semiconductor/electrolyte steady-state current–voltage relationship under illumination (Fig. 15), we can borrow the theories first developed by Gärtner for solid-state junctions (Gärtner, 1959) and later adoption by Butler (1977) for semiconductor/electrolyte junctions. Under dark condition (Fig. 15A), the free space charge in semiconductor can exchange with the redox pair, showing a net current flow. While for the illuminated condition (Fig. 15B), the current drastically increases due to the significantly increased minority charge density. The relationship is presented in Eq. (21), where the charge transfer rate across

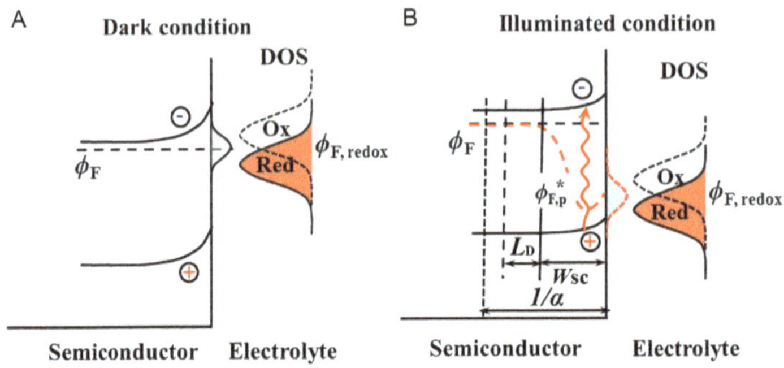

Fig. 15 Details of an n-type photoelectrode/electrolyte interface for the consideration of photocurrent calculations: (A) in dark and (B) under illumination. W_{SC} is the width of the space charge region and $1/\alpha$ is the wavelength-dependent penetration depth of the incident light, L_D is the diffusion length of the charge (hole).

the semiconductor/electrolyte is assumed to be sufficiently fast so that it is not a limiting factor in determining the current densities J_G (Reichman, 1980). We shall see later that such an assumption is not always reliable.

$$J_G = J_0 + eI_0\left(1 - \frac{\exp(-\alpha W_{SC})}{1 + \alpha L_D}\right) \tag{21}$$

where I_0 is the incident light flux and J_0 is the saturation dark current density ($J_0 = ep_0 L_D/\tau$), L_D is the diffusion length of minority charge, and α is the absorption coefficient.

Note that the above equation provides a quantitative measure on the upper limit of the current density by only assuming loses within the semiconductor (e.g., recombination and trapping of charges, as measured in L_D). This upper limit can be used to compute the upper limit of the incident photon to current conversion efficiency (IPCE):

$$\text{EQE} = \frac{j_{\text{photo}}}{eI_0} = 1 - \frac{\exp(-\alpha W_{SC})}{1 + \alpha L_D} \tag{22}$$

As aforementioned, the Gärtner prediction ignores the important fact that charge transfer across the semiconductor/electrolyte interface is also sluggish because of the nature of the chemical reactions (Fig. 16). The practical onset potential ϕ'_{onset} shift anodically for a photoanode compared with the theoretical ϕ_{onset}, where the measured photocurrent j_a is extremely low while much of the photoexcited charges contribute to the immeasurable recombination current j_{rec} in the low overpotential range. This is particularly true when it comes to complex reactions such as water oxidation which

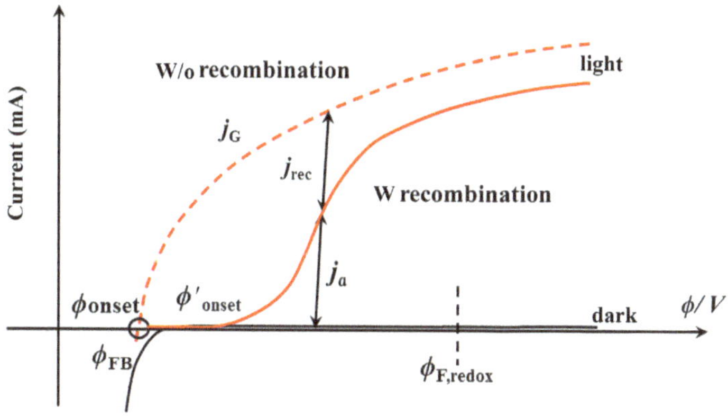

Fig. 16 Delayed turn-on of the photocurrent due to recombination, either in the space charge region or at the surface. j_G is the current predicted by the Gärtner equation, j_a is the experimental photocurrent, and j_{rec} is the recombination current. The ϕ_{onset} and ϕ'_{onset} correspond to the onset potentials of the photoelectrodes under the condition without recombination and with recombination, respectively. *Adapted from the book of Peter, L.M., 2016. Semiconductor electrochemistry. In: Giménez, S., Bisquert, J. (Eds.) Photoelectrochemical Solar Fuel Production: From Basic Principles to Advanced Devices. Cham: Springer International Publishing, with permission, Copyright © 2016, Springer International Publishing Switzerland.*

involves four protons and four electrons. The slow charge transfer on the one hand may act as a limiting factor to reduce the achievable photocurrent. On the other hand, it opens up doors for competing processes that contribute to the annihilation of the photogenerated charges. The latter processes may be broadly referred to as surface recombination. Worse still, if the competing processes involve chemical transformation of the semiconductor, photocorrosion of the electrode may become significant. We see now that speeding up charge transfer across the semiconductor/electrolyte interface is of paramount importance. It is within this context that significant research attention has been attracted to the development and study of catalysts, which is the topic we plan to address next.

4.1 Issues at the Semiconductor/Catalyst/Electrolyte Interface

To appreciate the complexities of the interface that involves catalyst, let us consider the overall processes on a combined practical semiconductor/electrolyte device (Fig. 17). In such a typical system, when the light is illuminated on the n-type semiconductor, light absorption leads to charge excitation which populates electrons in conduction band and holes in valence band. Ideally, electrons should be collected at the rear of the

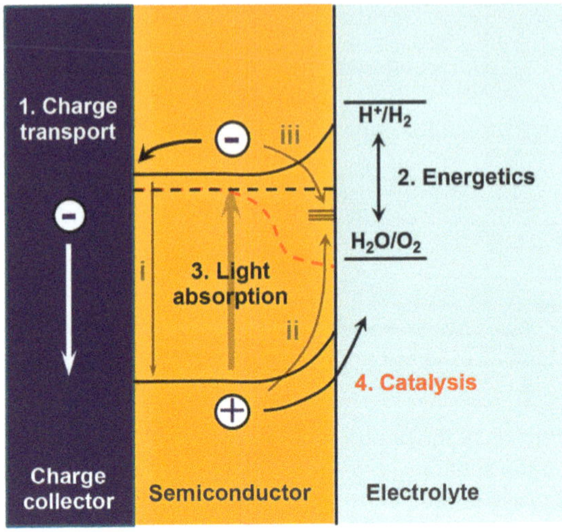

Fig. 17 Semiconductor-based photoelectrochemical H_2O splitting on an n-type electrode. Important processes to consider include (1) charge transport, (2) the energetics relative to H_2O redox potentials, (3) light absorption (spectral response), and (4) catalytic activities. Competing detrimental processes include (i) bulk recombination and (ii, iii) surface trapping and recombination. *Reprinted with permission from Mayer, M.T., Lin, Y., Yuan, G., Wang, D., 2013. Forming heterojunctions at the nanoscale for improved photoelectrochemical water splitting by semiconductor materials: case studies on hematite. Acc. Chem. Res. 46, 1558–1566, Copyright 2012 American Chemical Society.*

semiconductor, and holes should be concentrated on the surface, for eventual transfer to the electrolyte either directly or via cocatalysts. We see here that we are dealing with at least four somewhat independent issues, namely (1) charge collection at the rear of the photoelectrode; (2) matching of the energetics; (3) optoelectronic properties of the photoelectrode; and (4) charge transfer across the photoelectrode/electrolyte interface. It is reasonable to argue that the first three issues are shared by systems focused on solar-to-electricity conversion (such as solar cells). As such, they are not unique to the PEC systems. We therefore are interested in focusing on the fourth issue that concerns the surface chemical reactions.

4.2 Kinetics at Semiconductor/Catalyst/Electrolyte Interface

To help us further focus in developing an understanding of the system, let us only deal with water splitting reactions here. This way we can concentrate on the OER and HER reactions. It is well developed that the photophysical processes within a semiconductor are of significantly faster time scales than surface chemical reactions (Fujishima et al., 2008; Pendlebury et al., 2014). For example, charge generation owing to photoexcitation takes place in fs; charge transfer between bands and trap states is within hundreds of fs; charge recombination within the solid and on the surface is typically in tens of μs. By contrast, surface water splitting reactions mostly fall within the ms to s time scale (Le Formal et al., 2015). In other words, under majority circumstances, surface reaction is the rate-limiting step for water splitting reactions.

Under illumination and assume no nonradiative recombination on the surface, we obtain the quasi-equilibrium concentration of surface accumulated hoe concentration as follows (Pleskov, 1990; Thorne et al., 2015):

$$\frac{p_s}{N_{VB}} = \exp\left(-\frac{q\left(\phi^*_{F,p} - \phi_{VB}\right)}{kT}\right) = \exp\left[-\frac{q\left(\phi_{Appl.} - \phi_{VB} - V_{ph}\right)}{kT}\right] \quad (23)$$

where N_{VB} is the intrinsic hole concentration; $\phi_{Appl.}$, ϕ_{VB}, and V_{ph} correspond to the applied potential, valence band potential, and the photovoltage, respectively. The photovoltage is defined by the difference between the quasi-Fermi level of the holes $\phi^*_{F,p}$ under illumination and Fermi level under dark $\phi_{F,n}$ conditions, respectively. The photocurrent can be then be derived as (Thorne et al., 2015):

$$j_p = j_p^0\left(\frac{p_s}{N_{VB}}\exp\left(\frac{q\alpha_p\eta}{kT}\right) - \exp\left(\frac{-q\left(1-\alpha_p\right)\eta}{kT}\right)\right) \quad (24)$$

where j_p is the current due to the holes transferring, j_p^0 is the exchange current in the dark. q is the elementary charge of electrons, α_p is the hole transfer coefficient; η is the overpotential.

Eq. (24) has several important implications. First, it suggests that in order to maximize photocurrent, one desires to increase the quasi-equilibrium hole concentration. In other words, increasing the photovoltage is of great importance. An obvious route to do so is to reduce surface recombination, both by electrons from the conduction band and from parasitic chemical reactions. Second, it suggests that at the same applied potentials (and, hence, the same η), it is beneficial to increase α_p, which highlights the importance of catalysts that favor forward charge transfer, by reducing the activation energy (Wang et al., 2012). The understanding is illustrated in Fig. 18A and B. If we regard the photovoltage as the thermodynamic factor that defines the photocurrent–voltage dependence and the overpotential as the kinetic factor, Eq. (24) further implies that it is exceedingly difficult to accurately

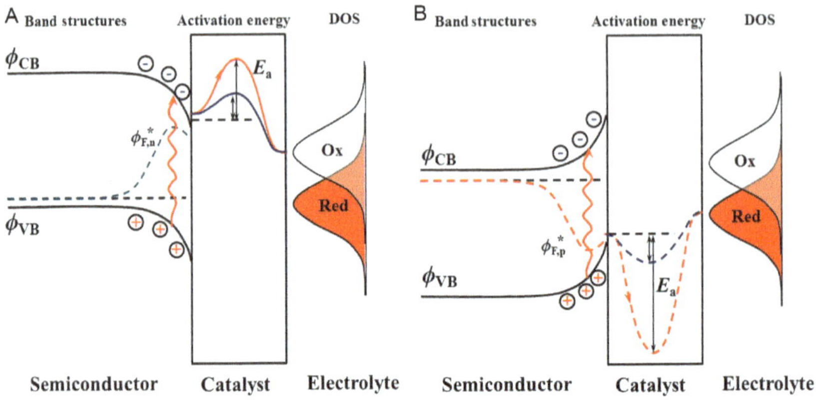

Fig. 18 Energy barriers at the semiconductor/catalyst/electrolyte interface for (A) reduction and (B) oxidation reactions.

understand what the distinctive reasons are (thermodynamic or kinetic) for a real system that is far from ideal in terms of performance. Conversely, the complexities also suggest that we have more than one way to improve the performance of a photoelectrode by applying cocatalysts with different functionalities.

4.3 Semiconductor/"Inhibitor"/Electrolyte

Simplistically, one way to improve the overall performance of a photoelectrode is to increase the photovoltage. Even without altering the charge transfer kinetics, we should be able to increase the measurable photocurrent at a given applied potential according to Eq. (24). Indeed, some catalysts have been found to act as a minority charge carrier reservoir. Under illumination, minority charges are first transferred to this reservoir and then further transferred into the electrolyte for the desired chemical reactions. Although the rate of chemical reactions is not altered, by quickly removing charges from the surface one can effectively reduce surface recombination and increase surface photovoltage (Barroso et al., 2011; Lin et al., 2012; Yang et al., 2014). The scenario is illustrated in Fig. 19A–C for a n-type photoelectrode system.

Following the same line of thoughts, we can next consider another condition in which no kinetic contribution is involved. That is, the same goal of increasing surface photovoltage can be achieved by forming buried junctions.

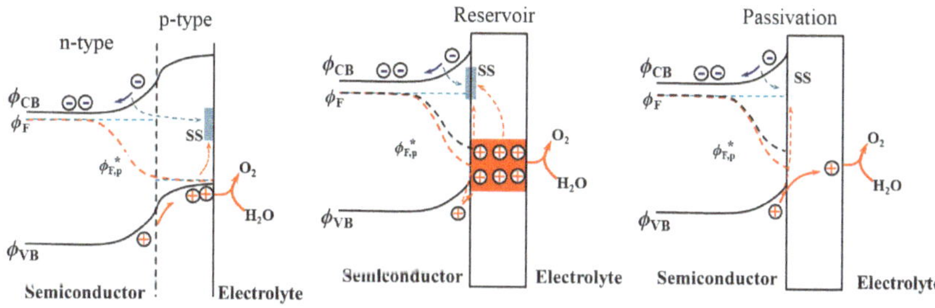

Fig. 19 "Inhibitor" modified n-type semiconductor for water oxidation by reducing recombination: (A) p–n junction, (B) reservoir effect, and (C) passivation effect.

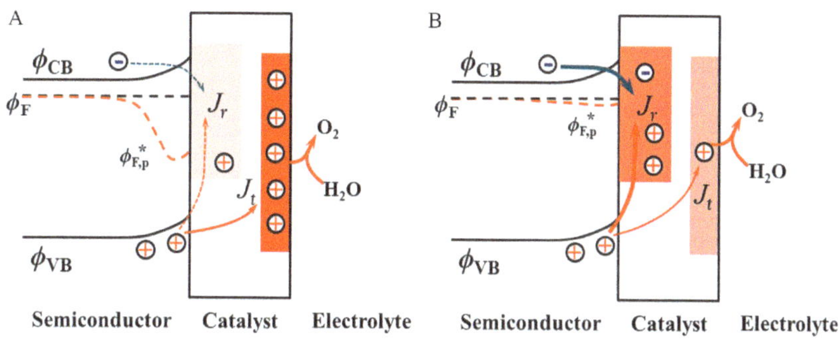

Fig. 20 "Promoter"-modified photoelectrode for water oxidation: (A) ideal semiconductor/catalyst interface and (B) inadvertent semiconductor/catalyst interface.

4.4 Semiconductor/"Promoter"/Electrolyte

Further examinations of Eq. (24) suggest that in addition to reduce surface recombination, promoting forward charge transfer is an equally effective way of improving the performance of a photoelectrode. That is, ideally the application of a "true" cocatalyst should only increase charge transfer rate without adversely impacting the surface photovoltage (Fig. 20A). In reality, however, the introduction of cocatalysts may have unintended effects. For instance, the cocatalyst may introduce additional surface states that promote surface recombination (in Fig. 20B). For any given system, these positive and negative effects should be studied separately for detailed understandings.

Next, let us consider MnO_x as an example to further illustrate this point. While a well-recognized OER catalyst, the application of MnO_x (prepared by atomic layer deposition) has been shown to almost completely diminish the performance of α-Fe_2O_3 (Yang et al., 2013). It was found that the high density of states introduced by MnO_x on the surface of α-Fe_2O_3 inadvertently promotes surface recombination. It produces an effect akin to almost complete Fermi level pinning, so that the achievable photovoltage is close to 0 V. In contrast, when amorphous $NiFeO_x$ OER catalyst was introduce to the surface of α-Fe_2O_3, the photovoltage was significantly increased, resulting in a greatly turn-on potential (Du et al., 2013). Interestingly, the negative effect of MnO_x is by no means absolute. For instance, Lewis et al. found that a thick MnO layer deposited on n-Si by ALD showed a clear positive effect in improving water oxidation performance (Strandwitz et al., 2013).

5. UNASSISTED WATER SPLITTING

The holy-grail of solar water splitting is a process that only requires sunlight as the energy input, and that the cost of the materials should be low, with good durability (Lewis, 2007). While it is a common practice to characterize (and compare) "cost" and "efficiency" of any given systems separately, these two considerations are intimately connected (Lewis, 2007, 2016). Ultimately, the golden standard with which any artificial system would have to be compared with is the natural photosystems. Within this context, we still have a long way to go. Existing systems that meet the efficiency benchmarks are invariably too expensive, and those that are cost-effective suffer low efficiencies. Below we use fundamental considerations for semiconductor-based PEC systems as a platform to present our view of the critical issues one has to take in account in designing and optimizing practical unassisted solar water splitting systems.

First and foremost is the consideration of the thermodynamic driving force. We need an overall photovoltage at the minimum of 1.23 V (more practically greater than 1.6 V). As such, single absorbers usually fail short to meet this requirement because only wide bandgap semiconductors would be possible for such a goal, which unfortunately cannot absorb broadly within the solar spectrum. Therefore, dual or triple absorber/junctions become necessary. Take a photocathode–photoanode configuration as an example. One way to achieve the goal is to use the two semiconductors separately, as shown in Fig. 21. Each semiconductor forms its own semiconductor/liquid junction, to provide its corresponding photovoltage. The photovoltage may be empirically measured as the difference between the turn-on potential and the corresponding thermodynamic equilibrium potential. For instance, suppose the turn-on potential of a photoanode is 0.4 V vs RHE; the photovoltage may be presented as 1.23–0.4 V = 0.83 V. Alternatively, one can combine the various absorbers into a single

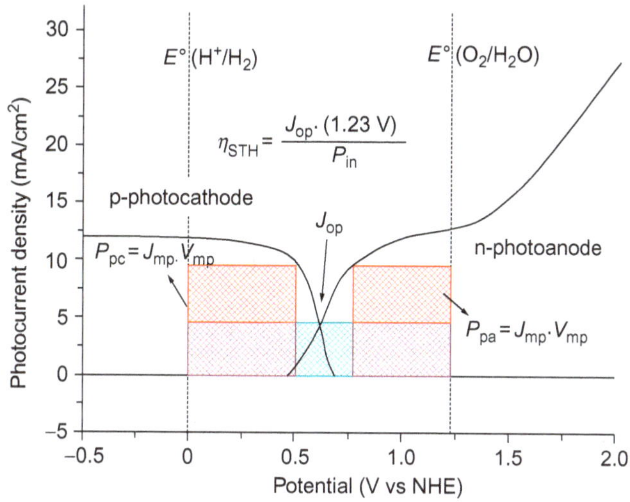

Fig. 21 Overlaid photocurrent density–potential behaviors for a p-type photocathode and an n-type photoanode, with the overall efficiency projected by the power generated $P_{STH}=J_{op}$ (1.23 V) by the cell for water splitting. *Reprinted with permission from Walter, M.G., Warren, E.L., Mckone, J.R., Boettcher, S.W., Mi, Q., Santori, E.A., Lewis, N.S., 2010. Solar water splitting cells. Chem. Rev. 110, 6446–6473 Copyright 2010 American Chemical Society.*

electrode by forming buried junctions. The latter approach benefits tremendously from efforts focused on developing better multijunction solar cells. It also means that this line of research would not be unique to that focused on PEC water splitting.

Second, similar to the need to match current densities by each individual layer within a multijunction solar cell, the balance of photocurrents by the photocathode and photoanode in a combined PEC system is of critical importance. Ideally, one would want to trade off the current densities so that both photoelectrodes function at the maximum power point. In reality, however, it is exceedingly difficult for such a goal. On the one hand, we need to consider the photophysical processes within the semiconductor for optimized light absorption, charge separation, and charge transfer. On the other hand, we must take into account of surface chemical reactions, to ensure that the kinetics match the influx of photogenerated charges. While it is generally accepted that cocatalysts are necessary to promote surface chemical reactions, how the presence of cocatalysts affects the behaviors of the semiconductor, and how does the semiconductor influence the chemical nature and reactivity of the cocatalyst are poorly understood.

Third, on top of all these issues, we need to be mindful that stability is of paramount importance. It may be the single most important distinguishing parameter that defines PEC water splitting. This consideration leads to the discussion on the balance of cost and stability (Lewis, 2007, 2016). When it comes to cost, we really mean at least two components, the cost of the composing elements (scarcity, ease of mining, etc.) and the cost of fabrication. Take Si as an example. As the second most abundant element on the crust of Earth, it is obviously a low-cost material. However, making solar-cell grade Si remains an expensive process because of the high purities needed. By and large, we view that the real hope of low-cost solar water splitting would have to be carried out by earth-abundant materials such as Cu_2O, Fe_2O_3, $BiVO_4$, and WO_3 (Yang et al., 2015). These materials also score as being relatively stable against photocorrosion. Nevertheless, we also agree that studies on high-performance, low-stability materials such as GaP are equally important as it represents a different route toward the final goal of economically competitive solar water splitting.

As a demonstration of unassisted solar water splitting by low-cost materials, we present a recent example developed by us (in Fig. 22A). Jang et al. employed $NiFeO_x$ on a two-step solution grown Fe_2O_3 nanofilms as the photoanode (V_{on} ~0.45 V), and a-Si as the photocathode (V_{on} ~0.8 V)

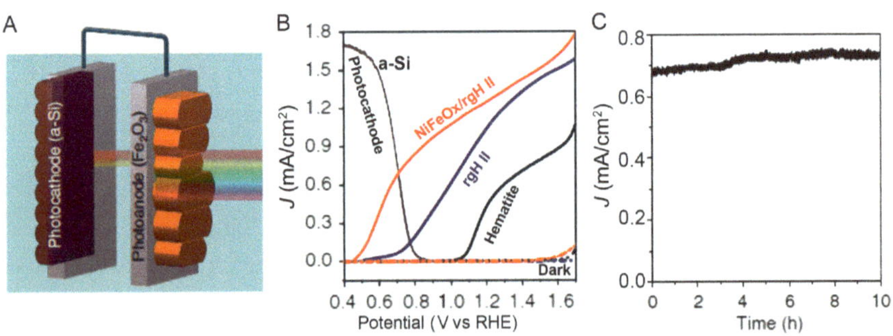

Fig. 22 An example of unassisted solar water splitting by earth-abundant materials, where Fe_2O_3 and Si are used as the photoanode and photocathode, respectively: (A) schematic PEC tandem cell under illumination, (B) *J–V* curves of various hematite photoanodes and a-Si photocathode, rgH II refers to hematite sample treated by twice growth condition, and (C) net photocurrent during the first 10 h of the tandem cell in 0.5 M phosphate solution (pH 11.8). *Adapted from the reference of Jang, J.-W., Du, C., Ye, Y., Lin, Y., Yao, X., Thorne, J., Liu, E., Mcmahon, G., Zhu, J., Javey, A., Guo, J., Wang, D., 2015. Enabling unassisted solar water splitting by iron oxide and silicon. Nat. Commun. 6, 7447, with permissions, Copyright © 2015, Rights Managed by Nature Publishing Group.*

in Fig. 22B (Jang et al., 2015). They achieved an unassisted PEC device to split water, where a photocurrent of 0.7 mA/cm^2 and total solar energy conversion efficiency of 0.9% was shown (Fig. 22C).

6. SUMMARY AND OUTLOOK

In this chapter, we summarized fundamental concepts that are important to semiconductor-based solar water splitting reactions. These principles not only govern photoelectrochemical processes, they are also applicable to semiconductor particle and nanoparticle-based photocatalysis because the light absorption, charge separation, and charge transfer details are shared by these processes (Hisatomi et al., 2014). From this perspective, photoelectrochemical studies may be regarded as a characterization tool which provides an opportunity to examine the detailed phenomenon of photocatalysis. Similar details would be exceedingly difficult to study separately on a particle-based system. For ease of reading, we presented the details of the system from a thermodynamic perspective, where how to maximize photovoltage is of great concerns. We also alluded to the importance of the kinetics at the photoelectrode/water interface, an area that has so far received undue attention. It is our assessment that detailed understandings of both the thermodynamic and kinetic factors will play key roles in advancing solar water splitting to a stage that is economically competitive. As of today, there is still a long way to go. Not only do we not have a consensus what materials would be the winners, we are also at a stage of debating what engineering designs should be adopted. Considering that research on this topic has lasted nearly five decades, it is easy for researchers and bystanders to be frustrated by the slow progress. Nevertheless, it is critically important to remember the magnitude of the issues we are facing, and the limited solutions we have to battle these challenges. Philosophically, there is a good reason why mother Nature developed natural photosynthesis as a solar energy storage solution. It is therefore hard to imagine a renewable-energy-powered future without water splitting technologies.

ACKNOWLEDGMENTS

The work at Boston College is supported by the National Science Foundation, Boston College, the Sloan Foundation and Massachusetts Clean Energy Center. And the work at Xuchang University is supported by National Natural Science Foundation of China (U1604121 and 21673200).

REFERENCES

B.P., 2014. BP Statistical Review of World Energy. British Petroleum Co, London.

Bard, A.J., Faulkner, L.R., 2000. Electrochemical Methods: Fundamentals and Applications, second ed. John Wiley & Sons, New York, USA.

Bard, A.J., Fox, M.A., 1995. Artificial photosynthesis: solar splitting of water to hydrogen and oxygen. Acc. Chem. Res. 28, 141–145.

Bard, A.J., Bocarsly, A.B., Fan, F.R.F., Walton, E.G., Wrighton, M.S., 1980. The concept of Fermi level pinning at semiconductor/liquid junctions. Consequences for energy conversion efficiency and selection of useful solution redox couples in solar devices. J. Am. Chem. Soc. 102, 3671–3677.

Barroso, M., Cowan, A.J., Pendlebury, S.R., Grätzel, M., Klug, D.R., Durrant, J.R., 2011. The role of cobalt phosphate in enhancing the photocatalytic activity of α-Fe_2O_3 toward water oxidation. J. Am. Chem. Soc. 133, 14868–14871.

Boddy, P.J., 1968. Oxygen evolution on semiconducting TiO_2. J. Electrochem. Soc. 115, 199–203.

Boekema, E.J., Kouřil, R., Dekker, J.P., Jensen, P.E., 2006. Association of photosystem I and light-harvesting complex II during state transitions. In: Golbeck, J.H. (Ed.), Photosystem I: The Light-Driven Plastocyanin:Ferredoxin Oxidoreductase. Springer Netherlands, Dordrecht.

Butler, M.A., 1977. Photoelectrolysis and physical properties of the semiconducting electrode WO_3. J. Appl. Phys. 48, 1914–1920.

Chattopadhyay, D., Rakshit, P.C., 2006. Electronics: Fundamentals and Applications. New Age International (P) Limited, New Delhi, India.

Dean, J.A., 1999. Lange's Handbook of Chemistry. McGraw-Hill, New York.

Demars, B.O.L., Gislason, G.M., Olafsson, J.S., Manson, J.R., Friberg, N., Hood, J.M., Thompson, J.J.D., Freitag, T.E., 2016. Impact of warming on CO_2 emissions from streams countered by aquatic photosynthesis. Nat. Geosci. 9, 758–761.

Doyle, R.L., Lyons, M.E.G., 2016. The oxygen evolution reaction: mechanistic concepts and catalyst design. In: Giménez, S., Bisquert, J. (Eds.), Photoelectrochemical Solar Fuel Production: From Basic Principles to Advanced Devices. Springer International Publishing, Cham.

Du, C., Yang, X., Mayer, M.T., Hoyt, H., Xie, J., Mcmahon, G., Bischoping, G., Wang, D., 2013. Hematite-based water splitting with low turn-on voltages. Angew. Chem. Int. Ed. Engl. 52, 12692–12695.

Fujishima, A., Honda, K., 1972. Electrochemical photolysis of water at a semiconductor electrode. Nature 238, 37–38.

Fujishima, A., Zhang, X., Tryk, D.A., 2008. TiO_2 photocatalysis and related surface phenomena. Surf. Sci. Rep. 63, 515–582.

Gärtner, W.W., 1959. Depletion-layer photoeffects in semiconductors. Phys. Rev. 116, 84–87.

Grätzel, M., 2001. Photoelectrochemical cells. Nature 414, 338–344.

Grätzel, M., 2009. Recent advances in sensitized mesoscopic solar cells. Acc. Chem. Res. 42, 1788–1798.

Hisatomi, T., Kubota, J., Domen, K., 2014. Recent advances in semiconductors for photocatalytic and photoelectrochemical water splitting. Chem. Soc. Rev. 43, 7520–7535.

Imanishi, A., Okamura, T., Ohashi, N., Nakamura, R., Nakato, Y., 2007. Mechanism of water photooxidation reaction at atomically flat TiO2 (rutile) (110) and (100) surfaces: dependence on solution pH. J. Am. Chem. Soc. 129, 11569–11578.

Jang, J.-W., Du, C., Ye, Y., Lin, Y., Yao, X., Thorne, J., Liu, E., Mcmahon, G., Zhu, J., Javey, A., Guo, J., Wang, D., 2015. Enabling unassisted solar water splitting by iron oxide and silicon. Nat. Commun. 6, 7447.

Klahr, B., Gimenez, S., Fabregat-Santiago, F., Hamann, T., Bisquert, J., 2012. Water oxidation at hematite photoelectrodes: the role of surface states. J. Am. Chem. Soc. 134, 4294–4302.
Krishnan, R., 2007. Fundamentals of semiconductor electrochemistry and photoelectrochemistry. Encyclopedia of Electrochemistry. Wiley-VCH Verlag GmbH & Co. KGaA, Weinheim, Germany.
Kumar, B., Llorente, M., Froehlich, J., Dang, T., Sathrum, A., Kubiak, C.P., 2012. Photochemical and photoelectrochemical reduction of CO_2. In: Johnson, M.A., Martinez, T.J. (Eds.), Annual Review of Physical Chemistry, vol. 63. Palo Alto, USA.
Le Formal, F., Pastor, E., Tilley, S.D., Mesa, C.A., Pendlebury, S.R., Grätzel, M., Durrant, J.R., 2015. Rate law analysis of water oxidation on a hematite surface. J. Am. Chem. Soc. 137, 6629–6637.
Lewis, N.S., 2007. Toward cost-effective solar energy use. Science 315, 798–801.
Lewis, N.S., 2016. Research opportunities to advance solar energy utilization. Science 351. aad1920-1–aad1920-9.
Lin, Y., Xu, Y., Mayer, M.T., Simpson, Z.I., Mcmahon, G., Zhou, S., Wang, D., 2012. Growth of p-type hematite by atomic layer deposition and its utilization for improved solar water splitting. J. Am. Chem. Soc. 134, 5508–5511.
Liu, R., Stephani, C., Han, J.J., Tan, K.L., Wang, D., 2013. Silicon nanowires show improved performance as photocathode for catalyzed carbon dioxide photofixation. Angew. Chem. Int. Ed. Engl. 52, 4225–4228.
Liu, R., Zheng, Z., Spurgeon, J., Yang, X., 2014. Enhanced photoelectrochemical water-splitting performance of semiconductors by surface passivation layers. Energy Environ. Sci. 7, 2504–2517.
Ma, G., Hisatomi, T., Domen, K., 2015. Semiconductors for photocatalytic and photoelectrochemical solar water splitting. In: Rozhkova, E.A., Ariga, K. (Eds.), From Molecules to Materials: Pathways to Artificial Photosynthesis. Springer International Publishing, Cham.
Memming, R., 2015. Solid–liquid interface. Semiconductor Electrochemistry. Wiley-VCH Verlag GmbH & Co. KGaA, Weinheim, Germany.
Morrison, S.R., 1980. Electrochemistry at Semiconductor and Oxidized Metal Electrodes. Plenum Press, New York and London.
Nocera, D.G., 2012. The artificial leaf. Acc. Chem. Res. 45, 767–776.
Pendlebury, S.R., Wang, X., Le Formal, F., Cornuz, M., Kafizas, A., Tilley, S.D., Grätzel, M., Durrant, J.R., 2014. Ultrafast charge carrier recombination and trapping in hematite photoanodes under applied bias. J. Am. Chem. Soc. 136, 9854–9857.
Peter, L.M., 2007. Characterization and modeling of dye-sensitized solar cells. J. Phys. Chem. C 111, 6601–6612.
Peter, L., 2013. Chapter 2 kinetics and mechanisms of light-driven reactions at semiconductor electrodes: principles and techniques. Photoelectrochemical Water Splitting: Materials, Processes and Architectures. The Royal Society of Chemistry, Cambridge, UK.
Pleskov, Y.V., 1990. Solar Energy Conversion: A Photoelectrochemical Approach. Springer-Verlag Berlin Heidelberg, Berlin.
Qi, B., Wang, J., 2012. Open-circuit voltage in organic solar cells. J. Mater. Chem. 22, 24315–24325.
Reichman, J., 1980. The current-voltage characteristics of semiconductor-electrolyte junction photovoltaic cells. Appl. Phys. Lett. 36, 574–577.
Rhoderick, E.H., Williams, R.H., 1988. Metal-Semiconductor Contacts. Clarendon Press, Oxford, UK.
Roger, I., Shipman, M.A., Symes, M.D., 2017. Earth-abundant catalysts for electrochemical and photoelectrochemical water splitting. Nat. Rev. Chem. 1, 0003.

Sato, N., 1998. Chapter 10—semiconductor photoelectrodes. Electrochemistry at Metal and Semiconductor Electrodes. Elsevier Science, Amsterdam.
Satoh, K., Wydrzynski, T.J., Govindjee, 2005. Introduction to photosystem II. In: Wydrzynski, T.J., Satoh, K., Freeman, J.A. (Eds.), Photosystem II: The Light-Driven Water:Plastoquinone Oxidoreductase. Springer Netherlands, Dordrecht.
Seeger, K., 2004. Energy band structure. Semiconductor Physics: An Introduction. Springer Berlin Heidelberg, Berlin, Heidelberg.
Srinivasan, S., 2006. Electrode/electrolyte interfaces: structure and kinetics of charge transfer. Fuel Cells: From Fundamentals to Applications. Springer US, Boston, MA.
Stephens, G.L., Li, J., Wild, M., Clayson, C.A., Loeb, N., Kato, S., L'ecuyer, T., Stackhouse, P.W., Lebsock, M., Andrews, T., 2012. An update on Earth's energy balance in light of the latest global observations. Nat. Geosci. 5, 691–696.
Strandwitz, N.C., Comstock, D.J., Grimm, R.L., Nichols-Nielander, A.C., Elam, J., Lewis, N.S., 2013. Photoelectrochemical behavior of n-type Si(100) electrodes coated with thin films of manganese oxide grown by atomic layer deposition. J. Phys. Chem. C 117, 4931–4936.
Thorne, J.E., Li, S., Du, C., Qin, G., Wang, D., 2015. Energetics at the surface of photoelectrodes and its influence on the photoelectrochemical properties. J. Phys. Chem. Lett. 6, 4083–4088.
Trasatti, S., 1986. The absolute electrode potential: an explanatory note. Pure Appl. Chem. 58, 955–966.
Van De Krol, R., 2012. Principles of photoelectrochemical cells. In: Van De Krol, R., Grätzel, M. (Eds.), Photoelectrochemical Hydrogen Production. Springer US, Boston, MA.
Wang, D.E., Li, R.G., Zhu, J., Shi, J.Y., Han, J.F., Zong, X., Li, C., 2012. Photocatalytic water oxidation on $BiVO_4$ with the electrocatalyst as an oxidation cocatalyst: essential relations between electrocatalyst and photocatalyst. J. Phys. Chem. C 116, 5082–5089.
Yang, X., Du, C., Liu, R., Xie, J., Wang, D., 2013. Balancing photovoltage generation and charge-transfer enhancement for catalyst-decorated photoelectrochemical water splitting: a case study of the hematite/MnO_x combination. J. Catal. 304, 86–91.
Yang, X., Liu, R., Du, C., Dai, P., Zheng, Z., Wang, D., 2014. Improving hematite-based photoelectrochemical water splitting with ultrathin TiO_2 by atomic layer deposition. ACS Appl. Mater. Interfaces 6, 12005–12011.
Yang, X., Liu, R., He, Y., Thorne, J., Zheng, Z., Wang, D., 2015. Enabling practical electrocatalyst-assisted photoelectron-chemical water splitting with earth abundant materials. Nano Res. 8, 56–81.

CHAPTER THREE

III–V Semiconductor Photoelectrodes

Georges Siddiqi, Zhenhua Pan, Shu Hu[1]
School of Engineering and Applied Sciences, Yale University, New Haven, CT, United States
[1]Corresponding author: e-mail address: shu.hu@yale.edu

Contents

Semiconductors and Semimetals, Volume 97
ISSN 0080-8784
http://dx.doi.org/10.1016/bs.semsem.2017.03.002

1. INTRODUCTION

Group III–V semiconductor photoelectrodes have been updating the records of photoelectrochemical (PEC) solar-to-fuel (STF) conversion efficiencies over the past decades. Their band gap tenability, versatile materials synthesis, and high radiative efficiency and quantum yield have been the major advantages that drive the continuous development and advances of III–V photoelectrodes. Compared with metal oxides, chalcogenides, etc., III–V materials for solar-fuel applications can realize tunable band gaps of 0.7–3.4 eV (Tiwari and Frank, 1992). These tunable band gaps can be realized in a monolithic and efficient multijunction configuration via epitaxial growth. III–Vs are one of the few materials systems promising >10% STF (Bolton et al., 1985; Hu et al., 2013; Nozik, 1977). New systems such as a membrane based, monolithically integrated can achieve STF > 10% for >40 h of continuous operation (Fig. 1), while via inverted metamorphic multijunction semiconductor architectures other systems have pushed STF efficiency to over 16% (Verlage et al., 2015a, and Deutsch, under review).

The major drawback is general instability of III–V/liquid interfaces and high cost of III–V materials growth. In this chapter, we will discuss through: (1) the historical context of III–V photoelectrode performance breakthroughs, (2) our current scientific understanding of the physical chemistry of III–V semiconductor/liquid electrolyte interfaces, and (3) the massive and growing applications of III–V photoelectrodes for light-driven chemical conversion, particularly enabled by protective coating strategies.

1.1 Brief History of III–V Photoelectrodes

Most reported applications of III–V photoelectrodes are PEC water splitting to produce solar hydrogen (H_2). To date, the record efficiency half-cell photocathode for H_2 evolution is Rh-hydride metalized p-type indium phosphide (p-InP) with 16.2% energy-conversion efficiency which is based on

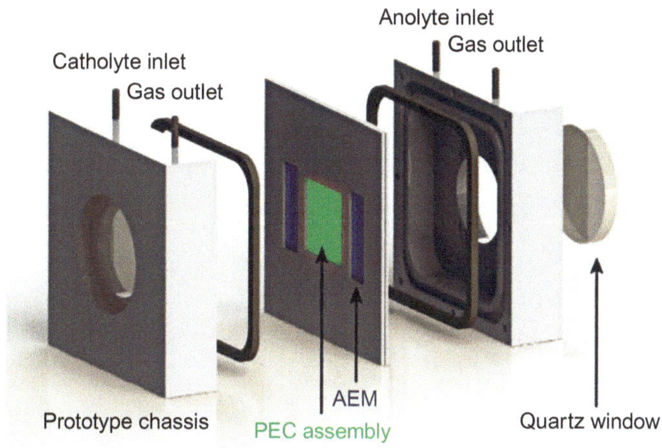

Fig. 1 Schematic illustration of a fully monolithically integrated intrinsically safe, solar hydrogen system prototype with >10% STF. *Figure adapted from Verlage, E., Hu, S., Liu, R., Jones, R.J.R., Sun, K., Xiang, C., Lewis, N.S., Atwater, H.A.A., 2015. Monolithically integrated, intrinsically safe, 10% efficient, solar-driven water-splitting system based on active, stable earth-abundant electrocatalysts in conjunction with tandem III–V light absorbers protected by amorphous TiO_2 films. Energy Environ. Sci. 8 (11), 3166–3172.*

solar-energy-conversion to H^+/H_2 redox couples (Aharon-Shalom and Heller, 1982). The record efficiency full-cell photoelectrodes for solar-driven water splitting so far is tandem two-junction GaInP/GaInAs with 16% solar-to-H_2 efficiency which is the efficiency of solar-energy stored to H_2 and O_2 bonds via water splitting (Deutsch, under review). Applying the Shockley–Queisser limit many groups have agreed upon a 25%–31% solar-to-hydrogen (STH) efficiency limit with two band gap, in which the current consensus is that III–V materials are the only system to at least providing the band gap and performance necessary to validate modeling (Hu et al., 2013; Pinaud et al., 2013). The quest for integration strategies that employ low-cost Si growth substrates and that address lattice-mismatched growth in a low-cost and scalable process remains ongoing.

Other than solar hydrogen, III–V photoelectrodes have versatile uses for regenerative PEC solar cells, water oxidation, and carbon dioxide reduction (CO_2R). There have also been solar-fuel reactor demonstrations for 10% solar-to-H_2 efficiency by water splitting and 10% solar-to-formate efficiency by CO_2 reduction and water oxidation, both utilizing two-junction III–V electrodes (Zhou et al., 2016). Note that III–V PV-based electrolysis have reported to achieve 31% STH (Jia et al., 2016). In this case, the III–V materials, which is a triple-junction InGaP/GaAs/GaInNAsSb tandem, were not submerged and thus have to be electrically wired to a separate electrolysis unit, which is two membrane-electrode-assemblies made of Pt/Nafion membrane/RuO_x. We will focus on immersed III–V photoelectrodes and their interfaces with liquids, because of potential advantages for low-cost solar H_2 production in the long term.

1.2 Motivation of III–V Photoelectrodes

Fujishima and Honda (1972) demonstrated solar-driven water oxidation under UV-illuminated n-TiO_2 photoelectrodes. Although later on it was discovered that single-crystal rutile TiO_2 can only split water into stoichiometric H_2 and O_2 under a chemical bias while anatase TiO_2 split water without bias, the quest for water splitting materials had started ever since. TiO_2, SnO_2, $SrTiO_3$, $KTaO_3$, and $KTa_{0.77}Nb_{0.23}O_3$ were discovered for unassisted water splitting, but only active under UV illumination (Wrighton et al., 1976). Beyond stable water splitting by oxides yet not efficient, significant efforts had shifted to reducing band gap of photoelectrodes from >3 eV to below 2 eV where visible light and near infrared has much more photon flux and thus much high predicted STF efficiency.

Nozik first introduced the two band gap concept in the form of a photochemical diode, which is back-to-back contacted two semiconductor absorbers (Nozik, 1977). Bolton later analyzed the absorption properties of semiconductors under two band gap concept (Bolton et al., 1985). Production of $^1H_2$ molecule, like the D2 (dual absorber, two photon) process in natural photosynthesis, utilizes D4 (2 + 2 photon excitation) in two-absorber tandem series for the needed free energy enough for water splitting.

For an applicable photocathode (Fig. 2A), the semiconductor as a light absorber should match several criteria: narrow enough band gap to harvest the solar spectrum; negative conduction band for H_2 reduction; sufficiently high photovoltage (V_{oc}) under illumination; and long-term stability in electrolyte solution. Group III–V materials, such as GaP and InP, fulfill most of the requirements and enable world-record STH efficiencies (Khaselev and Turner, 1998; Licht et al., 2001), but suffer from corrosion or degradation in photocatalytic reaction. Recent developments on surface protection technology greatly improve the stability of III–V materials based photoelectrodes (Hu et al., 2015; Liu et al., 2014), which facilitates their future application on nonbias photoelectrodechemical (PEC) solar water splitting cells.

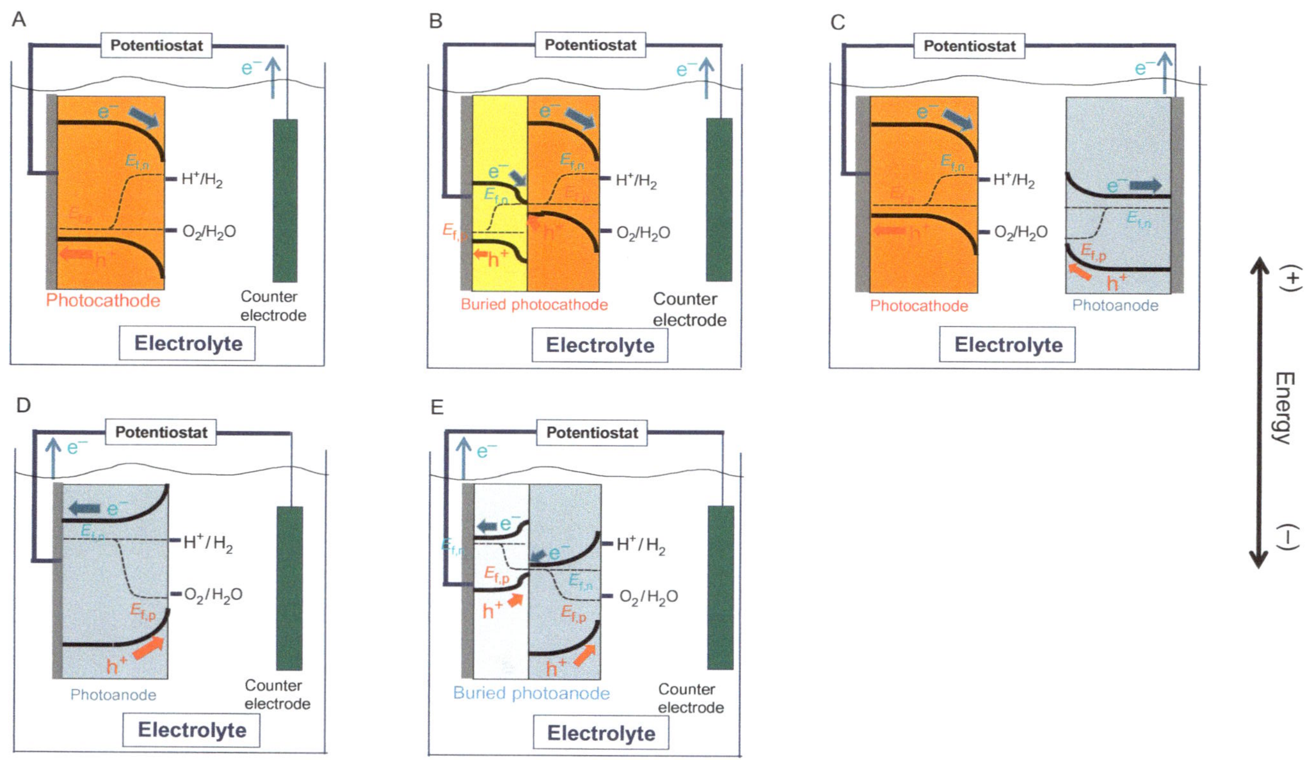

Fig. 2 Photoelectrochemical water splitting in various configurations: (A) single-absorber photocathode, (B) PV-assisted photocathode, (C) tandem photocathode + photoanode, (D) single-absorber photoanode, and (E) PV-assisted photoanode.

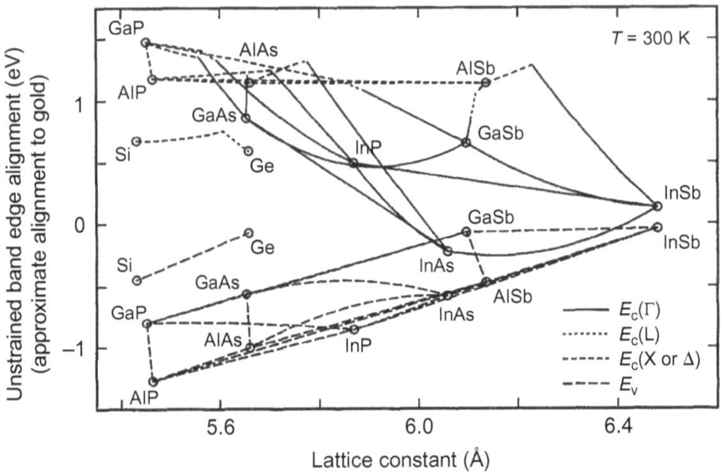

Fig. 3 Conduction- and valence-band edge energies plotted as a function of the lattice constant of conventional III–Vs. The zero energy point represents approximately the gold Schottky barrier position of any plotted III–V materials. *Solid lines* indicate direct transition, while *dashed lines* indicate indirect transition. *With permission, AIP from Tiwari, S., Frank, D.J., 1992. Empirical fit to band discontinuities and barrier heights in III–V alloy systems. Appl. Phys. Lett. 60 (5), 630–632.*

Through any possible materials synthesis pathway, the tandem photoelectrode configuration requires a band gap combination of 1.7 eV/1.1 eV for current matching under solar illumination, thus modeled to be >25% solar-to-H_2 efficiency. Tandem configuration and band gap tuning are the perfect alley of III–V semiconductors. Fig. 3 showed the band gap and theoretical band edge positions of conventional III–V materials and their ternary alloys, from GaAs, GaP to InP, GaSb, and InSb.

Furthermore, most III–V materials can be easily doped n-type and p-type. The ease of doping compared to oxides and chalcogenides made them model systems for basic PEC study. The synthesis of III–V materials largely leverages processes used in semiconductor manufacturing for logic, optical communication, and solid-state lighting. Their well-controlled materials properties, as well as surface chemistry, commensurate the basic study of semiconductor/liquid junctions (Fig. 4).

1.3 Challenges of III–V Photoelectrodes for Photochemistry

A decade before Fujishima and Honda (1972), Gerischer had observed the instability of an entire class of small band gap semiconductors in aqueous

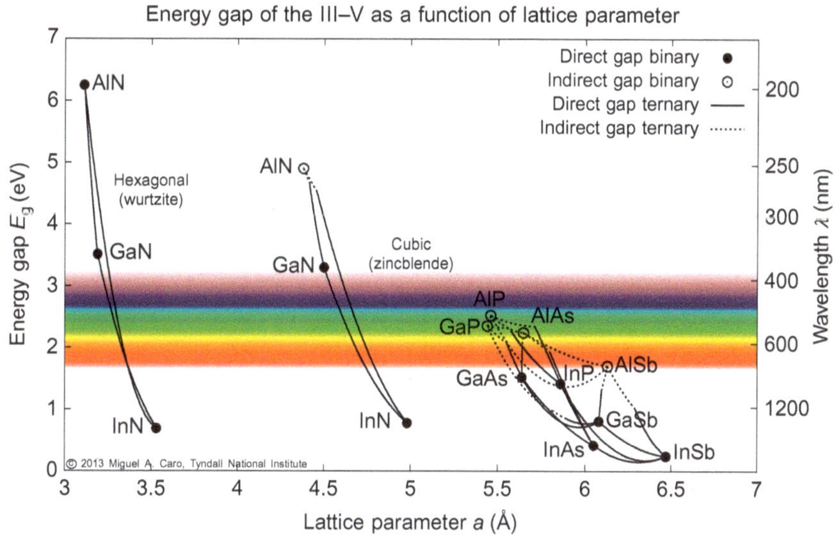

Fig. 4 Conduction and valence-band edge energies plotted as a function of the lattice constant of III-nitrides. *From Caro, M.A., 2013. Theory of elasticity and electric polarization effects in the group III-nitrides. PhD thesis. University College Cork.*

electrolytes, including Si, III–Vs such as GaAs, GaP, and InP, and II–VIs such as CdS, CdSe, and CdTe (Gerischer, 1977). The corrosion behavior of III–V electrodes and even their corrosion mechanism was well understood back then. Unlike III–Vs, Si cannot be further reduced so it is stable for H_2 production in acid; however, Si can be oxidized in water anodically to become SiO_2, which passivates Si surfaces as a charge transport barrier and limits further photocorrosion. Whereas for III–Vs, the III-elements can be further reduced while V-elements can be further oxidized in water. For example, Ga^- or In^- can be cathodically reduced to Ga and In metals which precipitates on surface. Furthermore, As^- or P^- can be anodically oxidized to make the oxides. Therefore, if the pH is acidic or basic, the (photo-) electrochemical corrosion products will either transform or dissolve, with corrosion continues until the entire immersed material disappears or until the transformation inhibits electrochemistry.

With practical goals of cost and performance considered, synthesis of high efficiency III–V materials one of the biggest challenge. So far, strategies including reuse of the III–V substrates, replacing III–V substrates with Si, and nanostructured morphology have all been investigated as promising approaches for further reducing materials cost. While metal-organic chemical vapor deposition (MOCVD) have achieved great performance-cost targets for logic and optoelectronic devices, solar-energy conversion for comparable price point to current energy sources requires further cost reduction to <\$200 m^{-2} (Energy, 2015). Therefore, low-cost synthesis such as higher growth rate metal-organic vapor phase epitaxy (MOVPE), liquid-phase epitaxy (LPE), closed space vapor transport (CSVT), and hydride vapor phase epitaxy (HVPE) specifically for solar-conversion materials are under development. In the next section, we will discuss this with greater details.

So far, combinations of III–V materials of 1.7 eV and Si of 1.1 eV grown monolithically in tandem have yet to achieve comparable photovoltaic performance as they were grown individually on a substrate with matching lattice sizes. Achieving this is a pre-requisite for high-efficiency STF conversion. Regardless, III–V materials are still the only potentially system for the predicted high efficiency and remains the holder of efficiency records.

2. SYNTHESIS OF III–V MATERIALS FOR PHOTOCHEMISTRY

To minimize the aforementioned problems, new growth techniques will be summarized: starting from a summary existing III–V semiconductor device fabrication techniques, this section will expand on these strategies and emphasize new research avenues toward minimizing lattice mismatch in III–V growth.

2.1 Liquid Phase Epitaxy

The first synthesis method discussed is that which is used for bulk III–V crystal growth, LPE. Unlike ionic and molecular solids, which can be crystallized from aqueous or organic solutions at relatively low temperatures, covalently bound solids such as III–V semiconductor are generally insoluble and must be grown either from a liquid melt (Arora, 1990). This mature field has been explored since the 1970s (Dawson, 1972; Ghez and Small, 1981; Kressel and Nelson, 1973; Nakajima, 1985) and is typically applicable for the synthesis of bulk III–V with high optical and electrical qualities (Arora, 1990; Saul and Lorimor, 1974). While it is experimentally simple to apply, it suffers from the drawback that it typically cannot grow thin epitaxial films. The basic principal of LPE typically is to use the low melting point III metal to serve as a "solvent" for the Group V metal. This saturated solution is then brought into contact with the substrate and allowed to cool. The composition and crystallinity can be controlled via the liquid melt temperature and the cooling rate. Although control of film thickness can be very challenging (Andreev et al., 1987; Kelting et al., 1986), the simplicity and applicability of LPE to a wide range of substrates and III–V compositions ensures it remains commonly used for III–V semiconductor synthesis (Liang et al., 2011; Lizhong et al., 2004; Mauk et al., 2001; Woo et al., 2008).

2.2 Metal-Organic Vapor Phase Epitaxy and Molecular Beam Epitaxy

An alternative for thin film III–V semiconductor growth is found in MOVPE and molecular beam epitaxy (MBE). Both of these methods can be used to grow epitaxial thin films of III–V semiconductors, with the main difference being how the metals are deposited onto the surface. As the names imply, crystal growth in MOVPE occurs via chemical reaction from organometallic compounds (Fig. 5), while MBE uses direct physical deposition from metal precursors (Arthur, 2002; McCray, 2007; Petroff and DenBaars, 1994; Pierson, 1999).

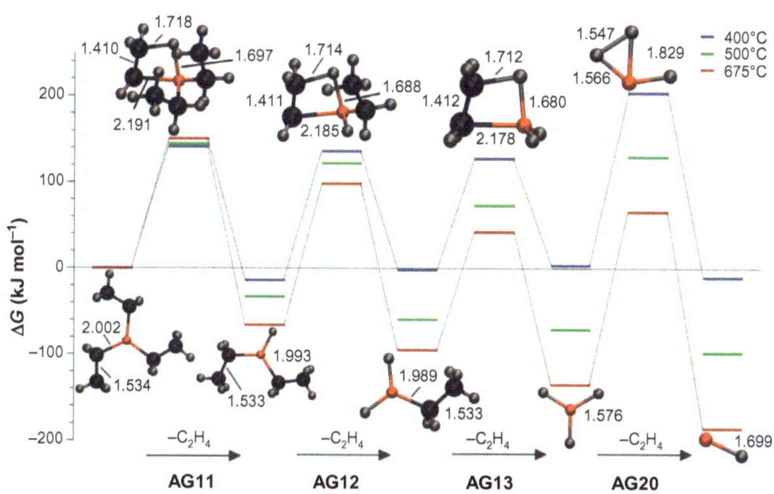

Fig. 5 Proposed mechanisms for the decomposition of triethylgallium under MOCVD conditions for the formation of GaP. Reaction pathway is thought to proceed via the gas phase decomposition of triethylgallium to Ga hydrides on the substrate surface followed by the reaction of *tert*-butylphosphine to form GaP. *Figure adapted from Stegmuller, A., Rosenow, P., Tonner, R.A., 2014. Quantum chemical study on gas phase decomposition pathways of triethylgallane (TEG, Ga(C2H5)3) and tert-butylphosphine (TBP, PH2(t-C4H9)) under MOVPE conditions. Phys. Chem. Chem. Phys. 16 (32), 17018–17029.*

Taking place under ultra-high vacuum, MBE uses direct sublimation of metal precursors, such as Ga or As, to serve as molecular "beams" aimed at the substrate surface. Once on the surface, the different metals can react to form alloys (Fig. 6A) (Arthur, 2002). Because no carrier gases are used, this technique results in ultra-pure films with highly tunable thicknesses and compositions. MOVPE uses a carrier gas to transport organometallic precursors across the substrate, where they can react, resulting in epitaxial growth (Fig. 6B).

Comparing the two, MOVPE offers higher film growth rates, and growth conditions closer to thermodynamic equilibrium, while MBE can give a faster switching of precursors for superior interfaces and can also be used to deposit materials far from equilibrium conditions. In addition to being useful for thin film synthesis, both of these methods have also been shown to be useful for deposition of III–V quantum dots (Petroff and DenBaars, 1994). The biggest advantage of these methods however is the fact that they can be achieved by epitaxial growth; part of the reason that these are the most commonly used III–V thin film deposition techniques for photoelectrodes and are responsible for some of the photoelectrodes with the highest recorded efficiencies (Green et al., 2012; Khaselev and Turner, 1998). Not only limited to thin films, MOVPE has been used to form GaAs pyramids (Bauhuis et al., 2002), and various III–V nanowires (Chu et al., 2010; Hersee et al., 2006; Johansson et al., 2006; Mandl et al., 2006a; Mårtensson et al., 2004; Seifert et al., 2004; Sköld et al., 2006; Wacaser et al., 2006). Similarly MBE has been used for the synthesis of a wide array of III–V nanostructures (Bertness et al., 2008; Dubrovskii et al., 2009; Li and Waag, 2012; Thelander et al., 2011).

2.3 Vapor–Liquid–Solid Synthesis

First discovered in the early 1960s as a method for the formation of single-crystal wires (Wagner and Ellis, 1964), the vapor–liquid–solid (VLS) synthesis was very rapidly adapted to the low-temperature synthesis of single-crystal III–V nanowires (Barns and Ellis, 1965; Ellis, 1968). The VLS process uses a nano-sized droplet of molten metal to serve as both a seed and catalyst, which is then contacted with a vapor of the metals desired to be deposited onto the surface, once the liquid droplet becomes supersaturated by the vapor, nucleation of the compounds in the vapor phase will begin at the solid–liquid interface (Fig. 7) (Xia et al., 2003).

The key requirements of the VLS process are that the metal catalyst must be able to form a liquid alloy with the gases in the vapor phase and that the vapor

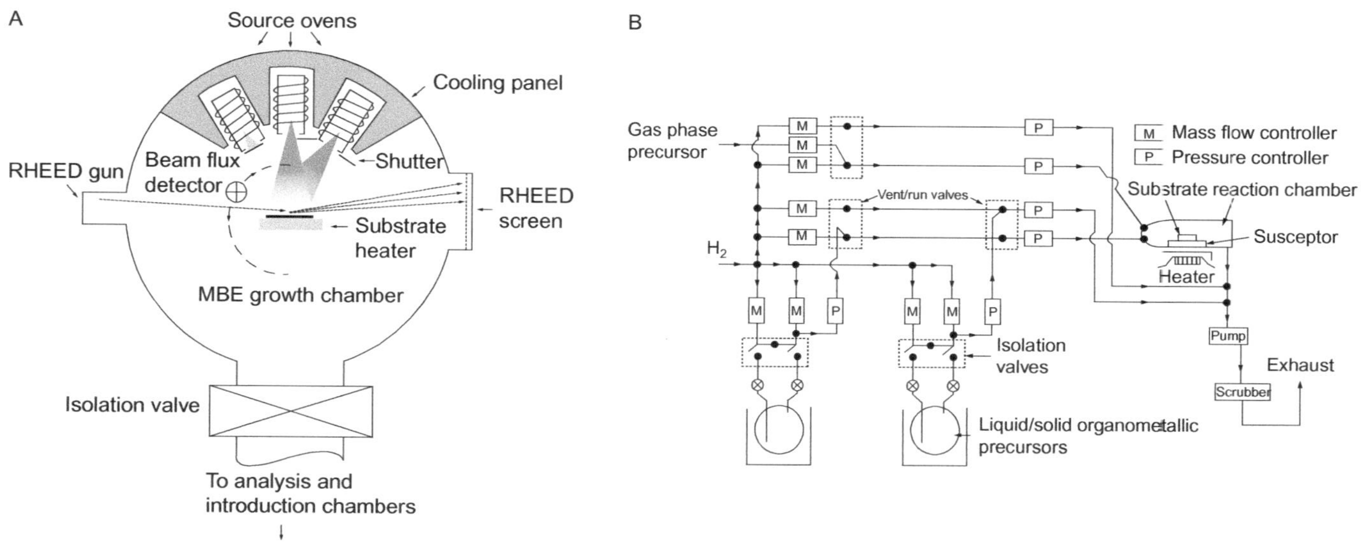

Fig. 6 Schematics of typical (A) MBE and (B) MOVPE setups. *Figures adapted from Arthur, J.R., 2002. Molecular beam epitcxy. Surf. Sci. 500 (1–3), 189–217 and Zilko, J.L., 2001. Metal organic chemical vapor deposition: technology and equipment A2. In: Seshan, K. (Ed.) Handbook of Thin Film Deposition Processes and Techniques, second ed., William Andrew Publishing, Norwich, NY, pp. 151–203, Chapter 4.*

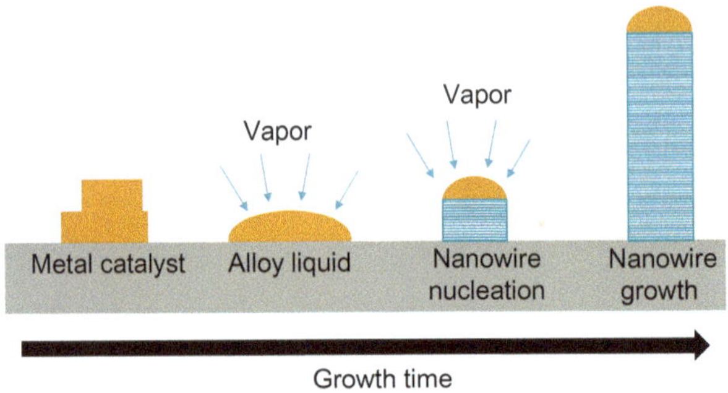

Fig. 7 Schematic of the phases of the VLS process; starting from a metal catalyst, upon heating and exposure to the precursor vapor, the metal catalyst melts, forming an alloy from which a solid nucleates then grows into a wire. *Adapted from Xia, Y., Yang, P., Sun, Y., Wu, Y., Mayers, B., Gates, B., Yin, Y., Kim, F., Yan, H., 2003. One-dimensional nanostructures: synthesis, characterization, and applications. Adv. Mater. 15 (5), 353–389.*

pressure of the gas must be sufficiently low that no secondary nucleation will occur in regions without the metal catalyst. In addition to the single-crystal nanowires formed by VLS, another significant advantage is that the source of the vapor phase gas does not matter; it can be used with well-established chemical processes such as thermal evaporation or chemical vapor transport. Additionally, the size of the nanowires formed can be very well controlled by simply controlling the size of the metal catalyst deposited onto the surface, giving great versatility to this process (Gudiksen and Lieber, 2000; Wu et al., 2002). A wide array of III–V semiconductors such as GaN, GaAs, InAs, and GaP have been grown using this method, and has as well as nanowire superlattices of GaAs and Gap, demonstrating its applicability (Chen and Yeh, 2000; Chen et al., 2001; Gudiksen et al., 2002; Hiruma et al., 1995; Jiang et al., 2012; Joyce et al., 2007; Shi et al., 2001; Yazawa et al., 1993; Zhang et al., 2001). While single crystals can be easily formed with this method, the disadvantages are that it is primarily limited to the nanowire geometry, and that the metal catalyst can serve as a potential source of contamination.

2.4 Selective-Area Growth of III–V Wires

Several strategies have been used to create selective-area growth of III–V nanowires. The majority of these methods rely on a self-catalyzed form of VLS growth, giving a twofold benefit of removing the metal catalyst as a source of contamination as well as providing selective growth, but some also are able to provide catalyst-free template growth.

Catalyst-free growth has been shown to synthesize InAs nanowires using a patterned wafer and carefully tuned MOVPE growth conditions. By either using a thin SiO_x ($x \approx 1$) layer without patterning to serve as a template or by simply patterning openings on a SiO_x or SiN_x ($x \approx 1$) mask, these methods can achieve vertical, patterned InAs growth directly on Si (Mandl et al., 2006b; Tomioka et al., 2008). Templating has also been used to direct GaN growth as well. The growth strategy was reported from early efforts using a patterned SiO_2 mask for direct GaN growth on Si or sapphire to recent work using a localized reversal of the surface polarity of growth substrates to create nanowires (Hiramatsu et al., 1998; Kawaguchi et al., 1998; Kitamura et al., 1995; Lekhal et al., 2016a,b; Nagahara et al., 1994; Okada et al., 2014; Schiaber et al., 2017; Seo et al., 2009; Sundaram et al., 2015; Wu et al., 2016a,b). A detailed study of the mass transport during selective-area MOCVD also presents results for general III-N growths (Coltrin and Mitchell, 2003). However, the exact mechanism of such selective-area growth is not yet elucidated, although in situ transmission electron microscopy has been used to observed III–V wire growth by Au-catalyzed or self-catalyzed growth in situ but not by selective-area growth particularly (Jacobsson et al., 2016). Regardless, it is possible the growth conditions of selective-area growth could be allowing for self-catalyzed growth, but this is inconclusive so far. Self-catalyzed growth uses a liquid seed of a Group III element, to catalyze VLS growth. The typical approach is to use patterning to provide the selective area, MBE to deposit the Group III seed, then MOVPE to grow the nanowires (Benedikt et al., 2010; Cirlin et al., 2010; Farrell et al., 2015; Gao et al., 2009; Krogstrup et al., 2010; Morral et al., 2008; Paek et al., 2009; Plissard et al., 2010, 2011; Scofield et al., 2011). Self-catalyzed growth can also be extended to grow two and three dimensions III–V structures such as nanosheets or pillars (Caroff et al., 2009; Chi et al., 2013).

2.5 Solution–Liquid–Solid Synthesis

Analogous to VLS, solution–liquid–solid (SLS) syntheses use a metal with a low melting point to serve as the catalyst, and the crystalline nanowire is grown by the decomposition of the organometallic precursors dissolved in an organic solvent (Fig. 8) (Trentler et al., 1995). The primary advantage of SLS compared to VLS is that the same highly crystalline III–V semiconducting nanowires can be grown at a relatively lower temperature (Trentler et al., 1995; Xia et al., 2003).

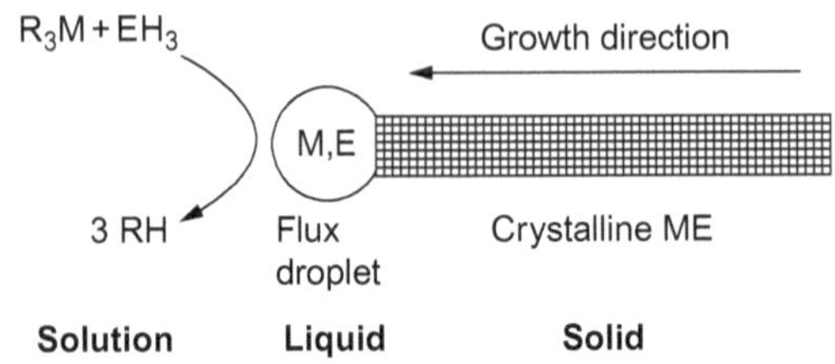

Fig. 8 Representation of the SLS process. *Figure adapted from Trentler, T.J., Hickman, K.M., Goel, S.C., Viano, A.M., Gibbons, P.C., Buhro, W.E., 1995. Solution-liquid–solid growth of crystalline III-V semiconductors: an analogy to vapor–liquid-solid growth. Science 270 (5243), 1791–1794.*

Precursors such as $t\text{-}Bu_2M[\mu\text{-}P(SiMe_3)_2]$ (M = In, Ga) can be coupled with simple hydrides such as AsH_3 or PH_3, mixed together in a solvent such as mesitylene with a flux of In or Ga particles to readily yield InP, InAs, or GaAs in a range of temperatures from 100°C to 200°C (Trentler et al., 1995). The key requirements and disadvantage of this method compared to VLS is the identification of the appropriate organometallic precursors and reaction solvents is not trivial. Nonetheless, this method has been successfully applied to create InN, InP, GaAs, or AlGaAs nanowires (Dingman et al., 2000; Markowitz et al., 2001; Trentler et al., 1995, 1997).

2.6 Nanoparticle III–V Synthesis

Synthesis of III–V nanoparticles presents an interesting alternative to synthesize high surface area III–V photoelectrodes. Compared to the 2D structures mainly synthesized by MOVPE and MBE, and the 1D structures synthesized by VLS and SLS, these 0D structures are typically referred to as quantum dots, with band gaps that can be tunable via particle size or surface chemistry control (Alivisatos, 1996; Burda et al., 2005; Peng et al., 1998; Vurgaftman and Meyer, 2003; Yoffe, 2001). Although these nanoparticles can be synthesized by method such as MOVPE or MBE (Petroff and DenBaars, 1994), this section will focus on solution phase syntheses, as these are the most widely used in current literature.

Nanoparticle growth requires an extremely rapid nucleation process followed by slower growth (Burda et al., 2005). This can be achieved by the very rapid injection of reagents into a hot coordinating solvent, with the exclusion of air (Fig. 9). The high temperature of the solution decomposes the reagents and creates a supersaturation of particles which gradually grow via reaction with the remaining precursor. Nanoparticle size can be controlled via reaction time or by increasing reaction temperature, and those

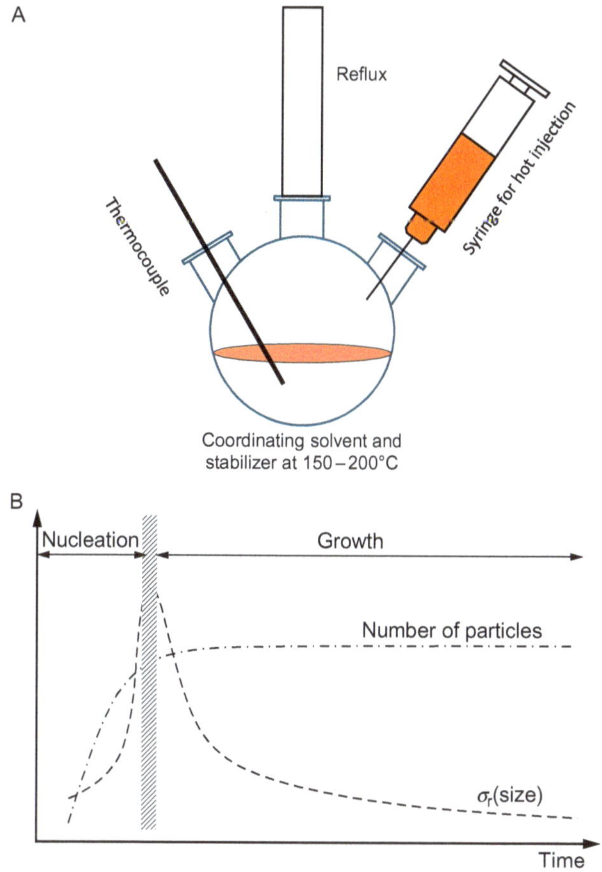

Fig. 9 (A) Schematics of a typical setup for injection of organometallic precursors to a hot solvent and (B) depiction of the nucleation and growth stages of nanoparticle synthesis.

parameters and also concentration, solvent, surfactant, and reagent selection can influence the particle shape and monodispersity of the particle size distributions obtained (Burda et al., 2005; Murray et al., 2000; Qu et al., 2001; Talapin et al., 2002).

Common precursors for GaN nanoparticles include gallium imide and gallium azides (Malik et al., 2010; Manz et al., 2000; Mićić et al., 1999), and adducts of metal chlorides and urea have been used to synthesize MN (M = Al, Ga, In) wurtzic nanoparticles (Sardar et al., 2005). InP and GaP have been synthesized using m diorganophosphides (Green and O'Brien, 1998, 2004), while InAs and GaAs have been synthesized from dimeric MAs (M = In, Ga) organometallic precursor (Malik et al., 2005, 2010). Although III–V nanoparticles can be synthesized with relative ease compared to the preceding methods of thin film or nanowire syntheses, they typically require surfactants to maintain their dispersion in solution. Upon impregnation to an electrode substrate, these surfactants can prove to be difficult to remove to create a pristine dispersion of III–V nanoparticles.

2.7 Electrodeposition of III–V Semiconductors

Instead of using heat to decompose precursors or induce crystallinity as shown in the previous synthetic methods, electrodeposition of semiconductors offers an interesting alternative via the use of potential to deposit metals or alloys onto surfaces. Using a solution with positive ions of the metals to be deposited, with a conducting substrate as the working electrode, potential can be applied between the two electrodes to reduce the metal ions in solution (Dharmadasa and Haigh, 2006). It should be noted that the nature of III–V semiconductor deposition necessitates that electrodeposition be done in nonaqueous conditions (Engelken and Van Doren, 1985; Wade et al., 2001). The properties of the deposited film can be controlled by varying the parameters such as ionic concentration, pH, temperature, or the deposition voltage to vary the composition.

The main advantage of electrodeposition compared to vacuum methods such as MOVPE or MBE is that is relatively low cost and can be easily scaled to large surface area substrates. However, it suffers from the drawback that the deposited films are usually amorphous or polycrystalline (Dharmadasa and Haigh, 2006; Lincot, 2005). One interesting application to overcome the polycrystalline nature of the electrodeposited films is to use self-limiting electrochemical reactions to achieve atomic layer epitaxy for the growth of III–V films of InAs, InSb, and $InAs_xSb_{1-x}$ (Wade et al., 2001). Since electrodeposition is not a line-of-sight process, 3D structures can also be grown by using a template, such as the deposition of GaAs mesoporous films via the use of a silica template by Lee et al (Lee et al., 2002). These applications illustrate the versatility of electrodeposition as a tool for III–V semiconductor synthesis.

2.8 Structured Materials, Tandem Junctions, and Reducing Lattice Mismatch

The following section focuses on using the preceding techniques to create photoelectrodes with deliberate structures and interfaces. One of the biggest challenges in the synthesis of III–V photoelelectrodes is the reduction of lattice mismatch between the semiconductor and substrate. One such way of achieving this is via lattice-mismatched epitaxy with a compositionally graded buffer to change the lattice constant slowly as a function of distance (Ahrenkiel et al., 2004; Geisz et al., 2006, 2007). This gradient layer allows the lattice constant to gradually change from that of the substrate to that of the deposited material, and coupled with methods such as MOVPE or MBE can yield high-performance photodevices (Geisz et al., 2008). Another strategy for minimizing lattice mismatch is to utilize the fact that in nanowires, this strain can be relaxed via the nanowire sidewall (Chuang et al., 2007; Ertekin et al., 2005; Glas, 2006; Sburlan et al., 2012). This has been utilized to create high efficiency tandem Si–III–V nanowire-based photodevices (Yan et al., 2009; Yao et al., 2014). Recent research has also found techniques to achieve ohmic contacts between Si and III–V semiconductors via direct fusion bonding. This technique minimizes lattice mismatch by using an amorphous layer to bond Si and III–V materials; even if the amorphous layer is an oxide, it can is sufficiently thin that it can function as an ohmic contact (Tanabe et al., 2012). The final strategy presented to minimize lattice mismatch is epitaxial liftoff. This method uses a sacrificial middle layer (such as AlAs) between a GaAs substrate and an epitaxially grown layer of GaAs (Cheng et al., 2013; Konagai et al., 1978). After the deposition of the final layer of GaAs, the sacrificial layer is removed via a wet chemical etch (such as HF or HCl), the yield the flexible deposited GaAs film, and allowing reuse of the original GaAs substrate. This strategy also minimizes the material costs associated with III–V semiconductor synthesis.

3. III–V PEC SOLAR CELLS WITH REDOX COUPLES

One prominent type of PEC solar cells is constructed by contacting III–V photoelectrodes with regenerative redox couples dissolved in electrolytes. Such a PEC solar cell utilizes the photovoltaic effect of semiconductor/liquid junctions: the photoelectrode oxidizes or reduces redox couples in the electrolyte, and the counter electrode (usually made of carbon or platinum) rereduces or reoxidizes to regenerate the redox couples. Such a PEC regenerative cell configuration outputs electrical energy and is distinctive to the other type of PEC solar cells, dye-sensitized solar cells. The main distinction is the charge separation process during energy conversion relies on semiconductor/liquid junctions in the regenerative cell.

3.1 Principles of PEC Solar Cells

Fig. 10 depicts a PEC solar cell by constructing a half-cell semiconductor/liquid junction between a semiconductor (e.g., GaAs) and redox couples (e.g., ferrocenium/ferrocene or $Fe(Cp)_2{}^{+/0}$). The regenerative PEC solar cell consists at least two electrodes, an illuminated photoelectrode and a counter electrode. Take an n-type semiconductor/liquid junction as an example. In dark before equilibration (Fig. 10A), electrons in the conduction band are injected into the redox energy levels by reducing the oxidized species in the contacting electrolyte. As the semiconductor loses free and mobile electrons to redox couples in the electrolyte, its Fermi energy level is lowered until equilibrates with the electrochemical potential of A/A^- and the immobilized dopant atoms remain in the semiconductor crystal lattice, forming a positively charged region. This is called the depletion region. As the mobile electrons were removed from the depletion region, it is the constant density of dopants and their constant fixed charges that build up linear electrical field and parabolic electrical potential across the depletion region. Therefore, the band edge energy appears to "bend up" spatially as it varies from bulk of semiconductor to the semiconductor/liquid interfaces. This is called "band bending," which is caused by fixed charges of ionized dopants left in the semiconductor's depletion region.

Under illumination (Fig. 10B), photogenerated hole charges are injected into the electrolyte (positive sign of the measured current) to oxidize A^- to A, while the electrons in the conduction band are also injected into the electrolyte (negative sign of the measured current) to reduce A to A^- just like the case in dark. Fig. 10C shows a typical current density–potential (J–E) curve, indicating an open-circuit potential (OCP) of the SC electrode, V_{oc}vs $E(A/A^-)$, a short-circuit photocurrent density J_{SC} measured at the SC electrode potential of $E(A/A^-)$, and a fill-factor measured at the point of maximum power output. Such asymmetrical "band bending" is the key for the photovoltaic effect by separation of photo-generated electron–hole pairs in semiconductor material-based PEC cell. At V_{oc}, the amount of positive photocurrents produced by hole injection is compensated with the negative electron-transfer current from the conduction band to the electrolyte.

One could recall a similar solid-state device configuration, Schottky diode, discussed in semiconductor device physics. Despite many analogies were drawn about liquid interface with respect to solid-state Schottky barrier, many fundamental characteristics of semiconductor/liquid junctions, including charge-transfer rates, surface chemistry, and interfacial

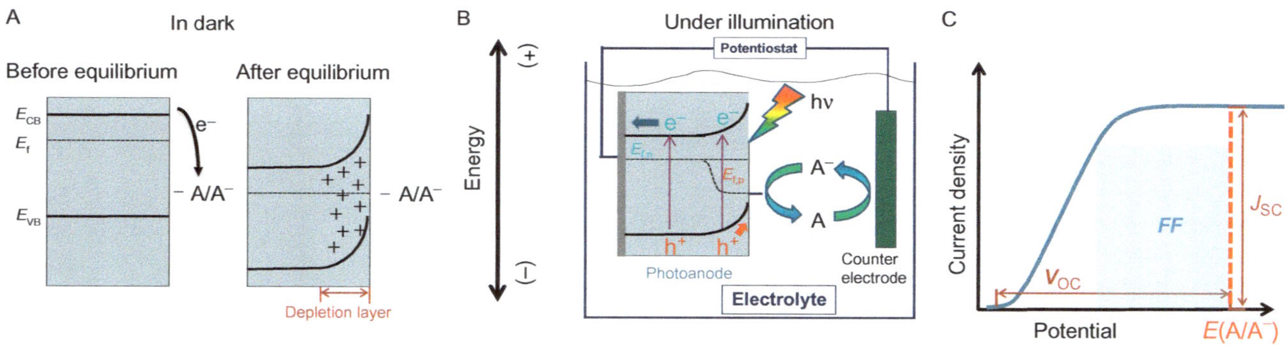

Fig. 10 Schematic configuration of (A) construction of an ideal semiconductor/liquid junction between an n-type SC and a generic A/A^- redox couple. (B) A regenerative photoelectrochemical solar cell with n-type semiconductor (SC) as the working electrode, and a Pt or carbon electrode as the counter electrode. (C) Typical current density–potential (*J–E*) curve of a regenerative cell of the n-type $SC/A/A^-$ redox junction.

electrochemical double layers (EDLs), are very much distinctive. We will discuss these differences later in details.

3.2 Stabilization of III–V Materials With Redox Systems

Studies of III–V semiconductors in contact with redox couples (in general, A/A^-) in an electrolyte dates back to 1977 when Miller and coworkers used fast poly-chalcogenide couples (e.g., Se_x^-/Se_2^{2-}) to contact n-type doped GaAs (n-GaAs) (Chang et al., 1977). In retrospect, the initial use of aqueous redox couples might be derived from early attempt to search for smaller band gap water splitting materials beyond TiO_2; 3.2–3.4 eV band gap of TiO_2 is too large to be efficient for water splitting despite the initial excitement started by Fujishima and Honda (Fujishima and Honda, 1972). Most metal oxides absorb light via oxygen 2p-to-metal d transition, and therefore varying the cations to lower the conduction band edge can reduce the band gap effectively, e.g., Fe_2O_3 in Fig. 11. What is really needed is raising O 2p level by replacing O with N, P, S, Se, etc., for band gap tuning.

However, none of these anionic compounds turned out to be stable for water oxidation to oxygen (O_2) so far after decades of new materials

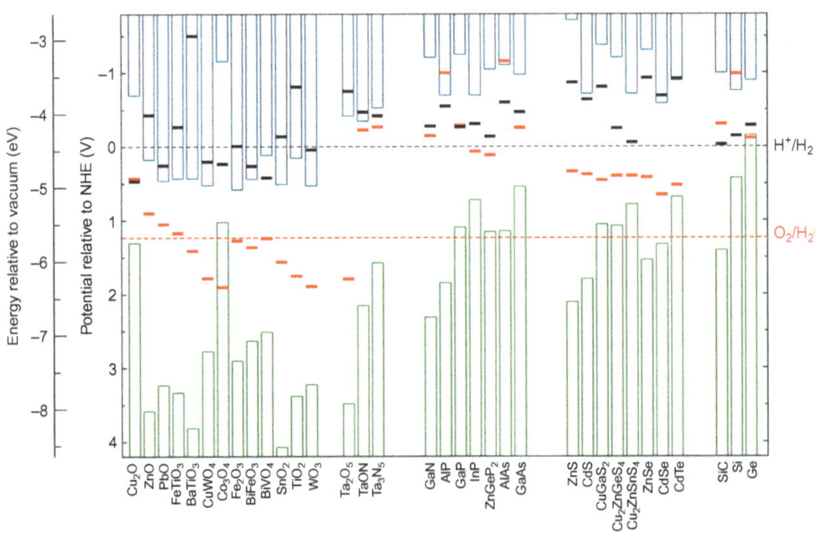

Fig. 11 Calculated oxidative corrosion potential (*red bars*) and reductive corrosion potential (*black bars*) relative to the NHE and the vacuum level for a series of semiconductors in aqueous solution of pH 0, a temperature of 298.15 K, and a pressure of 1 bar. *With permission, ACS from Chen, S., Wang, L.-W., 2012. Thermodynamic oxidation and reduction potentials of photocatalytic semiconductors in aqueous solution. Chem. Mater. 24 (18), 3659–3666.*

discovery. Furthermore, the cations of III–V materials can be reduced to Group III metals. In other words, metals are plated on III–V surfaces, and may be subsequently dissolved in electrolytes by corrosion or introduce surface states thus degrading performance. In parallel, another strategy then emerges as stabilization of these otherwise unstable compound semiconductors first, including II–VIs of CdS, CdSe, and CdTe, and III–Vs of GaAs, GaP, and InP.

Then the strategy in the late 1970s was: once these solid/liquid interfaces are stabilized, they will be adapted for fuel forming and water oxidation photocatalysis. Initially, sulfur/polysulfide $\left(S_x^-/S_2^{2-}\right)$ redox couples were found to chemisorb onto II–VI surfaces; these chalcogenide redox couples exchange anions with lattices to stabilize II–VIs in aqueous electrolytes despite dissolution. As a natural extension of this stabilization scheme, the chalcogenides were put in contact with III–Vs too and were found to stabilize III–Vs too. Therefore, a kinetic stabilization strategy was developed, where stabilizing n-GaAs/H_2O interface in regenerative PEC cells involves enhancing kinetic competition for photogenerated holes to oxidize redox couples vs crystal lattice by using electrolytes containing redox couples, such as S_x^-/S_2^{2-}(aq) (Ellis et al., 1977a) or I_3^-/I^-(aq) (Allongue et al., 1987; Verney et al., 1986) at high concentrations. But this semiconductor/liquid combination later on was found to be unstable in the long term. Furthermore, the kinetics was enhanced by adding metal ions that chemisorb to the electrode to increase the rate of transport of photogenerated holes away from the GaAs surface (Tufts et al., 1987). Kinetically, photogenerated holes were captured to oxidize redox couples rather than oxidize lattice anions to soluble species, thereby stabilizing III–V materials. The branching ratio of hole oxidation of lattices is reduced to once per million times of hole transfer to stabilize III–V/liquid junctions for hours to days.

While the efforts for semiconductor/aqueous electrolyte interfaces were to speed up kinetics for less corrosion for higher total photocurrents (Chang et al., 1977; Decker and Parkinson, 1980), later efforts on nonaqueous PEC cell is intended to reduce or remove water to improve stability. Nonaqueous electrolytes dissolved with various redox couples became main subject of investigation for decades since late 1970s and early 1980s. In presence of trace water and under illumination, both n-GaAs and p-GaAs surfaces easily generate trap states. This causes "Fermi-level pinning" and inefficiency which will be discussed later in this section.

Nonaqueous electrolytes through rigorous drying have produced most stable and least pinned surfaces for an efficient liquid junction cell. Even with nonaqueous solvents, preventing water impurities in nonaqueous electrolyte is the key (Ellis et al., 1977a); or otherwise, one consequently observes the band edges displacement ΔV to be shifting with respect to redox potential levels (indication of "Fermi-level pinning") as opposed to fixed band edge position of semiconductor/liquid junctions (Allongue and Cachet, 1988). So far, surface states still partially pin the GaAs Fermi level near the valence band.

3.3 Redox Couples for III–V Photoelectrodes

To start, n-GaAs has been reported to contact various aqueous couples of S_x^{2-}/S^{2-}, Se_x^{2-}/Se^{2-}, and Te_x^{2-}/Te^{2-} (Chang et al., 1977; Ellis et al., 1977a), TMPD (*N*,*N*,*N*′,*N*′-tetramethyl-*p*-phenylenediamine), metallocene family such as ferrocene, dimenthyl ferrocene, acetylferrocene, and colbaltocene (Ellis et al., 1977b; Gronet and Lewis, 1983; Heller et al., 1978), quinone family such as anthraquinone (Decker and Parkinson, 1980) and *p*-benzoquinone (Kohl and Bard, 1979a), I^-/I_3^-, Eu(II)/Er(III) (Fan and Bard, 1980), $TCNE^{0/-}$, $TCNQ^{0/-}$, methylviologen ($MV^{2+/+}$), thianthrene ($Th^{+/0}$) (Ba et al., 1994). Similarly, p-GaAs was reported to contact Eu(II)/Eu(III) at pH 1 (Memming, 1978), $Sn^{4+/2+}$, $Fe^{3+/2+}$ (Fan and Bard, 1980), $Cu^{2+/+}$, and nonaqueous metallocene couples including dimethylferrocene and cobaltocene (Nozik and Memming, 1996).

p-Type indium phosphide (p-InP) is similar: p-InP was investigated in contact with aqueous redox $V^{3+/2+}$ (Schefold and Vetter, 1994), $MV^{2+/+}$ (Feng and Cotton, 1988), as well as nonaqueous redox, such as anthracene and $Ru(TPTZ)_2^{3+/2+}$ (TPTZ, 2, 4,6-tripyridyl-*s*-triazine) (Kohl and Bard, 1979b), and later, $acetylferrocene^{+/0}$, $ferrocene^{+/0}$, $decamethylferrocene^{+/0}$, colbaltcene, and $decamethylcolbaltcene^{+/0}$ with well-defined surface chemistry (Pomykal et al., 1996; Segar et al., 1990). The surface states of InP pin the Fermi level near the conduction band. Whereas n-InP with those redox showed noticeably smaller open-circuit photovoltages: n-InP has been placed in aqueous electrolytes, contacting S_x^{2-}/S^{2-}, Se_x^{2-}/Se^{2-}, and Te_x^{2-}/Te^{2-} redox (with Pt as charge-transfer mediator) (Ellis et al., 1977c) for a photovoltage of up to 0.63 V, and in nonaqueous electrolytes contacting $Co(Cp)_2^{+/0}$ for a photovoltage of 0.715 V (Ellis et al., 1977c; Kohl and Bard, 1979b). Besides, sacrificial reagents of S^{2-} and SO_3^{2-} were also chosen for n-InP performance evaluation (Li et al., 2016a).

The ternary III–V compounds has also been in contact with aqueous couples such as Se_x^{2-}/Se^{2-} (Gronet and Lewis, 1984), and nonaqueous systems such as $Fe(Cp)_2{}^{+/0}$ (Casagrande et al., 2000). For $GaAs_xP_{1-x}$ materials in contact with Se_x^{2-}/Se^{2-}, the ratio of increasing open-circuit photovoltage (V_{oc}) vs increasing band gap energy (E_g) is very close and slight less than 1:1 for $0<x<0.4$. Similar V_{oc} vs E_g behavior was observed for n-$GaAs_xP_{1-x}$ in contact with $Fe(Cp)_2{}^{+/0}$ but for a wide range of $0<x<1$. For $Al_xGa_{1-x}As$ ($0<x<0.31$) in contact with $Fe(Cp)_2{}^{+/0}$, the activation energy, which is measured from temperature-dependent OCPs, varies at 1:1 ratio with E_g although not equal to E_g exactly. There are two possible explanations: (1) the position of conduction band edge, E_{CB}, is varying as E_g increases and therefore the barrier height and V_{oc} increase in excellent correlations; (2) while it is tempting to believe in the simple picture of varying E_{CB} positions, it is also possible that the simultaneous increases of barrier height and V_{oc} is predominately due to pinning of the Fermi level by defect levels or surfaces states of the ternary III–Vs.

Beyond conventional III-arsenides and III-phosphides, the emerging III-nitrides are recently studied as photoelectrode materials. n-GaN has been in contact with reversible Br redox couples. Luo et al. investigated the PEC properties of InGaN/GaN photoanode in HBr aqueous solution and the incident photon conversion efficiency (IPCE) was ca. 9% at 400–430 nm, but the absorption edge of InGaN was less than 500 nm for the low content of In in InGaN (Luo et al., 2008). A nanostructured InGaN photoanode with a small band gap (ca. 1.4 eV) was prepared by increasing the content of In and the IPCE reached 16% under 350 nm illumination (Alvi et al., 2014). AlOtaibi et al. reported a InGaN/GaN core/shell nanowire photoanode on Si substrate (AlOtaibi et al., 2015a). This unique core–shell structure worked as a double-band photoelectrode for a relatively high IPCE up to 27% under ultraviolet and visible light irradiation. Integrating InGaN nanowires with Si solar cell substrate, Fan et al. developed a monolithic tandem photoanode with a nearly ideal band gap configuration for maximum STH (Fan et al., 2016). The saturated photocurrent reached 16.3 mA cm^{-2} for HBr splitting, which is the highest value among PV cells and photoelectrodes with similar structure. Their PEC studies of GaN/electrolyte interfaces have just started and their photoelectrochemistry and surface chemistry is expected to be just as rich and versatile.

3.4 Characterization of III–V/Redox Liquid Junctions

Cyclic voltammetry (CV) of photoelectrodes in a quiescent solution was one of the very first characterization techniques, through which the V_{oc} of III–V liquid junctions were indicated as the shift of CV diagram under illumination versus in dark. One can also perform electrochemical impedance spectroscopy to plot Mott–Schottky diagram under illumination and in dark. Later, the OCP of photoelectrode with respect to the redox potential was often measured under varying redox species, varying light intensity, and varying temperatures. The photocurrent density, J_{SC}, was measured by rigorous stirring the solution to minimize the hysteresis of conventional CV. Therefore, the energy-conversion properties of semiconductor/liquid junctions are often plotted in current density vs potential curves. For further mechanistic study of semiconductor/liquid junctions, OCP vs log(J_{SC}) is plotted to extract the diode quality factor; OCP vs redox potential is plotted to study Fermi-level pinning and surface dipoles, and OCP vs temperature is plotted to extract excitation energy for photogenerated electron–hole recombination (Lewis, 1991; Tan et al., 1994).

Redox potentials in aqueous electrolytes with respect to the band edge positions of common III–V semiconductors are not well defined. Band edge position relative to individual redox energy levels ought to be measured either directly via operando photoelectron spectroscopy (Schwalbach et al., 2011) or indirectly via differential-capacitance (Mott–Schottky) measurements. For aqueous electrolytes with photo-corrosion or surface states (see Section 3.5), it is difficult to obtain frequency-independent Mott–Schottky data showing the relative band edge position with respect to redox levels. Photocorrosion produces time-varying semiconductor/electrolyte interfaces; and surface states trap and release charges at varying frequencies of the differential-capacitance measurements. Such processes contribute to a frequency-dependent surface-state capacitance and interfering with the depletion region capacitance of interests. Only for p-InP/$V^{3+/2+}$ junction, the flat-band potential was measured to be 0.520 V vs SCE in dark, according to its frequency-independent Mott–Schottky data (Schefold and Vetter, 1994). In this case, InP conduction-band (CB) edge is at −0.66 vs SCE and its valence-band (VB) edge is at 0.680 V vs SCE. Under illumination, the flat-band potential shifted to 0.300 V vs SCE, indicating a partial pinning of Fermi level to surface states. One should realize that such a number is not broadly applicable to any aqueous electrolyte; additionally, surface dipoles will play an important role in determining the band edge positions with respect to the redox levels. Therefore, surface chemistry of III–V materials in solid state as well as in contact with liquid electrolyte is a crucial part of consideration.

In nonaqueous electrolytes, because photocorrosion and surface states are effectively reduced but not yet completely eliminated, the Mott–Schottky analysis can reveal the band edge position with nonaqueous redox couple, e.g., $Fe(Cp)_2{}^{+/0}$ and $Co(Cp)_2{}^{+/0}$. From frequency-independent Mott–Schottky data, n-GaAs exhibits a barrier height of 1.053 V with $Fe(Cp)_2{}^{+/0} - MeCN$ and p-InP exhibits a barrier height of 1.16 V with $Co(Cp)_2{}^{+/0} - MeCN$ (Pomykal et al., 1996). Therefore, GaAs CB and VB edges are −1.053 and 0.367 V vs $Fe(Cp)_2{}^{+/0}$, whereas p-InP CB and VB edges are −1.511 and −0.171 V vs $Fe(Cp)_2{}^{+/0}$, respectively. Potential of $Co(Cp)_2{}^{+/0}$ used in the p-InP measurement is calculate to be −1.239 V vs the formal potential of $Fe(Cp)_2{}^{+/0}$. For $n - InP/\ Fe(Cp)_2{}^{+/0} - MeOH$ junction, the barrier height is measured to be 0.80 V vs $Fe(Cp)_2{}^{+/0}$ (Pomykal and Lewis, 1997). In this case, n-InP CB and VB edges are 0.80 and −0.54 V vs $Fe(Cp)_2{}^{+/0}$ in MeOH, respectively. Nonideality of III–V/liquid junctions, particularly partial Fermi-level pinning, causes various measured CB and VB positions with respect to different redox couples in different supporting electrolytes. Therefore, we hesitate to make a generic plot for band edge positions of III–V compound semiconductors aligning with redox energy levels.

3.5 Nonideality of III-V/Redox Junctions

Fig. 12 draws a nonideal semiconductor/liquid junction, e.g., $n - GaAs/\ Fe(Cp)_2{}^{+/0}$. Surface states, especially those near the mid-band gaps, are recombination centers. With surface states, the majority carrier electrons of n-GaAs can inject into either the empty traps on the surfaces (acceptor surface states), or the redox levels in the solution.(Fig. 12A) As illustrated in Fig. 12B, the positive charges fixed in the n-GaAs space-charge region balance with the negative charges of trapped electrons on surface states and net negative charges in the EDL. Fig. 12C indicates the ideal built-in voltage, V_{bi}, which is the energy difference between the Fermi level of n-GaAs and the electrochemical potential of the redox couple, $E_0 - E_F$. When there exist no surface trap states, one can observe that V_{bi} is almost

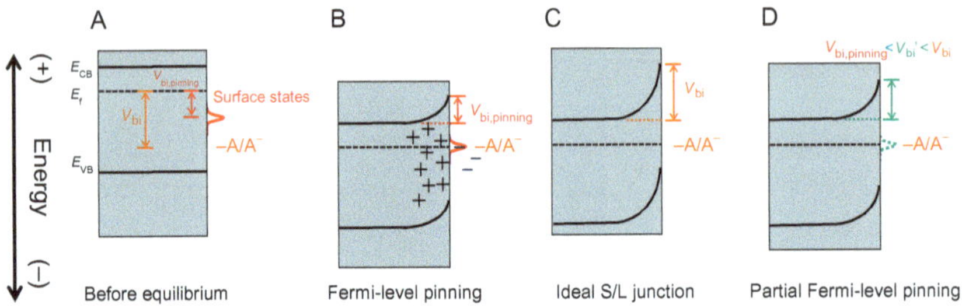

Fig. 12 Schematics of building up (A) a nonideal semiconductor/liquid (S/L) junction (B) and (D), and its comparison with an ideal semiconductor/liquid junction (C). Surface states are indicated in a *red* distribution drawn along the surface. $V_{bi,pinning}$ and V_{bi} are the built-in voltage of fully pinned and ideal junctions, respectively. The built-in voltage of nonideal junctions, V'_{bi}, should follow $V_{bi,pinning} < V'_{bi} < V_{bi}$.

the same as the potential difference ($E_0 - E_F$). The negligible potential drop across the EDL is because of the much smaller capacitance of the space-charge regions than the EDL. However with surface states, a portion of potential is dropped at the surface states, and therefore V_{bi} is always smaller than $E_0 - E_F$ with surface states, as shown in Fig. 12D. It was also observed that the recombination of photogenerated carriers via surface trap states is temperature dependent. For a $n-GaAs/Fe(Cp)_2{}^{+/0}-MeCN$ junction, a built-in voltage of 1.1 V with partial Fermi-level pinning (Fig. 12) can produce a V_{oc} of 1.05 V under 10-Sun at 200 K (Casagrande and Lewis, 1985a). At such a low temperature, the recombination between charges through surface trap states slowed down so that kinetically hole injection onto redox levels are faster. It behaves the semiconductor/liquid junction was close to an ideal one, as shown in Fig. 13.

3.5.1 Surface States of III–V/Redox Junctions

In aqueous electrolytes, GaAs easily forms surface states which causes nonradiative recombination and which pin the Fermi level of GaAs near the VB edge. Even the redox level of $Se_2{}^{2-}/Se^{2-}$ is near the conduction band of n-GaAs, a photovoltage of 0.68 V was observed (Gronet and Lewis, 1984). This suggests the built-in voltage, V_{bi} is 0.68 V out of a potential difference, $E_0 - E_F \sim = 0$. By replacing water with nonaqueous solvents, the issue of surface states was alleviated. Furthermore, one can systematically vary the electrochemical potential of redox couples in the nonaqueous electrolytes for basic studies of III–V/redox junctions.

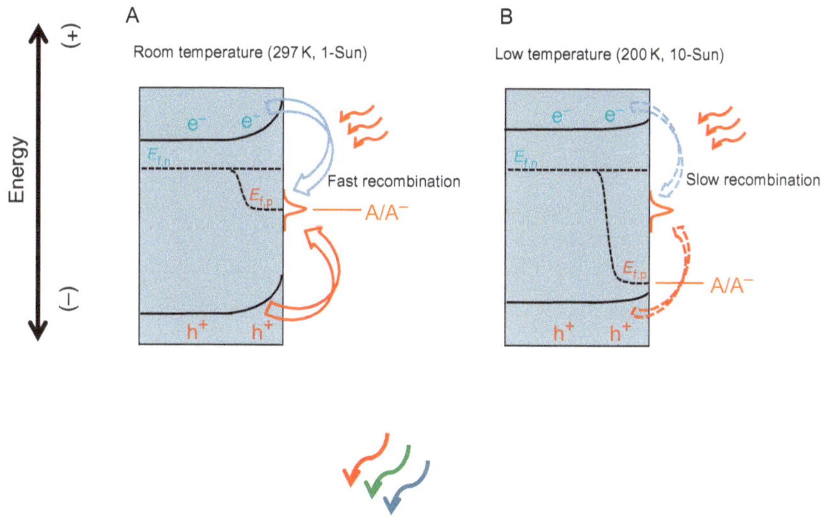

Fig. 13 Effect of temperature-dependent recombination on kinetics of redox charge transfer and thus observed increased photovoltages. At low temperatures, e.g., 200 K, splitting of quasi-Fermi levels of electrons and holes (B) further increases as compared to room temperatures (A). This effectively increases the barrier height. *Illustrated from illustrate data in Casagrande, L.G., Lewis, N.S., 1985. Measurement of the activation barrier for carrier transport at normal-gaas semiconductor liquid junctions. J. Am. Chem. Soc. 107 (20), 5793–5794.*

3.5.2 Fermi-Level Pinning

Band edge positions move together (1:1 ratio) with the redox potentials, if the Fermi level of III–V semiconductors is fully pinned at the surface states. For n-GaAs and p-InP in acetonitrile with various metallocene redox couples, a partial Fermi-level pinning was observed. In this case, the band edge position are not completely pined at the surface states. The band bending and the built-in voltage V'_{bi}, is partially dependent on the redox potential in the contacting electrolyte. The band edge positions of other semiconductors were plotted according to their measured work function in vacuum, and ought to be corrected by effects of surface dipoles or pinning.

It was reported that Mott–Schottky curve measured with p-GaAs in aqueous solution was shifted upon illumination of the semiconductor. This shift has been observed with many other semiconductors, such as bare p-InP/$V^{3+/2+}$ in H_2O (Schefold and Vetter, 1994). This shift of flat-band potential in dark and under illumination is due to Fermi-level pinning. The photovoltage of nonideal semiconductor/liquid junctions shifts the flat-band potential effectively. Therefore, the Mott–Schottky plot appears to shift negatively for p-type semiconductors and positively for n-type semiconductors.

3.5.3 V_{oc} vs Redox Potential

This partial Fermi-level pinning behavior was confirmed by the redox-dependent OCP measurements. Broadly, for n-GaAs and p-InP, per 1 V change of redox potential in the contacting electrolyte do not produce 1 V change of measured OCP (V_{oc}). As shown in Fig. 14A, per change of the redox potential of 0.40 V from −0.32 to 0.08 V vs SCE, the V_{oc} of n-GaAs/liquid junction is only increased by 0.20 V (Casagrande et al., 2000; Yang et al., 2016). Similarly, for n-GaAs/graphene/liquid junction, the change of redox levels of 0.40 V causes V_{oc} increase of 0.20 V. Compared with n-Si-MeCN with the same metallocene family, the ratio is ~1 for both n-Si (111):H surface and n-Si (111):–CH_3 surface. This suggests no Fermi-level pinning of Si in nonaqueous MeCN solvents (Grimm et al., 2012).

3.6 Evaluating Planar vs Nanostructured III–V/Redox Junctions

Despite the PEC solar cell acts as another version of photovoltaics, one can primarily use the semiconductor/liquid junction to evaluate the energy-conversion properties of III–V materials, particularly those porous III–Vs

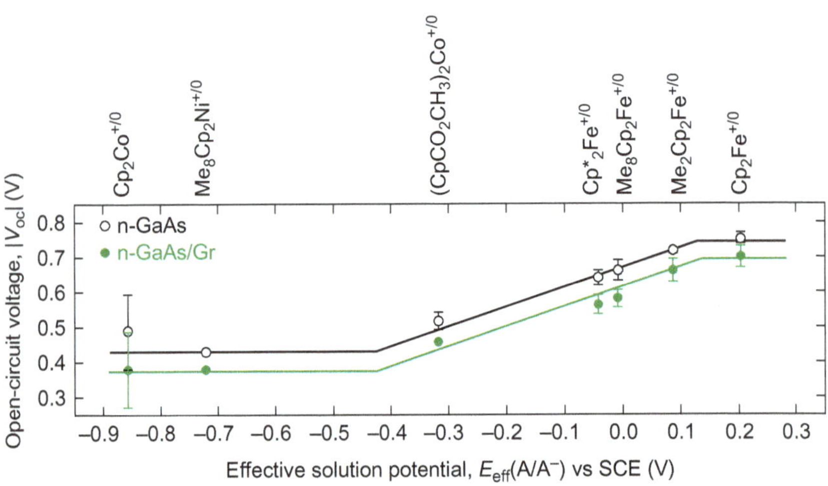

Fig. 14 Open-circuit voltage, V_{oc}, vs the effective solution redox potential, $E_{eff}(A/A^-)$, in 1.0 M $LiClO_4/CH_3CN$ for n-GaAs (*black, open circles*) and n-GaAs/Gr (*green, filled circles*). The corresponding lines indicate the observed trends in the different regions of V_{oc}, vs $E_{eff}(A/A^-)$. *Adapted from Yang, F., Nielander, A.C., Grimm, R.L., Lewis, N.S., 2016. Photoelectrochemical behavior of n-type GaAs(100) electrodes coated by a single layer of graphene. J. Phys. Chem. C 120 (13), 6989–6995.*

or III–V wires by constructing a conformal liquid contact with well-defined surface chemistry. In doing so, a PEC solar cell is constructed, as an intermediate step of solar-to-electricity conversion prior to STF conversion. One advantage is conformal junction formation without intensive fabrication processes.

Fig. 15 shows the photocurrent density–potential behavior of planar n-GaAs and n-GaAs nanowires in contact with $Fe(Cp)_2{}^{+/0}$ redox in acetonitrile. Figure insets of voltage vs log(J_{ph}) gives a diode quality factor of 1.1 for single-crystal $n-GaAs/Fe(Cp)_2{}^{+/0}$ junctions. This suggests a mild surface recombination, which dominates the overall recombination process in additional to bulk recombination. For n-GaAs NWs in contact with the same $MeCN-Fe(Cp)_2{}^{+/0}$ electrolyte, a diode quality factor of 1.8 was observed. While surface states of similar characteristics are expected to exist on both sidewall and top surfaces of GaAs NWs, the increase of diode quality

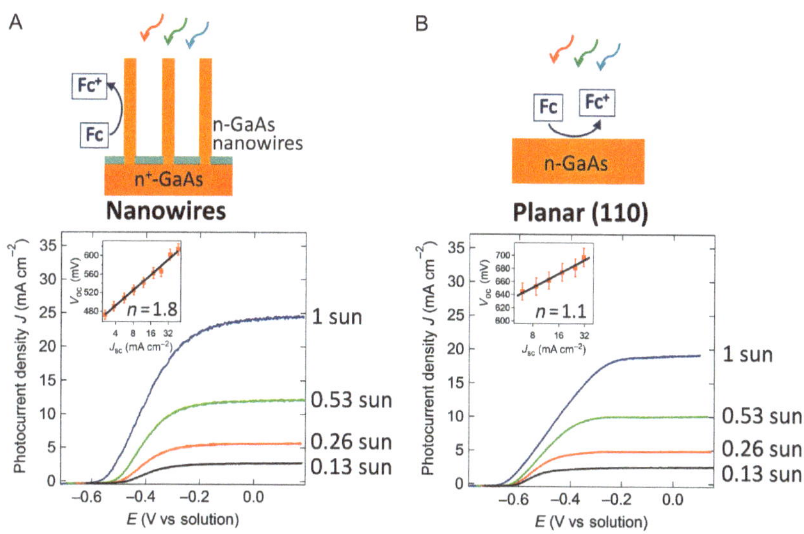

Fig. 15 *J–E* data of GaAs NW array and GaAs planar photoelectrodes as a function of the light illumination intensity. *Insets* are plotted light-dependent pairs of V_{oc} vs J_{SC} indicating the diode quality factor, *n*, of the GaAs/liquid junction. *Note*: The *I–V* curves are a subset of light-dependent measurements, and so the *I–V* curve was plotted with a photocurrent density of ~25 mA cm^{-2}, for up to 1-sun. The *insets* of V_{oc} vs ln(J_{sc}) show data points of above 30 mA cm^{-2}, which was measured under >1-sun. *Adapted from Hu, S., Chi, C.-Y., Fountaine, K. T., Yao, M., Atwater, H.A., Dapkus, P.D., Lewis, N.S., Zhou, C., 2013. Optical, electrical, and solar energy-conversion properties of gallium arsenide nanowire-array photoanodes. Energy Environ. Sci. 6 (6), 1879–1890.*

factor from 1.1 to 1.8 indicates additional recombination pathways in the depletion region. It could be the twin defects existing in depletion region of nanowires are causing depletion recombination. It could also be the increased surface-state density due to the nanowire geometry compared to planar case, increased the surface recombination. To separate these effects, temperature-dependent PEC measurements are needed to measure the activation barrier, the intersect of temperature-dependent V_{oc} extrapolated to $T=0$. The activation barrier of $\sim E_g/2$ prompts depletion recombination, while activation barrier of E_g prompts surface recombination (Casagrande and Lewis, 1985b).

4. INTERFACES OF III–V MATERIALS AND CATALYSTS WITH LIQUIDS

Given the critical role redox couples can play in the evaluation of electrode properties, it is clear that not only is a more detailed explanation of the semiconductor–liquid interface needed, especially with respect to the role that surface chemistry can play on the III–V electrode properties. This following section will present a chronological exploration of the role that surface chemistry plays in electrochemistry, starting from a theoretical overview of how surface chemistry can influence material properties, followed by experimental techniques to modify surface chemistry, ways to protect surfaces, and finally strategies to integrate catalysts.

4.1 Influence of Surface Chemistry on Bulk Material Properties

The general electronic structure of a semiconductor has the valence bands and conduction bands separated by a band gap of a certain energy, E_g. The energy at which the probability of having an electron or hole is identical is the Fermi level, E_F, and the Fermi level defines the electrochemical potential of the semiconductor. In an ideal intrinsic semiconductor, there would be no electronic states that exist in this band gap state; however, surface defects, surface reconstruction, or adsorbed materials can create localized surface states which can accept or donate electrons, in turn influencing the electronic properties of the semiconductor (Seker et al., 2000).

Two types of surface states can exist: extrinsic and intrinsic surface states. As the names imply, intrinsic surface states are due to the properties of the material when the lattice terminates a surface, and extrinsic states are caused by a foreign adsorbed material (Guo-Ping and Ruda, 1996; Ohno, 1991;

Ow and Wang, 1996; Seker et al., 2000). If electrons or holes become trapped by these surface states they can create a localized charge at the surface, changing the width of the depletion region, and in turn influencing the degree of band bending which occurs at the semiconductor–liquid interface (Mönch, 2001), or by shifting the E_F at the surface (Fig. 1) (Cierocki et al., 1992; Kiselev and Krylov, 1987; Rothschild et al., 2002).

Moving beyond just the electrical properties of the semiconductor, surface states can also directly influence the luminescent and electrochemical properties of the material, such as the recombination velocity of electrons and holes at the surface (Gmitter et al., 1988; Kelly and Memming, 1982; Lunt et al., 1991). On a more systems level, different types of etchants used to modify the intrinsic surface states or adsorbates to create extrinsic surface states can change the current densities which are achievable by III–V photovoltaic devices (Fig. 16) (Allongue and Cachet, 1984; Heller et al., 1978; Johnston et al., 1980; Miller et al., 1980).

4.2 Modification of III–V Surfaces Via Etching

Surface chemistry modification via etching can alter surface chemistry via the two routes previously discussed: the creation of intrinsic and extrinsic surface states (Aspnes and Studna, 1981; Stirland and Straughan, 1976). To modify intrinsic surface states, etching can remove adsorbates, passivating films, or even bulk material to reform intrinsic surface states on a freshly

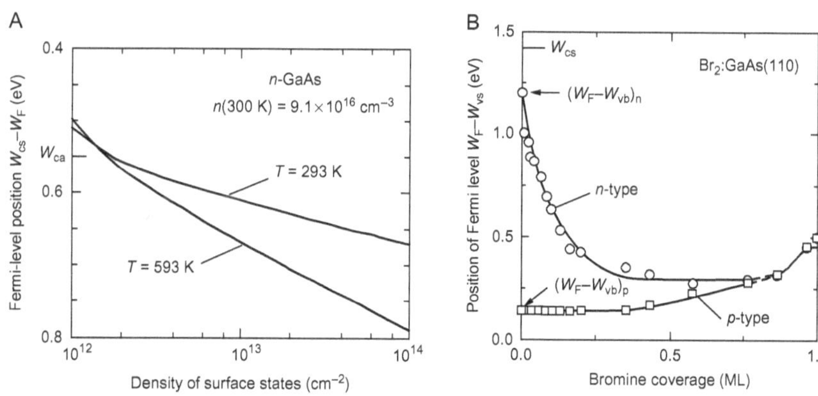

Fig. 16 Example of the effect of surface states on the electronic properties of semiconductors. (A) Calculation of the influence of the density of intrinsic surface states on the Fermi level and (B) effect of adsorbates (bromine) on the position of the Fermi level in n-type or p-type GaAs. *Figures adapted from Mönch, W., 2001. Semiconductor Surfaces and Interfaces. third ed. Springer-Verlag Berlin Heidelberg, 19 and 63.*

etched surface. But the etchant can leave behind adsorbants which can act as extrinsic surface states (Simpson and Yarmoff, 1996). The two most common etchants used are bromine solutions in alcohols and alkaline peroxides (Gomes, 2001).

Bromine-alcohol solutions are the most commonly used to create smooth, pristine GaAs surfaces. The mechanism of etching is thought to proceed via first the creation of an adsorbed surface layer of Br, followed by complexation of GaAs with Br to form $GaBr_3$ and $AsBr_3$, and finally dissolution via hydrolysis with the alcohol to create solution Ga and As oxohydroxides (McGhee et al., 1994). Numerous studies have categorically shown that etching III–V surfaces increases their photochemical properties (Gmitter et al., 1988; Lunt et al., 1991; Miller et al., 1980; Strubbe and Gomes, 1993; Tan et al., 1991).

Alkaline peroxides have also been extensively researched, but are comparatively less used in present research because of significant surface roughening can occur (Kelly and Reynders, 1987; Stirland and Straughan, 1976). In alkaline conditions, these etches dissolve the III–V semiconductor by forming soluble hydroxides. These hydroxides can however result in surface contamination with oxygen, and to reduce this, additives such as ammonia or EDTA can be added to the etch (Bryce and Berk, 1996; Kelly and Reynders, 1987).

4.3 Modification of III–V Surfaces Via Adsorption

Early investigations of electrodeposition of metals onto semiconductor surfaces and the influence of surface chemistry on nucleation were eventually followed by studies investigating the role of metal ions on the surface of GaAs photoelectrodes (Fig. 17) (Bindra et al., 1977; Johnston et al., 1980; Parkinson et al., 1978).

From there numerous studies have characterized the effects of numerous cations, the predominant among them being Ru^{III} (Allongue and Cachet, 1984; Miller et al., 1980; Nelson et al., 1980; Tan et al., 1991). However, metal ions are not the only adsorbates which can be used to improve photoelectrodes, the adsorption of anions can also be used to stabilize the surface of III–V electrodes (Bessolov and Lebedev, 1998; Bessolov et al., 1998; Cierocki et al., 1992; Guo-Ping and Ruda, 1996; Ohno, 1991; Simpson and Yarmoff, 1996). With these adsorbates, one would categorize their modification of surface chemistry as both adsorption and passivation.

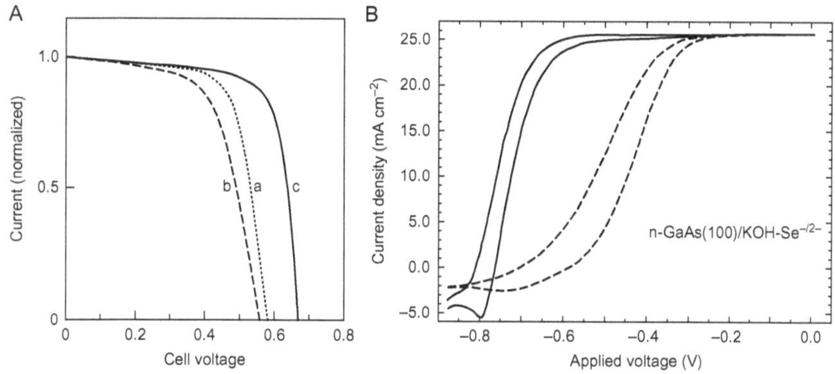

Fig. 17 (A) Studies of the effect of ions on the photocurrent produced by a n-GaAs photoelectrode; curve a is the electrode after an H_2O_2–H_2SO_4 etch, curve b is after adsorption of Bi^{III} ions, and curve c is after adsorption of Ru^{III} ions. (B) Effect of adsorption of Os^{III} on the surface of GaAs, *dashed line* the GaAs surface after etching, and *solid line* is the GaAs surface after etching and exposure to a $OsCl_3$ solution. *Figures adapted from Parkinson, B.A., Heller, A., Miller, B., 1978. Enhanced photoelectrochemical solar-energy conversion by gallium arsenide surface modification. Appl. Phys. Lett. 33 (6), 521–523.*

4.4 Nitrogen Ion Implantation on III–V Semiconductors

The "Turner Cell" was reported to get a record STH efficiency of 12.4% (Khaselev and Turner, 1998). However, $GaInP_2$ material is prone to photocorrosion due to P^{3-} reduction and dissolution in electrolytes. The GaN shows better stability than other III–V semiconductors, possibly due to the strong bonding between Ga and N. It's therefore proposed that incorporation of nitrogen into conventional III-phosphides or III-arsenide might improve their stabilization. Turner and coworkers successfully incorporated nitrogen into the surface layer of p-$GaInP_2$ films via $N_2{}^+$ ions implantation. This ion implantation increased corrosion resistance with little negative effects on the photocatalytic activity (Deutsch et al., 2012; Weir et al., 2012).

The nitrogen ion implantation method was extended to GaP, showing good stability against photocorrosion (Deutsch et al., 2006). However, incorporation of nitrogen into GaP induced lattice mismatch between the epilayer and the substrate. The lattice mismatch defects served as recombination centers and lowered the photoelectrode efficiency. For compensating the decreased lattice constant from nitrogen incorporation, arsenic was incorporated along with nitrogen. But the GaAsPN turned out to be prone to photocorrosion again. For the same reason of compensating lattice constants, indium was also investigated for an ideal nitride content and low lattice mismatch (Deutsch et al., 2008). The use of indium allowed the preparation of GaInPN epilayers with a nitrogen content over 2% and good lattice matching to a GaP substrate. Even though the stability of the nitrides was better than that of $GaInP_2$ and GaP alone, corrosion cannot be fully eliminated. Moreover, the poor electronic properties or recombination processes led to a low solar-energy-conversion efficiency.

4.5 Catalyst Integration

For efficient PEC water splitting, control over charge-transfer rates between semiconductors and liquid electrolytes is necessary. High turn-over frequency catalysts, like noble metal particles, are therefore always applied on III–V-based photoelectrodes. Early work by Heller et al. loaded Rh, Pt, and Ru on p-InP for hydrochloric acid and got an incident solar-energy-conversion efficiency of 12% (Heller and Vadimsky, 1981). Later, the PEC activities of p-InP were investigated by loading various metals (Pd, Cd, Co, Au, Ni, and Pt) in insular form. The variance of activity when loading different metals was attributed to the charge-transfer step at the metal–solution interface (Szklarczyk and Bockris, 1984). Goodman et al. drew a similar conclusion that the metals did not significantly affect the barrier height at the semiconductor/alkaline electrolyte interface but simply improved the kinetics of photoelectrolysis (Goodman and Wessels, 1986). However, Heller et al. reported that the work functions of metals could be adjusted by hydrogen alloying and such alloying increased the barrier height for a higher photovoltage (Heller et al., 1982). Using the Rh–H alloy and Re–H alloy loaded p-InP photocathode for H_2 generation, the solar-to-chemical energy-conversion efficiencies could reach 16.2% and 14.4%, respectively (Aharon-Shalom and Heller, 1982).

4.6 Protective Coating

It was pointed out that both cathodic and anodic corrosion of III–V semiconductors required multiple bond-breaking steps, which means that alleviation of corrosion is possible by facilitating charge transfer before enough bonds are broken to cause irreversible decomposition (Gerischer, 1978). Protection of III–V semiconductors by kinetic control of charge transfer is typically used by Turner's group (Deutsch et al., 2014). In their recent work, Pt–Ru cocatalyst was deposited on III–V semiconductors via flash sputtering to facilitate the hydrogen evolution reaction kinetics, improving the photocurrent onset potential and stability at the same time

(Deutsch, under review). However, the device was only stable for 20 min, then its performance starts to slowly decrease, indicating that this kind of protection is not ideal enough for long-term stability.

Neither N_2^+ implantation nor catalyst integration mentioned earlier can stabilize III–V semiconductors in PEC cells. Considering the ubiquity of corrosion (electrochemical, PEC, or chemical) for III–V semiconductors, several other methods have been proposed to mitigate it (Hu et al., 2015). Surface functionalization by organic reagents is another method for stabilization and manipulation of the interfacial energetics and kinetics of semiconductors. While such molecular functionalization may protect photocathodes, it is reluctant to apply it on photoanodes because the organic molecular is basically poor resist to oxidation. A primary approach for stabilization is the surface protection by thermodynamically stable films, which separate the semiconductors from direct contact with the electrolyte (Hu et al., 2015; Liu et al., 2014). An ideal protective film should be electrically conductive, electrochemically stable, and optically transparent. Sometimes the protective films also work as a cocatalyst for light-driven photochemistry. Candidates for protective films majorly come from metals and oxides, which can be prepared by physical-vapor deposition (PVD), chemical vapor deposition (CVD), atomic layer deposition (ALD), and electrodeposition. Surface protection on III–V semiconductors for PEC cells will be discussed in the later section in detail.

5. FUEL FORMING PHOTOELECTRODES

5.1 Photocathode for Hydrogen Evolution

A summary of the reported III–V photocathodes is provided in the following section.

5.1.1 GaP Photocathode

GaP, having a 2.25 eV indirect band gap and a 2.78 eV direct band gap, was investigated as a photocathode more than three decades ago (Nozik, 1976). Despite its promising band gap, GaP degrades (Nozik, 1976) during reactions and generates a low OCP (V_{oc}) of 1.2 V (Allen et al., 2011; Gronet and Lewis, 1982). One straightforward way to increase the V_{oc} of a GaP-based photocathode would be simply forming a junction on its surface. By sputtering n-TiO_2 or n-Nb_2O_5 on GaP, Malizia et al. reported that an ultimate V_{oc} of 0.710 V. GaP nanowires was also prepared to increase the light absorption and surface area using the VLS or SLS method (Liu et al., 2012; Standing et al., 2015; Sun et al., 2011). By further optimizing geometry to reduce electrical resistance and modifying the surface with a suitable cocatalyst, high PEC efficiencies of 2.9% with a V_{oc} over 0.75 V and a short-circuit current (I_{sc}) over 10 mA cm^{-2} were achieved (Standing et al., 2015).

5.1.2 InP Photocathode

InP with a band gap of 1.3 eV and a low-surface recombination velocity (ca. 10^5 cm s^{-1} for p-type InP) (Hoffman et al., 1978), has been widely studied as a photocathode, inspired by the attractive PEC performance with a STH conversion efficiencies of >13% (Aharon-Shalom and Heller, 1982; Heller and Vadimsky, 1981; Wallentin et al., 2010). Lee et al. created p-type InP nanopillars passivated with TiO_2 and Ru as a cocatalyst for a STH efficiency of approximately 14% (Lee et al., 2012). Gao et al. developed a similar nanostructured InP photocathode and via rational design of interface energetics and photon management a STH of 15.8% was obtained (Gao et al., 2016). A noble metal-free cocatalyst nanoparticle, MoS_3, were also tested on InP nanowires with a STH of 6.4% (Gao et al., 2014). Lee et al. employed p-type InP nanopillar photocathode for efficient water reduction in conjunction with a TiO_2 passivation layer and Ru as a cocatalyst with a STH of approximately 14% (Lee et al., 2012). Gao et al. developed a similar nanostructured InP photocathode with rational interface energetic design and photon management and a slightly higher STH (15.8%) was obtained (Gao et al., 2016). A noble-metal-free cocatalyst MoS_3 nanoparticles were also tested on InP nanowire with a STH of 6.4% (Gao et al., 2014). While most of the research focuses on nanostructured-InP, Javey and coworkers developed planar-based InP photocathode (Kapadia et al., 2013; Lobaccaro et al., 2014). In conjunction with a TiO_2 passivation layer and a Pt thin layer as cocatalyst, an onset potential over 800 mV and a STH of 11.6% was obtained (Hettick et al., 2015; Lin et al., 2015). The inspiring performance of InP photocathode makes it promising to be applied on tandem PEC cells for nonbias water splitting.

5.1.3 Ternary III–V Alloy Photocathode

Ternary alloys of III–V materials also attract much attention due to their band gap tunability. Incorporation of In into UV-responsive GaN forms a visible light-responsive InGaN ternary alloy and allows tuning of the band gap across nearly the entire solar spectrum, as importantly, straddling water redox potentials (Kuykendall et al., 2007). The PEC properties of InGaN was investigated by in situ electrochemical mass spectroscopy and its potential to work as a stable photocathode was confirmed (Kamimura et al., 2013). In most cases, InGaN was used in potential junction photoelectrodes, which will be discussed in the later section. Similar to InGaN, tuning the band gap of GaAsP is feasible. It is reported that the single $GaAs_{0.88}P_{0.12}$ (1.56 eV) cell exhibited high open-circuit voltages of 1.10–1.12 V in a solar cell, even though its PEC performance has not been investigated (Faucher et al., 2013). $GaInP_2$ was used as a photocathode unit in "Turner Cell" (Khaselev and Turner, 1998) and a highly active p-$GaInP_2$ photocathode protected through a TiO_2 layer functionalized by a cobaloxime molecular catalyst (Gu et al., 2016). The device presented a current density ca. 9 mA cm^{-2} at 0 V (vs RHE) and exhibited remarkable stability for 20 h compared with a bare $GaInP_2$ electrode.

5.1.4 Protection Layers for Photocathodes

The cathodic stability of most III–V semiconductors is mostly poor in reductive fuel-forming reaction (Chen and Wang, 2012). An important approach for solving this problem is the introduction of a protective layer. In addition to protecting the semiconductor from chemical corrosion, the protection layer commonly has other positive effects, like changing the reaction kinetics, passivating surface states, and improving charge separation (Liu et al., 2014).

Two types of protection layers are commonly coated on III–V-based photocathode: metals (Fig. 18A) and semiconductors (Fig. 18B and C). For metal protective layers, Britto et al. used Mo/MoS_2 as a protection layer and catalyst for $GaInP_2$ photocathode and achieved over 60 h of operation (Britto et al., 2016). The Mo metal layer isolated $GaInP_2$ from electrolytic solution and the MoS_2 was resistant to corrosion and possessed high activity for H_2 generation. A similar rational was applied to the modification of an InP photocathode with a Ti thin layers and amorphous MoS_x (Li et al., 2016b). The Ti thin layer reduced charge-transfer resistance and increased band bending of the InP photocathode, in addition to improving its stability. The common problem of using a metal layer for protection is that it reflects and/or absorbs light leading to reduced light-limited current.

A semiconductor protection layer with a large band gap is less likely to reflect or absorb the incoming light, compared with a metallic protection layer. This layer can conduct electrons from photocathode through well-aligned conduction bands (Fig. 18B). In this case, the valence band of the protection layer should be more positive than that of photocathode and

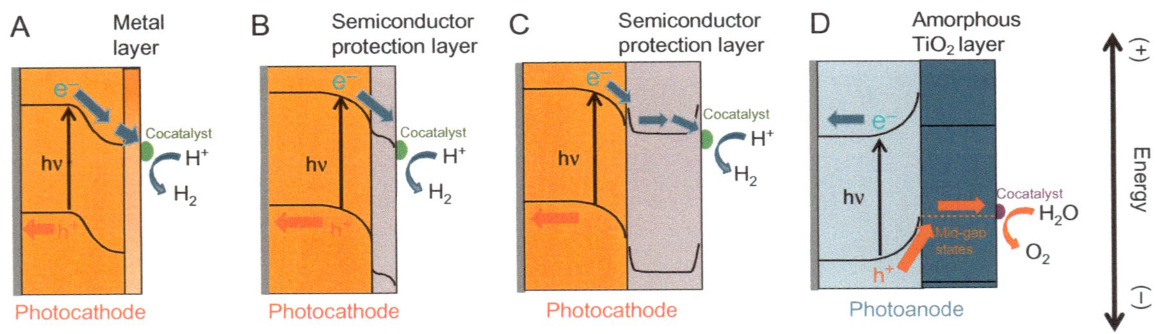

Fig. 18 Photocathode with metallic protection layer (A), semiconductor protection layer (thin with low doping) (B), and semiconductor protection layer (thick with high doping) (C); photoanode with "leaky" TiO_2 layer (D).

the conduction band should be close to the water reduction potential to form an accumulation mode (Seger et al., 2014). Moreover, the protection layer should be thinner than its depletion layer by reducing its thickness or/and donor density. Successful case is the application of TiO_2 layer on InP photocathode for both higher activity and stability.[120–121,124] The electrochemically produced oxide protection layer, Ga_2O_3, on GaP photocathode worked for similar effects (Standing et al., 2015). However, if the conduction bands are not well-aligned and the protection layer is relatively thick, the protection layer should highly doped for a sufficiently thin depletion layer for electronic tunneling (Fig. 18C). Malizia et al. deposited a n-TO_2 or n-Nb_2O_5 layer with high donor densities on a p-GaP photocathode (Malizia et al., 2014). The highly doped oxide layers endowed a large built-in potential drop in the GaP and contributed to a V_{oc} of 0.71 V under illumination. The photocathode was stabilized for hours, but longer term stability is difficult to achieve. Moreover, too heavy doping on the protection layer will easily form detrimental electron–hole recombination centers.

5.2 Photoanode for Water Oxidation

The III–V semiconductors are less investigated as photoanode for water oxidation compared with that as photocathode. Except GaN with a large band gap, the valence bands of most III–V semiconductors are too negative for water oxidation and a high bias is usually required by the photoanodes. What is more, the issue of stability in photoanode is much more serious than that of photocathode. Until surface protection strategies were developed, activity by III–V-based photoanodes in aqueous solution remained low. Hu et al. reported GaAs and GaP photoanodes for efficient and stable water oxidation by novel amorphous TiO_2 coatings. The GaAs and GaP photoanodes exhibited V_{oc} of 0.81 and 0.59 V and saturated photocurrent densities of 14.3 and 3.4 mA cm^{-2}, respectively. A p-type transparent conducting oxides was applied on InP photoanode to function both as a selective hole contact and corrosion protection layer (Chen et al., 2015). A high saturation photocurrent density was obtained, but the onset potential was as positive as 1.55 V (vs RHE). Sputtering NiO_x films on InP photoanodes was reported to shown similar result and the onset potential can be negatively shifted to 0.86 V (vs RHE) by incorporating a thin p^+-doped emitter layer (Sun et al., 2015). Compared with other photoanodes, like $BiVO_4$ (Zhong et al., 2016), the V_{oc} of III–V-based photoanodes are still too low and further research is necessary to improve its future applications in tandem cells.

5.2.1 *Protective Layer for Photoanodes*

The issue of stability is always encountered for an applicable III–V-based photoanode because corrosion through surface traps is often thermodynamically and kinetically favorable than water oxidation (Gerischer, 1990; Hu et al., 2015). Currently, it could be protected by a p-type oxide or a new "leaky" TiO_2 layer (Fig. 18D).

Similar to photocathodes, an oxide semiconductor is commonly used as a protective layer for photoanodes. However, the quest for suitable candidates is much more challenging. The valence band of most oxides are primarily composed of O 2p orbital, which is too deep for well-aligned valence bands with photoanode. Moreover, the oxides are rarely p-type. One of the few cases is naturally p-type NiO_x with a large band gap and a low valence band (Bai et al., 2014; Zhong et al., 2015). Sun et al. stabilized p^+n-InP photoanode for over 48 h in both alkaline and neutral electrolyte by depositing NiO_x films with comparable antireflective properties to TiO_2 coatings, high conductivity and catalytic activity for oxygen evolution (Sun et al., 2015). Chen et al. developed a p-type transparent conducting oxide ($NiCo_2O_4$) with a higher hole conductivity than NiO and formed an efficient and stable photoanode heterostructure with n-InP (Chen et al., 2015).

Recently, Hu et al. developed a "leaky" amorphous TiO_2 layer to stabilize GaAs and GaP photoanodes for more than 100 h (Hu et al., 2014). It was pointed out that the unannealed ALD-TiO_2 layer was electronically defective and thus highly conductive for hole transportation by intermixing of a Ni film. Further analysis indicated that holes passed the amorphous TiO_2 layer though the mid-gap states and an ohmic contact was formed at the interface between the photoanode and the protection layer (Hu et al., 2016).

5.3 Tandem Electrodes for Overall Water Splitting

5.3.1 *Single Cell for Potential Tandem Electrode*

Some single cells were developed for PEC/photocatalytic water splitting, but they can potentially work as a buried tandem cell (Fig. 1B), with a Si substrate being commonly used. Wu et al. reported a p–n homojunction 1.7 eV GaAsP core–shell nanowire photocathode on a Si substrate yielded a STH of 0.5% and was found to be relatively resistant to photocorrosion (Wu et al., 2014). InGaN ternary alloy is one of the few III–V semiconductor applicable in single-junction cell for water splitting. GaN with a large band gap has been reported to achieved one-step water splitting with suitable cocatalysts (Wang et al., 2011). Its activity can be remarkably improved by tuning the Fermi level on the nonpolar surfaces by Mg doping (Kibria et al., 2014). Incorporation of In into GaN for InGaN can narrow its band gap while keeping the band structure straddling the water redox potentials as mention above. Kibria et al. prepared InGaN/GaN nanowire heterostrucutres on a Si substrate, using Rh/Cr_2O_3 nanoparticles as a cocatalyst for one-step water splitting. The device can absorb light irradiation up to 560 nm and achieved an internal quantum efficiency of ca. 13% at 440–450 nm (Kibria et al., 2013). They further improved the performance of the device by p-type dopant incorporation, similar to that for GaN, and a STH of ca. 1.8% under concentrated sun light (Kibria et al., 2015). This value is still far from 5% which is estimated for industrial application (Pinaud et al., 2013), but it very promising to get by adjusting the amount of In in InGaN for harvesting more light irradiation or incorporating the Si substrate as a tandem cell for a higher driving force.

5.3.2 Dual-Junction Tandem Cell

For increasing the electrochemical driving force for water splitting, an integrated system with dual junction (Fig. 2B and C) is commonly considered and it is possible to achieve a STH of >25% (Hu et al., 2013). In an early study, p-GaP was used as a photocathode in tandem cells combining with n-TiO_2, n-$SrTiO_3$, or n-GaP as a photoanode (Nozik, 1976; Ohashi et al., 1977). Both the efficiencies and stability reported in this early work were quite low. Later, Kainthla et al. reported an efficient tandem with a STH up to 8.2% using InP as a photocathode and GaAs as a photoanode decorated with Pt and MnO_x, respectively. A breakthrough was make Khaselev et al. using an buried $GaInP_2$(p)/GaAs(pn) photocathode for a STH of 12.4% (Khaselev and Turner, 1998). Recently, an integrated GaInP/GaInAs photocathode was reported to enable a STH of 14% for unbiased solar water splitting, which is more efficient than "Turner Cell" in a similar structure, but the activity of the device kept degrading during reaction (May et al., 2015). Verlage et al. developed a buried GaAs/InGaP photoanode decorated with earth-abundant electrocatalysts and coated with an amorphous TiO_2 layer for stable water splitting over 40 h with a STH of 10.5% (Verlage et al., 2015b). Considering that the amorphous protection layer is effective for more than 2200 h (Shaner et al., 2015), the device is the only one with long-term operational stability and a STH of >10%. Photoelectrode with nanowire arrays is also tried on tandem cell. InP photocathodes were combined with a nanoporous $BiVO_4$ photoanode for unassisted water splitting, but the STH is only 0.5% currently (Kornienko et al., 2016).

5.3.3 Triple-Junction Tandem Cell

Integrated triple-junction devices are feasible to provide a relatively high photovoltage with rational interface engineering between semiconductors but more photons are required to driven the same water splitting reaction. Yamane et al. reported a complex n-Si/p-CuI/n–i–p a-Si/n–p-GaP/RuO_2 photoanode (n–i–p stands for n-type/intrinsic/p-type) based the model of "crystalline Si/a-Si/GaP." (Yamane et al., 2009) The V_{oc} was as high as 1.4 V was achieved (Yamane et al., 2009). The V_{oc} was as high as 1.4 V but the saturated photocurrent was less than 2 mA cm^{-2} and the STH was calculated to be 2.3%. Fujii et al. developed a "concentrated photovoltaic electrochemical cell" by GaInP(pn)/InGaAs(pn)/Ge(pn), which operated stably and achieved a STH of >12% and claimed it to be a stand-alone renewable energy storage system (Fujii et al., 2013). The device was applied on a membrane-separated PEC system for water splitting in near neutral pH with a controlled recirculating stream, which yield continues nearly pure H_2 stream with a STH above 6.2% (Modestino et al., 2014).

The tandem cell with a STH of 10% was estimated to generate H_2 at a cost about 10 \$/kg, which is still much higher than the targeted threshold cost of \$2–\$4/kg. Several tandem cells mentioned earlier reach the STH of 10% but it is still necessary to pursue a higher activity (Pinaud et al., 2013). What is more, the targeted device is designed to operate for several years. However, the currently available tandem cells are unable to achieve such long-term stability, which calls for advanced protection technologies.

5.4 PEC CO_2 Reduction

Hydrogen generation by solar-driven water splitting systems has attracted much attention since the report of the Honda–Fujishima effect. However, hydrogen has a number of drawbacks as a medium of energy storage: high flammability, explosiveness, and high diffusivity through commonly used metals and materials (Goeppert et al., 2014). Due to these reasons, more attention is being paid to PEC CO_2 reduction to easily stored and transported chemicals such as methanol.

5.4.1 Photocathodes for CO_2 Reduction

Pioneer work for PEC reduction of CO_2 on conducted by Halmann several decades ago on p-GaP (Halmann, 1978). The selectivity for products was poor, generating formic acid, formaldehyde and methanol simultaneously and a very negative bias (−1.2 V vs SCE) was necessary to obtain a low photocurrent (0.3 mA cm^{-2}). Photoreduction of CO_2 for high selectivity and activity by a single photocatalyst lacking reaction sites without a suitable cocatalyst will be challenging, similar to that for H_2 generation. Parkinson et al. proposed to combine a semiconductor photoelectrode (p-InP) with a biological catalyst (a formate dehydrogenase enzyme) by an electron mediator (methylviologen/methylviologen$^+$) for formic acid generation (Parkinson and Weaver, 1984). High selectivity for formic acid formation with a low overpotential loss was achieved, but the performance of the system was threatened by the stability of the electron mediator and enzyme and the backward reaction of product oxidation. In this stage, CO_2 reduction by III–V semiconductors was generally limited to relatively a low photocurrent density (activity) (<10 mA cm^{-2}), which was attributed to the CO_2 mass transport limitations and competition with H_2 production (Hirota et al., 1998). Hirota et al. applied a high-pressure (40 atm) CO_2 and methanol medium for PEC CO_2 reduction using p-InP. The current densities reached 200 mA cm^{-2} with faradic efficiency of $>90\%$ for production and the stability of p-InP was greatly improved (Hirota et al., 1998).

Most previously discussed devices only functioned at high applied potentials of -0.97 to -1.37 V vs RHE in CO_2-saturated aqueous electrolytes. Barton et al. pointed out that if the CO_2 reduction process can be carried out at a metal electrode employing an overpotential that was lower than need at a photoelectrode, one do not expect to use such a PEC process to convert light energy to chemical energy (Barton et al., 2008). They attached the soluble pyridinium component on a p-GaP photocathode to reduce CO_2 to methanol with near 100% faradaic efficiency at -0.22 V (vs SCE) at pH of 5.2 without degrading pyridine. Moreover, the cell produced sustained cathodic currents as high as $\sim$0.2 mA cm^{-2} under illumination with no applied bias. Recently developed passivation layers were applied to CO_2 reduction photoelectrodes to improve the photoconversion efficiency and stability. Qiu et al. used a TiO_2-passivated InP nanopillar photocathode to produce methanol from CO_2 (Qiu et al., 2015). Based on theoretical calculations, the oxygen vacancies in TiO_2 layer serve as catalytically active sites in CO_2 reduction process. However, the Faraday efficiency was still as low as 8.7% even though Cu nanoparticles were deposited on TiO_2 as a cocatalyst.

5.4.2 Full Cell for CO_2 Reduction and Water Oxidation

Similar to one-step water splitting, Al-Otaibi et al. used GaN nanowire with Rh/Cr_2O_3 core/shell or Pt cocatalyst for nonbias CO_2 reduction under ultraviolet light irradiation (AlOtaibi et al., 2015b). The device exhibited relatively high conversion rate for CH_4 and CO and stability over 24 h. This only ultraviolet light responsive device was turned to a visible light responsive one but using InGaN nanowires instead of GaN nanowires (AlOtaibi et al., 2016). The CO_2 molecules were revealed to be spontaneously adsorbed and deformed on the nonpolar surfaces of InGaN based on calculations. A Mg dopant effectively reduced the surface potential barrier and enhanced adsorption of CO_2 on nanowire, leading to a photocatalytic activity by nearly 50-fold.

Application of PEC tandem cells is another approach for nonbias CO_2 reduction. Sato et al. screened the Ru complex polymer catalysts for CO_2 to formate reduction sites and loaded it on p-InP (or p-GaP) as a photocathode. Then, combining it with Pt/TiO_2 photoanode for water oxidation for a tandem cell (Sato et al., 2011). Under no external bias, the system operated with the selectivity for formate over 70% and the conversion efficiency of solar energy to chemical energy was about 0.03%. The latter value was improved to 0.14% using a reduced-$SrTiO_3$ photoanode instead of a TiO_2 photoanode. Using Pd/C instead of Ni–Mo as a cathode, Zhou et al. developed the previous reported NiMo/GaAs/InGaP/TiO_2/Ni water splitting system to a CO_2 reduction system with the help of a bipolar membrane (Verlage et al., 2015b; Zhou et al., 2016). This system was demonstrated to reduce CO_2 to formate selectively at a STF energy-conversion efficiency of 10% and kept stable for more than 2 h.

REFERENCES

Aharon-Shalom, E., Heller, A., 1982. Efficient p-InP (Rh-H alloy) and p-InP (Re-H alloy) hydrogen evolving photocathodes. J. Electrochem. Soc. 129 (12), 2865–2866.

Ahrenkiel, S.P., Wanlass, M.W., Carapella, J.J., Gedvilas, L.M., Keyes, B.M., Ahrenkiel, R.K., Moutinho, H.R., 2004. Characterization survey of $Ga_xIn_{1-x}As/InAs_yP_{1-y}$ double heterostructures and $InAsyP_{1-y}$ multilayers grown on InP. J. Electron. Mater. 33 (3), 185–193.

Alivisatos, A.P., 1996. Semiconductor clusters, nanocrystals, and quantum dots. Science 271 (5251), 933–937.

Allen, C.R., Woodall, J.M., Jeon, J.H., 2011. Results of a gallium phosphide photovoltaic junction with an AR coating under concentration of natural sunlight. Sol. Energy Mater. Sol. Cells 95 (9), 2655–2658.

Allongue, P., Cachet, H., 1984. Photodissolution kinetics of n-GaAs in 1 M KOH and calculation of the stabilization by Se^{2-}: effect of the Ru^{3+} surface treatment. J. Electrochem. Soc. 131 (12), 2668–2861.

Allongue, P., Cachet, H., 1988. Charge-transfer and stabilization at illuminated N-GaAs aqueous-electrolyte junctions. Electrochim. Acta 33 (1), 79–87.

Allongue, P., Cachet, H., Clechet, P., Froment, M., Martin, J.R., Verney, E., 1987. Stabilization of normal-GaAs in acidic concentrated iodide electrolytes. J. Electrochem. Soc. 134 (3), 620–625.

AlOtaibi, B., Fan, S., Vanka, S., Kibria, M.G., Mi, Z., 2015a. A metal-nitride nanowire dual-photoelectrode device for unassisted solar-to-hydrogen conversion under parallel illumination. Nano Lett. 15 (10), 6821–6828.

AlOtaibi, B., Fan, S., Wang, D., Ye, J., Mi, Z., 2015b. Wafer-level artificial photosynthesis for CO_2 reduction into CH4 and CO using GaN nanowires. ACS Catal. 5 (9), 5342–5348.

AlOtaibi, B., Kong, X., Vanka, S., Woo, S.Y., Pofelski, A., Oudjedi, F., Fan, S., Kibria, M.G., Botton, G.A., Ji, W., Guo, H., Mi, Z., 2016. Photochemical carbon dioxide reduction on Mg-doped Ga(in)N nanowire arrays under visible light irradiation. ACS Energy Lett. 1 (1), 246–252.

Alvi, N.H., Soto Rodriguez, P.E.D., Kumar, P., Gómez, V.J., Aseev, P., Alvi, A.H., Alvi, M.A., Willander, M., Nötzel, R., 2014. Photoelectrochemical water splitting and hydrogen generation by a spontaneously formed InGaN nanowall network. Appl. Phys. Lett. 104 (22), 223104.

Andreev, V.M., Vodnev, A.A., Mintairov, A.M., Rumyantsev, V.D., Khvostikov, V.P., 1987. Photoluminescence of quantum-well layers in AlGaAs heterostructures formed by low-temperature liquid-phase epitaxy. Sov. Phys. Semicond. 21 (7), 736–738.

Arora, B.M., 1990. Growth of III–V compounds by liquid phase epitaxy. Bull. Mater. Sci. 13 (1), 11–26.

Arthur, J.R., 2002. Molecular beam epitaxy. Surf. Sci. 500 (1–3), 189–217.

Aspnes, D.E., Studna, A.A., 1981. Chemical etching and cleaning procedures for Si, Ge, and some III-V compound semiconductors. Appl. Phys. Lett. 39 (4), 316–318.

Ba, B., Cachet, H., Fotouhi, B., Gorochov, O., 1994. Orientation-dependent electronic properties of n-type GaAs in contact with various redox systems in acetonitrile and methanol. Semicond. Sci. Technol. 9 (8), 1529–1534.

Bai, S., Cao, M., Jin, Y., Dai, X., Liang, X., Ye, Z., Li, M., Cheng, J., Xiao, X., Wu, Z., Xia, Z., Sun, B., Wang, E., Mo, Y., Gao, F., Zhang, F., 2014. Low-temperature combustion-synthesized nickel oxide thin films as hole-transport interlayers for solution-processed optoelectronic devices. Adv. Energy Mater. 4 (6), 1301460.

Barns, R.L., Ellis, W.C., 1965. Whisker crystals of gallium arsenide and gallium phosphide grown by the vapor–liquid–solid mechanism. J. Appl. Phys. 36 (7), 2296–2301.

Barton, E.E., Rampulla, D.M., Bocarsly, A.B., 2008. Selective solar-driven reduction of CO_2 to methanol using a catalyzed p-GaP based photoelectrochemical cell. J. Am. Chem. Soc. 130 (20), 6342–6344.

Bauhuis, G.J., Mulder, P., van Kempen, H., 2002. Tip formation of micrometer scale GaAs pyramid structures grown by MOCVD. J. Cryst. Growth 240 (1–2), 104–111.

Benedikt, B., Andreas, R., Marcello, S., Anna Fontcuberta i, M., Josef, Z., Dieter, S., Elisabeth, R., 2010. Position controlled self-catalyzed growth of GaAs nanowires by molecular beam epitaxy. Nanotechnology 21 (43), 435601.

Bertness, K.A., Roshko, A., Mansfield, L.M., Harvey, T.E., Sanford, N.A., 2008. Mechanism for spontaneous growth of GaN nanowires with molecular beam epitaxy. J. Cryst. Growth 310 (13), 3154–3158.

Bessolov, V.N., Lebedev, M.V., 1998. Chalcogenide passivation of III–V semiconductor surfaces. Semiconductors 32 (11), 1141–1156.

Bessolov, V.N., Lebedev, M.V., Ivankov, A.F., Bauhofer, W., Zahn, D.R.T., 1998. Electronic properties of GaAs(100) surface passivated in alcoholic sulfide solutions. Appl. Surf. Sci. 133 (1–2), 17–22.

Bindra, P., Gerischer, H., Kolb, D.M., 1977. Electrolytic deposition of thin metal films on semiconductor substrates. J. Electrochem. Soc. 124 (7), 1012–1018.

Bolton, J.R., Strickler, S.J., Connolly, J.S., 1985. Limiting and realizable efficiencies of solar photolysis of water. Nature 316 (6028), 495–500.

Britto, R.J., Benck, J.D., Young, J.L., Hahn, C., Deutsch, T.G., Jaramillo, T.F., 2016. Molybdenum disulfide as a protection layer and catalyst for gallium indium phosphide solar water splitting photocathodes. J. Phys. Chem. Lett. 7 (11), 2044–2049.

Bryce, C., Berk, D., 1996. Kinetics of GaAs dissolution in H_2O_2–NH_4OH–H_2O solutions. Ind. Eng. Chem. Res. 35 (12), 4464–4470.

Burda, C., Chen, X., Narayanan, R., El-Sayed, M.A., 2005. Chemistry and properties of nanocrystals of different shapes. Chem. Rev. 105 (4), 1025–1102.

Caroff, P., Dick, K.A., Johansson, J., Messing, M.E., Deppert, K., Samuelson, L., 2009. Controlled polytypic and twin-plane superlattices in iii-v nanowires. Nat. Nanotechnol. 4 (1), 50–55.

Casagrande, L.G., Lewis, N.S., 1985a. Measurement of the activation barrier for carrier transport at normal-gaas semiconductor liquid junctions. J. Am. Chem. Soc. 107 (20), 5793–5794.

Casagrande, L.G., Lewis, N.S., 1985b. Measurement of the activation barrier for carrier transport at n-gallium arsenide semiconductor/liquid junctions. J. Am. Chem. Soc. 107 (20), 5793–5794.

Casagrande, L.G., Juang, A., Lewis, N.S., 2000. Photoelectrochemical behavior of n-GaAs and n-Al_x Ga_{1-x} as in CH_3CN. J. Phys. Chem. B 104 (23), 5436–5447.

Chang, K.C., Heller, A., Schwartz, B., Menezes, S., Miller, B., 1977. Stable semiconductor liquid junction cell with 9 percent solar-to-electrical conversion efficiency. Science (New York, N.Y.) 196 (4294), 1097–1099.

Chen, S., Wang, L.-W., 2012. Thermodynamic oxidation and reduction potentials of photocatalytic semiconductors in aqueous solution. Chem. Mater. 24 (18), 3659–3666.

Chen, C.C., Yeh, C.C., 2000. Large-scale catalytic synthesis of crystalline gallium nitride nanowires. Adv. Mater. 12 (10), 738–741.

Chen, C.-C., Yeh, C.-C., Chen, C.-H., Yu, M.-Y., Liu, H.-L., Wu, J.-J., Chen, K.-H., Chen, L.-C., Peng, J.-Y., Chen, Y.-F., 2001. Catalytic growth and characterization of gallium nitride nanowires. J. Am. Chem. Soc. 123 (12), 2791–2798.

Chen, L., Yang, J., Klaus, S., Lee, L.J., Woods-Robinson, R., Ma, J., Lum, Y., Cooper, J.K., Toma, F.M., Wang, L.-W., Sharp, I.D., Bell, A.T., Ager, J.W., 2015. p-type transparent conducting oxide/n-type semiconductor heterojunctions for efficient and stable solar water oxidation. J. Am. Chem. Soc. 137 (30), 9595–9603.

Cheng, C.-W., Shiu, K.-T., Li, N., Han, S.-J., Shi, L., Sadana, D.K., 2013. Epitaxial lift-off process for gallium arsenide substrate reuse and flexible electronics. Nat. Commun. 4, 1577.

Chi, C.-Y., Chang, C.-C., Hu, S., Yeh, T.-W., Cronin, S.B., Dapkus, P.D., 2013. Twin-free GaAs nanosheets by selective area growth: implications for defect-free nanostructures. Nano Lett. 13 (6), 2506–2515.

Chu, H.-J., Yeh, T.-W., Stewart, L., Dapkus, P.D., 2010. Wurtzite InP nanowire arrays grown by selective area MOCVD. Phys. Status Solidi C 7 (10), 2494–2497.

Chuang, L.C., Moewe, M., Chase, C., Kobayashi, N.P., Chang-Hasnain, C., Crankshaw, S., 2007. Critical diameter for III-V nanowires grown on lattice-mismatched substrates. Appl. Phys. Lett. 90 (4), 043115.

Cierocki, K., Troost, D., Koenders, L., Mönch, W., 1992. Bromine-induced surface states on cleaved GaAs(110) surfaces: experiment and tight-binding model. Surf. Sci. 264 (1), 23–32.

Cirlin, G.E., Dubrovskii, V.G., Samsonenko, Y.B., Bouravleuv, A.D., Durose, K., Proskuryakov, Y.Y., Mendes, B., Bowen, L., Kaliteevski, M.A., Abram, R.A., Zeze, D., 2010. Self-catalyzed, pure zincblende GaAs nanowires grown on Si(111) by molecular beam epitaxy. Phys. Rev. B 82 (3), 035302.

Coltrin, M.E., Mitchell, C.C., 2003. Mass transport and kinetic limitations in MOCVD selective-area growth. J. Cryst. Growth 254 (1–2), 35–45.

Dawson, L.R., 1972. Liquid phase epitaxy. Prog. Solid State Chem. 7, 117–139.

Decker, F., Parkinson, B.A., 1980. The suppression of GaAs photocorrosion in aqueous-solutions by sulfonated anthraquinones. J. Electrochem. Soc. 127 (11), 2370–2374.

Deutsch, T.G., Koval, C.A., Turner, J.A., 2006. III-V nitride epilayers for photoelectrochemical water splitting: GaPN and GaAsPN. J. Phys. Chem. B 110, 25297–25307.

Deutsch, T.G., Head, J.L., Turner, J.A., 2008. Photoelectrochemical characterization and durability analysis of GaInPN epilayers. J. Electrochem. Soc. 155 (9), B903–B907.

Deutsch, T.G., Welch, A., Bär, M., Weinhardt, L., Weir, M.G., George, K.E., Heske, C., Turner, J.A., 2012. Photoelectrolysis on p-GaInP2; extended durability by nitrogen ion implantation. In: Meeting Abstracts. MA2012-02 (14), 1743.

Deutsch, T., Turner, J., Leisch, J., Wang, H., Welch, A., Lindeman, A., O'neill, K., Pinkard, A., Dameron, A., Heske, C., 2014. Stable photoelectrode surfaces and methods. Google Patents: 2014.

Dharmadasa, I.M., Haigh, J., 2006. Strengths and advantages of electrodeposition as a semiconductor growth technique for applications in macroelectronic devices. J. Electrochem. Soc. 153 (1), G47–G52.

Dingman, S.D., Rath, N.P., Markowitz, P.D., Gibbons, P.C., Buhro, W.E., 2000. Low-temperature, catalyzed growth of indium nitride fibers from azido-indium precursors. Angew. Chem. Int. Ed. 39 (8), 1470–1472.

Dubrovskii, V.G., Sibirev, N.V., Cirlin, G.E., Soshnikov, I.P., Chen, W.H., Larde, R., Cadel, E., Pareige, P., Xu, T., Grandidier, B., Nys, J.P., Stievenard, D., Moewe, M., Chuang, L.C., Chang-Hasnain, C., 2009. Gibbs-Thomson and diffusion-induced contributions to the growth rate of Si, InP, and GaAs nanowires. Phys. Rev. B 79 (20), 205316.

Ellis, W., 1968. Morphology of gallium phosphide crystals grown by VLS mechanism with gallium as liquid-forming agent. J. Cryst. Growth 2 (2), 61–68.

Ellis, A.B., Bolts, J.M., Kaiser, S.W., Wrighton, M.S., 1977a. Study of n-type gallium arsenide- and gallium phosphide-based photoelectrochemical cells. Stabilization by kinetic control and conversion of optical energy to electricity. J. Am. Chem. Soc. 99 (9), 2848–2854.

Ellis, A.B., Bolts, J.M., Wrighton, M.S., 1977c. Characterization of n-type semiconducting indium phosphide photoelectrodes. J. Electrochem. Soc. 124, 1603–1607.

Energy, U.D.O., 2015. Multi-Year Research, Development, and Demonstration Plan—Section 3.1 Hydrogen Production, Department of Energy, Office of Energy Efficiency and Renewable Energy; Washington, DC, https://energy.gov/eere/fuelcells/downloads/fuel-cell-technologies-office-multi-year-research-development-and-22.

Engelken, R.D., Van Doren, T.P., 1985. Ionic electrodeposition of II–VI and III–V compounds: I. Development of a simple, Butler-Volmer equation-based kinetic model for electrodeposition. J. Electrochem. Soc. 132 (12), 2904–2909.

Ertekin, E., Greaney, P.A., Chrzan, D.C., Sands, T.D., 2005. Equilibrium limits of coherency in strained nanowire heterostructures. J. Appl. Phys. 97 (11), 114325.

Fan, F.R.F., Bard, A.J., 1980. Semiconductor electrodes. 24. Behavior of photoelectrochemical cells based on p-type gallium arsenide in aqueous solutions. J. Am. Chem. Soc. 102 (11), 3677–3683.

Fan, S., Shih, I., Mi, Z., 2016. A monolithically integrated InGaN nanowire/Si tandem photoanode approaching the ideal bandgap configuration of 1.75/1.13 eV. Adv. Energy Mater. 7, 1600952.

Farrell, A.C., Lee, W.-J., Senanayake, P., Haddad, M.A., Prikhodko, S.V., Huffaker, D.L., 2015. High-quality InAsSb nanowires grown by catalyst-free selective-area metal–organic chemical vapor deposition. Nano Lett. 15 (10), 6614–6619.

Faucher, J., Gerger, A., Tomasulo, S., Ebert, C., Lochtefeld, A., Barnett, A., Lee, M.L., 2013. Single-junction GaAsP solar cells grown on SiGe graded buffers on Si. Appl. Phys. Lett. 103 (19), 191901.

Feng, Q., Cotton, T.M., 1988. A study of charge-transfer at an illuminated P-InP electrode in aqueous-solution. J. Electrochem. Soc. 135 (3), 591–598.

Fujii, K., Nakamura, S., Sugiyama, M., Watanabe, K., Bagheri, B., Nakano, Y., 2013. Characteristics of hydrogen generation from water splitting by polymer electrolyte electrochemical cell directly connected with concentrated photovoltaic cell. Int. J. Hydrogen Energy 38 (34), 14424–14432.

Fujishima, A., Honda, K., 1972. Electrochemical photolysis of water at a semiconductor electrode. Nature 238 (5358), 37–38.

Gao, L., Woo, R.L., Liang, B., Pozuelo, M., Prikhodko, S., Jackson, M., Goel, N., Hudait, M.K., Huffaker, D.L., Goorsky, M.S., Kodambaka, S., Hicks, R.F., 2009. Self-catalyzed epitaxial growth of vertical indium phosphide nanowires on silicon. Nano Lett. 9 (6), 2223–2228.

Gao, L., Cui, Y., Wang, J., Cavalli, A., Standing, A., Vu, T.T.T., Verheijen, M.A., Haverkort, J.E.M., Bakkers, E.P.A.M., Notten, P.H.L., 2014. Photoelectrochemical hydrogen production on InP nanowire arrays with molybdenum sulfide electrocatalysts. Nano Lett. 14 (7), 3715–3719.

Gao, L., Cui, Y., Vervuurt, R.H.J., van Dam, D., van Veldhoven, R.P.J., Hofmann, J.P., Bol, A.A., Haverkort, J.E.M., Notten, P.H.L., Bakkers, E.P.A.M., Hensen, E.J.M., 2016. High-efficiency InP-based photocathode for hydrogen production by interface energetics design and photon management. Adv. Funct. Mater. 26 (5), 679–686.

Geisz, J.F., Olson, J.M., Romero, M.J., Jiang, C.S., Norman, A.G., 2006. Lattice-mismatched GaAsP solar cells grown on silicon by OMVPE. In: 2006 IEEE 4th World Conference on Photovoltaic Energy Conference, May, pp. 772–775.

Geisz, J.F., Kurtz, S., Wanlass, M.W., Ward, J.S., Duda, A., Friedman, D.J., Olson, J.M., McMahon, W.E., Moriarty, T.E., Kiehl, J.T., 2007. High-efficiency GaInP/GaAs/InGaAs triple-junction solar cells grown inverted with a metamorphic bottom junction. Appl. Phys. Lett. 91 (2), 023502.

Geisz, J.F., Friedman, D.J., Ward, J.S., Duda, A., Olavarria, W.J., Moriarty, T.E., Kiehl, J.T., Romero, M.J., Norman, A.G., Jones, K.M., 2008. 40.8% efficient inverted triple-junction solar cell with two independently metamorphic junctions. Appl. Phys. Lett. 93 (12), 123505.

Gerischer, H., 1977. On the stability of semiconductor electrodes against photodecomposition. J. Electroanal. Chem. Interfacial Electrochem. 82 (1–2), 133–143.

Gerischer, H., 1978. Electrolytic decomposition and photodecomposition of compound semiconductors in contact with electrolytes. J. Vac. Sci. Technol. 15 (4), 1422–1428.

Gerischer, H., 1990. The impact of semiconductors on the concepts of electrochemistry. Electrochim. Acta 35 (11), 1677–1699.

Ghez, R., Small, M.B., 1981. Growth and dissolution of ternary alloys of III–V compounds by liquid phase epitaxy and the formation of heterostructures. J. Cryst. Growth 52, 699–709.

Glas, F., 2006. Critical dimensions for the plastic relaxation of strained axial heterostructures in free-standing nanowires. Phys. Rev. B 74 (12), 121302.

Gmitter, T.J., Yablonovitch, E., Heller, A., 1988. Recombination velocities on GaAs <100> surfaces immersed in aqueous solutions. J. Electrochem. Soc. 135 (9), 2391–2392.

Goeppert, A., Czaun, M., Jones, J.-P., Surya Prakash, G.K., Olah, G.A., 2014. Recycling of carbon dioxide to methanol and derived products—closing the loop. Chem. Soc. Rev. 43 (23), 7995–8048.

Gomes, W.P., 2001. Wet etching of III–V semiconductors. In: Willardson, R., Nalwa, H.S. (Eds.), Semiconductors and Semimetals. vol. 73. Elsevier, London, UK, pp. 215–295. Chapter 6.
Goodman, C.E., Wessels, B.W., 1986. Function of cobalt and platinum on p-InP in the photoevolution of hydrogen from alkaline solutions. Appl. Phys. Lett. 49 (13), 829–830.
Green, M., O'Brien, P., 1998. A novel metalorganic route for the direct and rapid synthesis of monodispersed quantum dots of indium phosphide. Chem. Commun. 22, 2459–2460.
Green, M., O'Brien, P., 2004. The synthesis of III-V semiconductor nanoparticles using indium and gallium diorganophosphides as single-molecular precursors. J. Mater. Chem. 14 (4), 629–636.
Green, M.A., Emery, K., Hishikawa, Y., Warta, W., Dunlop, E.D., 2012. Solar cell efficiency tables (version 39). Prog. Photovolt. Res. Appl. 20 (1), 12–20.
Grimm, R.L., Bierman, M.J., O'Leary, L.E., Strandwitz, N.C., Brunschwig, B.S., Lewis, N.S., 2012. Comparison of the photoelectrochemical behavior of H-terminated and methyl-terminated Si(111) surfaces in contact with a series of one-electron, outer-sphere redox couples in CH3CN. J. Phys. Chem. C 116 (44), 23569–23576.
Gronet, C.M., Lewis, N.S., 1982. Design of a 13% efficient n-$GaAs_{1-x}P_x$ semiconductor–liquid junction solar cell. Nature 300 (5894), 733–735.
Gronet, C.M., Lewis, N.S., 1983. n-type GaAs photoanodes in acetonitrile: design of a 10.0% efficient photoelectrode. Appl. Phys. Lett. 43 (1), 115–117.
Gronet, C.M., Lewis, N.S., 1984. Systematic studies of the semiconductor/liquid junction: n-gallium arsenide phosphide anodes in aqueous Se2-/Se22- solutions. J. Phys. Chem. 88 (8), 1310–1317.
Gu, J., Yan, Y., Young, J.L., Steirer, K.X., Neale, N.R., Turner, J.A., 2016. Water reduction by a p-GaInP2 photoelectrode stabilized by an amorphous TiO2 coating and a molecular cobalt catalyst. Nat. Mater. 15 (4), 456–460.
Gudiksen, M.S., Lieber, C.M., 2000. Diameter-selective synthesis of semiconductor nanowires. J. Am. Chem. Soc. 122 (36), 8801–8802.
Gudiksen, M.S., Lauhon, L.J., Wang, J., Smith, D.C., Lieber, C.M., 2002. Growth of nanowire superlattice structures for nanoscale photonics and electronics. Nature 415 (6872), 617–620.
Guo-Ping, J., Ruda, H.E., 1996. Ab initio studies of S chemisorption on GaAs(100). J. Appl. Phys. 79 (7), 3758–3762.
Halmann, M., 1978. Photoelectrochemical reduction of aqueous carbon dioxide on p-type gallium phosphide in liquid junction solar cells. Nature 275 (5676), 115–116.
Heller, A., Vadimsky, R.G., 1981. Efficient solar to chemical conversion: 12% efficient photoassisted electrolysis in the [p-type InP(Ru)]/HCl-KCl/Pt(Rh) cell. Phys. Rev. Lett. 46 (17), 1153–1156.
Heller, A., Chang, K.C., Miller, B., 1978. Photocurrent spectroscopy of semiconductor electrodes in liquid junction solar cells. J. Am. Chem. Soc. 100 (3), 684–688.
Heller, A., Aharon-Shalom, E., Bonner, W.A., Miller, B., 1982. Hydrogen-evolving semiconductor photocathodes: nature of the junction and function of the platinum group metal catalyst. J. Am. Chem. Soc. 104 (25), 6942–6948.
Hersee, S.D., Sun, X., Wang, X., 2006. The controlled growth of GaN nanowires. Nano Lett. 6 (8), 1808–1811.
Hettick, M., Zheng, M., Lin, Y., Sutter-Fella, C.M., Ager, J.W., Javey, A., 2015. Non-epitaxial thin-film InP for scalable and efficient photocathodes. J. Phys. Chem. Lett. 6 (12), 2177–2182.

Hiramatsu, K., Matsushima, H., Shibata, T., Sawaki, N., Tadatomo, K., Okagawa, H., Ohuchi, Y., Honda, Y., Matsue, T., 1998. Selective area growth of GaN by MOVPE and HVPE. Mater. Res. Soc. Symp. Proc. 482, 257–268.

Hirota, K., Tryk, D.A., Yamamoto, T., Hashimoto, K., Okawa, M., Fujishima, A., 1998. Photoelectrochemical reduction of CO_2 in a high-pressure CO^{2+} methanol medium at p-type semiconductor electrodes. J. Phys. Chem. B 102 (49), 9834–9843.

Hiruma, K., Yazawa, M., Katsuyama, T., Ogawa, K., Haraguchi, K., Koguchi, M., Kakibayashi, H., 1995. Growth and optical properties of nanometer-scale GaAs and InAs whiskers. J. Appl. Phys. 77 (2), 447–462.

Hoffman, C.A., Jarašiūnas, K., Gerritsen, H.J., Nurmikko, A.V., 1978. Measurement of surface recombination velocity in semiconductors by diffraction from picosecond transient free-carrier gratings. Appl. Phys. Lett. 33 (6), 536–539.

Hu, S., Xiang, C., Haussener, S., Berger, A.D., Lewis, N.S., 2013. An analysis of the optimal band gaps of light absorbers in integrated tandem photoelectrochemical water-splitting systems. Energ. Environ. Sci. 6 (10), 2984.

Hu, S., Shaner, M.R., Beardslee, J.A., Lichterman, M., Brunschwig, B.S., Lewis, N.S., 2014. Amorphous TiO_2 coatings stabilize Si, GaAs, and GaP photoanodes for efficient water oxidation. Science 344 (6187), 1005–1009.

Hu, S., Lewis, N.S., Ager, J.W., Yang, J., McKone, J.R., Strandwitz, N.C., 2015. Thin-film materials for the protection of semiconducting photoelectrodes in solar-fuel generators. J. Phys. Chem. C 119 (43), 24201–24228.

Hu, S., Richter, M.H., Lichterman, M.F., Beardslee, J., Mayer, T., Brunschwig, B.S., Lewis, N.S., 2016. Electrical, photoelectrochemical, and photoelectron spectroscopic investigation of the interfacial transport and energetics of amorphous TiO2/Si heterojunctions. J. Phys. Chem. C 120 (6), 3117–3129.

Jacobsson, D., Panciera, F., Tersoff, J., Reuter, M.C., Lehmann, S., Hofmann, S., Dick, K.A., Ross, F.M., 2016. Interface dynamics and crystal phase switching in GaAs nanowires. Nature 531 (7594), 317–322.

Jia, J., Seitz, L.C., Benck, J.D., Huo, Y., Chen, Y., Ng, J.W., Bilir, T., Harris, J.S., Jaramillo, T.F., 2016. Solar water splitting by photovoltaic-electrolysis with a solar-to-hydrogen efficiency over 30. Nat. Commun. 7, 13237.

Jiang, N., Parkinson, P., Gao, Q., Breuer, S., Tan, H.H., Wong-Leung, J., Jagadish, C., 2012. Long minority carrier lifetime in Au-catalyzed GaAs/AlxGa_{1-x}As core-shell nanowires. Appl. Phys. Lett. 101 (2), 023111.

Johansson, J., Karlsson, L.S., Svensson, C.P.T., Mårtensson, T., Wacaser, B.A., Deppert, K., Samuelson, L., Seifert, W., 2006. Structural properties of <111> B-oriented III-V nanowires. Nat. Mater. 5 (7), 574–580.

Johnston, W.D., Leamy, H.J., Parkinson, B.A., Heller, A., Miller, B., 1980. Effect of ruthenium ions on grain boundaries in gallium arsenide thin film photovoltaic devices. J. Electrochem. Soc. 127 (1), 90–95.

Joyce, H.J., Gao, Q., Tan, H.H., Jagadish, C., Kim, Y., Zhang, X., Guo, Y., Zou, J., 2007. Twin-free uniform epitaxial GaAs nanowires grown by a two-temperature process. Nano Lett. 7 (4), 921–926.

Kamimura, J., Bogdanoff, P., Lähnemann, J., Hauswald, C., Geelhaar, L., Fiechter, S., Riechert, H., 2013. Photoelectrochemical properties of (In, Ga)N nanowires for water splitting investigated by in situ electrochemical mass spectroscopy. J. Am. Chem. Soc. 135 (28), 10242–10245.

Kapadia, R., Yu, Z., Wang, H.-H.H., Zheng, M., Battaglia, C., Hettick, M., Kiriya, D., Takei, K., Lobaccaro, P., Beeman, J.W., Ager, J.W., Maboudian, R., Chrzan, D.C., Javey, A., 2013. A direct thin-film path towards low-cost large-area III-V photovoltaics. Sci. Rep. 3, 2275.

Kawaguchi, Y., Honda, Y., Matsushima, H., Yamaguchi, M., Hiramatsu, K., Sawaki, N., 1998. Selective area growth of GaN on Si substrate using SiO2 mask by metalorganic vapor phase epitaxy. Jpn. J. Appl. Phys. 37 (Pt. 2 No. 8b), L966–L969.

Kelly, J.J., Memming, R., 1982. The influence of surface recombination and trapping on the cathodic photocurrent at p-type III-V electrodes. J. Electrochem. Soc. 129 (4), 730–738.

Kelly, J.J., Reynders, A.C., 1987. A study of GaAs etching in alkaline H_2O_2 solutions. Appl. Surf. Sci. 29 (2), 149–164.

Kelting, K., Koehler, K., Zwicknagl, P., 1986. Luminescence of Ga1-xAlxAs/GaAs single quantum wells grown by liquid phase epitaxy. Appl. Phys. Lett. 48 (2), 157.

Khaselev, O., Turner, J.A., 1998. A monolithic photovoltaic-photoelectrochemical device for hydrogen production via water splitting. Science 280 (5362), 425–427.

Kibria, M.G., Nguyen, H.P.T., Cui, K., Zhao, S., Liu, D., Guo, H., Trudeau, M.L., Paradis, S., Hakima, A.-R., Mi, Z., 2013. One-step overall water splitting under visible light using multiband InGaN/GaN nanowire heterostructures. ACS Nano 7 (9), 7886–7893.

Kibria, M., Zhao, S., Chowdhury, F., Wang, Q., Nguyen, H., Trudeau, M., Guo, H., Mi, Z., 2014. Tuning the surface Fermi level on p-type gallium nitride nanowires for efficient overall water splitting. Nat. Commun. 5, 3825.

Kibria, M.G., Chowdhury, F.A., Zhao, S., AlOtaibi, B., Trudeau, M.L., Guo, H., Mi, Z., 2015. Visible light-driven efficient overall water splitting using p-type metal-nitride nanowire arrays. Nat. Commun. 6, 6797.

Kiselev, V.F., Krylov, O.V., 1987. Electronic Phenomena in Adsorption and Catalysis. Springer-Verlag New York Inc., New York, NY, United States.

Kitamura, S., Hiramatsu, K., Sawaki, N., 1995. Fabrication of gan hexagonal pyramids on dot-patterned gan/sapphire substrates via selective metalorganic vapor-phase epitaxy. Jpn. J. Appl. Phys. 34 (Pt. 2 No. 9b), L1184–L1186.

Kohl, P.A., Bard, A.J., 1979a. Semiconductor electrodes. 18. Liquid junction photo-voltaic cells based on n-GaAs electrodes and acetonitrile solutions. J. Electrochem. Soc. 126 (4), 603–608.

Kohl, P.A., Bard, A.J., 1979b. Semiconductor electrodes. 17. Electrochemical behavior of N-type and P-type InP electrodes in acetonitrile solutions. J. Electrochem. Soc. 126 (4), 598–603.

Konagai, M., Sugimoto, M., Takahashi, K., 1978. High efficiency GaAs thin film solar cells by peeled film technology. J. Cryst. Growth 45, 277–280.

Kornienko, N., Gibson, N.A., Zhang, H., Eaton, S.W., Yu, Y., Aloni, S., Leone, S.R., Yang, P., 2016. Growth and photoelectrochemical energy conversion of wurtzite indium phosphide nanowire arrays. ACS Nano 10, 5525–5535.

Kressel, H., Nelson, H., 1973. Properties and applications of III–V compound films deposited by liquid phase epitaxy. In: Georg Hass, M.H.F., Richard, W.H. (Eds.), Physics of Thin Films. vol. 7. Academic Press, New York, pp. 115–256.

Krogstrup, P., Popovitz-Biro, R., Johnson, E., Madsen, M.H., Nygård, J., Shtrikman, H., 2010. Structural phase control in self-catalyzed growth of GaAs nanowires on silicon (111). Nano Lett. 10 (11), 4475–4482.

Kuykendall, T., Ulrich, P., Aloni, S., Yang, P., 2007. Complete composition tunability of InGaN nanowires using a combinatorial approach. Nat. Mater. 6 (12), 951–956.

Lee, Y.-C., Kuo, T.-J., Hsu, C.-J., Su, Y.-W., Chen, C.-C., 2002. Fabrication of 3D macroporous structures of II − VI and III − V semiconductors using electrochemical deposition. Langmuir 18 (25), 9942–9946.

Lee, M.H., Takei, K., Zhang, J., Kapadia, R., Zheng, M., Chen, Y.-Z., Nah, J., Matthews, T.S., Chueh, Y.-L., Ager, J.W., Javey, A., 2012. p-Type InP nanopillar photocathodes for efficient solar-driven hydrogen production. Angew. Chem. Int. Ed. 51 (43), 10760–10764.

Lekhal, K., Bae, S.Y., Lee, H.J., Mitsunari, T., Tamura, A., Deki, M., Honda, Y., Amano, H., 2016a. Controlled morphology of regular GaN microrod arrays by a selective area growth with HVPE. J. Cryst. Growth 447, 55–61.

Lekhal, K., Bae, S.Y., Lee, H.J., Mitsunari, T., Tamura, A., Deki, M., Honda, Y., Amano, H., 2016b. Selective-area growth of GaN microrods on strain-induced templates by hydride vapor phase epitaxy. Jpn. J. Appl. Phys. 55 (5), 05FF03.

Lewis, N.S., 1991. An analysis of charge transfer rate constants for semiconductor/liquid interfaces. Annu. Rev. Phys. Chem. 42 (1), 543–580.

Li, S., Waag, A., 2012. GaN based nanorods for solid state lighting. J. Appl. Phys. 111 (7), 071101. 071101-23.

Li, Q., Zheng, M., Ma, L., Zhong, M., Zhu, C., Zhang, B., Wang, F., Song, J., Ma, L., Shen, W., 2016a. Unique three-dimensional InP nanopore arrays for improved photoelectrochemical hydrogen production. ACS Appl. Mater. Interfaces 8 (34), 22493–22500.

Li, Q., Zheng, M., Zhong, M., Ma, L., Wang, F., Ma, L., Shen, W., 2016b. Engineering MoS(x)/Ti/InP hybrid photocathode for improved solar hydrogen production. Sci. Rep. 6, 29738.

Liang, X., Zhang, H., Zhao, Y., 2011. Peelable GaAs mircotips grown by one-step selective liquid phase epitaxy. In: 2011 International Conference on Multimedia Technology, pp. 2279–2281.

Licht, S., Wang, B., Mukerji, S., Soga, T., Umeno, M., Tributsch, H., 2001. Over 18% solar energy conversion to generation of hydrogen fuel; theory and experiment for efficient solar water splitting. Int. J. Hydrogen Energy 26 (7), 653–659.

Lin, Y., Kapadia, R., Yang, J., Zheng, M., Chen, K., Hettick, M., Yin, X., Battaglia, C., Sharp, I.D., Ager, J.W., Javey, A., 2015. Role of TiO_2 surface passivation on improving the performance of p-InP photocathodes. J. Phys. Chem. C 119 (5), 2308–2313.

Lincot, D., 2005. Electrodeposition of semiconductors. Thin Solid Films 487 (1–2), 40–48.

Liu, C., Sun, J., Tang, J., Yang, P., 2012. Zn-doped p-type gallium phosphide nanowire photocathodes from a surfactant-free solution synthesis. Nano Lett. 12 (10), 5407–5411.

Liu, R., Zheng, Z., Spurgeon, J., Yang, X., 2014. Enhanced photoelectrochemical water-splitting performance of semiconductors by surface passivation layers. Energ. Environ. Sci. 7 (8), 2504.

Lizhong, H., Hongzhi, Z., Zhijun, W., Jie, S., Yu, Z., Xiuping, L., 2004. GaAs pyramidal microtips grown by selective liquid-phase epitaxy. J. Cryst. Growth 271 (1–2), 46–49.

Lobaccaro, P., Raygani, A., Oriani, A., Miani, N., Piotto, A., Kapadia, R., Zheng, M., Yu, Z., Magagnin, L., Chrzan, D.C., Maboudian, R., Javey, A., 2014. Electrodeposition of high-purity indium thin films and its application to indium phosphide solar cells. J. Electrochem. Soc. 161 (14), D794–D800.

Lunt, S.R., Ryba, G.N., Santangelo, P.G., Lewis, N.S., 1991. Chemical studies of the passivation of GaAs surface recombination using sulfides and thiols. J. Appl. Phys. 70 (12), 7449–7467.

Luo, W., Liu, B., Li, Z., Xie, Z., Chen, D., Zou, Z., Zhang, R., 2008. Stable response to visible light of InGaN photoelectrodes. Appl. Phys. Lett. 92 (26), 262110.

Malik, M.A., O'Brien, P., Helliwell, M., 2005. A new synthesis of InAs quantum dots from [tBu2AsInEt2]2. J. Mater. Chem. 15 (14), 1463–1467.

Malik, M.A., Afzaal, M., O'Brien, P., 2010. Precursor chemistry for main group elements in semiconducting materials. Chem. Rev. 110 (7), 4417–4446.

Malizia, M., Seger, B., Chorkendorff, I., Vesborg, P.C.K., 2014. Formation of a p-n heterojunction on GaP photocathodes for H2 production providing an open-circuit voltage of 710 mV. J. Mater. Chem. A 2 (19), 6847–6853.

Mandl, B., Stangl, J., Mårtensson, T., Mikkelsen, A., Eriksson, J., Karlsson, L.S., Bauer, G.U.N., Samuelson, L., Seifert, W., 2006a. Au-free epitaxial growth of InAs nanowires. Nano Lett. 6 (8), 1817–1821.

Manz, A., Birkner, A., Kolbe, M., Fischer, R.A., 2000. Solution synthesis of colloidal gallium nitride at unprecedented low temperatures. Adv. Mater. 12 (8), 569–573.
Markowitz, P.D., Zach, M.P., Gibbons, P.C., Penner, R.M., Buhro, W.E., 2001. Phase separation in AlxGa1-xAs nanowhiskers grown by the solution–liquid–solid mechanism. J. Am. Chem. Soc. 123 (19), 4502–4511.
Mårtensson, T., Svensson, C.P.T., Wacaser, B.a., Larsson, M.W., Seifert, W., Deppert, K., Gustafsson, A., Wallenberg, L.R., Samuelson, L., 2004. Epitaxial III–V nanowires on silicon. Nano Lett. 4 (10), 1987–1990.
Mauk, M.G., Tata, A.N., Feyock, B.W., 2001. Selectively-grown InGaP/GaAs on silicon heterostructures for application to photovoltaic–photoelectrolysis cells. J. Cryst. Growth 225 (2–4), 359–365.
May, M.M., Lewerenz, H.-J., Lackner, D., Dimroth, F., Hannappel, T., 2015. Efficient direct solar-to-hydrogen conversion by in situ interface transformation of a tandem structure. Nat. Commun. 6, 8286.
McCray, W.P., 2007. MBE deserves a place in the history books. Nat. Nanotechnol. 2 (5), 259–261.
McGhee, L., McMeekin, S.G., Nicol, I., Robertson, M.I., Winfield, J.M., 1994. Chemo-mechanical polishing of gallium arsenide and cadmium telluride to subnanometre surface finish. Evaluation of the action and effectiveness of hydrogen peroxide, sodium hypochlorite and dibromine as reagents. J. Mater. Chem. 4 (1), 29–34.
Memming, R., 1978. Role of energy-levels in semiconductor-electrolyte solar-cells. J. Electrochem. Soc. 125 (1), 117–123.
Mićić, O.I., Ahrenkiel, S.P., Bertram, D., Nozik, A.J., 1999. Synthesis, structure, and optical properties of colloidal GaN quantum dots. Appl. Phys. Lett. 75 (4), 478–480.
Miller, B., Heller, A., Menezes, S., Lewerenz, H.J., 1980. Surface modification in semiconductor liquid-junction cells. Faraday Discuss. Chem. Soc. 70, 223–232.
Modestino, M.A., Walczak, K.A., Berger, A., Evans, C.M., Haussener, S., Koval, C., Newman, J.S., Ager, J.W., Segalman, R.A., 2014. Robust production of purified H2 in a stable, self-regulating, and continuously operating solar fuel generator. Energy Environ. Sci. 7 (1), 297–301.
Mönch, W., 2001. Semiconductor Surfaces and Interfaces, third ed. Springer-Verlag Berlin Heidelberg, 200B, New York, 19 and 63.
Morral, A.F.i., Colombo, C., Abstreiter, G., Arbiol, J., Morante, J.R., 2008. Nucleation mechanism of gallium-assisted molecular beam epitaxy growth of gallium arsenide nanowires. Appl. Phys. Lett. 92 (6), 063112.
Murray, C.B., Kagan, C.R., Bawendi, M.G., 2000. Synthesis and characterization of monodisperse nanocrystals and close-packed nanocrystal assemblies. Annu. Rev. Mater. Sci. 30 (1), 545–610.
Nagahara, M., Miyoshi, S., Yaguchi, H., Onabe, K., Shiraki, Y., Ito, R., 1994. Selective growth of cubic gan in small areas on patterned GaAs(100) substrates by metalorganic vapor-phase epitaxy. Jpn. J. Appl. Phys. 33 (Pt. 1 No. 1b), 694–697.
Nakajima, K., 1985. The liquid-phase epitaxial growth of InGaAsP. Chapter 1, In: Tsang, W.T. (Ed.), Semiconductors and Semimetals. vol. 22 (Pt.A). Elsevier, Orlando, FL, pp. 1–94.
Nelson, R.J., Williams, J.S., Leamy, H.J., Miller, B., Casey, H.C., Parkinson, B.A., Heller, A., 1980. Reduction of GaAs surface recombination velocity by chemical treatment. Appl. Phys. Lett. 36 (1), 76–79.
Nozik, A.J., 1976. p-n Photoelectrolysis cells. Appl. Phys. Lett. 29 (3), 150–153.
Nozik, A.J., 1977. Photochemical diodes. Appl. Phys. Lett. 30 (11), 567–569.
Nozik, A.J., Memming, R., 1996. Physical chemistry of semiconductor-liquid interfaces. J. Phys. Chem. 100 (31), 13061–13078.
Ohashi, K., McCann, J., Bockris, J.O.M., 1977. Stable photoelectrochemical cells for the splitting of water. Nature 266 (5603), 610–611.

Ohno, T., 1991. Sulfur passivation of GaAs surfaces. Phys. Rev. B 44 (12), 6306–6311.
Okada, S., Miyake, H., Hiramatsu, K., Enatsu, Y., Nagao, S., 2014. Selective-area growth of GaN on non- and semi-polar bulk GaN substrates. Jpn. J. Appl. Phys. 53 (5), 05FL04.
Ow, K.N., Wang, X.W., 1996. First-principles pseudopotential calculations of passivated GaAs(001) surfaces. Phys. Rev. B 54 (24), 17661–17666.
Paek, J.H., Nishiwaki, T., Yamaguchi, M., Sawaki, N., 2009. Catalyst free MBE-VLS growth of GaAs nanowires on (111)Si substrate. Phys. Status Solidi C 6 (6), 1436–1440.
Parkinson, B.A., Weaver, P.F., 1984. Photoelectrochemical pumping of enzymatic CO_2 reduction. Nature 309 (5964), 148–149.
Parkinson, B.A., Heller, A., Miller, B., 1978. Enhanced photoelectrochemical solar-energy conversion by gallium arsenide surface modification. Appl. Phys. Lett. 33 (6), 521–523.
Peng, X., Wickham, J., Alivisatos, A.P., 1998. Kinetics of II-VI and III-V colloidal semiconductor nanocrystal growth: "focusing" of size distributions. J. Am. Chem. Soc. 120 (21), 5343–5344.
Petroff, P.M., DenBaars, S.P., 1994. MBE and MOCVD growth and properties of self-assembling quantum dot arrays in III-V semiconductor structures. Superlattice. Microst. 15 (1), 15.
Pierson, H.O., 1999. Metallo-organic CVD (MOCVD). In: Handbook of Chemical Vapor Deposition (CVD), second ed. William Andrew Publishing, Norwich, NY, pp. 84–107. Chapter 4.
Pinaud, B.A., Benck, J.D., Seitz, L.C., Forman, A.J., Chen, Z., Deutsch, T.G., James, B.D., Baum, K.N., Baum, G.N., Ardo, S., Wang, H., Miller, E., Jaramillo, T.F., 2013. Technical and economic feasibility of centralized facilities for solar hydrogen production via photocatalysis and photoelectrochemistry. Energ. Environ. Sci. 6 (7), 1983–2002.
Plissard, S., Dick, K.A., Wallart, X., Caroff, P., 2010. Gold-free GaAs/GaAsSb heterostructure nanowires grown on silicon. Appl. Phys. Lett. 96 (12), 121901.
Plissard, S., Larrieu, G., Wallart, X., Caroff, P., 2011. High yield of self-catalyzed GaAs nanowire arrays grown on silicon via gallium droplet positioning. Nanotechnology 22 (27), 275602.
Pomykal, K.E., Lewis, N.S., 1997. Measurement of interfacial charge-transfer rate constants at n-type InP/CH 3 OH junctions. J. Phys. Chem. B 101 (14), 2476–2484.
Pomykal, K.E., Fajardo, A.M., Lewis, N.S., 1996. Theoretical and experimental upper bounds on interfacial charge-transfer rate constants between semiconducting solids and outer-sphere redox couples. J. Phys. Chem. 100 (9), 3652–3664.
Qiu, J., Zeng, G., Ha, M.-A., Ge, M., Lin, Y., Hettick, M., Hou, B., Alexandrova, A.N., Javey, A., Cronin, S.B., 2015. Artificial photosynthesis on TiO_2-passivated InP nanopillars. Nano Lett. 15 (9), 6177–6181.
Qu, L., Peng, Z.A., Peng, X., 2001. Alternative routes toward high quality CdSe nanocrystals. Nano Lett. 1 (6), 333–337.
Rothschild, A., Komem, Y., Ashkenasy, N., 2002. Quantitative evaluation of chemisorption processes on semiconductors. J. Appl. Phys. 92 (12), 7090–7097.
Sardar, K., Dan, M., Schwenzer, B., Rao, C.N.R., 2005. A simple single-source precursor route to the nanostructures of AlN, GaN and InN. J. Mater. Chem. 15 (22), 2175–2177.
Sato, S., Arai, T., Morikawa, T., Uemura, K., Suzuki, T.M., Tanaka, H., Kajino, T., 2011. Selective CO_2 conversion to formate conjugated with H_2O oxidation utilizing semiconductor/complex hybrid photocatalysts. J. Am. Chem. Soc. 133 (39), 15240–15243.
Saul, R.H., Lorimor, O.G., 1974. Liquid phase epitaxy processes for GaP led's. J. Cryst. Growth 27, 183–192.
Sburlan, S., Daniel Dapkus, P., Nakano, A., 2012. Critical dimensions of highly lattice mismatched semiconductor nanowires grown in strain-releasing configurations. Appl. Phys. Lett. 100 (16), 163108.

Schefold, J., Vetter, M., 1994. Solar-energy conversion at the p-InP/vanadium3+/2+ semiconductor electrolyte contact—a study based on differential capacitance and current–voltage data. J. Electrochem. Soc. 141 (8), 2040–2048.

Schiaber, Z.D., Calabrese, G., Kong, X., Trampert, A., Jenichen, B., da Silva, J.H.D., Geelhaar, L., Brandt, O., Fernandez-Garrido, S., 2017. Polarity-induced selective area epitaxy of GaN nanowires. Nano Lett. 17 (1), 63–70.

Schwalbach, E.J., Davis, S.H., Voorhees, P.W., Wheeler, D., Warren, J.a., 2011. Liquid droplet dynamics and complex morphologies in vapor–liquid–solid nanowire growth. J. Mater. Res. 26, 2186–2198.

Scofield, A.C., Shapiro, J.N., Lin, A., Williams, A.D., Wong, P.-S., Liang, B.L., Huffaker, D.L., 2011. Bottom-up photonic crystal cavities formed by patterned III–V nanopillars. Nano Lett. 11 (6), 2242–2246.

Segar, P.R., Koval, C.A., Koel, B.E., Gebhard, S.C., 1990. A comprehensive investigation of HCl- and Br_2/NH_3(aq)-etched p-InP interfaces. J. Electrochem. Soc. 137 (2), 544–552.

Seger, B., Castelli, I.E., Vesborg, P.C.K., Jacobsen, K.W., Hansen, O., Chorkendorff, I., 2014. 2-Photon tandem device for water splitting: comparing photocathode first versus photoanode first designs. Energ. Environ. Sci. 7 (8), 2397–2413.

Seifert, W., Borgström, M., Deppert, K., Dick, K.A., Johansson, J., Larsson, M.W., Mårtensson, T., Sköld, N., Svensson, C.P.T., Wacaser, B.A., Reine Wallenberg, L., Samuelson, L., 2004. Growth of one-dimensional nanostructures in MOVPE. J. Cryst. Growth 272 (1–4), 211–220.

Seker, F., Meeker, K., Kuech, T.F., Ellis, A.B., 2000. Surface chemistry of prototypical bulk II–VI and III–V semiconductors and implications for chemical sensing. Chem. Rev. 100 (7), 2505–2536.

Seo, J., Hasegawa, S., Asahi, H., 2009. Selective area growth of GaN nanorods on patterned W/SiO2/Si substrates by RF-MBE. J. Cryst. Growth 311 (19), 4437–4441.

Shaner, M.R., Hu, S., Sun, K., Lewis, N.S., 2015. Stabilization of Si microwire arrays for solar-driven H_2O oxidation to O_2(g) in 1.0 M KOH(aq) using conformal coatings of amorphous TiO_2. Energy Environ. Sci. 8, 203–207.

Shi, W.S., Zheng, Y.F., Wang, N., Lee, C.S., Lee, S.T., 2001. Synthesis and microstructure of gallium phosphide nanowires. J. Vac. Sci. Technol. B 19 (4), 1115–1118.

Simpson, W.C., Yarmoff, J.A., 1996. Fundamental studies of halogen reactions with III-V semiconductor surfaces. Annu. Rev. Phys. Chem. 47 (1), 527–554.

Sköld, N., Wagner, J.B., Karlsson, G., Hernán, T., Seifert, W., Pistol, M.-E., Samuelson, L., 2006. Phase segregation in AlInP shells on GaAs nanowires. Nano Lett. 6 (12), 2743–2747.

Standing, A., Assali, S., Gao, L., Verheijen, M.A., van Dam, D., Cui, Y., Notten, P.H.L., Haverkort, J.E.M., Bakkers, E.P.A.M., 2015. Efficient water reduction with gallium phosphide nanowires. Nat. Commun. 6, 7824.

Stirland, D.J., Straughan, B.W., 1976. A review of etching and defect characterisation of gallium arsenide substrate material. Thin Solid Films 31 (1), 139–170.

Strubbe, K., Gomes, W.P., 1993. Bromine-methanol as an etchant for semiconductors: a fundamental study on GaP: I. Etching behavior of n- and p-type. J. Electrochem. Soc. 140 (11), 3294–3300.

Sun, J., Liu, C., Yang, P., 2011. Surfactant-free, large-scale, solution–liquid–solid growth of gallium phosphide nanowires and their use for visible-light-driven hydrogen production from water reduction. J. Am. Chem. Soc. 133 (48), 19306–19309.

Sun, K., Kuang, Y., Verlage, E., Brunschwig, B.S., Tu, C.W., Lewis, N.S., 2015. Sputtered NiOx films for stabilization of p+n-InP photoanodes for solar-driven water oxidation. Adv. Energy Mater. 5 (11), 1402276.

Sundaram, S., El Gmili, Y., Puybaret, R., Li, X., Bonanno, P.L., Pantzas, K., Patriarche, G., Voss, P.L., Salvestrini, J.P., Ougazzaden, A., 2015. Nanoselective area growth and characterization of dislocation-free InGaN nanopyramids on AlN buffered Si(111) templates. Appl. Phys. Lett. 107 (11), 113105.

Szklarczyk, M., Bockris, J.O.M., 1984. Photoelectrochemical evolution of hydrogen on p-indium phosphide. J. Phys. Chem. 88 (22), 5241–5245.

Talapin, D.V., Rogach, A.L., Shevchenko, E.V., Kornowski, A., Haase, M., Weller, H., 2002. Dynamic distribution of growth rates within the ensembles of colloidal II–VI and III–V semiconductor nanocrystals as a factor governing their photoluminescence efficiency. J. Am. Chem. Soc. 124 (20), 5782–5790.

Tan, M.X., Newcomb, C., Kumar, A., Lunt, S.R., Sailor, M.J., Tufts, B.J., Lewis, N.S., 1991. Chemical modification of n-gallium arsenide photoanodes with group VIIIB metal ions: stability in contact with 1.0 M potassium hydroxide(aq)-0.10 M dipotassium selenide(aq) solutions and I-V properties in contact with 1.0 M potassium hydroxide(aq)-0.3 M dipotassium telluride(aq) electrolytes. J. Phys. Chem. 95 (24), 10133–10142.

Tan, M.X., Laibinis, P.E., Nguyen, S.T., Kesselman, J.M., Stanton, C.E., Lewis, N.S., 1994. Principles and applications of semiconductor photoelectrochemistry. In: Progress in Inorganic Chemistry, vol. 41 (K. D. Karlin, Ed.), John Wiley & Sons, Inc., Hoboken, NJ, 21–144. http://dx.doi.org/10.1002/9780470166420.ch2.

Tanabe, K., Watanabe, K., Arakawa, Y., 2012. III-V/Si hybrid photonic devices by direct fusion bonding. Sci. Rep. 2, 349.

Thelander, C., Caroff, P., Plissard, S., Dey, A.W., Dick, K.A., 2011. Effects of crystal phase mixing on the electrical properties of InAs nanowires. Nano Lett. 11 (6), 2424–2429.

Tiwari, S., Frank, D.J., 1992. Empirical fit to band discontinuities and barrier heights in III–V alloy systems. Appl. Phys. Lett. 60 (5), 630–632.

Tomioka, K., Motohisa, J., Hara, S., Fukui, T., 2008. Control of InAs nanowire growth directions on Si. Nano Lett. 8 (10), 3475–3480.

Trentler, T.J., Hickman, K.M., Goel, S.C., Viano, A.M., Gibbons, P.C., Buhro, W.E., 1995. Solution-liquid–solid growth of crystalline III-V semiconductors: an analogy to vapor–liquid-solid growth. Science 270 (5243), 1791–1794.

Trentler, T.J., Goel, S.C., Hickman, K.M., Viano, A.M., Chiang, M.Y., Beatty, A.M., Gibbons, P.C., Buhro, W.E., 1997. Solution–liquid–solid growth of indium phosphide fibers from organometallic precursors: elucidation of molecular and nonmolecular components of the pathway. J. Am. Chem. Soc. 119 (9), 2172–2181.

Tufts, B.J., Abrahams, I.L., Santangelo, P.G., Ryba, G.N., Casagrande, L.G., Lewis, N.S., 1987. Chemical modification of N-GaAs electrodes with OS-3+ gives a 15-percent efficient solar-cell. Nature 326 (6116), 861–863.

Verlage, E., Hu, S., Liu, R., Jones, R.J.R., Sun, K., Xiang, C., Lewis, N.S., Atwater, H.A., 2015a. A monolithically integrated, intrinsically safe, 10% efficient, solar-driven water-splitting system based on active, stable earth-abundant electrocatalysts in conjunction with tandem III–V light absorbers protected by amorphous TiO_2 films. Energy Environ. Sci. 8 (11), 3166–3172.

Verney, E., Martin, J.R., Jaffrezic-Renault, N., Clechet, P., 1986. Photoelectrochemical and chemical corrosion inhibition of n-type GaAs in highly concentrated iodide solutions. J. Electroanal. Chem. Interfacial Electrochem. 209 (1), 219–222.

Vurgaftman, I., Meyer, J.R., 2003. Band parameters for nitrogen-containing semiconductors. J. Appl. Phys. 94 (6), 3675.

Wacaser, B.a., Deppert, K., Karlsson, L.S., Samuelson, L., Seifert, W., Reuter, M.C., Khayyat, M.M., Wen, C.-y., Haight, R., Guha, S., Ross, F.M., 2006. Growth and characterization of defect free GaAs nanowires. J. Cryst. Growth 287 (2), 504–508.

Wade, T.L., Vaidyanathan, R., Happek, U., Stickney, J.L., 2001. Electrochemical formation of a III–V compound semiconductor superlattice: InAs/InSb. J. Electroanal. Chem. 500 (1–2), 322–332.

Wagner, R.S., Ellis, W.C., 1964. Vapor-liquid-solid mechanism of single crystal growth. Appl. Phys. Lett. 4 (5), 89–90.

Wallentin, J., Ek, M., Wallenberg, L.R., Samuelson, L., Deppert, K., Borgström, M.T., 2010. Changes in contact angle of seed particle correlated with increased zincblende formation in doped InP nanowires. Nano Lett. 10, 4807–4812.

Wang, D., Pierre, A., Kibria, M.G., Cui, K., Han, X., Bevan, K.H., Guo, H., Paradis, S., Hakima, A.-R., Mi, Z., 2011. Wafer-level photocatalytic water splitting on GaN nanowire arrays grown by molecular beam epitaxy. Nano Lett. 11 (6), 2353–2357.

Weir, M.G., George, K.E., Deutsch, T.G., Welch, A., Wilks, R.G., Hanks, D.C., Blum, M., Yang, W., Bar, M., Weinhardt, L., Turner, J.A., Heske, C., 2012. Impact of nitrogen treatment on the electronic and chemical structure of GaInP2 thin-film surfaces. In: Meeting Abstracts. MA2012-02 (14), 1797.

Woo, R.L., Xiao, R., Kobayashi, Y., Gao, L., Goel, N., Hudait, M.K., Mallouk, T.E., Hicks, R.F., 2008. Effect of twinning on the photoluminescence and photoelectrochemical properties of indium phosphide nanowires grown on silicon (111). Nano Lett. 8 (12), 4664–4669.

Wrighton, M.S., Ellis, A.B., Kaiser, S.W., 1976. Photoelectrochemical cells—conversion of intense optical energy, In: Proceedings of the International Symposium on Solar Energy, Washington, DC, May 5–7, 1976 (A76-47051 24-44). Electrochemical Society, Inc., Princeton, NJ, pp. 66–91. NASA-supported research.

Wu, Y., Yan, H., Huang, M., Messer, B., Song, J.H., Yang, P., 2002. Inorganic semiconductor nanowires: rational growth, assembly, and novel properties. Chem. A Eur. J. 8 (6), 1260–1268.

Wu, J., Li, Y., Kubota, J., Domen, K., Aagesen, M., Ward, T., Sanchez, A., Beanland, R., Zhang, Y., Tang, M., Hatch, S., Seeds, A., Liu, H., 2014. Wafer-scale fabrication of self-catalyzed 1.7 eV GaAsP core–shell nanowire photocathode on silicon substrates. Nano Lett. 14 (4), 2013–2018.

Wu, C.H., Lee, P.Y., Chen, K.Y., Cheng, K.Y., 2016a. Selective area growth of high-density GaN nanowire arrays on Si(111). In: 2016 IEEE Photonics Society Summer Topical Meeting Series (SUM), pp. 169–170.

Wu, C.H., Lee, P.Y., Chen, K.Y., Tseng, Y.T., Wang, Y.L., Cheng, K.Y., 2016b. Selective area growth of high-density GaN nanowire arrays on Si(111) using thin AlN seeding layers. J. Cryst. Growth 454, 71–81.

Xia, Y., Yang, P., Sun, Y., Wu, Y., Mayers, B., Gates, B., Yin, Y., Kim, F., Yan, H., 2003. One-dimensional nanostructures: synthesis, characterization, and applications. Adv. Mater. 15 (5), 353–389.

Yamane, S., Kato, N., Kojima, S., Imanishi, A., Ogawa, S., Yoshida, N., Nonomura, S., Nakato, Y., 2009. Efficient solar water splitting with a composite "n-Si/p-CuI/n-i-p a-Si/n-p GaP/RuO_2" semiconductor electrode. J. Phys. Chem. C 113 (32), 14575–14581.

Yan, R., Gargas, D., Yang, P., 2009. Nanowire photonics. Nat. Photonics 3 (10), 569–576.

Yang, F., Nielander, A.C., Grimm, R.L., Lewis, N.S., 2016. Photoelectrochemical behavior of n-type GaAs(100) electrodes coated by a single layer of graphene. J. Phys. Chem. C 120 (13), 6989–6995.

Yao, M., Huang, N., Cong, S., Chi, C.-Y., Seyedi, M.A., Lin, Y.-T., Cao, Y., Povinelli, M.L., Dapkus, P.D., Zhou, C., 2014. GaAs nanowire array solar cells with axial p–i–n junctions. Nano Lett. 14 (6), 3293–3303.

Yazawa, M., Koguchi, M., Muto, A., Hiruma, K., 1993. Semiconductor nanowhiskers. Adv. Mater. 5 (7–8), 577–580.

Yoffe, A.D., 2001. Semiconductor quantum dots and related systems: electronic, optical, luminescence and related properties of low dimensional systems. Adv. Phys. 50 (1), 1–208.

Zhang, J., Peng, X.S., Wang, X.F., Wang, Y.W., Zhang, L.D., 2001. Micro-Raman investigation of GaN nanowires prepared by direct reaction Ga with NH3. Chem. Phys. Lett. 345 (5–6), 372–376.

Zhong, M., Hisatomi, T., Kuang, Y., Zhao, J., Liu, M., Iwase, A., Jia, Q., Nishiyama, H., Minegishi, T., Nakabayashi, M., Shibata, N., Niishiro, R., Katayama, C., Shibano, H., Katayama, M., Kudo, A., Yamada, T., Domen, K., 2015. Surface modification of CoOx loaded BiVO4 photoanodes with ultrathin p-type NiO layers for improved solar water oxidation. J. Am. Chem. Soc. 137 (15), 5053–5060.

Zhong, M., Hisatomi, T., Minegishi, T., Nishiyama, H., Katayama, M., Yamada, T., Domen, K., 2016. Bulky crystalline BiVO4 thin films for efficient solar water splitting. J. Mater. Chem. A 4 (25), 9858–9864.

Zhou, X., Liu, R., Sun, K., Chen, Y., Verlage, E., Francis, S.A., Lewis, N.S., Xiang, C., 2016. Solar-driven reduction of 1 atm of CO_2 to formate at 10% energy-conversion efficiency by use of a TiO_2-protected III–V tandem photoanode in conjunction with a bipolar membrane and a Pd/C cathode. ACS Energy Lett. 1 (4), 764–770.

CHAPTER FOUR

III-Nitride Semiconductor Photoelectrodes

Katsushi Fujii[1]
Institute of Environmental Science and Technology, The University of Kitakyushu, Kitakyushu, Fukuoka, Japan
RIKEN Center for Advanced Photonics, Wako, Saitama, Japan
[1]Corresponding author e-mail address: katsushi.fujii@riken.jp

Contents

1. GENERAL PROPERTIES OF III-NITRIDE SEMICONDUCTORS

The nitride (N) compounds of aluminum (Al), gallium (Ga), and indium (In) and their mixed phases are usually called as III-nitride semiconductors. III-Nitride semiconductors are widely used for ultraviolet (UV), blue, and green light-emitting diodes and UV and blue laser diodes. The semiconductors are also expected to be high power transistors and field-effect transistors. These facts show that, from the semiconductor point of view,

Semiconductors and Semimetals, Volume 97
ISSN 0080-8784
http://dx.doi.org/10.1016/bs.semsem.2017.03.003

III-nitride semiconductors can be obtained with high-quality (relatively low dislocation density, low impurity incorporation, and high controllability of doping) single-phase crystals that is enough for fabricating real devices.

III-Nitrides do not exist in nature and are usually grown by metal-organic vapor phase epitaxy (MOVPE), molecular beam epitaxy (MBE), hydride vapor phase epitaxy (HVPE), and high-pressure bulk growth method like ammonothermal growth. The semiconductors are usually used as single crystal and heteroepitaxial thin film on a different substrate like silicon (Si) or sapphire (Al_2O_3) single crystal. The hetero-substrate is commonly used since the bulk single crystal of III-nitride is only possible with GaN, which is not so easy to grow as a substrate size. The equilibrium phase of III-nitrides is wurtzite, but cubic phase is also found commonly. The cubic phase is easily grown by MBE with Ga-rich condition. However, 100% cubic phase is difficult; that is, some wurtzite stacking faults are usually mixed. Commonly used doping element for n-type is silicon (Si) and that for p-type is magnesium (Mg). The p-type materials are relatively difficult to grow because the amount of dopant is close to its solubility limit due to its deep acceptor level (over 100 meV). In addition, especially for the III-nitride semiconductors grown by the MOVPE method, which is the most common growth technique for III-nitride semiconductors, low-temperature growth is needed compared with its undoped or n-type materials to increase the Mg doping. Annealing process after epitaxial growth is also required to electrically activate by removing hydrogen (H) from Mg–H bonding. III-Nitride semiconductors containing In element require much lower growth temperature than Mg-doped p-type GaN due to the difficulty in In incorporation. Thus, the quality of crystallinity of these crystals is not so high. The details of growth and characteristics for III-nitride semiconductors are described in Yao and Hong (2009) and Nakamura and Chichibu (2000).

The band gap of III-nitride semiconductors is a direct gap and has a wide range from IR (InN: 0.65 eV) to UV (GaN: 3.4 eV, AlN: 6.2 eV). The lattice parameters and band gaps are summarized in Table 1. This is a suitable characteristic for photoelectrodes because the visible range photoabsorption can be covered by III-nitride semiconductors. The dependence of band gap on composition of group III materials has the bowing term for the lattice parameter (Marques et al., 2003); thus, the relationship between the lattice constant and the band gap is not linear as shown in Fig. 1.

The dry plasma (reactive ion) etching process is required for GaN device fabrication because GaN is chemically stable even in strong acidic and basic aqueous solutions. The AlN is also chemically stable, but InN and its

Table 1 Basic Semiconductor and Electrochemical Properties of III-Nitride Materials

			Dopant	
	Band Gap (eV)	**Structure (nm)**	**p-Type**	**n-Type**
AlN	6.2	*a*: 0.3111	Mg	Si
		c: 0.4978		
GaN	3.4	*a*: 0.3189	Mg	Si
		c: 0.5185		
InN	0.65	*a*: 0.3544	Mg	Si
		c: 0.5718		

The band gaps are direct and the structures are wurtzite.
The structure of zinc blende is also observed.

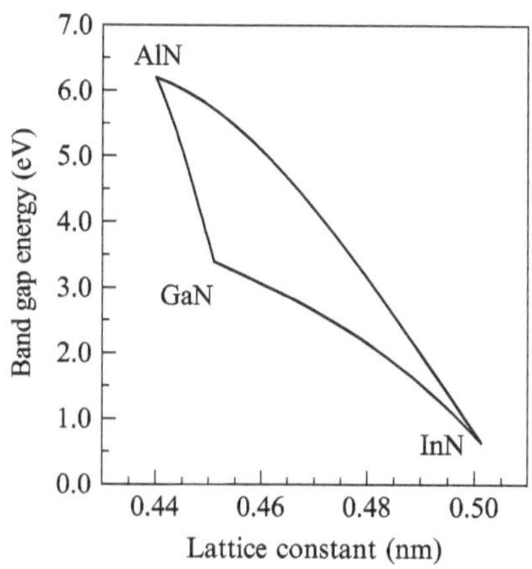

Fig. 1 The band gaps of III-nitride semiconductors as functions of lattice parameters using the band gap energy bowing parameters (Marques et al., 2003). The band gaps and lattice parameters were used the value listed in Table 1. The lattice parameter in this graph is $c\sqrt{2}$ in order to compare with other cubic lattice parameters easily.

alloys are not chemically stable probably due to their low crystal quality. The chemical stability of GaN was believed to be good for a water-oxidizing photoanode even when the water-oxidized reaction was performed under highly oxidized condition. The photoanode stability of GaN is, however, not good probably because oxide is more stable in water-oxidized condition than nitride. This instability is expected from the fact that the valence of cation in IIInitrides is close to zero, whereas those of oxides are +3. The cation valence probably tends to change to more positive ones in strong anodic condition.

For photoelectrochemical usage of III-nitride semiconductors, the contact electrode for III-nitride semiconductors is required at the photo-irradiated surface when sapphire is used as the III-nitride substrate because sapphire is an insulator. The back-contact electrode is available when Si or GaN substrate is used due to its electrical conductivity. Even when Si substrate is used, since the heterojunction between III-nitride semiconductor and Si exists, an additional voltage drop due to the heterojunction must be considered.

2. GENERAL PROPERTIES FOR PHOTOELECTROCHEMICAL REACTIONS OF SEMICONDUCTOR

There are several important parameters to understand in discussing the photoelectrochemical activity of the semiconductor. In this section, basic and important properties of electrochemistry are briefly introduced. The details of the fundamental electrochemical properties are found elsewhere (Bard et al., 1980; Fujii, 2016a,b; Memming, 2015).

Firstly, it is necessary to note that electrochemical reactions are performed by one pair. That is, the oxidation and reduction reactions occur at different electrodes simultaneously. Since water-splitting reaction is the most common for photoelectrochemical reaction, water oxidation and reduction in acidic electrolyte are selected here as an example:

$$2H_2O(\ell) = O_2(g) + 4H^+(\text{in}\ell) + 4e^-(\text{ins}), \quad (1)$$

$$2H^+(\text{in}\ell) + 2e^-(\text{ins}) = H_2(g), \quad (2)$$

where ℓ is the liquid phase, g is the gas phase, inℓ exists in liquid (electrolyte), and ins exists in solid (electrode). The energy difference of these two reactions is calculated from the standard Gibbs energy of the formation of H_2 and O_2 and is +237 kJ/mol, i.e., +1.23 V.

In the electrochemical reaction, the electrode potential of hydrogen formation (Eq. (2)) at pH 0 and under all of the activities unity is set as a reference with 0.0 V and is called standard hydrogen electrode (SHE) or normal hydrogen electrode (NHE). That is, $\phi(H_2O/H_2) = 0.00$ (V vs SHE or NHE) at pH 0. From the semiconductor point of view, the potential locates almost at −4.5 eV below from the vacuum level of electron energy ($E(H_2O/H_2) = -4.50$ [eV]) (Memming, 2015). In addition, the positive and negative directions of the electrode potential and those of electron energy are opposite. From this definition, the electrode potential of oxygen formation of water $\phi(O_2/H_2O)$ is +1.23 V vs SHE (the electrode energy: $E(O_2/H_2O)$ is −5.73 eV). For the water splitting, the water-oxidized and -reduced reactions in basic aqueous solution are changed as the dominant ion is changed from H^+ to OH^-. That is, instead of Eqs. (1) and (2),

$$4OH^-(\text{in}\ell) = O_2(g) + 2H_2O(\ell) + 4e^-(\text{ins}), \tag{3}$$

$$2H_2O(\ell) + 2e^-(\text{ins}) = H_2(g) + 2OH^-(\text{in}\ell). \tag{4}$$

When the equilibrium of the electrode and the electrolyte is considered, the Fermi level of an electrode E_F and the chemical potential of an electrolyte μ_{el} are the same energy. That is,

$$E_F = \mu_{el}. \tag{5}$$

The Fermi level in a solid is defined as a hypothetical energy level of an electron, such that at thermodynamic equilibrium, this energy level would have a 50% probability of being occupied at any given time. The definition of the chemical potential of an electrolyte is a bit complicated. Since the Gibbs energy of an electrolyte is expressed as the sum of the all elements of "the amounts of substance (mole) times the chemical potential" for each element, the chemical potential of the electrolyte can be regarded as the Gibbs energy of the electrolyte divided by the sum of the amounts of substance.

Expanding this chemical potential definition to electrochemistry, the electrochemical potential of the ith element under standard condition (25°C, 101.3 kPa, shown with the "^{0}" mark meaning with "standard") in electrochemistry $\widetilde{\mu}_i^0$ can be expressed as

$$\widetilde{\mu}_i^0 = \mu_i^{0\phi} + RT \ln a_i + z_i F \phi_i, \tag{6}$$

where $\mu_i^{0\phi}$ is the standard chemical potential of the ith element calculated from the Gibbs energy, R is the gas constant, T is the absolute temperature, a_i is the activity of the ith element, z_i is the valence of the ith ion, F is the Faraday constant, and ϕ_i is the potential for the ith element. The electrochemical potential for the electron is

$$\widetilde{\mu}_e^0 = \mu_e^{0\phi} + RT \ln a_e - F\phi_e. \tag{7}$$

For equilibrium condition, the potential is the same for all elements and is expressed as ϕ^0. From the equilibrium of oxide and reductant with the nth electron,

$$pP + qO + \cdots + ne = xX + yY + \cdots, \tag{8}$$

$$\begin{aligned}\phi^0 &= -\frac{n\left(\mu_e^{0\phi} + RT\ln a_e\right)}{nF} \\ &= \phi^0(\mathrm{Ox/Red}) + \frac{RT}{nF}\ln\frac{a_p^p a_q^q\cdots(\mathrm{oxide})}{a_x^x a_y^y\cdots(\mathrm{reductant})}.\end{aligned} \tag{9}$$

This relationship is known as the Nernst relationship. For water oxidation (oxygen evolution) and reduction (hydrogen evolution), the equilibriums are

$$\phi^0(\mathrm{O_2/H_2O}) = +1.23 - 0.059\mathrm{pH}\quad [\mathrm{V\ vs\ SHE}] \tag{10}$$

$$\phi^0(\mathrm{H_2O/H_2}) = +0.00 - 0.059\mathrm{pH}\quad [\mathrm{V\ vs\ SHE}] \tag{11}$$

where $\phi^0(O_2/H_2O)$ is the equilibrium potential for water oxidation at the pH, and $\phi^0(H_2O/H_2)$ is the equilibrium potential for water reduction at the pH.

The other important phenomenon at the electrode and electrolyte interface is electrochemical kinetics relating the rate of an electrochemical reaction to the overpotential. The Tafel equation was proposed experimentally

$$\eta = a + b\ln|j|, \tag{12}$$

where η is the overpotential, j is the current density, and a and b are constants. Under the conditions of overpotential being dominant (difficult to reach equilibrium state) and the bias region with reaction activation being dominant (not diffusion limited), the Tafel equation can be comparable to the Butler–Volmer equation, which is the theoretical electrochemical kinetics based on the transition state theory. For the case, the bias slope of the $\ln|j|$ has the relationship:

$$b = \frac{k_B T}{\alpha z q}\ln 10 = \frac{RT}{\alpha z F}\ln 10, \tag{13}$$

where k_B is the Boltzmann constant, T is the absolute temperature, z is the number of electron related to the reaction, q is the elementary charge, R is the gas constant, and F is the Faraday constant. The α is called as the charge transfer coefficient, which is in between 0 and 1. From this relationship, the number of electrons related to the reaction can be obtained experimentally.

Electrochemical reaction kinetics are also explained from the Marcus theory under nonadiabatic condition and the Gerischer model under adiabatic condition (Memming, 2015). Adiabatic condition is usually established under one-electron outer sphere (no chemical interaction between the electrode and the electrolyte with monolayer of the Helmholtz layer in between) reaction, whereas it is established at least one-electron reaction with the shape changes of electron donor and acceptor molecules (chemical interaction existing between the electrode and the electrolyte) for nonadiabatic condition. Although both theoretical models are very important to analyze electrochemical reactions, the models are not enough to explain real electrochemical reactions especially for multielectron reactions with intermediate active states.

Second, the relationship between the conduction and valence band edge energies and reduced and oxidized electron energies of targeted electrochemical reactions are discussed. The electrochemical reaction is performed by electron flow even using the metal electrode. The electron flow must be from high to low energy in the electron energy diagram. In the case of using the semiconductor electrode, the electron can move freely at the conduction band edge, and the hole can move freely at the valence band edge. Thus, the minimum condition of a targeted electrochemical reaction must be satisfied that the conduction band edge electron energy of the semiconductor electrode must be above the targeted reduction reaction redox energy in case of reduction, and the valence band edge electron energy of semiconductor electrode must be below the targeted oxidation reaction redox energy in case of oxidation.

Third, the conduction and valence band edge electron energies are usually fixed at the interface with the electrolyte when the electrolyte is the same. This is explained by the semiconductor surface adsorbed molecules, and ions are fixed when the kinds of semiconductor and electrolyte are decided. For this situation, the electric field is formed by the absorbed ions in the electrolyte and the ionized trap levels in the depletion layer of the semiconductor. The band edge electron energies are moved when, for example, the molecular adsorption structure on the electrode changes due to a change in the pH of the electrolyte.

The band edge energy of the semiconductor can be calculated from the experimentally obtained the Mott–Schottky plot, which uses the depletion layer capacitance (Nozik and Memming, 1996). When the free electric particle (ions, electrons, and holes) concentration of electrolyte is higher enough than that of semiconductor, the capacitance of the semiconductor is sufficient smaller than that of the electrolyte. For this case, the capacity of the depletion layer W below the semiconductor surface for n-type semiconductor C_{SC}, which is the same as the capacitance in the semiconductor at the Schottky barrier of the semiconductor–metal interface, is expressed

$$\frac{1}{C_{SC}^2} = \frac{2}{\varepsilon\varepsilon_0 q N_D} |\phi_{SC}|, \tag{14}$$

and the depletion layer thickness W is

$$W = \sqrt{\frac{2\varepsilon\varepsilon_0}{qN_D} |\phi_{SC}|}, \tag{15}$$

where ε is the relative dielectric constant, ε_0 is the permittivity of free space, q is the elemental charge, N_D is the activated donor concentration, and ϕ_{SC} is the potential difference as a bending of the semiconductor surface. For p-type semiconductor, N_D is replaced by the activated acceptor concentration N_A. The ϕ_{SC} is

$$U_E = \phi_{SC} + \phi_H + C, \tag{16}$$

where U_E is the electrode potential measured as the potential of semiconductor electrode with respect to the reference electrode, ϕ_H is the potential difference in the Helmholtz double layer, and C is a constant depending on the reference electrode. Since the depletion layer capacitance C_{SC} is much smaller than the Helmholtz double layer capacitance C_H due to their thickness, $1/C_{SC}^2$ can easily be obtained experimentally. Since the Mott–Schottky plot indicates the relationship between $1/C_{SC}^2$ and U_E, an extrapolation of $1/C_{SC}^2 = 0$ shows the point at which the space charge potential at the depletion layer becomes nearly zero ($\phi_{SC} \to 0$). Accordingly,

$$\phi_{SC} = 0 \quad \text{and} \quad U_E = U_{fb} \quad \text{at } 1/C_{SC}^2 = 0, \tag{17}$$

where U_{fb} ($=\phi_{fb}$) is called the flatband potential. Since the discussion of the potentials is based on the energy level of the semiconductor Fermi level and electrolyte chemical potential, this flatband potential shows the Fermi level of semiconductor when the depletion layer potential (ϕ_{SC}) is close to zero. The band edge energy for the n-type semiconductor requires that

$$\phi_C = \phi_{fb} - \frac{k_B T}{q} \ln\left(\frac{N_C^*}{n}\right), \tag{18}$$

where k_B is the Boltzmann constant, T is the absolute temperature, ϕ_C is the conduction band edge potential, n is the free electron concentration, and the conduction band-effective density of states. The N_C^* is expressed as

$$N_C^* = 2\left(\frac{2\pi m_e^* k_B T}{h^2}\right)^{3/2}, \tag{19}$$

where m_e^* is the electron effective mass, and h is the Plank's constant (Fujii and Ohkawa, 2006a). When the semiconductor is p-type, it is changed to obtain the valence band edge potential ϕ_V, and the sign of the second term in Eq. (18) changed from "−" to "+" and ϕ_C, n, N_C^*, m_e^* are replaced by ϕ_V, p, N_V^*, m_h^*, which are the valence band edge potential, the free hole concentration, the valence band-effective density of states, and the hole effective mass, respectively.

The schematic relationships between the applied bias and the current density of n- and p-type semiconductor electrode are shown in Fig. 2. The band diagram including the flatband conditions is also shown in the graph. Oxidation current for n-type and reduction current for p-type semiconductor electrodes are only observed under light irradiation and the

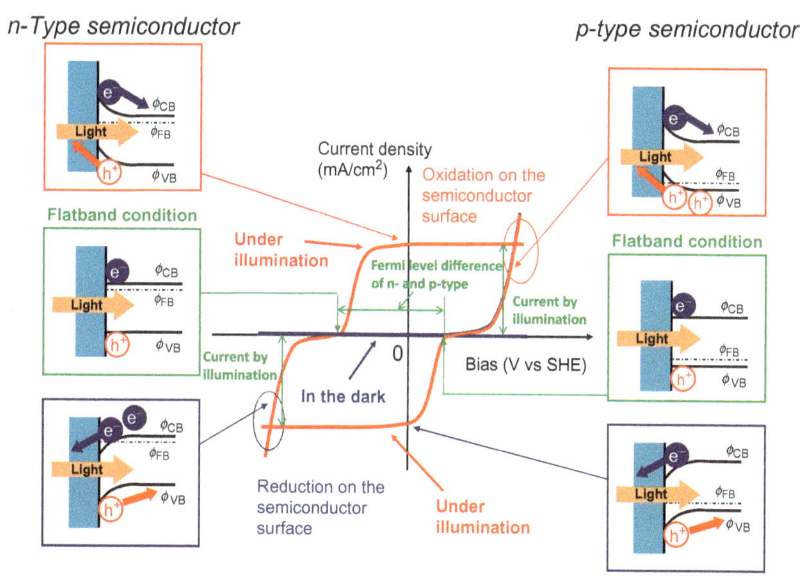

Fig. 2 Schematic relationship between the bias and current density of n- and p-type semiconductors. The band structures for each bias are also illustrated. The turn-on of the photocurrent occurs theoretically at the flatband condition. The photocurrent can be observed just under illuminated condition of the semiconductor electrode.

current densities are limited by the absorbed light power by semiconductor. This is because the carrier source is the electron–hole pair generated by the absorbed light. When the semiconductor materials are the same for n- and p-type, the difference of the turn-on biases of each photocurrent shows the difference of the Fermi level (correctly, quasi-Fermi level) of n- and p-type. These conditions are also shown in the graph.

Finally, the decomposition of semiconductor in an electrolyte must be discussed. The reliability of photoelectrode is not so much discussed, but this is very important for the real use. Since electrochemical oxidation and reduction conditions are so harsh for the materials, the electrode is usually oxidized in the oxidation condition and reduced in the reduction condition. It is easy to understand when it is considered that almost room temperature of 300K is about 26 meV, that is, 2.4 kJ/mol or 0.6 kcal/mol. The additional voltage is usually extremely large compared to this energy of room temperature; that is, the amount is at least subvolt order. Theoretically, the stability can be calculated when the electrode material has a perfect structure (Chen and Wang, 2012; Gerischer, 1977). The corrosion is not easy to discuss in reality because some of the defects can be reaction centers when the material has defects and because some materials can react chemically with electrolyte.

For the III-nitride materials, the semiconductors are believed to be stable in reduced condition and nonstable in oxidized condition because the nitride semiconductor bonds are basically covalent ones; that is, the valences of cation metals (Al, Ga, and In) are almost zero and not oxidized. The oxides of metals, in which valences of metals are already oxidized and +3 for III oxides, are expected to be more stable in oxidized condition of electrodes. It is well known that n-type III-nitride semiconductors are much more high quality than that for p-type because the crystal growth temperature of p-type materials is lower than that for n-type and because the amount of the doping material of Mg in III-nitride semiconductors is very high and close to the solubility limit. Thus, n-type nitride semiconductors are often used as photoelectrodes (photoanodes) of photoelectrochemical reactions. Still, keeping its stability is not easy under the reaction conditions.

3. BAND EDGE ENERGY AND WATER SPLITTING OF III-NITRIDES IN PHOTOELECTROCHEMISTRY

3.1 Band Edge Energy of GaN

As discussed until here, one of the important parameters for photoelectrochemical water splitting is the relationship between conduction and valence band edge energies (potentials) of semiconductors and water oxidation and reduction redox energy (potential). The conduction band edge energy (potential) must be higher (more negative in potential) than the redox energy (potential) of water reduction (hydrogen evolution), and the valence band edge energy (potential) must be lower (more positive in potential) than the redox energy (potential) of water oxidation (oxygen evolution) for the semiconductor to split water completely.

The first report of pH-dependent flatband potential of n-type GaN was 1995 and it was performed from the Mott–Schottky plot (Kocha et al., 1995). The flatband potential ϕ_{fb} as a function of pH is

$$\phi_{\mathrm{fb}} = -0.75 - 0.055\mathrm{pH} \quad [\mathrm{V\ vs\ SHE}] \tag{20}$$

where SHE shows the standard hydrogen electrode. The carrier concentration (n) was $1 \times 10^{19}\ \mathrm{cm}^{-3}$. Since the band gap of GaN is 3.4 eV and the flatband potential of n-type material is close to the conduction band edge, it is expected that GaN can split water.

Similar evaluation of n-type GaN using the Mott–Schottky plot was performed (Huygens et al., 2000), and the flatband potential ϕ_{fb} is

$$\phi_{\mathrm{fb}} = -0.78 - 0.060\mathrm{pH} \quad [\mathrm{V\ vs\ SHE}] \tag{21}$$

The carrier concentration (n) was $2 \times 10^{18}\ \mathrm{cm}^{-3}$.

The next report was that the conduction and valence band edge energies were obtained using the Mott–Schottky plot for n-type GaN and using the turn-on voltage of the current–voltage relationship for n-type and p-type GaN (Beach et al., 2003). The turn-on voltage was regarded as the flatband potential for this case. The carrier concentration of n-type GaN (n) was $7 \times 10^{17}\ \mathrm{cm}^{-3}$ and that for p-type GaN (p) was $1 \times 10^{17}\ \mathrm{cm}^{-3}$. The conduction band edge potential ϕ_{cb} and the valence band edge potential ϕ_{vb} for n-type measured by the Mott–Schottky plots are

$$\phi_{\mathrm{cb}} = -0.816 - 0.047\mathrm{pH} \quad [\mathrm{V\ vs\ SCE}] \tag{22}$$

$$\phi_{\mathrm{vb}} = +2.594 - 0.047\mathrm{pH} \quad [\mathrm{V\ vs\ SCE}] \tag{23}$$

where SCE is the standard calomel electrode (SCE = SHE + 0.2444 [V]). The electron effective mass of 0.18 and the band gap value of 3.41 eV were used in the calculations from the flatband potential to band edge potentials. The potentials for n-type measured by turn-on voltage are

$$\phi_{\mathrm{cb}} = -0.538 - 0.046\mathrm{pH} \quad [\mathrm{V\ vs\ SCE}] \tag{24}$$

$$\phi_{vb} = +2.872 - 0.046\text{pH} \quad [\text{V vs SCE}] \tag{25}$$

The potentials for p-type measured by turn-on voltage are

$$\phi_{cb} = -1.372 - 0.063\text{pH} \quad [\text{V vs SCE}] \tag{26}$$

$$\phi_{vb} = +2.048 - 0.063\text{pH} \quad [\text{V vs SCE}] \tag{27}$$

The hole effective mass of 0.8 and the band gap value of 3.41 eV were used in the calculations. The discrepancy of the band edge potentials for n-type and p-type GaN semiconductors was explained by the fact that the measurement values of turn-on voltages contained the overpotentials for the photoelectrochemical reactions. The values are the overpotentials of water oxidations for n-type GaN photoanode and that of water reduction for p-type photocathode.[a]

The band edge potentials were obtained from the Mott–Schottky plot of n-type GaN (Fujii and Ohkawa, 2006a) and compared with those of p-type GaN (Fujii and Ohkawa, 2005). The band edge potentials are

$$\phi_{cb} = -0.73 - 0.055\text{pH} \quad [\text{V vs Ag/AgCl/NaCl}] \tag{28}$$

$$\phi_{vb} = +2.69 - 0.055\text{pH} \quad [\text{V vs Ag/AgCl/NaCl}] \tag{29}$$

where Ag/AgCl/NaCl is sodium-chloride-saturated silver-chloride electrode (Ag/AgCl/NaCl = SHE + 0.212 [V]). The electron and hole effective masses of 0.25 and 0.80, respectively and the band gap value of 3.42 eV were used in the calculations from the flatband potential to band edge potentials.

The summary of the flatband and band edge potentials is shown in Fig. 3. The slopes of the band edge potentials show a similar value of −0.059 pH, which is the pH dependence of Nernst equation of water. This probably shows that the surface of GaN is mostly covered by the water molecules, H^+ and OH^- ions. In addition, all the results show that the conduction band edge is more negative value in the electrode potential (higher electron energy) than that of hydrogen evolution from water and the valence band edge is more positive value in the electrode potential (lower electron energy) than that of oxygen evolution from water in whole pH region. Thus, GaN can be expected to split water without bias.

[a] The original paper of calculation for Eqs. (23) and (26) is incorrect due to its mistaken band gap energy of GaN; thus, the values were recalculated here.

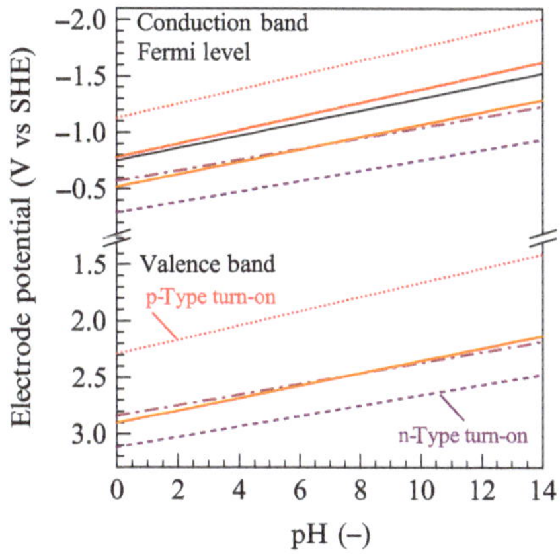

Fig. 3 The pH dependences of flatband potential of n-type GaN by Kocha et al. (*solid line*) (Kocha et al., 1995) and by Huygens et al. (*chained line*) (Huygens et al., 2000), conduction and valence band edge potentials of n-type GaN obtained by flatband potentials (*bold chained lines*), conduction and valence band edge potentials of n-type GaN obtained by turn-on voltage (*bold break lines*), conduction and valence band edge potentials of p-type GaN obtained by turn-on voltage (*bold dotted lines*) (Beach et al., 2003), and conduction and valence band edge of n- and p-type GaN obtained by flatband potentials (*bold lines*) (Fujii and Ohkawa, 2006a). The potentials were obtained by the Mott–Schottky plot except for obtained by turn-on voltage. The potentials were converted to "V vs SHE."

3.2 Band Edge Energy of III-Nitrides

The flatband potentials for the other nitrides were evaluated using with $Al_yGa_{1-y}N$ and $Ga_{1-x}In_xN$ (Fujii et al., 2007c). The summary of the band edge potential is shown in Fig. 4. It was summarized that the conduction band edge changed with band gap, but the valence band edge remained almost constant. This result is also expected from the oxide semiconductors, in which the valence band edge does not change when the cation element is changed with the same anion of oxygen. From the result, the conduction band edge potential is expected to approach the hydrogen evolution potential with an increase in the In content of $Ga_{1-x}In_xN$. The flatband potential of $Ga_{1-x}In_xN$ ($x=0.12$) was measured to be −0.12 V vs Ag/AgCl/NaCl in the later study (Fujii et al., 2011b). This value has also good agreement with this band edge potential change with the In content. From these results, the conduction band edge is calculated to cross the hydrogen evolution potential when the In composition x is 0.20 (410 nm in wavelength), where the band gap is 3.0 eV. This means that $Ga_{1-x}In_xN$ has a possibility of water splitting by photoelectrochemical reaction with blue visible-light illumination.

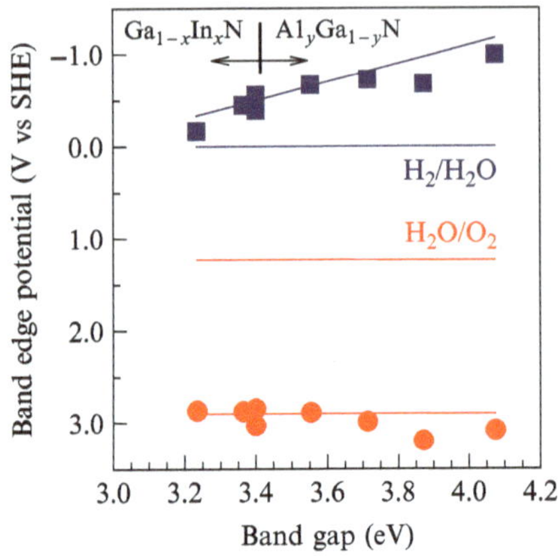

Fig. 4 Conduction and valence band potentials of $Al_yGa_{1-y}N$ ($y=0.00$, 0.05, 0.10, 0.15, and 0.21) and $Ga_{1-x}In_xN$ ($x=0.00$, 0.02, 0.09) and water oxidation and reduction potentials. The lines of conduction band and valence band are calculated from Fig. 1 band gap values with the estimation of the valence band edge potential being constant with GaN valence band potential of +2.90 V vs SHE. The potential was converted from the potentials of $Ga_{1-x}In_xN$ at pH 0.1 and the potentials of $Al_yGa_{1-y}N$ at pH 2.7 with assumption of the slope with −0.055 pH (Fujii and Ohkawa, 2006a). Water oxidation and reduction potentials are also indicated.

4. PHOTOELECTROCHEMICAL WATER SPLITTING BY N-TYPE GaN

As discussed until here, GaN is expected to split water without bias but has the possibility of dissolution into the electrolyte when high-quality n-type GaN is used as a photoelectrode. The first report of photocurrent densities without bias at pH 0 and 14 showed that the densities were not significantly different considering the difference of pH under the RuO_2 colloid catalyst loaded on the GaN photoanode surface (Kocha et al., 1995). They also pointed out the hysteresis at the turn-on and turn-off regions of photocurrent in the current–voltage characteristics (cyclic voltammetry) for an as-grown n-type GaN and summarized that the existence of surface state of GaN affects the hysteresis, which was similar with the other III–V semiconductors.

The anodic corrosion of GaN was evaluated with the detection of Ga ion in electrolyte after photoelectrochemical reaction (Huygens et al., 2000). The Ga ion was detected in the electrolytes of 1 mol/L KOH and 1 mol/L H_2SO_4, but the ion was not detected in the 1.2 mol/L HCl electrolyte.

The surfaces were roughened by the electrochemical reaction in 1 mol/L KOH and 1 mol/L H_2SO_4, but the smooth surface remained even after the reaction in 1.2 mol/L HCl observed by the atomic force microscope (AFM). From the calculation of the electrical equivalence relationship of Faraday's law,

$$n = It/FcV, \tag{30}$$

where n is the electrical equivalence, I is the current of electrochemical reaction in A, t is the reaction time in s, F is the Faraday's constant, c is the dissolved Ga in mol/L, and V is the volume of the electrolyte in L. The n of 3.54 ± 0.35 for 1 mol/L KOH and 3.06 ± 0.37 for 1 mol/L H_2SO_4 was found. From this relationship, the corrosion reactions were proposed in the KOH aqueous electrolyte

$$GaN(s) + 6OH^- + 3h^+(in\,GaN) \rightarrow GaO_3{}^{3-}(in\,\ell) + 0.5N_2(g) + 3H_2O, \tag{31}$$

and in the H_2SO_4 aqueous electrolyte

$$GaN(s) + 3h^+(in\,GaN) \rightarrow Ga^{3+}(in\,\ell) + 0.5N_2(g). \tag{32}$$

They also pointed out important properties of III-nitride semiconductors that only holes in valence band involved the corrosion and no electron contribution. Electron contribution for the corrosion process often observes in the case of the photoelectrochemical water splitting with the other III–V materials. In 1.2 mol/L HCl electrolyte, Cl^- oxidation was occurred instead of GaN oxidation. That is,

$$2Cl^-(in\,\ell) + 2h^+(in\,GaN) \rightarrow Cl_2(g). \tag{33}$$

Thus, these hydrogen halide aqueous electrolytes are often used for the III-nitride photoelectrochemical reaction to prevent the anodic corrosion of n-type GaN photoanode after this result.

Clear hydrogen evolution from the counterelectrode with n-type GaN photoelectrochemical reaction was observed in 1.0 mol/L KOH aqueous electrolyte with +1.0 V vs counterelectrode bias applied to working electrode (Fujii et al., 2005a). The overpotential for water reduction was higher at neutral aqueous electrolyte of 1.0 mol/L NaCl used than that at acidic 1.0 mol/L HCl and basic 1.0 mol/L KOH used (Fujii and Ohkawa, 2006b). These situations were the same when the acidic, neutral, and basic aqueous solutions were 0.5 mol/L Na_2SO_4, 0.5 mol/L H_2SO_4, and

1.0 mol/L NaOH, respectively. This is explained by the fact that the electric conductivity of the neutral electrolyte is lower than those of acidic and basic electrolytes due to the dominant moving ions not being H^+ or OH^-.

Hydrogen generation without bias was achieved with n-type GaN on the sapphire substrate at relatively low GaN carrier concentration of around 1×10^{17} cm^{-3} in 1.0 mol/L HCl aqueous solution (Ono et al., 2007). The photocurrent density had the maximum point as a function of n-type GaN carrier concentration from the analysis. This is explained as follows. High electric resistivity of GaN is a serious problem for electron transfer since the GaN layer is usually grown on insulating sapphire substrate and the thickness is around a few μm. Thus, the current density decreases with the carrier concentration of n-type GaN especially at the carrier concentration below 1×10^{17} cm^{-3}. In contrast, the depletion layer thickness is shortened with proportional to reverse square root of donor concentration as shown in Eq. (15). Since the electron–hole pair generated outside of this depletion layer recombines shortly and does not contribute photocurrent, the photocurrent decreases with increasing carrier concentration. Thus, the optimized carrier concentration exists to maximize the photocurrent density of n-type GaN grown on the sapphire substrate.

The electrolyte dependence was also evaluated and summarized that NaOH aqueous electrolyte was relatively good except for nonwater-oxidized reaction case like in the HCl aqueous electrolyte (Koike et al., 2014). The overpotential, which was evaluated from the comparison between the flatband potential obtained from the Mott–Schottky plot and the turn-on voltage of photocurrent, under H_2SO_4 acidic solution, was larger than that under NaOH basic solution. This is probably explained from the fact that oxidation of water (H_2O) molecule is more difficult than the oxidation of OH^- ion considered from the reactions of Eqs. (1) and (3). As a result, the anodic oxidized corrosion of GaN is expected to be difficult in the basic aqueous electrolyte. This is coincident with the experimental results.

The anodic corrosion rates of GaN were, however, different even between the basic aqueous electrolytes KOH and NaOH. The corrosion rate in KOH electrolyte was larger than that in NaOH (Koike et al., 2014). The difference of anodic corrosion between KOH and NaOH electrolytes is probably explained by the difference of cation ion radius. Since the K^+ ion radius (0.137–0.164 nm) is larger than that of the Na^+ ion radius (0.099–0.139 nm) (the radius is in solid and changes with coordination number) (Shannon, 1976), the protection of OH^- ion approaching to n-type GaN photoanode by K^+ ion is much stronger than that by Na^+ ion. As a result, the OH^- oxidation at the GaN electrode surface in the KOH electrolyte is much less likely than that in NaOH aqueous solution. This diffusion speed change of reactant is probably the role of cation in the electrolyte near photoanode. Since NaOH aqueous electrolyte is good for reducing anodic corrosion rate, the evaluation of n-type GaN anodic corrosion mechanism is basically performed in the NaOH aqueous electrolyte.

5. ANODIC CORROSION AND WATER OXIDATION MECHANISMS OF N-TYPE GaN

5.1 Anodic Corrosion of n-Type GaN Photoanode

Anodic corrosion of n-type photoanode is a well-known phenomenon in III–V and II–VI semiconductors, but the mechanism is not well understood. It is because that some materials show the anodic corrosion with chemical corrosion and/or are difficult to distinguish between the anodic corrosion and corrosion due to the material properties or due to low crystal quality. In contrast, high-quality oxide crystals hardly show anodic corrosion. It is believed that the chemical bond of oxides is relatively ionic and not covalent ones, that is, the cation atom in the oxide crystal already oxidized. Thus, the cation elements cannot oxidize any more, which means that the corrosion of oxides is difficult. In contrast, the bonds of usual III–V and II–VI crystals are covalent ones and the cation atom valence is almost zero. As a result, since the cation element can oxidize, the compound semiconductors may be prone to anodic corrosion. From the discussion here, usually III–V and II–VI crystals are expected to be stable in the reduced condition because the cations are already reduced; that is, p-type materials for photocathode are expected to be stable.

Anodic corrosion is also a problem for n-type GaN photoanode although GaN is one of chemically stable materials. As discussed until here, n-type GaN shows a stability difference with different electrolytes and relatively stable in the NaOH aqueous electrolyte. Even in the 1.0 mol/L NaOH electrolyte, however, n-type GaN shows anodic corrosion and the color turns from transparent (due to the GaN band gap is UV region of 3.4 eV) to frost mirror-like white due to the GaN surface roughening. Even though n-type GaN is suitable material for the analysis of anodic corrosion mechanisms since GaN is relatively stable in chemically and high-quality materials. Detail mechanism of GaN anodic corrosion phenomenon, which mechanism is expected to be applied the other III–V and II–VI semiconductors, is discussed here.

The first important results for n-type GaN anodic photocorrosions were taken with bias application of +1.0 V vs counterelectrode (Fujii et al., 2007a). The electrolytes used in the experiments were 0.5 mol/L H_2SO_4 (pH 0.4), a mixture of 0.005 mol/L H_2SO_4 and 0.2 mol/L Na_2SO_4 (pH 2.7), 0.2 mol/L Na_2SO_4 (pH 6.2), a mixture of 0.01 mol/L NaOH and 0.2 mol/L Na_2SO_4 (pH 11.9), and 1.0 mol/L NaOH (pH 13.6). The irradiated light was 150 W Xe lamp.

The photocurrent densities were over 1.0 mA/cm^2 at the beginning and started to decrease after about 50 min in pH 0.4 and about 70 min in pH 13.6. In contrast, the densities for the others of pH 2.7, 6.2, and 11.9 showed not much changes and the current densities were below 0.5 mA/cm^2. The corrosion amounts after 180 min photo-irradiations with 150 W Xe lamp were calculated from the weight losses of GaN and compared with the expected corrosion amounts from the sums of the current, which showed 100% GaN corrosion from Eq. (31). The amounts of corrosion weights were almost the same value expected from the currents at the electrolyte pH below 11.9, but the amount was lower than the expected value from the current at the pH 13.6 as shown in Fig. 5. It is summarized that the oxygen formation from water is easier in the higher pH electrolytes from the analysis

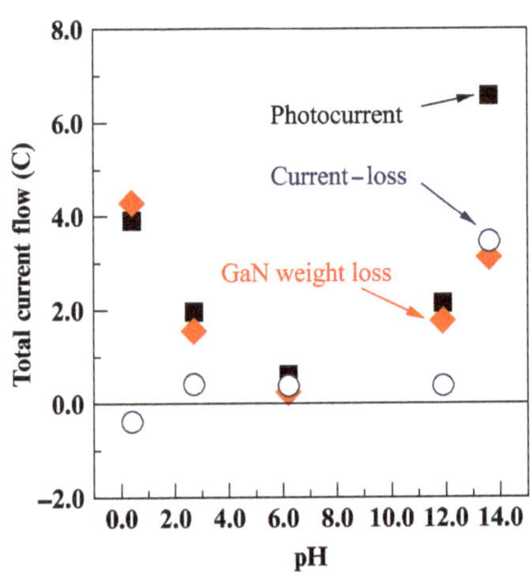

Fig. 5 Total current flows (coulomb) after 180 min photoelectrochemical reaction of n-type GaN with +1.0 V vs counterelectrode dependent on pH calculated from the photocurrent and GaN weight loss. The differences between photocurrent and weight loss are also shown as "current–loss." *Quoted from fig. 2 of Fujii, K., Ito, T., Ono, M., Iwaki, Y., Yao, T., Ohkawa, K., 2007a. Investigation of surface morphology of n-type GaN after photoelectrochemical reaction in various solutions for H2 gas generation. Phys. Status Solidi C 4, 2650–2653.*

of the corrosion amounts. This probably shows that OH^- is absorbed at the surface of n-type GaN at the higher pH and this adsorbed OH^- is easier to form oxygen than H_2O molecule, which is oxidized reactant at the lower pH. The reactants are expected from the water oxidation process as shown in Eqs. (1)–(4). The explanations of the results are similar to the previous discussions of the GaN overpotential difference for water splitting in the difference of electrolytes (Koike et al., 2014).

The surface morphologies after the reactions observed by a scanning electron microscope (SEM) were different with the pH of electrolyte as shown in Fig. 6. The surfaces changed from smooth ones to high-density needle-like holes with whisker-like materials on the surface after the reaction at the electrolyte pH below 6.2. Especially, the large and clear whisker-like shapes were observed in pH 6.2. After the reaction in the electrolyte pH 11.9 and 13.6, the surfaces were completely different from that in lower pH and relatively large-sized hexagonal-shaped columns without whisker-like material were observed. Similar results of hexagonal-shaped columns were obtained from the others (Ko et al., 2002).

From the energy-dispersive X-ray spectroscopy (EDS) analysis with SEM, the whisker-like material was found to be Ga oxide. These results are consistent with the stable form of Ga in solution with different pH

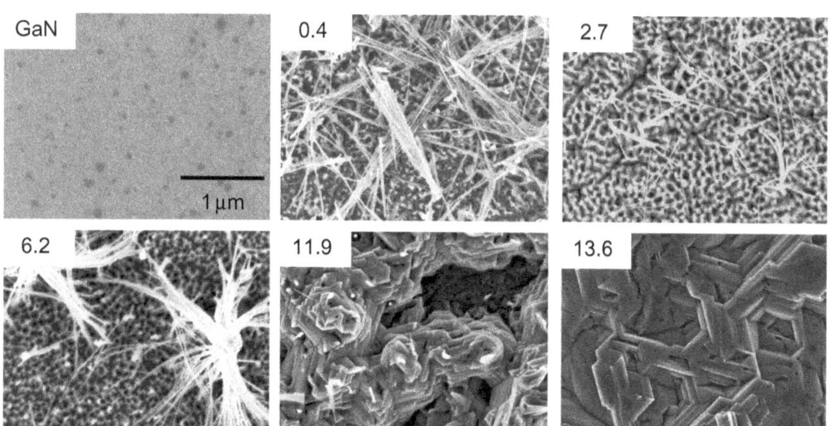

Fig. 6 Surface SEM images before and after 180 min photoelectrochemical reaction of n-type GaN with +1.0 V vs counterelectrode. The *GaN* and *numbers* denote the surface before photoelectrochemical reaction and the pH values of the electrolyte. *Quoted from fig. 3 of Fujii, K., Ito, T., Ono, M., Iwaki, Y., Yao, T., Ohkawa, K., 2007a. Investigation of surface morphology of n-type GaN after photoelectrochemical reaction in various solutions for H2 gas generation. Phys. Status Solidi C 4, 2650–2653.*

(Pourbaix diagram). These results show that oxidation of GaN is easier than oxygen generation from water. Anodic corrosion and Ga^{3+} formation probably occurred in the lower pH solutions. The Ga^{3+} formation is easier than oxygen generation from water, and the formation of Ga^{3+} changing to the formation of GaO_3^{3-} in the higher pH solutions followed by the stable form of Ga in solution pH. The amount of dissolved Ga^{3+} or GaO_3^{3-} ion in the solution is controlled by Ga solubility in an electrolyte; thus, the over-saturated Ga ions probably formed Ga_2O_3 in the lower pH solutions (especially in the middle range of (neutral) pH). This Ga_2O_3 was observed as whisker-like Ga oxides.

As discussed here, basic NaOH aqueous electrolyte is better to prevent n-type GaN anodic corrosion and to enhance water oxidation at the surface of GaN than acidic ones and KOH aqueous electrolyte. Therefore, the conditions of NaOH aqueous electrolyte and without bias application are usually used for the stability discussion when the reaction is water splitting (and does not oxidize the halogen ion in hydrogen halide like oxidizing Cl^- in HCl for example). This condition of NaOH aqueous electrolyte and without bias used for photoelectrochemical reaction is expected to be close to the real use.

In order to prevent the corrosion, the sacrificial reagent was considered first because water oxidation was a four-electron transfer process and was generally believed difficult (Kudo, 2003). The sacrificial reagents here were alcohols of CH_3OH and C_2H_5OH. The alcohols are easy to be oxidized compared with water. Thus, the alcohol-containing electrolytes were evaluated using an n-type GaN photocathode (Fujii et al., 2008; Fujii et al., 2009a).

Photocurrent densities in 0.5 mol/L H_2SO_4, 0.5 mol/L Na_2SO_4, and 1.0 mol/L NaOH aqueous electrolytes were increased with 1.0 mol/L alcohol additions in current–voltage characteristics. The current density decreasing with time using the electrolytes with alcohols was more moderate than that without alcohols in 300 min time-dependent measurement as shown in Fig. 7.

The surface oxide was observed only when the Na_2SO_4 aqueous electrolyte was used by an energy-dispersive X-ray spectroscope with a scanning electron microscope (SEM-EDS), and the oxygen peak decreased when alcohol was added. The surface roughness also decreased with the addition of alcohol observed by the root mean square (RMS) value of AFM measurements as shown in Table 2.

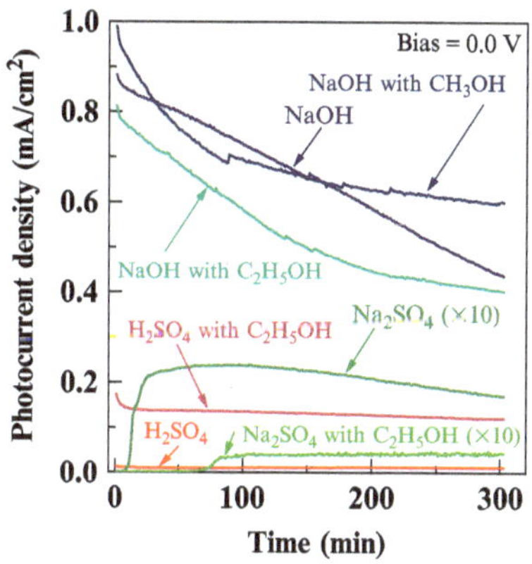

Fig. 7 Photocurrent dependence on reaction time without bias under irradiation. The photoanodes are n-type GaN and the electrolytes are 0.5 mol/L H_2SO_4, 0.5 mol/L Na_2SO_4, and 1.0 mol/L NaOH electrolytes with and without 1.0 mol/L C_2H_5OH and 1.0 mol/L CH_3OH (for NaOH electrolyte only) (Fujii et al., 2008, 2009a).

Table 2 The RMS Values of the n-Type GaN Surfaces Obtained From AFM Measurements After the Photoelectrochemical Treatment

nm	H_2SO_4	Na_2SO_4	NaOH
Without C_2H_5OH	90.1	19.1	20.0
With C_2H_5OH	20.4	11.3	9.8

The electrolytes were 0.5 mol/L H_2SO_4, 0.5 mol/L Na_2SO_4, and 1.0 mol/L NaOH without C_2H_5OH (*upper row*) and with 1.0 mol/L C_2H_5OH (*lower row*) (Fujii et al., 2009a).

These results show that the GaN oxidation is suppressed with alcohol additions in aqueous electrolytes but not enough. This probably indicates that the electron transfer rate of the electrolyte is relatively fast, but the transfer near GaN surface is slow and has a rate-limiting problem.

In order to evaluate this electron transfer rate limitation, the crystal quality, which meant different growth methods here, was changed from the common growth method of MOVPE to HVPE (Ga- and N-face free standing wafer) and sodium-flux (Na-flux) growth (single crystal, large- and small-grain polycrystals) (Fujii et al., 2009b, 2010). This is because the crystal qualities of HVPE and Na-flux grown n-type GaN evaluated by X-ray diffraction are much better than MOVPE-grown n-type GaN. (For example, the full width of half maximum (FWHM) of (0002) X-ray rocking curve (XRC: ω scan) for MOVPE-grown GaN is around 200–300 arcsec.

In contrast, those for HVPE and Na-flux grown GaN are usually less than 50 arcsec.) This indicates that the dislocation densities of HVPE and Na-flux grown GaN are less than that of MOVPE-grown GaN (Kaganer et al., 2005).

In the 1.0 mol/L NaOH aqueous electrolyte, the photocurrent density of HVPE-grown Ga-face GaN was less than that of MOVPE-grown n-type GaN. The photocurrent density of HVPE-grown N-face GaN was, however, more than that of MOVPE-grown n-type GaN. In addition, the surfaces of both HVPE-grown GaN became rough after 600 min photoelectrochemical reaction. These results mean that both of the surfaces show photocorrosion. Similar evaluation was performed in later, and almost similar results were obtained (Bae et al., 2013).

The photocurrent density of Na-flux GaN single crystal was higher than that of MOVPE in the 1.0 mol/L HCl aqueous electrolyte. The photocurrent densities for polycrystal samples showed, however, large overpotential at the turn-on and less-saturated photocurrent density at the positive bias application. The overpotential increased and the saturated photocurrent decreased with decreasing the grain size.

These results indicate that the dislocation density decreasing of GaN does not affect the anodic corrosion but decreases the electron–hole recombination loss generated by light irradiation. Since the photocurrent density of the Ga-face HVPE sample was lower than that of MOVPE, which had the same surface orientation of HVPE, the electron transfer limitation cannot be explained only by the surface structure at the semiconductor electrolyte interface.

5.2 Water Oxidation Mechanisms by n-Type GaN

The effects of the light irradiation intensity and applied bias were also evaluated with using MOVPE-grown n-type GaN samples with the carrier concentration of 1.0×10^{18} cm^{-3} in 1.0 mol/L NaOH aqueous electrolyte (Koike et al., 2010). The current density dependences with photoelectrochemical reaction times were evaluated with changing irradiated light intensities and applied biases.

The time dependence became faster like a kind of acceleration reliability test when the light irradiation intensity was increased as shown in Fig. 8. The current density dependence with light irradiation was, however, changed from proportional to the light intensity at 100 min photoelectrochemical reaction to almost no relationship to the light intensity at 750 min.

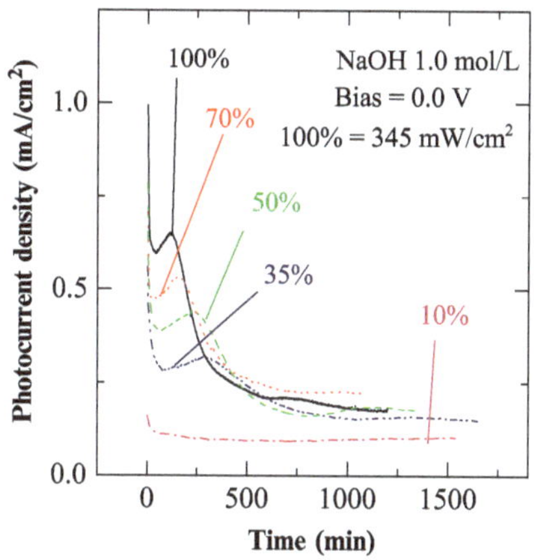

Fig. 8 Time dependences of photocurrent densities at zero bias. The electrolyte was 1.0 mol/L NaOH. Light intensities were 100% (345 mW/cm^2), 70%, 50%, 35%, and 10%. *Quoted from fig. 1 of Koike, K., Sato, K., Fujii, K., Goto, T., Yao, T., 2010. Time variation of GaN photoelectrochemical reactions affected by light intensity and applied bias. Phys. Status Solidi C 7, 2221–2223.*

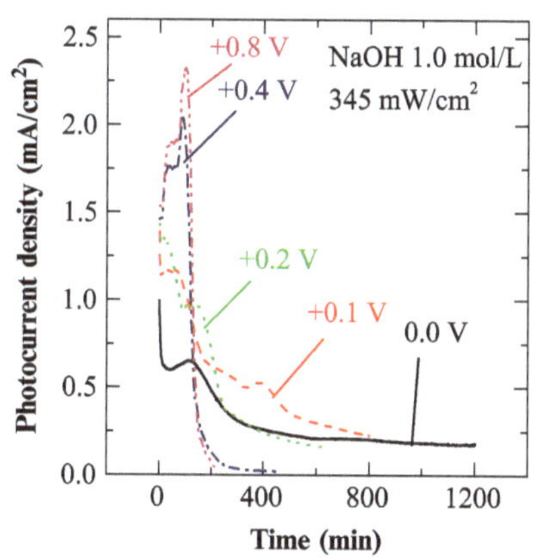

Fig. 9 Time dependences of the photocurrent densities under 345 mW/cm^2 light intensity. The electrolyte was 1.0 mol/L NaOH. The biases were 0.0, +0.1, +0.2, +0.4, and +0.8 V vs counterelectrode. *Quoted from fig. 4 of Koike, K., Sato, K., Fujii, K., Goto, T., Yao, T., 2010. Time variation of GaN photoelectrochemical reactions affected by light intensity and applied bias. Phys. Status Solidi C 7, 2221–2223.*

The time dependence was drastically changed when the applied bias was changed as shown in Fig. 9. The time dependences became faster like the dependence of light irradiation change at the applied bias from 0.0 to +0.2 V vs counterelectrode. The time dependences with the applied bias of +0.4 and +0.8 V vs counterelectrode were completely changed, and the photocurrent densities became nearly 0 mA/cm^2 at 200–400 min after the reaction started. Even after the photoelectrochemical reactions with +0.4 and +0.8 V vs counterelectrode biases applied, the surfaces became rough but were still the GaN layers remaining on sapphire substrates. The surface after the reaction consisted of many grooves and plateaus at the bias between 0.0 and +0.2 V vs counterelectrode also changed to many deep pits at the bias between +0.4 and +0.8 V vs counterelectrode observed by SEM. This probably indicates that the surface corrosion reaction changes in between +0.2 and +0.4 V vs counterelectrode.

The detailed analysis of anodic photocorrosion of n-type GaN was expected because the corrosion was estimated to be related to point defects worked as reaction centers in n-type GaN from these analyses. Thus, the time dependence of photocurrent density, surface morphology, and photoluminescence (PL) was evaluated (Sato et al., 2009). The samples were n-type GaN with the carrier concentration of 1.0×10^{18} cm^{-3}. The experiments were performed in 1.0 mol/L NaOH aqueous electrolyte under 170 mW/cm^2 irradiations by Xe lamp.

The photocurrent density dependence of reaction time showed strange behavior as shown in Fig. 10. The photocurrent density changing with time showed a "w" shape, i.e., the photocurrent density decreased rapidly

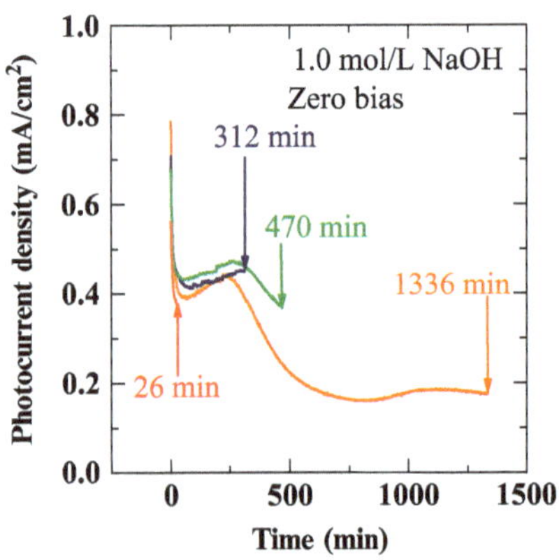

Fig. 10 Time dependence of photocurrent density without bias. The times show the end of the reactions. The electrolyte and light irradiation were 1.0 mol/L NaOH aqueous solution and 170 mW/cm^2, respectively. *Quoted from fig. 2 of Sato, K., Fujii, K., Koike, K., Goto, T., Yao, T., 2009. Anomalous time variation of photocurrent in GaN during photoelectrochemical reaction for H2 gas generation in NaOH aqueous solution. Phys. Status Solidi C 6, S635–S638.*

followed by its increase, and the density decreased again. In order to evaluate the reason of this strange behavior, five samples were evaluated with different photo-irradiated times in the electrolyte. The sample (A) was as grown (shown as 0 min). The next sample (B) was stopped the reaction at the almost end of first rapid photocurrent decrease of 26 min reaction. The sample (C) was stopped the reaction at 312 min, where was the almost end of slightly photocurrent increasing. The sample (D) was stopped the reaction at the middle of its decreasing of 470 min. The last sample (E) was stopped the reaction at the photocurrent decreasing almost stopped of 1336 min. As also can be seen here, the repeatability of the photocurrent density was almost maintained.

It is interesting that the surface roughness increased with the treatment time linearly from the observation of RMS evaluation of AFM measurements. This was coincident with the PL measurement at 10K. As shown in the wide range PL spectra of Fig. 11A, the deep-level emissions of samples (A) and (B) have intensity oscillation, but the samples (C), (D), and (E) have no oscillation. The oscillation is the interference between the GaN surface and the interface between the sapphire and GaN. Thus, this result also indicates that the surface roughness of GaN increases with the treatment time.

Fig. 11B with PL energy from 3.45 to 3.50 eV shows the peak energy shift of the exciton-bound neutral donor (D_0X_A) depending on the photoelectrochemical reaction time. The D_0X_A peak of strain-free GaN was reported to be located at 3.471 eV (Reshchikov and Morkoç, 2005). As the treatment time increases, the D_0X_A emission peak gradually shifted from 3.486 eV of sample (A) to the lower energy of 3.473 eV of sample (E). The peak of sample (D) seemed to be double peaked. These two peaks probably indicate the mixture of the strained and almost strain-free peaks. The peak position of sample (E) was very close to the position of the strain-free D_0X_A emission of 3.471 eV. This result can explain that the strain of GaN layer is relaxed by the roughing of the surface.

In addition, new peaks (3.402 eV (Y2), 3.381 eV (Y4), 3.353 eV (Y5), 3.322 eV (Y6), and 3.246 eV (Y7) (Reshchikov and Morkoç, 2005)) appeared as shown in the figure of the energy range of 3.22–3.44 eV in Fig. 11C. These new peaks were observed only from the sample (B), which treatment time was 26 min and the photocurrent density showed the first minimum. This result shows that some new defects are formed at the surface of the GaN.

The UVL peaks (2.8–3.3 eV) of sample (A), that is the as-grown sample, were relatively strong compared with the others. This UVL peak decreased

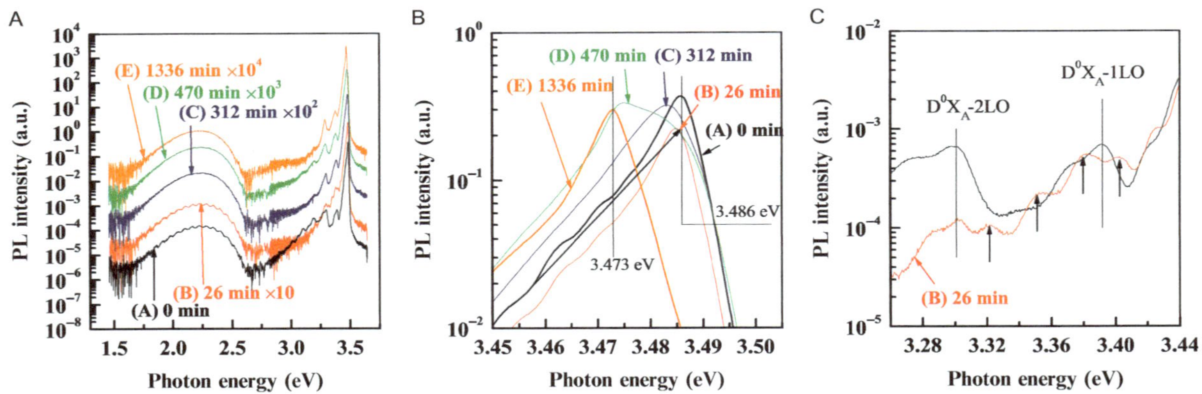

Fig. 11 PL spectra at 10K of samples (A) 0, (B) 26, (C) 312, (D) 470, and (E) 1336 min of photoelectrochemical reaction times without bias. The electrolyte and light irradiation were 1.0 mol/L NaOH aqueous solution and 170 mW/cm^2, respectively. The figures show (A) the wide range spectra between 1.4–3.6 eV, (B) the spectra near band edge between 3.45–3.51 eV, and (C) the spectrum of sample (B) 26 min and that of sample (A) 0 min at around photon energy of 3.22–3.44 eV range. *Quoted from fig. 4 of Sato, K., Fujii, K., Koike, K., Goto, T., Yao, T., 2009. Anomalous time variation of photocurrent in GaN during photoelectrochemical reaction for H2 gas generation in NaOH aqueous solution. Phys. Status Solidi C 6, S635–S638.*

rapidly from the as-grown sample (A) to the 26-min treated sample (B). These UVL peaks were reported to be a donor–acceptor pair transitions or a conduction band to acceptor (e–A) transitions and their LO phonon replicas (Reshchikov and Morkoç, 2005). From this definition, the result shows that the acceptors near the GaN surface disappear during the initial stage of the treatment.

In summary of these evaluations, the photocurrent density changing with time is related to the defect disappearance and formation until the sample (C), in which the reaction time is about 300 min. After that time, the surface morphology change, which means the change in surface index and surface chemical properties, is the dominant factor for the decrease in the photocurrent density. The time dependence of photocurrent density is expected to be different when the sample defect structure is different, like the case of change in the crystal growth condition. One of the good examples of this surface property change is changing the surface direction. The photoelectrochemical properties were observed to change with surface direction (surface index) as expected here (Bae et al., 2016a; Fujii et al., 2007b).

Point defects worked as reaction centers are strongly expected to affect the photocurrent density time dependence of n-type GaN from the discussions until here. Thus, the photoelectrochemical property evaluation with changing of the Si doping amount during the MOVPE crystal growth, which was the origin of the donor for n-type GaN, was evaluated (Fujii et al., 2011a, 2012). The properties of used samples are summarized in Table 3. The donor concentration of Si-doped samples ranged from 10^{16} to 10^{18} cm^{-3} and that of the undoped GaN sample below 10^{16} cm^{-3} were obtained from the Mott–Schottky plot (Kocha et al., 1995). The dependence of the carrier concentration on the Si doping level obtained from the Hall measurements (van der Pauw method) also showed a similar trend. The full width at half maximum (FWHM) of the XRC (ω scan) and 2θ–ω scan of the (0002) ranged from 200 to 300 arcsec for all samples. The FWHMs of the $(10\bar{1}2)$ XRC (ω scan) and 2θ–ω scan were different between the samples. These results are also summarized in Table 3.

The time dependences of photocurrent densities with changing FWHM of XRC (ω scan) and donor concentration are shown in Fig. 12. The photocurrent densities between (1012) XRC (ω scan) FWHM of 320 and 690 arcsec for Si-doped GaN with the similar donor concentrations about 1×10^{18} cm^{-3} showed a similarity in time dependences and showed a

Table 3 The Samples of n-Type Ga-Face (0001) GaN Grown by the Metal-Organic Chemical Vapor Deposition (MOVPE) Method on (0001) Sapphire Substrate Used for the Evaluations of Photocurrent Changes (Fujii et al., 2011a, 2012)

n-Type GaN (0001)+c	Carrier Conc. (cm^{-3})	Donor Conc. (cm^{-3})	X-Ray 2θ–ω (0002) FWHM (arcsec)	X-Ray ω (0002) FWHM (arcsec)	X-Ray 2θ–ω ($10\bar{1}2$) FWHM (arcsec)	X-Ray ω ($10\bar{1}2$) FWHM (arcsec)
Undoped	4.6×10^{16}	Not available	340	310	1230	690
Si-doped (1.5)	5.0×10^{17}	1.9×10^{18}	240	230	1300	690
Si-doped (1.5)	8.4×10^{17}	1.1×10^{18}	230	190	550	320
Si-doped (1.0)	1.9×10^{17}	3.3×10^{17}	250	190	530	300
Si-doped (0.5)	5.5×10^{16}	1.3×10^{16}	240	190	530	290

Carrier concentrations by the van der Pauw method, donor concentrations by the Mott–Schottky plot, and (0002) and ($10\bar{1}2$) X-ray rocking curve (XRC: ω scan) and X-ray diffraction of 2θ–ω scan are summarized. The *number in bracket* in the *left column* is the flow rate ("sccm" unit) of 100 ppm SiH_4 diluted by H_2 at the MOVPE growth.

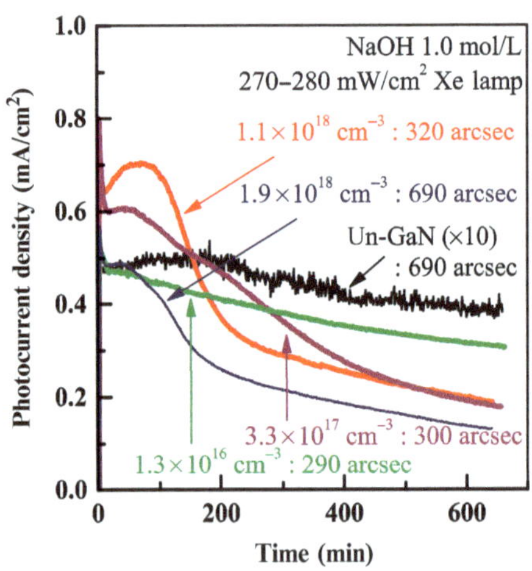

Fig. 12 Time dependences of photocurrent densities for the Si-doped and -undoped GaN without bias. The electrolyte and light irradiations were 1.0 mol/L NaOH aqueous solution and 270–280 mW/cm^2, respectively. The *numbers* in the graph are the donor concentrations and ($10\bar{1}2$) X-ray rocking curve (XRC: ω scan) (Fujii et al., 2011a, 2012). The detail properties of samples are shown in Table 3.

difference in absolute values. The peak photocurrent density with the smaller $(10\bar{1}2)$ XRC (ω scan) FWHM (0.70 mA/cm^2 at 69 min) was higher than that with the larger $(10\bar{1}2)$ XRC (ω scan) FWHM (0.49 mA/cm^2 at 39 min). This shows that defects observed by $(10\bar{1}2)$ XRC (ω scan) in GaN acting as recombination center and prevent from the photo-generated electrons and holes acting as photoreaction sources. This result is similar to the results from the comparison of carrier recombination with GaN growth methods as discussed (Fujii et al., 2009b, 2010). In contrast, the time-dependent trends of relative photocurrent densities for these Si-doped samples were similar. Thus, both of the photocurrent densities after 600 min from the photo-irradiation started were 28% of the peaks. These results indicate that the crystal quality observed by the FWHM of $(10\bar{1}2)$ XRC (ω scan) is one of the key characteristics for effective use of photo-generated electron and hole as the electrochemical reaction, but is independent of the photoreaction stability.

The time dependences of photocurrent densities with changing donor concentration are shown in Fig. 12. Although the FWHMs of $(10\bar{1}2)$ XRC (ω scan) were similar and around 300 arcsec for Si-doped GaN, the time dependence was drastically changed with the donor concentration. The photocurrent peak decreased with the donor concentration and almost disappeared for the sample with the donor concentration of 1.3×10^{16} cm^{-3}. In addition, the photocurrent density at the initial stage decreased with the donor concentration. The photocurrent density of the sample with 1.3×10^{16} cm^{-3} was much stable; thus, it was the highest after 600 min from the start of light irradiation as shown in Fig. 12. This indicates that the lower donor concentration sample is better for the stable photocurrent density. Since the photocurrent of the undoped sample, which had a high FWHM of $(10\bar{1}2)$ XRC (ω scan), was relatively stable, the key factor in a stable photocurrent is apparently the impurity concentration of GaN and not to be the FWHM of $(10\bar{1}2)$ XRC (ω scan).

SEM photographs of GaN surfaces after photoelectrochemical reactions of about 650 min are shown in Fig. 13. The surface morphology of the undoped sample rarely changed from that of the as-grown sample. The surface morphology of the Si-doped sample with a donor concentration of 1.3×10^{16} cm^{-3} was relatively smooth, while many pits were observed in the surface of the other Si-doped samples. The scratch-like line on the Si-doped sample with a donor concentration of 1.3×10^{16} cm^{-3} probably reflected some surface damage.

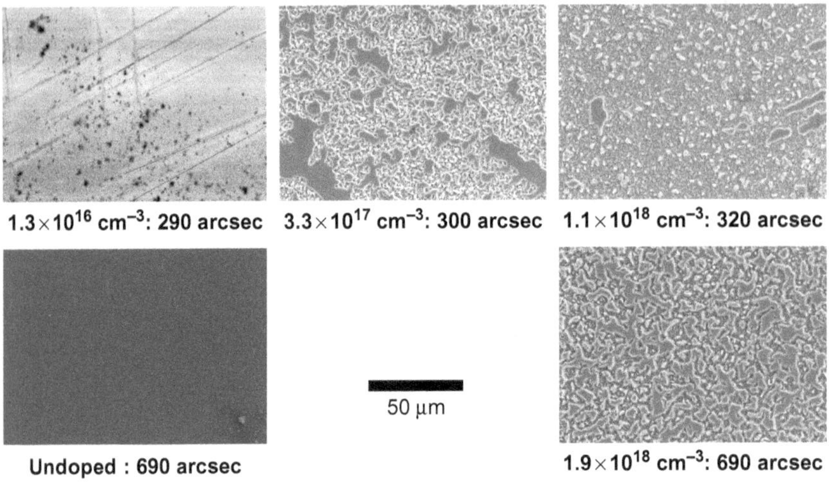

Fig. 13 Scanning electron microscope (SEM) photographs of the n-type GaN samples after photoelectrochemical reactions in 1.0 mol/L NaOH electrolyte under 270–280 mW/cm^2 Xe-lamp irradiation without bias. The numbers in the graph are the donor concentrations and $(10\bar{1}2)$ X-ray rocking curve (XRC: ω scan) (Fujii et al., 2011a, 2012). The detailed properties of samples are shown in Table 3.

From the low-temperature PL measurements, the peaks related to stacking faults (3.24–3.44 eV) were clearly observed from all of the samples after 600 min photoelectrochemical reactions. These results show that even if ones reduce the surface anodic corrosion for n-type GaN samples, the surface of the samples is still affected by the photoelectrochemical reaction.

From these evaluations throughout this section, point defect-like defects probably relate to the photoanodic reactions at the n-type GaN surface. The defects are affected not only by the crystal growth procedure and the doping amount of Si but also by the photoelectrochemical reaction time. This shows that the defects probably decide the number of reaction sites also. These results strongly suggest that the photoelectrochemical water oxidation at the surface of n-type GaN happens at a hole trap near the surface. This is very close to the model of the surface trap as an oxidation reaction center in the electrolyte reactant (Memming, 2015). This reaction model via a surface trap has a similar model of the Gerischer model. Since water oxidation is a four-electron reaction, the reaction is probably not the same as the one-electron outer sphere transfer model, which is the Gerischer model (adiabatic reaction). However, when the point defect-like defects are the reaction center of the photoelectrochemical oxidations, the reactions are probably similar to the Gerischer model especially for each electron transfer.

In addition, the reaction has probably some modifications like Marcus theory, which means that electron paths exist between the electrode and the electrolyte with the reactant chemisorption or physisorption at the electrode surface and changing the molecule shape after each electron transfer (nonadiabatic reaction).

5.3 Role of Electrocatalyst Loading on n-Type GaN

It is difficult to prevent n-type GaN photoanodic corrosion with changing the GaN properties as discussed. In order to prevent the photoanodic corrosion, electrocatalyst loading was proposed (Hayashi et al., 2012). The NiO electrocatalyst was selected as the candidate on the n-type GaN photoanode surface considered from the analogy of metal oxide photocatalyst. The NiO electrocatalysts were small particles and were dispersed on the GaN surface. The effect of NiO electrocatalyst on n-type GaN was excellent comparing with that of the property changes of GaN as shown in Fig. 14, and the anodic photocorrosion was completely suppressed. The effect of NiO is amazing that the photocurrent density was still similar at the beginning even after 500 h of photoelectrochemical reactions (Ohkawa et al., 2013).

The NiO particle loading procedure was clarified to use *n*-butyl acetate diluted NiO metal-organic decomposition (MOD) solution and annealed at over 500°C for 15 min after 120°C for 10 min solvent-removing process (Kim et al., 2014). The efficiency of the NiO MOD solution dilution ratio

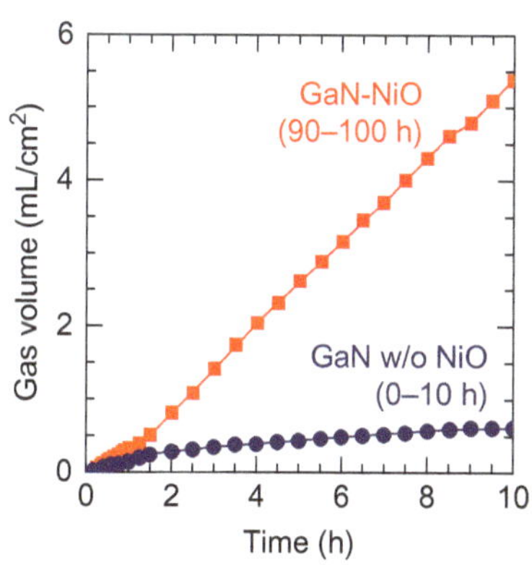

Fig. 14 Time dependence of hydrogen evolution at the Pt counterelectrode using working electrodes of NiO-loaded GaN (GaN-NiO) and as-grown GaN (GaN without NiO). The reactions were performed in 1.0 mol/L NaOH electrolyte under 100 mW/cm^2 Xe-lamp irradiation. *Quoted from fig. 2 in Hayashi, T., Deura, M., Ohkawa, K., 2012. High stability and efficiency of GaN photocatalyst for hydrogen generation from water. Jpn. J. Appl. Phys. 51, 112601. Copyright 2012 The Japan Society of Applied Physics.*

from 1:50 to 1:400 on n-type GaN with carrier concentration of 5×10^{18} cm^{-3} was evaluated. The optimized dilution ratio of 1:200 was observed from the experiment and the reason was summarized the fast collection of holes into the NiO region of optimized thickness and morphology.

The other NiO particle formation procedure on n-type GaN was proposed by using $Ni(OH)_2$ dispersed in ethanol as 0.02 mol/L (Koike et al., 2016). The $Ni(OH)_2$ dispersed ethanol was coated on GaN by spin coating and annealed at 280°C for 1 h. Basic photoelectrochemical properties were the same as the diluted MOD solution used. The time dependences of the photocurrent density showed a difference in the $Ni(OH)_2$ synthetic ways although the photocurrent densities and the stabilities of all NiO-loaded n-type GaN were improved from that of the as-grown GaN sample. The step and terrace structures on GaN surface, which were formed during the crystal growth, were observed even after photoelectrochemical reactions for NiO-loaded GaN. This means that the surface of GaN is kept the original structure and does not corrode.

The n-type GaN sample coated by 3 wt% NiO MOD solution showed low current density even compared with the as-grown GaN sample. The MOD-coated GaN surface had a NiO layer structure, whereas the $Ni(OH)_2$-coated samples had NiO particles on the GaN surface. This probably suggests that the NiO layer has high electric resistivity and prevents the electrochemical reaction on the surface.

The NiO electrocatalyst was reported to be more effective when the carrier concentration of n-type GaN was higher as the order of 10^{19} cm^{-3} when the carrier concentration was between 3×10^{18} and 2×10^{19} cm^{-3} (Kang et al., 2014). However, the NiO electrocatalytic effect was observed even in the case of relatively low carrier concentration of 10^{17} cm^{-3} order (Hayashi et al., 2012; Ohkawa et al., 2013). This probably means that the NiO-loading effect is independent of carrier concentration of n-type GaN.

The pH dependence of NiO-loaded n-type GaN clearly shows the role of NiO on GaN surface (Koike et al., in press). The electrolytes used during the evaluation were 0.5 mol/L H_2SO_4, 1.0 mol/L KOH, and 1.0 mol/L NaOH. In acidic electrolyte of H_2SO_4, the NiO particles corroded into the electrolyte and the morphology of GaN surface was the same as the non-treated as-grown GaN sample after photoelectrochemical reactions. In contrast, the anodic corrosion was not observed in basic solutions of KOH and NaOH. This clearly reflects the solubility of NiO pH dependence of solutions and the GaN surface showed anodic corrosion when the NiO particles were corroded in the electrolyte.

The mechanisms of the NiO particle loaded on n-type GaN were discussed but still obscure. The Ni metal of NiO is the transition metal and has 3d electrons. This NiO is a semiconductor and undoped NiO is said to have p-type electric properties. Since the NiO band is affected by the 3d electrons, the band structure is complicated with the isolated band in between so-called conduction and valence bands (Cox, 2010). The possible band alignment was proposed in 2014 (Kang et al., 2014), which shows that the valence band (or the isolated band related to 3d electrons) edge of NiO is close to the water oxidation redox potential with a small amount of high hole energy as shown in Fig. 15. This close band edge position to water oxidation redox potential is believed to drive the water oxidation more effectively compared with n-type GaN, in which the valence band edge is far from the water oxidation redox potential. The electron trap density for NiO is considered to be higher than that of intermediate state (point defect-like defects) of GaN, which is believed to be the electrochemical reaction center for water oxidation of GaN. This difference of density is probably related to the reaction speed of water oxidation.

In addition, since NiO is reported to be easily oxidized to NiOOH (Sialvi et al., 2013), the role of NiO electrocatalyst is also affected by this valence change from 2+ to 3+. The NiOOH probably is easy to make hydrogen bond with water molecule and H^+, OH^- ions in electrolyte; thus, the reaction mechanism is expected to be much more complicated.

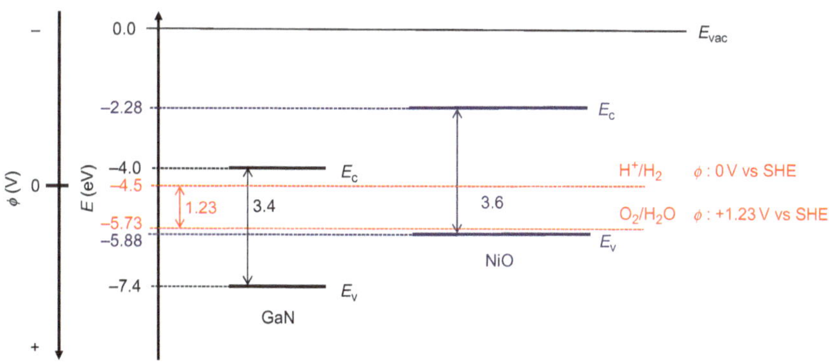

Fig. 15 Band alignments of GaN and NiO with water reduction (hydrogen evolution) and oxidation (oxygen evolution) redox potentials. The ϕ, E, E_{vac}, E_c, and E_v show electrode potential, electron energy vacuum level, conduction band edge energy, and valence band edge energy, respectively. The GaN band edge energies were evaluated from the hydrogen evolution energy. The band edge energy of NiO was obtained from literature (Toroker et al., 2011). Since the type of NiO is difficult to evaluate, the Fermi level of NiO is considered to be in the middle of the band gap.

6. CO_2 REDUCTION USING III-NITRIDE SEMICONDUCTORS

CO_2 reduction by III-nitride semiconductor is also proposed because of its high conduction band energy of electron. Since some of the redox energies of CO_2 reduction are more positive (redox potentials are more negative) than that of water reduction (hydrogen generation), the higher energy of electron at the conduction band edge energy was believed to be suitable for CO_2 reduction. Since electron energy used for reduction is defined by its quasi-Fermi level, the real mechanism is more complicated, but it can be used as an indicator. Thus, n-type GaN and $Al_xGa_{1-x}N$ were considered to be used as the photoanode for CO_2 reduction.

The first report of CO_2 reduction was performed by using n-type GaN as a photoanode in 1.0 mol/L NaOH aqueous electrolyte and Cu plate cathode in 0.1 mol/L $KHCO_3$ aqueous electrolyte (Yotsuhashi et al., 2011, 2012b). The electrolytes of anode and cathode sides were connected with a 0.18-mm cation exchange membrane; thus, the reaction was accelerated by the chemical bias induced from the difference of the electrolyte pH even though the anode and cathode were electrically connected directly. The production materials were CO, CH_4, HCOOH, and parasitic production of water-reduced H_2. The Faradic ratio of H_2 was dominant followed by HCOOH. The Faradic efficiency of HCOOH formation was 3.2% after 10 C electron charge flow. In addition, the HCOOH production was proportional with the amount of electron charge.

The Faradic efficiency was improved when the III-nitride structure changed from n-type GaN layer to 100 nm undoped GaN or 100 nm undoped $Al_{0.1}Ga_{0.9}N$ with NiO particles loading on the surfaces. The other experimental conditions were the same as the previous ones except for the concentration of $KHCO_3$ aqueous electrolyte changed from 0.1 to 0.5 mol/L. The product materials were H_2, CO, CH_4, C_2H_4, C_2H_6, and HCOOH evaluated by gas and liquid chromatography. The Faradic efficiency of HCOOH was the highest except for the parasitic production of H_2. The Faradic efficiencies were reported to be 14.15% for $Al_{0.1}Ga_{0.9}N$ and 8.88% for GaN after 20 C electron charge. The photoanode surfaces were not damaged due to the NiO particle-loaded effects (Yotsuhashi et al., 2012a,c).

Next improvement was performed with using a GaN substrate instead of conventional sapphire substrate to reduce the total electric resistance. The other conditions except for the electrolytes were the same as the NiO particle-loaded 100 nm undoped $Al_{0.1}Ga_{0.9}N$ electrode case. The electrolytes for undoped $Al_{0.1}Ga_{0.9}N$ electrode and Cu cathode were: (A) (on sapphire) and (B) (on GaN) 1.0 mol/L NaOH and 0.5 mol/L $KHCO_3$, (C) 5.0 mol/L NaOH and 1.0 mol/L $KHCO_3$, and (D) 5.0 mol/L NaOH and 3.0 mol/L KCl, respectively. A cation exchange membrane was used for the separation of the anodic and cathodic reaction chambers. The HCOOH production ratio was increased with the order of A, B, and C. The main products were H_2 and HCOOH for condition C and CO, C_2H_4, HCOOH, C_2H_5OH, and CH_3CHO for condition D. The produced materials for condition D were different from those for other conditions. The Faradic efficiency was greatly increased when KCl aqueous solution was used as the electrolyte for the cathode side (condition D) and the total CO_2 reduction was 75% (Deguchi et al., 2013).

The NiO-loaded undoped $Al_{0.1}Ga_{0.9}N$ on n-type GaN electrically connected on the p/i/n-type Si photovoltaic structure was also used as photoanode for CO_2 reduction (Yotsuhashi et al., 2014). The experiment was performed with an irradiated light area of 1.6 in. in diameter under AM1.5 condition. The electrolytes for GaN on the Si photoanode side was 5 mol/L NaOH aqueous solution and In cathode side was 0.5 mol/L $KHCO_3$. This condition was compared with the electrolytes for GaN on the Si photoanode side was 5 mol/L NaOH aqueous solution and Cu cathode and 3 mol/L KCl aqueous electrolyte. The cation exchange membrane was used for the separation of the anodic and cathodic reaction chambers. The product of HCOOH was dominant when the In cathode was used and the ratio of HCOOH for cathode and O_2 for anode was 2:1. When the cathode material changed from In to Cu, the products were also changed from HCOOH to hydrocarbons.

These results show that the electron located at the conduction band edge of III-nitrides can be used as the CO_2 reduction promoter.

7. $In_xGa_{1-x}N$, III-NITRIDE PHOTOANODES AND POROUS MATERIALS

7.1 $In_xGa_{1-x}N$

High In content $In_xGa_{1-x}N$ is one of the expected materials in III-nitrides due to it has a possibility to absorb visible light for water splitting. However, the first report was just evaluating its band edge energies (Fujii et al., 2005b, 2007c).

The photoelectrochemical properties of n-type $In_xGa_{1-x}N$ ($x \sim 0.2$ and 0.4) were observed in 1 mol/L HCl aqueous electrolyte with white light irradiation (Li et al., 2008). Preliminary results were obtained, and the photocurrent density as a function of applied bias showed that the photocurrent density for $x \sim 0.4$ was clearly higher than that of $x \sim 0.2$, although the crystal quality of $x \sim 0.4$ was lower than that of $x \sim 0.2$.

The photoelectrochemical properties of $In_xGa_{1-x}N/GaN$ ($0 \leq x \leq 0.20$) were evaluated and found that the flatband potential of $In_xGa_{1-x}N$ is shifted to more positive voltage positions with increasing indium incorporation. In HBr aqueous solution, $In_{0.20}Ga_{0.80}N$ has good stability under photo-irradiation even when the reaction is a photoanodic one. Moreover, $In_{0.20}Ga_{0.80}N$ shows the highest visible-light response and the photon conversion efficiency is about 9% at 400–430 nm light irradiation in HBr aqueous solution (Luo et al., 2008).

7.2 p-Type GaN and $In_xGa_{1-x}N$

The p-type III-nitrides can be used as photocathode; thus, it is expected to be stable under photoelectrochemical condition. However, since the crystal qualities of p-type GaN are not high due to its crystal growth condition, which requires being not the same as the optimized crystal growth condition, photoelectrochemical properties of p-type III-nitrides have not been reported so much (Fujii and Ohkawa, 2005). The effects of the flow rate of Mg doping to grow p-type GaN by metal MOVPE were reported about a decade after the first report (Bae et al., 2016b). The optimum flow rate of the Mg precursor was found to exist and the photocathode was stable over 2500 s under the condition with −1.0 V vs Ag/AgCl/NaCl (the redox potential of Ag/AgCl/NaCl is +0.212 V vs SHE) in the 1.0 mol/L aqueous electrolyte.

The photoelectrochemical properties of 0.25 μm thick p-type $In_xGa_{1-x}N$ ($0 \leq x \leq 0.22$) on semiinsulating (0001) GaN/AlN grown on the sapphire substrate were evaluated (Aryal et al., 2010). The electric contact was deposited of Ni (30 nm)/Au (120 nm) and annealed 550°C for 90 s in the air. This anneal process was probably also used for the p-type activation process. The light source, electrolyte, and counterelectrode used were AM1.5 solar simulator light source (the light intensity at the sample was 132 mW/cm^2), Pt, and 1 mol/L HBr, respectively. The measured photocurrent densities were much higher in p-type $In_xGa_{1-x}N$ than in p-type GaN as shown in Fig. 16. However, the dependence of photocurrent density on the In composition was not seen. This is probably due to the fact that the material quality of p-type $In_xGa_{1-x}N$ dominates over all other factors; thus, the advantage of

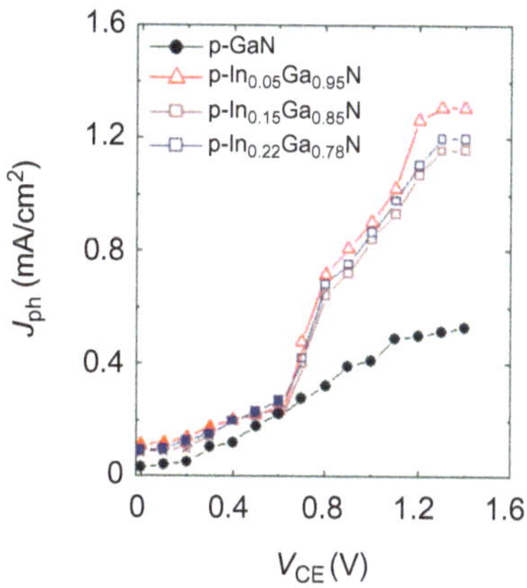

Fig. 16 Photocurrent densities (J_{ph}) as a function of V_{CE} (voltage applied between working and counterelectrodes) under white light irradiation using a standard AM1.5 solar simulator. The light intensity at the sample surface was about 132 mW/cm^2. The electrolyte was 1 mol/L HBr aqueous solution. *Open symbols* indicate generation of H_2 gas, while *solid symbols* indicate no H_2 gas generation. *Quoted from fig. 1 in Aryal, K., Pantha, B.N., Li, J., Lin, J.Y., Jiang, H.X., 2010. Hydrogen generation by solar water splitting using p-InGaN photoelectrochemical cells. Appl. Phys. Lett. 96, 052110.*

the small band gap is not observed in present p-type $In_xGa_{1-x}N$ materials. This was also explained by the relationship between photocurrent density and hole mobility of p-type $In_xGa_{1-x}N$. The photocurrent density was almost proportional to the hole mobility.

The photocurrent densities decreased at the beginning, but the values kept constant for about 24 h after 10 min at the beginning as shown in Fig. 17. Any etching effects occurring on the surface of the p-type $In_xGa_{1-x}N$ working photocathode were not observed. The reason is that p-type conductivity provides a reduction reaction and prevents the photoanodic corrosion.

As discussed here, the crystal qualities of p-type III-nitrides are not good compared with those of undoped or n-type ones. New idea with using nitride crystal polarity was proposed for photocathodic electrode (Nakamura et al., 2014). The water-splitting photocathode structure was an undoped GaN/a few-nm-thick undoped AlN/an n-type GaN (undoped GaN/AlN/n-type GaN). Although this structure does not contain any p-type semiconductors, the polarization-induced field in the AlN layer makes this structure work as a photocathode.

In a GaN/AlN/GaN structure grown in the direction of (0001) (Ga-face), for example, the charge density of the polarization is observed and its sign is positive at the bottom AlN/GaN and negative at the top GaN/AlN

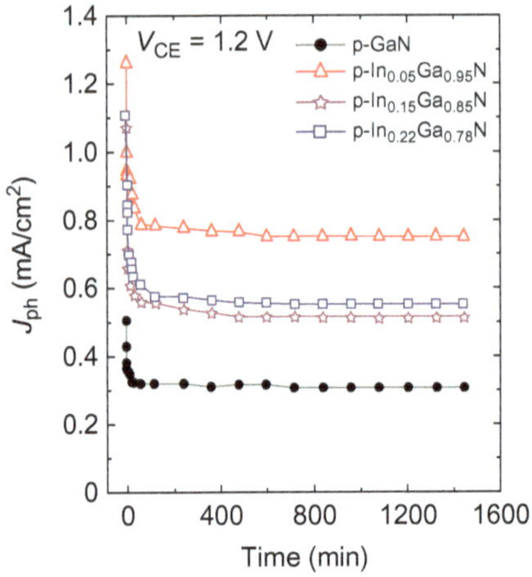

Fig. 17 Photocurrent densities (J_{ph}) of p-type $In_xGa_{1-x}N$ electrodes at $V_{CE}=1.2$ V as a function of the measurement time (t). The light intensity at the sample surface was about 132 mW/cm^2. The electrolyte was 1 mol/L HBr aqueous solution. *Quoted from fig. 2 in Aryal, K., Pantha, B.N., Li, J., Lin, J.Y., Jiang, H.X., 2010. Hydrogen generation by solar water splitting using p-InGaN photoelectrochemical cells. Appl. Phys. Lett. 96, 052110.*

interface. The calculated band diagrams under thermal equilibrium conditions of the p-type GaN/AlN/n-type GaN structure with doping concentrations of acceptor for p-type GaN$=1\times10^{19}$ cm^{-3}, donor for AlN$=3\times10^{17}$ cm^{-3}, and donor for n-type GaN$=2\times10^{18}$ cm^{-3} are shown in Fig. 18. The AlN layer was 4.5 nm thick. Since the huge polarization charge with different signs exists at the two interfaces, a tunnel junction is formed at the thin AlN layer between a p-type and an n-type GaN. When the top layer is replaced from the p-type GaN layer to undoped GaN, the electric field appears in the undoped GaN layer. Entire electric field is observed in the surface-undoped GaN due to the entire depletion by the bias when the thickness and doping concentration of donor for undoped GaN are 150 nm and 3×10^{16} cm^{-3} as also shown in Fig. 18.

The real polarized-engineering structures were also grown by MOVPE. The structures were 150 nm undoped GaN/thin undoped AlN/1.3 μm 2.0×10^{17} cm^{-3} n-type GaN/undoped GaN on (0001) sapphire substrates. The thicknesses of the thin AlN layers were 3.4, 4.4, 5.4, and 6.3 nm. The characteristics of photocurrent densities vs bias compared with that for p-type GaN were shown in Fig. 19. Clear photocathodic properties were observed for the polarization-engineered structures, and the properties were

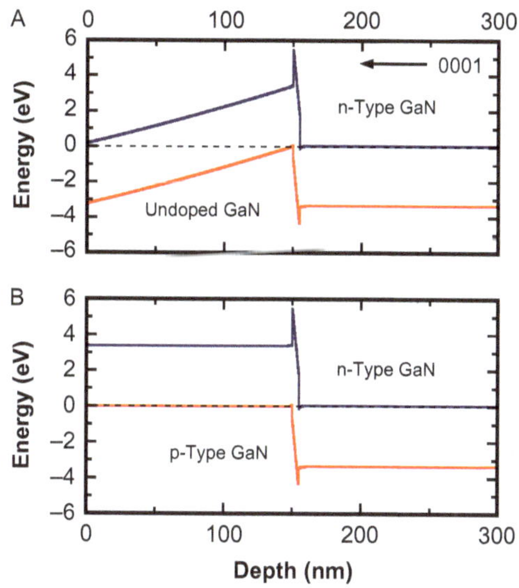

Fig. 18 Calculated energy band diagrams of (A) the p-type GaN/AlN/n-type GaN polarization-induced tunneling junction and (B) the undoped GaN/AlN/n-type GaN proposed photocathode in thermal equilibrium with a 4.5 nm thick AlN layer. *Quoted from fig. 1 of Nakamura, A., Fujii, K., Sugiyama, M., Nakano, Y., 2014. A nitride based polarization-engineered photocathode for water splitting without a p-type semiconductor. Phys. Chem. Chem. Phys. 16, 15326–15330.*

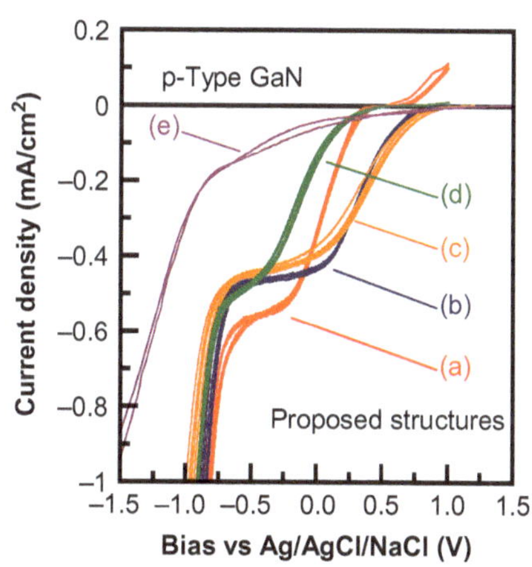

Fig. 19 Photocurrent density vs bias characteristics under 110 mW/cm^2 light irradiation of the proposed structures of 150 nm undoped GaN/AlN/n-type GaN with (a) 3.4, (b) 4.4, (c) 5.4, and (d) 6.3 nm thick AlN layers. The electrolyte was 0.5 mol/L H_2SO_4. Photoelectrochemical properties of p-type GaN are also plotted in (e). *Quoted from fig. 6 of Nakamura, A., Fujii, K., Sugiyama, M., Nakano, Y., 2014. A nitride based polarization-engineered photocathode for water splitting without a p-type semiconductor. Phys. Chem. Chem. Phys. 16, 15326–15330.*

better than that of p-type GaN. This is probably because of the better crystal quality and longer depletion layer thickness for the proposed structure. From the comparison of turn-on voltage for the proposed polarization-engineered structure, the AlN layer thickness of 5.4 nm was the most positive. This shows that the AlN layer has optimized thickness.

The stability of the proposed polarized-engineered sample with −0.5 V vs Ag/AgCl/NaCl under 330 mW/cm^2 irradiation compared with n-type GaN with +1.0 V vs Ag/AgCl/NaCl under 330 mW/cm^2 is shown in Fig. 20. The current of GaN surface was cathodic for proposed sample and anodic for n-type GaN. The stability of the proposed sample was similar to the p-type III-nitrides and was stable for 200 min under photoelectrochemical reaction, whereas the photocurrent density of n-type GaN drastically decreased and approached to zero at around 100 min. This indicates that the proposed polarized-engineered sample shows p-type photoelectrochemical properties and the properties are expected to be better than the p-type III-nitrides.

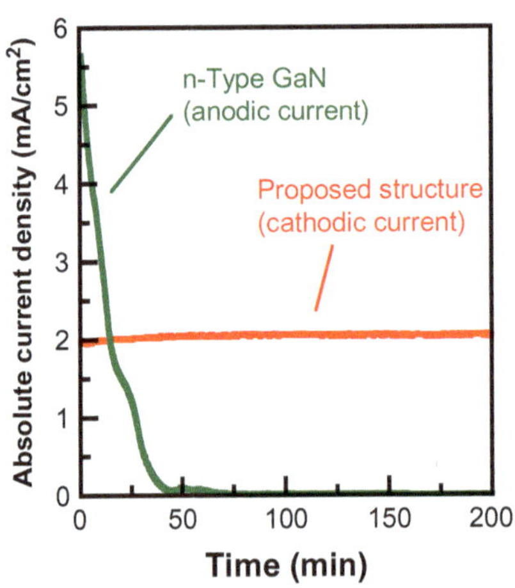

Fig. 20 Photocurrent density of the 3.4 nm AlN polarization-engineered photocathode and n-type bulk GaN photoanode under light irradiation as a function of reaction time. Biases on the proposed structure of 150 nm undoped GaN/AlN/n-type GaN and the n-type GaN were −0.5 and +1.0 V vs Ag/AgCl/NaCl, respectively. The light intensity at the sample position was about 330 mW/cm^2. The electrolyte was 0.5 mol/L H_2SO_4. *Quoted from fig. 5 of Nakamura, A., Fujii, K., Sugiyama, M., Nakano, Y., 2014. A nitride based polarization-engineered photocathode for water splitting without a p-type semiconductor. Phys. Chem. Chem. Phys. 16, 15326–15330.*

7.3 Nanoporous GaN Photoelectrochemical Properties

Photoelectrochemical properties of nanoporous n-type GaN are expected to improve from the planer structure since the nanoporous structure increased the reaction surface area (Ryu et al., 2012). The nanoporous GaN sample was 2 μm thick highly doped n-type GaN with the carrier concentration of 5×10^{18} cm^{-3} grown after 1.8 μm thick undoped GaN on (0001) sapphire substrate by MOVPE. The top surface of GaN was processed by electrochemical etching at 15 V for 10 min in 0.3 mol/L oxalic acid at room temperature.

The properties of photocurrent density vs bias and stability were better when nanoporous sample was used in 1.0 mol/L NaOH aqueous electrolyte under 200 W Hg lamp irradiation. The biases were observed in the working electrode biases vs counterelectrode of Pt. This fact was explained by the enhanced the amount of hole transport from photoanode to electrolyte due to the enlargement of the reaction surface area of nanoporous n-type GaN from planer one This fact was explained by the enhanced the amount of hole transport from photoanode to electrolyte due to the enlargement of the reaction surface area of nanoporous n-type GaN from planer one as expected. This behavior would be beneficial for photoelectrochemical water-splitting system.

8. SUMMARY

Photoelectrochemical properties of III-nitride semiconductors were discussed based on the semiconductor photoelectrochemical evaluation technique, and some applications of the III-nitride photoelectrodes were introduced.

The III-nitride semiconductors, especially for the GaN, have suitable properties for water splitting. The band edge energy of GaN straddles the water-splitting energies, GaN is chemically stable, and GaN has high quality, which can be used for light-emitting devices. $In_xGa_{1-x}N$ is also expected to absorb visible light due to the band gap decreasing with increasing the In content.

The pH dependence of the band edge potential of GaN shows almost the same of the Nernst relationship for water oxidation and reduction reactions, which suggests that the surface of GaN was covered by water, H^+, and OH^- ions. Photoanodic corrosion is observed when n-type GaN is used although GaN is chemically stable. In contrast, photocorrosion is not observed when p-type III-nitrides are used as photocathodes. This probably indicated that the bond structure of III-nitrides is covalent bond and can be oxidized under oxidized condition, which is anodic corrosion. This means that, since the Ga valence of III-nitrides is close to zero, the surface is stable when the reaction condition is cathodic.

In order to suppress the photoanodic corrosion, NiO particle loading on n-type GaN was proposed. The effect of GaN loading is excellent and no surface change is observed even after 500-h reaction. This can be explained by the assumption that the hole near the n-type GaN surface is collected into the loaded NiO particles and the carrier transfer at the NiO is much faster than that at the GaN surface.

$Al_xGa_{1-x}N$ photoanode was applied for CO_2 reduction due to its higher conduction band edge energy than that for GaN. The higher conduction band edge energy is believed to be the easier CO_2 reduction at the counter-electrode. The CO_2 reduction was observed in addition to chemical bias produced by pH or in addition to Si solar cell bias application.

Photoelectrochemical water splitting by p-type GaN, $In_xGa_{1-x}N$, and polarized-engineered GaN (the surface-undoped GaN works like p-type semiconductor due to AlN thin-layer introduction) showed excellent stability compared with those by n-type III-nitrides even though the p-type III-nitrides require bias for operation. Although the photoelectrochemical properties of p-type III-nitrides are not high compared with those for n-type ones due to its low crystal qualities, the polarized-engineered sample shows a comparable performance of n-type GaN.

Nanoporous n-type GaN showed better photoelectrochemical properties than that for planer one. This indicates that nanostructure GaN is expected to show good photoelectrochemical performance.

REFERENCES

Aryal, K., Pantha, B.N., Li, J., Lin, J.Y., Jiang, H.X., 2010. Hydrogen generation by solar water splitting using p-InGaN photoelectrochemical cells. Appl. Phys. Lett. 96, 052110.

Bae, H., Park, J., Jung, K.C., Nakamura, A., Fujii, K., Park, H.J., Jeong, T., Lee, H.J., Moon, Y.B., Ha, J.S., 2013. The polarity effect on the photoelectrochemical properties of Ga- and N-face free-standing GaN substrate. Jpn. J. Appl. Phys. 52, 4.

Bae, H., Kim, E., Park, J.-B., Kang, S.-J., Fujii, K., Lee, S.H., Lee, H.-J., Ha, J.-S., 2016a. Effect of polarity on photoelectrochemical properties of polar and semipolar GaN photoanode. J. Electrochem. Soc. 163, H213–H217.

Bae, H., Kim, E., Park, J.B., Fujii, K., Lee, S., Lee, H.J., Ryu, S.W., Lee, J.K., Ha, J.S., 2016b. Influence of Mg flux on the photoelectrochemical properties of p-type GaN for hydrogen production. J. Nanosci. Nanotechnol. 16, 10635–10638.

Bard, A.J., Faulkner, L.R., Leddy, J., Zoski, C.G., 1980. Electrochemical Methods: Fundamentals and Applications.Wiley, New York.

Beach, J.D., Collins, R.T., Turner, J.A., 2003. Band-edge potentials of n-type and p-type GaN. J. Electrochem. Soc. 150, A899–A904.

Chen, S., Wang, L.-W., 2012. Thermodynamic oxidation and reduction potentials of photocatalytic semiconductors in aqueous solution. Chem. Mater. 24, 3659–3666.

Cox, P.A., 2010. Transition Metal Oxides: An Introduction to Their Electronic Structure and Properties. Oxford University Press, Oxford, UK.

Deguchi, M., Yotsuhashi, S., Hashiba, H., Yamada, Y., Ohkawa, K., 2013. Enhanced capability of photoelectrochemical CO2 conversion system using an AlGaN/GaN photoelectrode. Jpn. J. Appl. Phys. 52, 08JF07.

Fujii, K., 2016a. Non-oxide materials (nitrides, chalcogenides, and arsenides). In: Giménez, S., Bisquert, J. (Eds.), Photoelectrochemical Solar Fuel Production: From Basic Principles to Advanced Devices. Springer International Publishing, Cham, Switzerland.

Fujii, K., 2016b. Thermodynamics for electrochemistry and photoelectrochemistry. In: Sugiyama, M., Fujii, K., Nakamura, S. (Eds.), Solar to Chemical Energy Conversion: Theory and Application. Springer International Publishing, Cham, Switzerland.

Fujii, K., Ohkawa, K., 2005. Photoelectrochemical properties of p-type GaN in comparison with n-type GaN. Jpn. J. Appl. Phys. Pt 2 44, L909–L911.

Fujii, K., Ohkawa, K., 2006a. Bias-assisted H2 gas generation in HCl and KOH solutions using n-Type GaN photoelectrode. J. Electrochem. Soc. 153, A468–A471.

Fujii, K., Ohkawa, K., 2006b. Hydrogen generation from aqueous water using n-GaN by photoassisted electrolysis. Phys. Status Solidi C 3, 2270–2273.

Fujii, K., Karasawa, T., Ohkawa, K., 2005a. Hydrogen gas generation by splitting aqueous water using n-type GaN photoelectrode with anodic oxidation. Jpn. J. Appl. Phys. 44, L543.

Fujii, K., Kusakabe, K., Ohkawa, K., 2005b. Photoelectrochemical properties of InGaN for H-2 generation from aqueous water. Jpn. J. Appl. Phys. Pt 1 44, 7433–7435.

Fujii, K., Ito, T., Ono, M., Iwaki, Y., Yao, T., Ohkawa, K., 2007a. Investigation of surface morphology of n-type GaN after photoelectrochemical reaction in various solutions for H2 gas generation. Phys. Status Solidi C 4, 2650–2653.

Fujii, K., Iwaki, Y., Masui, H., Baker, T.J., Iza, M., Sato, H., Kaeding, J., Yao, T., Speck, J.S., Denbaars, S.P., Nakamura, S., Ohkawa, K., 2007b. Photoelectrochemical properties of nonpolar and semipolar GaN. Jpn. J. Appl. Phys. 46, 6573.

Fujii, K., Ono, M., Ito, T., Iwaki, Y., Hirako, A., Ohkawa, K., 2007c. Band-edge energies and photoelectrochemical properties of n-type AlxGa1-xN and InyGa1-yN alloys. J. Electrochem. Soc. 154, B175–B179.

Fujii, K., Nakayama, H., Sato, K., Kato, T., Cho, M.-W., Yao, T., 2008. Improvement of hydrogen generation efficiency using GaN photoelectrochemical reaction in electrolytes with alcohol. Phys. Status Solidi C 5, 2333–2335.

Fujii, K., Sato, K., Kato, T., Koike, K., Yao, T., 2009a. Surface changes by GaN photoelectrochemical treatment using various electrolytes with and without C2H5OH. Phys. Status Solidi C 6, S627–S630.

Fujii, K., Sato, K., Kato, T., Minegishi, T., Yao, T., 2009b. Improvement of photoelectrochemical reaction for hydrogen generation from water using N-face GaN. MRS Proc. 1202, 1202–I07–03.

Fujii, K., Kato, T., Minegishi, T., Yamada, T., Yamane, H., Yao, T., 2010. Photoelectrochemical properties of single crystalline and polycrystalline GaN grown by the Na-flux method. Electrochemistry 78, 136–139.

Fujii, K., Koike, K., Atsumi, M., Goto, T., Itoh, T., Yao, T., 2011a. Time dependence of water-reducing photocurrent with change of the characteristics of n-type GaN photoilluminated working electrodes. Phys. Status Solidi C 8, 2457–2459.

Fujii, K., Nakamura, S., Yokojima, S., Goto, T., Yao, T., Sugiyama, M., Nakano, Y., 2011b. Photoelectrochemical properties of $In_xGa_{1-x}N$/GaN multiquantum well structures in depletion layers. J. Phys. Chem. C 115, 25165–25169.

Fujii, K., Koike, K., Atsumi, M., Goto, T., Itoh, T., Yao, T., 2012. Photoluminescence changes in n-type GaN samples after photoelectrochemical treatment. Phys. Status Solidi C 9, 715–718.

Gerischer, H., 1977. On the stability of semiconductor electrodes against photodecomposition. J. Electroanal. Chem. Interfacial Electrochem. 82, 133–143.
Hayashi, T., Deura, M., Ohkawa, K., 2012. High stability and efficiency of GaN photocatalyst for hydrogen generation from water. Jpn. J. Appl. Phys. 51, 112601.
Huygens, I.M., Strubbe, K., Gomes, W.P., 2000. Electrochemistry and photoetching of n-GaN. J. Electrochem. Soc. 147, 1797–1802.
Kaganer, V.M., Brandt, O., Trampert, A., Ploog, K.H., 2005. X-ray diffraction peak profiles from threading dislocations in GaN epitaxial films. Phys. Rev. B 72, 045423.
Kang, J.-H., Kim, S.H., Ebaid, M., Lee, J.K., Ryu, S.-W., 2014. Efficient photoelectrochemical water splitting by a doping-controlled GaN photoanode coated with NiO cocatalyst. Acta Mater. 79, 188–193.
Kim, S.H., Ebaid, M., Kang, J.-H., Ryu, S.-W., 2014. Improved efficiency and stability of GaN photoanode in photoelectrochemical water splitting by NiO cocatalyst. Appl. Surf. Sci. 305, 638–641.
Ko, C.H., Su, Y.K., Chang, S.J., Lan, W.H., Webb, J., Tu, M.C., Cherng, Y.T., 2002. Photo-enhanced chemical wet etching of GaN. Mater. Sci. Eng., B 96, 43–47.
Kocha, S.S., Peterson, M.W., Arent, D.J., Redwing, J.M., Tischler, M.A., Turner, J.A., 1995. Electrochemical investigation of the gallium nitride-aqueous electrolyte interface. J. Electrochem. Soc. 142, L238–L240.
Koike, K., Sato, K., Fujii, K., Goto, T., Yao, T., 2010. Time variation of GaN photoelectrochemical reactions affected by light intensity and applied bias. Phys. Status Solidi C 7, 2221–2223.
Koike, K., Nakamura, A., Sugiyama, M., Nakano, Y., Fujii, K., 2014. Surface stability of n-type GaN depending on carrier concentration and electrolytes under photoelectrochemical reactions. Phys. Status Solidi C 11, 821–823.
Koike, K., Yamamoto, K., Ohara, S., Sugiyama, M., Nakano, Y., Fujii, K., 2016. Photoelectrochemical property differences between NiO dots and layer on n-type GaN for water splitting. J. Electrochem. Soc. 163, H1091–H1095.
Koike K., Yamamoto K., Ohara S., Kikitsu T., Ozasa K., Nakamura S., Sugiyama M., Nakano Y. and Fujii K., Effects of NiO-loading on n-type GaN photoanode for photoelectrochemical water splitting using different aqueous electrolytes, *Int. J. Hydrogen Energ.* in press.
Kudo, A., 2003. Photocatalyst materials for water splitting. Catal. Surveys Asia 7, 31–38.
Li, J., Lin, J.Y., Jiang, H.X., 2008. Direct hydrogen gas generation by using InGaN epilayers as working electrodes. Appl. Phys. Lett. 93, 162107.
Luo, W., Liu, B., Li, Z., Xie, Z., Chen, D., Zou, Z., Zhang, R., 2008. Stable response to visible light of InGaN photoelectrodes. Appl. Phys. Lett. 92, 262110.
Marques, M., Teles, L., Scolfaro, L., Leite, J., Furthmuller, J., Bechstedt, F., 2003. Lattice parameter and energy band gap of cubic AlxGayIn1-x-yN quaternary alloys. Appl. Phys. Lett. 83, 890–892.
Memming, R., 2015. Semiconductor Electrochemistry, second ed. Wiley-VCH, Weinheim, Germany.
Nakamura, S., Chichibu, S.F., 2000. Introduction to Nitride Semiconductor Blue Lasers and Light Emitting Diodes. Taylor & Francis, London, UK.
Nakamura, A., Fujii, K., Sugiyama, M., Nakano, Y., 2014. A nitride based polarization-engineered photocathode for water splitting without a p-type semiconductor. Phys. Chem. Chem. Phys. 16, 15326–15330.
Nozik, A.J., Memming, R., 1996. Physical Chemistry of Semiconductor–Liquid Interfaces. J. Phys. Chem. 100, 13061–13078.
Ohkawa, K., Ohara, W., Uchida, D., Deura, M., 2013. Highly stable GaN photocatalyst for producing H2 gas from water. Jpn. J. Appl. Phys. 52, 08JH04.
Ono, M., Fujii, K., Ito, T., Iwaki, Y., Hirako, A., Yao, T., Ohkawa, K., 2007. Photoelectrochemical reaction and H2 generation at zero bias optimized by carrier concentration of n-type GaN. J. Chem. Phys. 126, 054708.

Reshchikov, M.A., Morkoç, H., 2005. Luminescence properties of defects in GaN. J. Appl. Phys. 97, 061301.

Ryu, S.-W., Zhang, Y., Leung, B., Yerino, C., Han, J., 2012. Improved photoelectrochemical water splitting efficiency of nanoporous GaN photoanode. Semicond. Sci. Technol. 27, 015014.

Sato, K., Fujii, K., Koike, K., Goto, T., Yao, T., 2009. Anomalous time variation of photocurrent in GaN during photoelectrochemical reaction for H2 gas generation in NaOH aqueous solution. Phys. Status Solidi C 6, S635–S638.

Shannon, R., 1976. Revised effective ionic radii and systematic studies of interatomic distances in halides and chalcogenides. Acta Crystallogr. A 32, 751–767.

Sialvi, M.Z., Mortimer, R.J., Wilcox, G.D., Teridi, A.M., Varley, T.S., Wijayantha, K.G., Kirk, C.A., 2013. Electrochromic and colorimetric properties of nickel(II) oxide thin films prepared by aerosol-assisted chemical vapor deposition. ACS Appl. Mater. Interfaces 5, 5675–5682.

Toroker, M.C., Kanan, D.K., Alidoust, N., Isseroff, L.Y., Liao, P., Carter, E.A., 2011. First principles scheme to evaluate band edge positions in potential transition metal oxide photocatalysts and photoelectrodes. Phys. Chem. Chem. Phys. 13, 16644–16654.

Yao, T., Hong, S.-K., 2009. Oxide and Nitride Semiconductors: Processing, Properties, and Applications. Springer, Berlin, Heidelberg, Germany.

Yotsuhashi, S., Deguchi, M., Zenitani, Y., Hinogami, R., Hashiba, H., Yamada, Y., Ohkawa, K., 2011. Photo-induced CO2 reduction with GaN electrode in aqueous system. Appl. Phys. Expr. 4, 117101.

Yotsuhashi, S., Deguchi, M., Hashiba, H., Zenitani, Y., Hinogami, R., Yamada, Y., Ohkawa, K., 2012a. Enhanced CO2 reduction capability in an AlGaN/GaN photoelectrode. Appl. Phys. Lett. 100, 243904.

Yotsuhashi, S., Deguchi, M., Zenitani, Y., Hinogami, R., Hashiba, H., Yamada, Y., Ohkawa, K., 2012b. CO2 conversion with light and water by GaN photoelectrode. Jpn. J. Appl. Phys. 51, 02BP07.

Yotsuhashi, S., Hashiba, H., Deguchi, M., Zenitani, Y., Hinogami, R., Yamada, Y., Deura, M., Ohkawa, K., 2012c. Highly efficient photochemical HCOOH production from CO2 and water using an inorganic system. AIP Adv. 2, 042160.

Yotsuhashi, S., Deguchi, M., Yamada, Y., Ohkawa, K., 2014. Effect of inserted Si p-n junction on GaN-based photo-electrochemical CO2 conversion system. AIP Adv. 4, 067135.

CHAPTER FIVE

Rare-Earth-Containing Materials for Photoelectrochemical Water Splitting Applications

Jennifer Leduc, Yakup Gönüllü, Aida Raauf, Thomas Fischer, Sanjay Mathur[1]
Institute of Inorganic Chemistry, University of Cologne, Cologne, Germany
[1]Corresponding author: e-mail address: sanjay.mathur@uni-koeln.de

Contents

1. INTRODUCTION

Finitude of carbon-based energy resources and the sociopolitical dilemma on the future of nuclear power explains the technological thrust in the demonstration of advanced concepts for renewable energy production. Solar hydrogen produced by photoelectrochemical (PEC) splitting of water carries the potential of becoming an important pillar of the future energy economy based on renewable sources especially for decentralized small-scale applications. Feasibility of artificial photosynthesis driven by the development of large number of advanced water oxidation photocatalysts has been successfully demonstrated at the laboratory; however, implementation of these concepts and materials into technologically relevant device structures is currently hampered by several engineering challenges

Semiconductors and Semimetals, Volume 97
ISSN 0080-8784
http://dx.doi.org/10.1016/bs.semsem.2017.05.001

associated with the complex interfacial (liquid–solid) processes and lack of material compositions unifying both high reactivity and long-term stability to operate under field conditions, which on one hand fuels the quest for new water oxidation catalysts and on the other hand points out the necessity of exploring new material compositions.

Rare earth elements are known to play an irreplaceable role in important catalytic and energy production processes such as three-phase catalysts for reducing automotive emissions (e.g., cerium-doped zirconium oxide), high temperature conductors (e.g., yttrium stabilized zirconia), and redox reactions (e.g., lanthanum manganate in solid oxide fuel cells). Given their electronic structures, the lanthanide elements can act as catalysts and cocatalysts to improve the efficiency of the catalytic system by (i) tuning the acid–base properties of the catalyst surface, (ii) enhancing the thermal stability of catalytic oxides, (iii) improving the catalytic efficiency due to their redox capabilities and conductivity enhancement, as well as (iv) augmenting oxygen uptake and release properties of the catalyst materials. Several fundamental questions need to be answered in order to understand the role of rare earth elements in chemical reactions such as the influence of 4f electrons and the interaction of rare earth oxides both as dopants and in conjunction with other metals and oxides in metal–oxide and oxide–oxide heterostructures. This chapter presents a brief account of the status and perspectives of rare earth materials in photocatalytic and photoelectrochemical processes and discusses the potential of rare earth materials as active catalysts or support materials. After a brief introduction on the electronic structure and properties of rare earth ions and oxide materials, the examples of their application in catalytic processes are discussed.

Rare earths (REs), unlike their name, are rather commonly occurring elements in the earth crust with atomic numbers 57–71. Light rare earth elements (atomic numbers 57–63) such as cerium or lanthanum even have a higher abundance (67 and 39 ppm) than standard industrial metals such as copper (60 ppm) or tin (2.3 ppm). The heavy rare earth elements (atomic numbers 64–71) such as terbium or lutetium have lower abundances of 1.2 and 0.5 ppm which is comparable to tungsten (1.25 ppm) but still orders of magnitude higher than gold and platinum. However, due to their less concentrated appearance and distribution over several continents, the cost-effective mining, extraction, and refining process proves rather difficult. The major resources for light rare earth elements are the minerals bastnäsite ($(Ce,La,Nd,Y)[(F,OH)CO_3]$) and monazite (e.g., $(Ce,La,Nd,Th)[PO_4]$), whereas the so-called lateritic ore is dominated by yttrium and heavy rare earth elements. Deposits of bastnäsite are predominantly found in Mountain Pass in California, whereas lateritic ore is prevalent in southern China (Jha, 2014).

Although more than 50% of the total rare earths world reserves (~110,000,000 tons) belong to the United States, Russia, Australia, India, and other countries, 97% of the annual mining capacity (130,000 tons) was assigned to Chinese production facilities in 2010 (Table 1). This market dominance was mainly owed to the geographical distribution of rare earths in the southern Tibetan region and low mining and processing costs in China. However, since the world demand for rare earths (136,000 tons) was already higher than the mine output in 2010 and was expected to rise to 160,000 tons per year by 2016, the prices for rare earth elements have continuously increased (Fig. 1). In order to ensure a safe long-term supply with a stable price for rare earth materials, countries such as the United States, Australia, and Japan were pursuing the establishment of their own mining facilities. For instance, Japanese researchers have discovered a new potential source for rare earth elements in the deep-sea mud of the Pacific Ocean (Kato et al., 2011). However, the recovery of these resources from deep sea (4000–5000 m) still requires further technological development to

Table 1 World Rare Earth Materials Reserves and Their Mining Production in 2010 and 2013

	Rare Earth Oxide Reserves (tons)		**Mine Output of Rare Earth Oxides (tons)**	
Country	**2010**	**2013**	**2010**	**2013**
China	55,000,000	55,000,000	130,000	100,000
Other countries	22,000,000	41,000,000	NA	NA
Russia	19,000,000	[a]	NA	2400
USA	13,000,000	13,000,000	0	4000
India	3,100,000	3,100,000	2700	2900
Australia	1,600,000	2,100,000	0	2000
Brazil	48,000	22,000,000	550	140
Malaysia	30,000	30,000	350	100
World total (rounded)	110,000,000	140,000,000	130,000	110,000

[a]Included with "other countries."

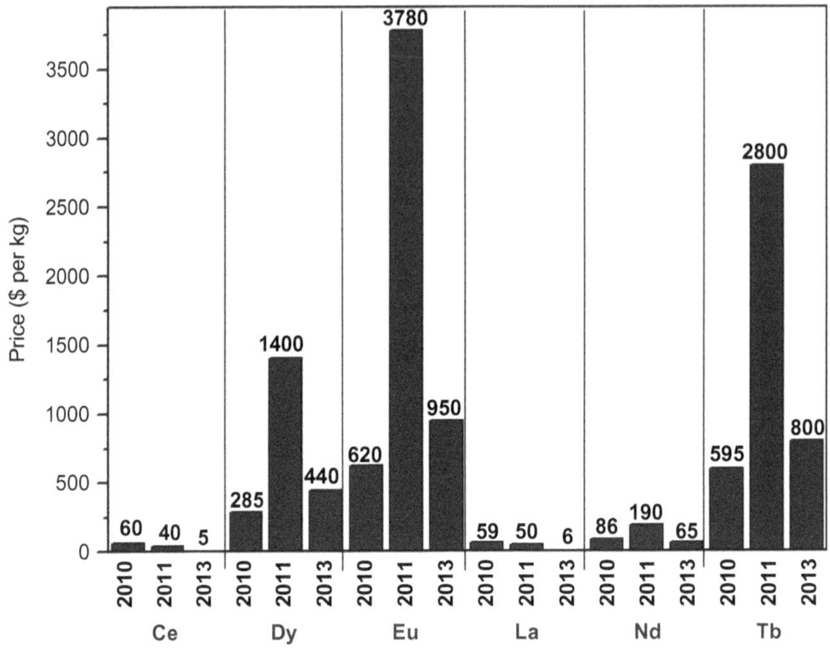

Fig. 1 Prices ($/kg) of selected rare earth oxides containing cerium (Ce), dysprosium (Dy), europium (Eu), lanthanum (La), neodymium (Nd), and terbium (Tb) for the years 2010, 2011, and 2013.

make the extraction economically viable. The increased output outside China led to a price reduction of more than 50% compared to 2011, where the costs for popular rare earth elements (such as europium and terbium) had reached their maximum (Lehmann, 2014).

The increasing demand of rare earth materials is due to their implementation in various commercial products, such as mobile phones, hybrid electrical vehicles, high capacity batteries, infrared lasers, light bulbs, glass additives, or permanent magnets for electric motors (Jha, 2014). In addition, high purity rare earth materials are used for defense and high precision military system applications. Moreover, the presence of rare earth materials in catalysts knowingly enhances the catalytic properties, chemically regulates surface and thermal stabilities, and enhances the ionic conductivity that are important features for catalytic processes. In comparison to the prominent role of lanthanide compounds in optical materials and applications, their potential in energy production and utilization processes is hitherto less explored.

2. RELEVANT PROPERTIES OF RARE EARTH IONS AND COMPOUNDS FOR PHOTO(ELECTRO)CATALYSIS

In general, the rare earth series consists of 17 elements containing all lanthanides as well as scandium and yttrium, which mostly occur together due to their chemically similar nature. One of their common properties is the preference for trivalent oxidation states with the electronic structure [Xe] $4f^n$ and their minor crystallochemical resemblance to other trivalent elements (Riedel and Janiak, 2007).

Relevant properties of lanthanides for the application in photo(electro) catalysis are their size (ionic radii), electronic structure (band gap), redox capability, as well as their optical properties. The ionic radii of the lanthanides (+3 oxidation state, sixfold coordination) range from 1.03 (La) to 0.86 (Lu) Å (Table 2) and are substantially larger than their transition metal analogs. When incorporated as dopant into a transition metal oxide matrix, the lattice gets distorted which can either result in the formation of pseudo intermediate bands or the generation of a dipole (Abe et al., 2004a). The latter increases the charge carrier separation and thus the conductivity in the matrix material, whereas the additional intermediate bands shift the light absorption toward lower energy (Liu et al., 2011). In addition, due to their ionic radii, lanthanides form characteristic bimetallic oxides (such as pyrochlores or perovskites) with other transition metals. These crystal structures are of substantial interest for solar water splitting reactions due to (a) cooperative effects emerging from the presence of different metallic centers in the crystal structure and (b) a high photochemical stability. For examples, rare earth metal titanates ($Ln_2Ti_2O_7$) and stannates ($Ln_2Sn_2O_7$), build up by an interwoven 3d network of corner-linked MO_6 octahedra and LnO_8 edge-sharing cubes, are attractive candidates for catalytic applications due to their high temperature stability and defect-rich structure. As shown in Fig. 2, $La_2Ti_2O_7$ forms a layered assembly which enables the use of different reaction sites for water oxidation/reduction (Zhang et al., 2007).

The high affinity of lanthanides toward oxygen is reflected in the existence of various Ln:O stoichiometries (LnO_2, Ln_2O_3, LnO, as well as nonstoichiometric oxides with the composition $LnO_{1.5-2.0}$) with different

Table 2 Effective Ionic Radii in pm for six-Coordinated Ln^{3+} Ions (Sastri et al., 2003; Shannon, 1976)

La	Ce	Pr	Nd	Pm	Sm	Eu	Gd	Tb	Dy	Ho	Er	Tm	Yb	Lu
103	101	99	98	97	96	95	94	92	91	90	89	88	87	86

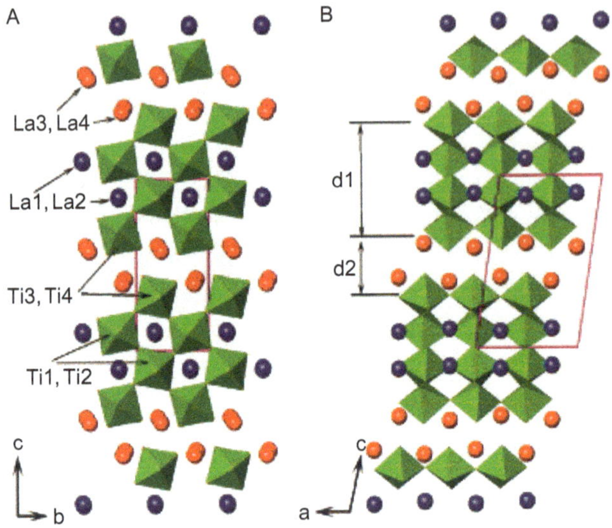

Fig. 2 View along the (A) [100] and (B) [010] directions of $La_2Ti_2O_7$. The structure shows four layers of corner-sharing TiO_6 octahedra which are separated by two layers of La cations (La3 and La4). *Reprinted from Zhang, F.X., et al., 2007. Structural change of layered perovskite La2Ti2O7 at high pressures. J. Solid State Chem. 180, 571–576.*

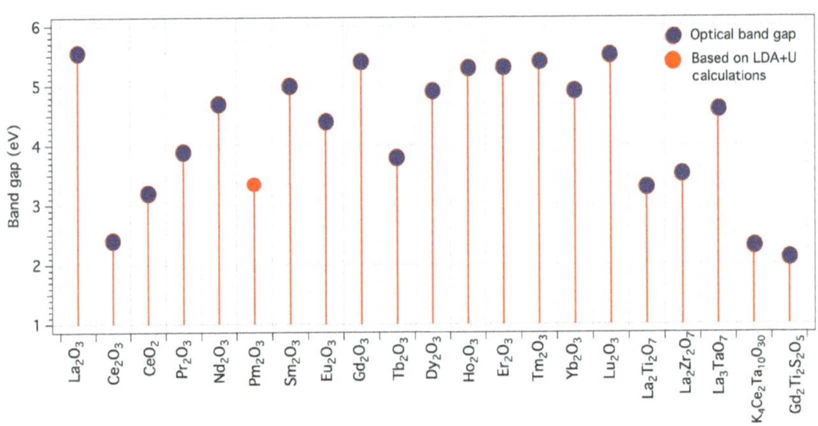

Fig. 3 Experimentally determined optical band gaps (in eV) of binary and multimetallic RE oxides and oxysulfides (He et al., 2013; Hebbink et al., 2002; Heer et al., 2003; Idriss, 2010; Jiang et al., 2012; Prokofiev et al., 1996; Zhang et al., 2015).

optical band gap values (Fig. 3). In general, the Ln_2O_3 compounds are insulators with wide band gaps (3.8–5.55 eV) except for the metastable Ce_2O_3, which shows a deep minimum in the E_g value (2.4 eV) (Prokofiev et al., 1996). The conductivity and photocatalytic activity of pristine lanthanide oxides directly correlate with the band gap and redox capability of the material. Thus, CeO_2 is most widely studied due to its band gap of 3.2 eV and reversible oxygen uptake-release properties. It crystallizes in the fluorite structure and can be reduced at high temperatures as well as at low partial pressures of oxygen to a nonstoichiometric form, CeO_{2-x} ($0 < x < 0.28$), that retains the fluorite lattice and exhibits high electronic conductivity. It forms solid solutions with other broad-band semiconductors such as TiO_2 and has been investigated for several photocatalytic and PEC water splitting applications (Kundu et al., 2012; Luo et al., 2015; Primo et al., 2011; Wu et al., 2015; You et al., 2016; Zhao et al., 2015). Due to the valence dynamics in the CeO_x system, it is particularly used for solar thermochemical hydrogen generation applications (Al-Shankiti et al., 2013; Cotton, 2006).

The extraordinary optical properties of lanthanide ions (e.g., sharp and narrow f–f transitions bands) derive from the effective shielding of low-energy 4f orbitals by the outer filled 5s and 5p orbitals (Fig. 4) (Bunzli and Piguet, 2005; Cotton, 2006). Due to their low transition probabilities with long-lived excited states, lanthanide-doped host materials (e.g., $NaYF_4{:}Er^{3+}$)

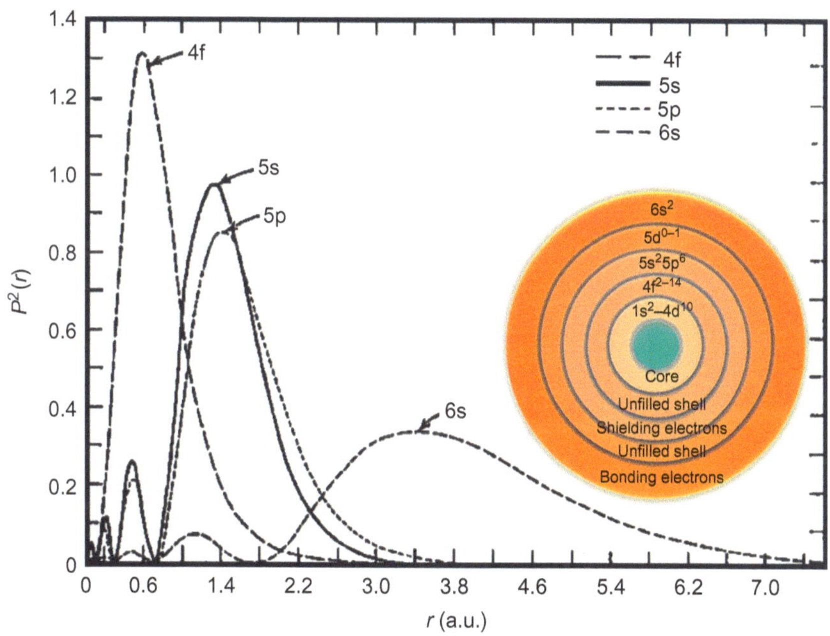

Fig. 4 Electronic radial distribution functions for the 4f, 5s, 5p, and 6s energy levels on a Gd^{3+} ion. The electronic density is plotted as a function of the distance from the nucleus. The picture *inset* schematically shows that the partially filled 4f orbitals are buried below the filled 5s and 5p orbitals (Bunzli and Piguet, 2005; Rivera et al., 2012). *Reprinted from Rivera, V.A.G., Ferri, F.A., Marega, E., 2012. Localized Surface Plasmon Resonances: Noble Metal Nanoparticle Interaction with Rare-Earth Ions.*

are of particular interest for upconversion (UC) processes in combination with solar cells. The UC process is a nonlinear optical process with successive absorption of at least two photons via intermediate long-lived energy states followed by the emission of photons with shorter wavelengths. Thus, IR radiation (which represents about 52% of the solar spectrum) can be converted into energetically higher visible light leading to a more efficient absorption of sunlight (Auzel, 2004). Proof-of-principle experiments for c-Si solar cells in combination with the $NaYF_4:Er^{3+}$ system have shown higher external quantum efficiencies (EQEs) than the intrinsic absorption of silicon in this wavelength region (Shalav et al., 2007).

3. APPLICATION IN SOLAR ENERGY GENERATION

In the following sections we present a comprehensive overview of rare-earth-doped materials, rare-earth-based oxide–oxide composites and mixed-metal and ternary oxide materials in the fields of photoelectrochemical and photocatalytic water splitting (Fig. 5). Herein, the emphasis will be on the influence of 4f electrons on the catalytic activity and the direct comparison of rare-earth-based materials to transition metal-modified semiconductors.

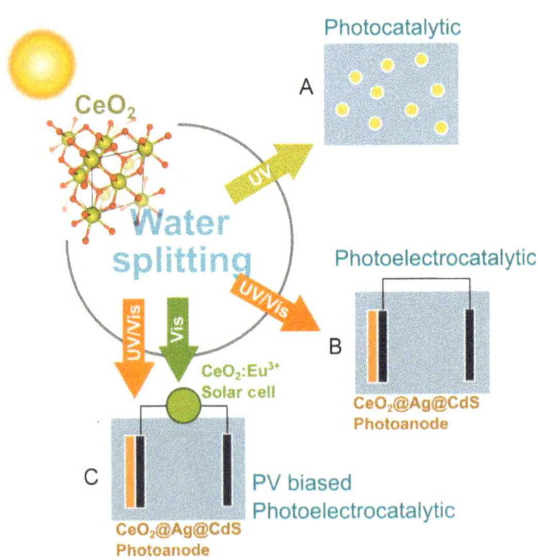

Fig. 5 Graphical outline on the applications of rare earth materials in solar energy generation: (A) photocatalytic water splitting, (B) photoelectrochemical water splitting, (C) PEC-PV tandem cell approach.

In addition, the status and perspectives of rare earth materials in photoelectrochemical tandem cells will be briefly discussed underlining the high potential of this material class.

3.1 Photoelectrochemical Water Splitting

Rare-earth-modified materials for solar energy production are rather unexplored compared to common materials such as Fe_2O_3, TiO_2, or WO_3. Though their potential in UC processes has been described already in 1973, most reports on application in solar hydrogen generation were published after 2010. Although there are many examples for photocatalysis, e.g., the degradation of dye molecules, there are only a few examples on photoelectrochemical water splitting applications. A proof of concept for rare-earth-doped ceramics has been published by Méndez-Ramos et al. (2013). They reported that the PEC performance of large band gap materials (e.g., TiO_2) can be enhanced by harvesting long wavelength light of solar irradiation using RE-doped oxyfluoride nanoglass ceramics such as $NaYF_4$, YF_3, and KYF_4. This material class is capable of mimicking the green chlorophyll antenna of plants due to highly efficient IR to UV-blue upconversion. However, these UC powered PEC devices have only been proposed but not yet tested. Later reports on rare-earth-doped TiO_2 have shown that this approach is valid. Ce doping of TiO_2, for instance, shifts the absorption edge to 550 nm, which equals a 170 nm red shift compared to pristine TiO_2 (Tian et al., 2014). Therefore, under visible light irradiation a photocurrent density of 300 $\mu A\,cm^{-2}$ was observed, whereas without doping no photocurrent was detected.

Gonell et al. have shown that mesoporous TiO_2 structures sensitized with CdSe displayed enhanced light harvesting efficiencies when loaded with Er^{3+}/Yb^{3+}-codoped Y_2O_3 submicrometric particles. These particles upconvert IR photons into visible radiation and thereby lead to effective H_2 generation of the overall semiconductor material (Gonell et al., 2014). The UC particles were synthesized by a simple precipitation method, added to the mesoporous TiO_2 paste, and then sensitized by the successive ionic layer adsorption and reaction method. The EQE of this material at 980 nm was estimated to be $3.75 \times 10^{-4}\%$, and the photocurrent density was determined to be 0.8 $\mu A\,cm^{-2}$ at 0 V vs Ag/AgCl under IR illumination. These values indicate that further optimization of the material is required to be useful in nonacademic technological applications. Rare earth doping of TiO_2 cannot only enhance the PEC performance by UC mechanisms but also by surface states passivation. Gd doping of mesoporous TiO_2 has been shown to prevent charge carrier recombination at the electrode/electrolyte interface and thereby resulted in a 180% higher generation of H_2 ($2.34\,mL\,h^{-1}\,cm^{-2}$) compared to pristine TiO_2 ($1.28\,mL\,h^{-1}\,cm^{-2}$) under UV light illumination (Sudhagar et al., 2014). Gd doping of WO_3 nanoplates has been reported to combine both effects. On the one hand, the absorption edge of WO_3 is red-shifted, and on the other hand, Gd dopants, which are homogeneously allocated at the surface of the nanoplates in the form of gadolinia, function as a passivation layer and thus decrease the recombination of electron–hole pairs. XRD, TEM, EDX, as well as XPS measurements have shown that Gd is both incorporated into the WO_3 lattice (via displacement of W atoms and occupation of interstitial positions) as well as distributed at the surface. Due to the large ionic radius of gadolinium, the WO_3 lattice is expanded upon Gd doping resulting in a shift of peak positions toward smaller angles in the XRD pattern. XPS and HRTEM analysis indicate the formation of hexagonal Gd_2O_3 at the surface which inhibits the charge carrier recombination at the electrode/electrolyte interface. This double-effect leads to an increase of photocurrent density by 153% ($2.28\,mA\,cm^{-2}$ at 1 V vs Ag/AgCl) compared to undoped WO_3 nanoplates. In addition, the calculated light energy to chemical energy conversion efficiencies of the Gd–WO_3 films (0.84%) was 2.47 times higher than that of the pristine WO_3 film (Liu et al., 2015a).

Not only TiO_2 and WO_3 but also Fe_2O_3 films have been reported to yield higher solar water splitting efficiencies when modified with rare earth UC nanomaterials (Zhang et al., 2012). Zhang et al. have prepared this composite material by drop casting a solution of hexagonal disk-shaped rare earth nanocrystals (RENs, $NaYF_4$:Yb,Er) in hexane onto highly uniform hematite films deposited via ALD (atomic layer deposition). These RENs absorb IR radiation and emit photons at 550 and 670 nm which are subsequently absorbed by the hematite film underneath (Fig. 6). Chronoamperometry measurements under IR irradiation display a current density of $130\,nA\,cm^{-2}$ at an applied potential of 1.43 V compared to $<10\,nA\,cm^{-2}$ for pristine hematite films. However, the incident photon to current conversion efficiency (IPCE) of this system was calculated to be 1.24×10^{-4}% at 980 nm. Thus, this work can be considered as a proof of concept, which needs further investigation for improving the efficiency of the hematite-based PEC device.

Pristine CeO_2 and Eu_2O_3 materials have been examined for photoelectrochemical water splitting applications, although their overall performance is very low compared to common transition metal oxides.

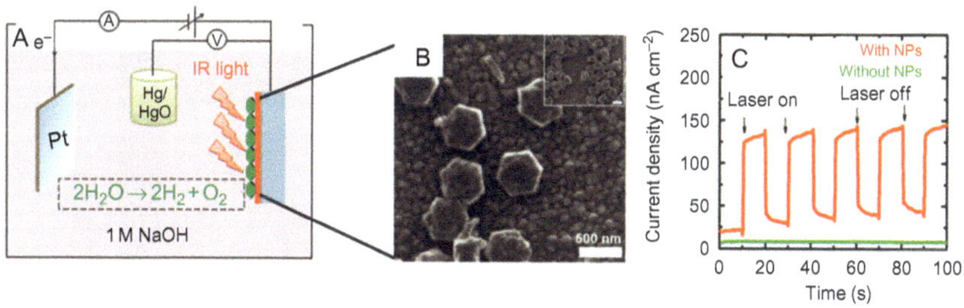

Fig. 6 (A) Schematic diagram of the photoelectrochemical cell setup using IR irradiation and REN/hematite composite film electrodes; (B) HRSEM image of RENs on a hematite film (the *inset* shows a zoom-out image); (C) Chronoamperometry measurements of hematite films coated with (*red*) and without RENs (*green*), in which a 980 nm laser was turned on and off. *Reprinted and modified from Zhang, M., et al., 2012. Improving Hematite's solar water splitting efficiency by incorporating rare-earth upconversion nanomaterials. J. Phys. Chem. Lett. 3, 3188–3192.*

Zeng et al. have reported the synthesis of porous Eu_2O_3 nanoparticles (NPs) and the investigation of their photoelectrochemical performance under white light irradiation. Due to the very large band gap of 4.0–4.4 eV (Fig. 3), the Eu_2O_3 NPs only yield a photocurrent density of 40 μA cm^{-2} at 0.6 V vs Ag/AgCl (Zeng et al., 2015). Free-standing CeO_2 nanorods (NRs) which were grown on Ti substrates via electrodeposition exhibit a smaller band gap (2.9–3.2 eV; Fig. 3) then Eu_2O_3 but display an even lower photocurrent density of 1.1 μA cm^{-2} under visible light irradiation (Lu et al., 2010). However, when CeO_2 NRs are combined with CdS NPs, the photocurrent density is increased to 120 μA cm^{-2} (without bias) vs Ag/AgCl under visible light irradiation in comparison to bare CdS which only exhibits a photocurrent density of ~15 μA cm^{-2}. This composite material was fabricated by a two-step hydrothermal method and was also used to examine photocatalytic H_2 evolution activity with water and lactic acid as sacrificial agent (see Section 3.2) (You et al., 2016). The CeO_2@Ag@CdS composite material shows a higher photocurrent density of 2.14 mA cm^{-2} at 0.2 V (vs Ag/AgCl) under visible light irradiation compared to CdS/CeO_2. This three-component system consists of CeO_2 nanotube arrays (NTAs) which were electrodeposited on FTO substrates and then covered with Ag NPs via successive electrodeposition using a solution of $AgNO_3$. In the final step, the CeO_2@Ag system was entirely coated by a CdS layer (Fig. 7). The increase in photocurrent density and high incident photon to electron conversion efficiency (IPCE) of 69% at 360 nm is dedicated to the cooperative effect of enhanced visible light absorption and better charge carrier transport at

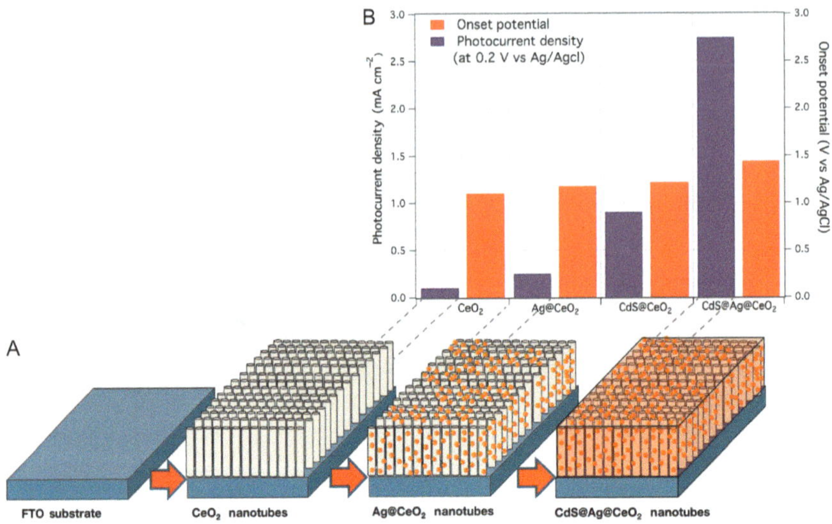

Fig. 7 (A) Schematic illustration of the fabrication process of the three-component system CeO_2@Ag@CdS; (B) Photocurrent densities (mA cm^{-2}) and onset potentials (V vs Ag/AgCl) of the CeO_2, CeO_2@Ag, CeO_2@CdS, and CeO_2@Ag@CdS systems (Zhao et al., 2015).

the interface of this composite due to the Ag NPs (Zhao et al., 2015). This has also been confirmed by Khan et al. who synthesized Ag@CeO_2 nanocomposites using electrochemically active biofilms as reducing agents (Khan et al., 2014).

There are only three publications on mixed phase rare earth materials for PEC application (Uno et al., 2005, 2006) that deal with bimetallic oxides such as lanthanum titanium oxide ($La_2Ti_2O_7$) and lanthanum zirconium oxide ($La_2Zr_2O_7$). In contrast to the perovskite-related layered $La_2Ti_2O_7$, $La_2Zr_2O_7$ exhibits a pyrochlore structure. Not only the photocatalytic activities of both materials were explored by hydrogen evolution measurements, but also their band structures were elucidated. However, the amount of hydrogen gas evolved was low with values of 10.5 μmol/10 h ($La_2Ti_2O_7$) and 9.0 μmol/10 h ($La_2Zr_2O_7$) under the illumination of a 500 W Xenon lamp which is congruent with their calculated band gap energies of 3.29 and 3.52 eV (Fig. 3).

Jamil et al. (2016) have recently published the application of the pyrochlore $Nd_2Sn_2O_7$ for solar water splitting. A thin film of $Nd_2Sn_2O_7$ was fabricated by spin coating a solution of the bimetallic alkoxide precursor [NdSn(O^{t}Bu)$_5$Py] onto a hematite layer deposited by plasma-enhanced chemical vapor deposition (Fig. 8A and B). The heterojunctioned

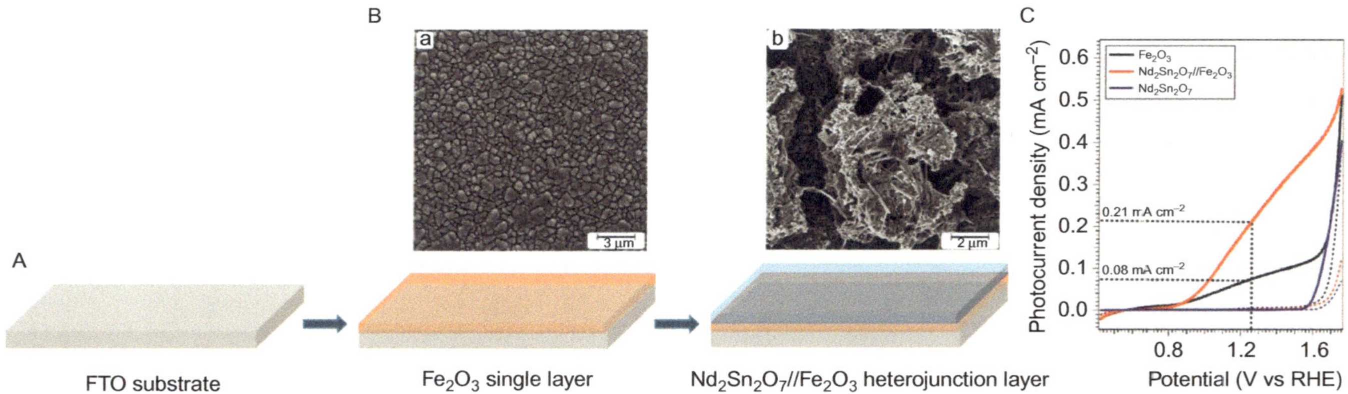

Fig. 8 (A) Schematic illustration of the fabrication process to generate $Nd_2Sn_2O_7//Fe_2O_3$ bilayers; (B) *Top view* SEM images of (a) Fe_2O_3 and (b) $Nd_2Sn_2O_7//Fe_2O_3$; (C) Photoelectrochemical measurements of single (Fe_2O_3, $Nd_2Sn_2O_7$) and bilayered ($Nd_2Sn_2O_7//Fe_2O_3$) photoanodes (Jamil et al., 2016).

$Nd_2Sn_2O_7//Fe_2O_3$ bilayers showed an almost threefold enhancement in photocurrent density (0.21 mA cm^{-2}) when compared to the bare Fe_2O_3 layer (0.08 mA cm^{-2}) manifesting the importance of tightly coupled phase boundaries and an unintentional doping of lanthanide ions in the absorber iron oxide film (Fig. 8C). The improved PEC performance was explained by the photocatalytic effect and the large band gap of the $Nd_2Sn_2O_7$ material (3.5 eV) (Bradley et al., 1973), which inhibited possible backward migration of holes upon illumination.

To conclude, rare-earth-modified compounds are rather novel but emerging materials for photoelectrochemical water splitting applications. However, since the PEC efficiencies are not high enough at this stage, further investigations are required to achieve industrial/technological usability. PEC performance improvement, for instance, might be achieved by implementing diverse UC dopants to well-investigated materials in order to cover a broader range of the solar spectrum (Zhang et al., 2012).

3.2 Photocatalytic Water Splitting

Rare-earth-containing compounds are already widely used as catalysts in various industrial applications, e.g., petroleum refining (lanthanum chloride), automobile emission reduction (cerium oxide), organic polymerization (neodymium versatate), hydrogen and synthesis gas production (ceria, lanthania), plastics decomposition, and dechlorination (samarium iodide) (Sastri et al., 2003). In addition, solar thermochemical water splitting (Al-Shankiti et al., 2013; Dey et al., 2015; Le Gal and Abanades, 2012; Le Gal et al., 2013) and photocatalytic reactions have been investigated thoroughly, especially for La-, Ce-, Eu-, and Gd-modified materials. The review by Weber et al. on "Lanthanide-modified semiconductor photocatalysts" discusses several RE-modified TiO_2 materials for the photocatalytic degradation of organic molecules, such as benzene, *p*-nitrophenol, aniline, or organic dyes (e.g., methyl orange, rhodamine B, methylene blue) (Weber et al., 2012). The enhancement of activity (compared to pristine TiO_2) is explained by (a) the increasing adsorption of organic molecules, (b) the decreased charge carrier recombination, (c) the enhanced formation of hydroxyl groups and thus hydroxyl radicals which initiate the photocatalytic degradation, and (d) the shift in the onset of absorption into the visible region. Besides TiO_2 and its composites (e.g., Fe_3O_4–TiO_2:La) (Shi et al., 2011), further RE-doped materials such as Bi_2MoO_6:Er (Zhou et al., 2011), $BiVO_4$:RE (Xu et al., 2009; Zhang and Zhang, 2010),

$AgNbO_3$:La, and $SrTiO_3$:La (Jia et al., 2010; Wang et al., 2005) are well investigated for photocatalytic applications.

In contrast to the photocatalytic degradation of organic molecules, publications on photocatalytic water splitting using lanthanide-based materials have been rather scarce until recently. Catalysts for water oxidation (Kundu et al., 2012; Liu et al., 2015b; Primo et al., 2011; Wu et al., 2015; Xie et al., 2015), such as calcium-doped lanthanum nickelate layered perovskites (Liu et al., 2015b) or ZnO/Eu_2O_3 (Xie et al., 2015), have been recently reported. In addition, there are several examples of RE-doped titanium oxide materials, such as platinized TiO_2:Ce,Gd,Eu,Yb,Ho (Luo et al., 2015; Zalas, 2014; Zalas and Laniecki, 2005) or cerium-doped TiO_2 NTAs assembled by Ag NPs (Fan et al., 2015) for application in photocatalytic hydrogen generation.

The latter were synthesized by potentiostatic anodization of titanium in a cerium(III) nitrate containing glycerol solution to obtain TiO_2:Ce NTAs followed by microwave-assisted chemical reduction of $AgNO_3$ to generate the desired Ag–Ce/TiO_2 NTs. The measured maximum hydrogen evolution rate of this catalyst was determined to be 1.47 μmol/(cm^2 h) compared to 0.47 and 1.17 μmol/(cm^2 h) for pure TiO_2 and Ag/TiO_2 NTs (with a measured time of 6 h under UV and visible light illumination). The mechanism for the enhanced photocatalytic activity is described schematically in Fig. 9. Upon UV light irradiation electrons are either raised to the conduction band of TiO_2 or to the Ce 4f level. In case of the latter, the electrons are transferred (via the Ce^{3+}/Ce^{4+} redox couple) to oxygen molecules at the surface of TiO_2 to result in negatively charged O_2 radicals. These radicals react with protons of the aqueous solution to give hydroxide radicals which are consumed by the sacrificial agent (ethanol). When raised to the conduction band of TiO_2 instead, electrons are transferred to the silver NP, since their Fermi level is energetically lower. These electrons diffuse to the surface of the Ag NP where they reduce protons to give H_2. The photogenerated holes accumulate in the valence band of TiO_2 until the water oxidation reaction occurs. This Ag–Ce/TiO_2 system is claimed to decrease the charge carrier recombination rate, as well as to enhance visible absorption due to the local surface plasmon resonance effect of the silver NPs.

Besides TiO_2, tantalates and niobates are of interest for photocatalytic water splitting applications despite their relatively wide band gaps: $NaTaO_3$ and $NaNbO_3$, for instance, display band gap energies of 4.0 (Kato et al., 2003; Kudo and Kato, 2000) and 3.5 eV, respectively (Liu et al., 2011). However, due to their high photochemical stabilities and photocatalytic

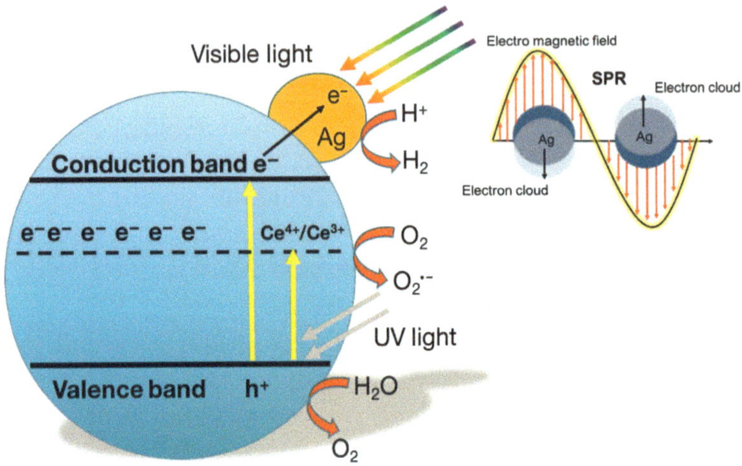

Fig. 9 Schematic drawing of the photocatalytic water splitting mechanism using Ag–Ce/TiO_2 NTs (Fan et al., 2015).

activities even without cocatalysts, several working groups are actively investigating these materials. Kudo et al. have reported on NiO loaded, lanthanide-doped $NaTaO_3$ (Kato et al., 2003; Kudo and Kato, 2000). These materials were prepared in a straightforward fashion by calcining a mixture of Na_2CO_3, Ta_2O_5, and lanthanide oxides with subsequent NiO loading via impregnation from $Ni(NO_3)_2$. Band gap energies of the Ln-doped materials were determined by UV–vis–NIR spectroscopy, and their surface areas were estimated by BET measurements. These experiments have shown that the band gap values of the undoped and Ln-doped materials are almost equal, whereas the surface areas of rare-earth-doped materials are much higher compared to pure NiO/$NaTaO_3$. In addition, the hydrogen and oxygen evolution for different Ln-doped NiO/$NaTaO_3$ materials under UV irradiation were investigated. The La-doped catalyst thereby showed the highest photoactivity with a H_2 evolution rate of 19.8 mmol h^{-1} (nine times higher than nondoped NiO/$NaTaO_3$) and a quantum yield of ~56% at 270 nm. This is explained by the decreasing particle size and change in morphology which enhances the probability of photogenerated electrons and holes to react with water. More recent publications have shown that Y-doped $NaTaO_3$ exhibits a higher photocatalytic water splitting performance (using methanol as sacrificial agent) than $NaTaO_3$:La (Jana et al., 2014).

In contrast to $NaTaO_3$, Ln-doping of $NaNbO_3$ effects the band gap width as well as the band edge positions (Liu et al., 2011). Lanthanum doping thus reduces the band gap energy of $NaNbO_3$ by over 50% to 1.53 eV.

Liu et al. have performed first principle calculations to investigate the electronic structure of La-doped $NaNbO_3$ and (La + N)-codoped $NaNbO_3$. They have shown that doping enlarges the O–Nb–O angles of the NbO_6 octahedra which results in a splitting of the Nb 4d orbital. This splitting induces the formation of a pseudointermediate band which enables wide band gap semiconductors to absorb visible light photons. Due to the favorable band edge positions (Fig. 10) relative to the electrochemical potentials of H_2/H_2O and O_2/H_2O, (La + N)-codoped $NaNbO_3$ is assumed to be a promising material for photocatalytic water splitting applications.

A further approach for enhancing the water splitting efficiency of photocatalysts has been presented by Yu et al. They have shown that lanthanum doping into the depletion layer of $Cd_{0.6}Zn_{0.4}S$ (in the form of La–S–Zn bonds) enhances the charge carrier separation and thereby enables higher performances (1.39 mmol h^{-1}) without the use of noble metal cocatalysts (Yu et al., 2013).

Very recently, Wu et al. (2017) studied the reaction mechanism of photocatalytic water splitting using rare-earth (Sm, Ce, Eu, Pr, Er)-doped carbon dots. They found that in contrast to their transition metal (Cr, Cd, Fe, Zn)-doped analogs, which split water through a two-electron pathway, the photocatalytic reaction occurs via a four-electron pathway (Fig. 11). This was explained by the favorable formation of oxygen at the catalyst surface resulting from the defect structure, oxygen storage-release capacity, and valence band positions of the rare-earth-doped materials. In addition, the photocatalytic activity of Sm-doped carbon dots exhibited the highest

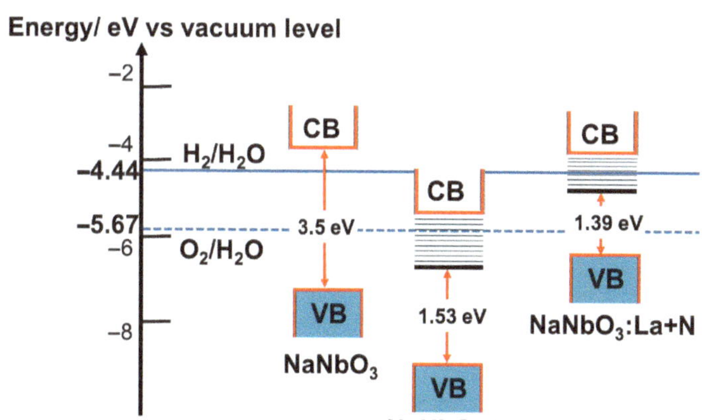

Fig. 10 Schematic illustration of band edge positions of La-doped and (La + N)-codoped $NaNbO_3$ relative to the redox potentials of the water splitting reaction (Liu et al., 2011).

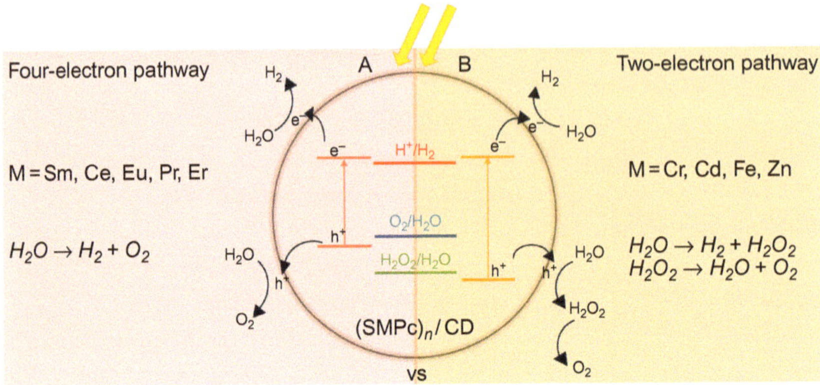

Fig. 11 Schematic illustration of the water splitting pathway using either rare earth- (*left*) or transition metal-doped (*right*) carbon dots (Wu et al., 2017).

photocatalytic activity among all systems (H_2 rate of 1.04 μmol h^{-1}), which was explained by a decreased charge carrier recombination rate resulting from the orbital hybridization of 4f electrons.

Lanthanide oxide composite materials for photocatalytic water splitting mainly consist of CeO_2, which is either implemented in the form of NPs (Zhang et al., 2015) or 1D structures such as nanowires (NWs) (Li et al., 2013). Lu et al. found that CeO_2 nanorods with a diameter of ~200 nm and a length of ~800 nm exhibit a smaller band gap (2.85 eV) than originally reported (3.2 eV) (Zhang et al., 2014). This red-shift of the absorption edge is attributed to the existence of Ce^{3+} ions. The special redox capability of CeO_2 is not only beneficial to light harvesting but also to the charge separation in the photocatalytic water splitting process using Na_2S–Na_2SO_3 as sacrificial agents. Photogenerated holes are consumed by Ce^{3+} ions which are subsequently oxidized to Ce^{4+}. However, as stronger reductants, the sacrificial agents get oxidized in a second step and Ce^{3+} is regenerated. This redox cycle promoted photocatalytic water splitting yields a hydrogen evolution rate of 741 μmol g^{-1} (Lu et al., 2011a).

In combination with a narrow direct band gap semiconductor such as CdS in the form of CdS/CeO_x heterostructured NWs, the hydrogen production is enhanced to 1290.5 μmol g^{-1} h^{-1} under white light irradiation. Pristine CeO_x NWs and commercial CdS NPs, in comparison, only show rates of 603.3 and 579.2 μmol g^{-1} h^{-1}. Since CdS exhibits a band gap of 2.4 eV and band edges that straddle the redox potential of the water splitting reaction, it acts as a photosensitizer for CeO_x. In addition, the energy band structures of CdS and CeO_2 match for the charge carrier transport (Fig. 12). This enables a

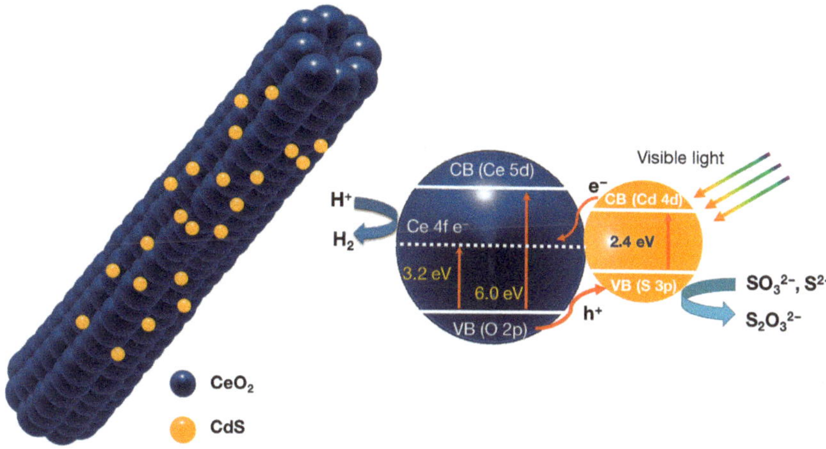

Fig. 12 Schematic drawing of the photocatalytic hydrogen generation mechanism under visible light irradiation using CdS/CeO_x heterostructured nanowires (Li et al., 2013; You et al., 2016).

photocatalytic performance of 473.6 $\mu mol\,g^{-1}\,h^{-1}$ under visible light irradiation (Li et al., 2013). A very recently published article on CdS NPs/CeO_2 NRs composites by You et al. demonstrates an even higher H_2 evolution rate of 8.4 $mmol\,g^{-1}\,h^{-1}$ under visible light illumination using lactic acid as sacrificial agent (You et al., 2016). As the morphologies of this two-component system are inverted (CdS NWs–CeO_2 NPs), a rate of 549.5 $\mu mol\,g^{-1}\,h^{-1}$ under analogous conditions is observed (Zhang et al., 2015).

A more complex system to achieve water splitting under visible light irradiation without the use of a sacrificial agent has been reported by Yoshida et al. They deposited different lanthanide oxide layers onto a rhodium metal NP loaded $(Ga_{1-x}Zn_x)(N_{1-x}O_x)$ photocatalyst with the composition $x=0.12$. In general, noble metals are efficient cocatalysts due to their low overvoltage for hydrogen evolution. They, however, permit the backward water formation reaction which is the reason for the very low water splitting activity of the pure $Rh/(Ga_{1-x}Zn_x)(N_{1-x}O_x)$ system. When modified with cerium or europium oxide still no enhancement in activity was observed. In contrast, when covered with lanthanum oxide a hydrogen evolution rate of 49 $\mu mol\,h^{-1}\,g^{-1}$ under visible light irradiation was detected (Fig. 13) (Yoshida et al., 2013). This increase in water splitting performance is attributed to the redox-inactive La(III) species which is incapable of accepting electrons from the $Rh/(Ga_{1-x}Zn_x)(N_{1-x}O_x)$ photocatalyst and thereby suppresses the back reaction of water generation.

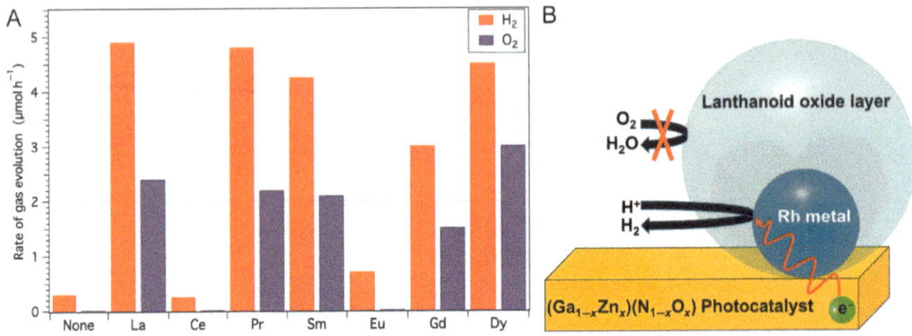

Fig. 13 Illustration of (A) the hydrogen evolution reaction on the $(Ga_{1-x}Zn_x)(N_{1-x}O_x)$/rhodium metal system which is covered by a lanthanide oxide layer to suppress the backward water formation reaction and (B) the photocatalytic activity of the composite material with different lanthanide oxides (Yoshida et al., 2013).

In addition, mixed-phase lanthanide oxysulfides $Ln_2Ti_2S_2O_5$ (Ishikawa et al., 2004) and lanthanide oxides such as $Ln_2Ti_2O_7$ (Khan et al., 2014) and $K_2Ln_2Ti_3O_{10}$ (Tai et al., 2004; Takata et al., 1998) have been investigated for the photocatalytic generation of hydrogen (Abe et al., 2004b; Hu et al., 2014; Meng et al., 2013). $Gd_2Ti_2S_2O_5$/Pt, for instance, displays an optical bandgap of 2.1 eV and a H_2 evolution rate of 120 μmol g^{-1} h^{-1} under visible light irradiation using a Na_2S/Na_2SO_3 containing electrolyte. $La_2Ti_2O_7$ exhibits a perovskite-type structure with a high chemical stability and an interlayer spatial assembly which enables the use of different reaction sites for the hydrogen and oxygen evolution reaction. However, $La_2Ti_2O_7$ displays a band gap of 3.3–3.8 eV (Fig. 3) and therefore shows a very low PEC performance (Section 3.1) under visible light illumination. Meng et al. have shown that nitrogen doping of $La_2Ti_2O_7$ nanosheets results in an absorption edge shift from 330 nm (3.8 eV) to 560 nm which corresponds to a band gap energy of 2.2 eV. Due to their special synthetic approach (hydrothermal processing, wet impregnation, and heat treatment in a NH_3 flow) they were able to increase the visible light absorption while retaining a high specific surface area (Meng et al., 2013). The hydrogen production under visible light irradiation without the use of sacrificial agents amounts to 10 μmol g^{-1} h^{-1} which is in the order of $Rh/(Ga_{1-x}Zn_x)(N_{1-x}O_x)$ but still very low compared to the CeO_x–TiO_2/Pt (14.1 mmol h^{-1} g^{-1}) and the CdS–CeO_2 (8.4 mmol h^{-1} g^{-1}) materials. However, the photoactivity of these systems was measured in the presence of methanol and lactic acid as sacrificial agents which makes a comparison with the other photocatalysts difficult. In addition, the photocatalytic hydrogen formation of the CeO_x–TiO_2/Pt catalyst was measured under UV light irradiation. The performance of $La_2Ti_2O_7$

nanosheets is enhanced when combined with In_2O_3 NPs (hydrogen evolution rate of 68.14 μmol g^{-1} h^{-1} under 1 sun illumination using methanol as sacrificial agent). In_2O_3 is an indirect semiconductor material with a band gap of 2.8 eV and displays a H_2 production rate of 2.3 μmol g^{-1} h^{-1} when used without further modifications. As shown in Fig. 14C and D, the In_2O_3 NPs with a size of ~50 nm are homogeneously dispersed on the $La_2Ti_2O_7$ nanosheets and thereby promote the separation of photogenerated charge carriers resulting in a higher H_2 formation activity of the composite.

Tantalates and niobates are interesting materials for photocatalytic water splitting. The NiO/$NaTaO_3$:La system exhibits the highest hydrogen evolution rate (19.8 mmol h^{-1}) reported for rare-earth-modified materials until today. However, this value has been achieved under UV light illumination using NiO as cocatalyst. In addition, rare earth (La_3TaO_7) and mixed rare earth/alkaline ($K_4Ce_2Ta_{10}O_{30}$) or rare earth/alkaline earth ($H_{1-x}La_xCa_{2-x}Nb_3O_{10}$) tantalates and niobates have been tested as catalysts

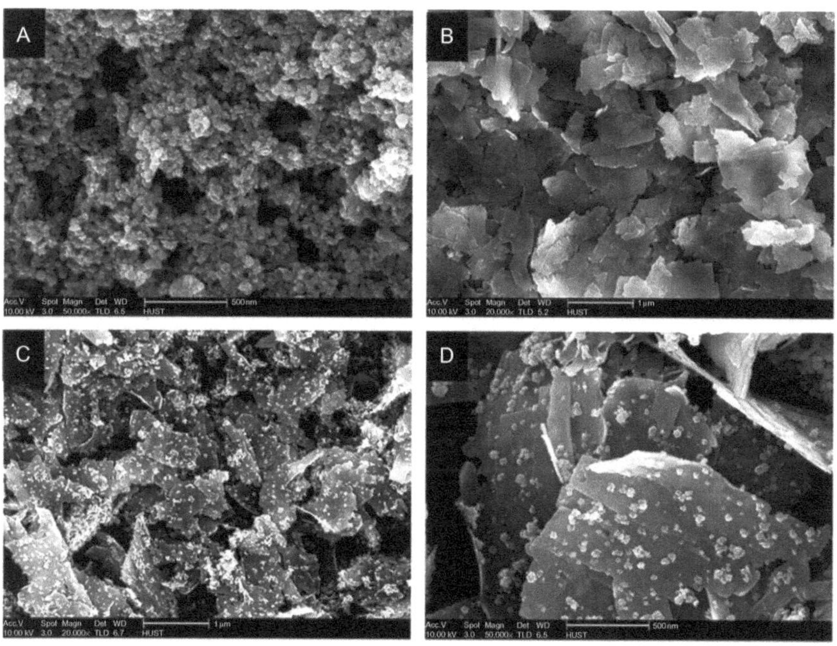

Fig. 14 SEM images of (A) In_2O_3 nanocrystals, (B) $La_2Ti_2O_7$ nanosheets and the In_2O_3/$La_2Ti_2O_7$ nanocomposite with a molar ratio of 1.5:1 at a magnification of (C) 20k× and (D) 50k×. *Reprinted from Hu, S., Chi, B., Pu, J., Jian, L., 2014. Novel heterojunction photocatalysts based on lanthanum titanate nanosheets and indium oxide nanoparticles with enhanced photocatalytic hydrogen production activity. J. Mater. Chem. A 2, 19260–19267.*

for solar hydrogen production. La_3TaO_7 exhibits a weberite-orthorhombic structure with a band gap energy of 4.6 eV (in comparison to $NaTaO_3$ with a band gap of 4 eV). This distorted structure shows a higher activity than the ordered cubic structure of La_3TaO_7 which is explained by the change in resistance of the material as well as by the enhanced charge carrier separation due to the dipole moment which is generated by the distortion in the crystal structure (Abe et al., 2004a). However, the H_2 production activity of La_3TaO_7/NiO under UV irradiation amounts to 158 μmol g^{-1} h^{-1} which is 125 times smaller than the activity of NiO/$NaTaO_3$:La. Several mixed rare earth/alkaline ($K_4Ln_2Ta_{10}O_{30}$) materials have been investigated by Kudo et al. (2000). Among these, only $K_4Ce_2Ta_{10}O_{30}$ showed absorption in the visible light region and therefore has been further investigated for photocatalytic water splitting applications (Tian et al., 2006). $K_4Ce_2Ta_{10}O_{30}$ displays a band gap energy of 2.3 eV and a H_2 evolution rate of 30 μmol g^{-1} h^{-1} (without Pt) and 55 μmol g^{-1} h^{-1} (with Pt) under visible light illumination. In order to understand why the latter value is ~50% lower than the activity of $Gd_2Ti_2S_2O_5$/Pt which exhibits a similar optical band gap, the electronic structure of $K_4Ce_2Ta_{10}O_{30}$ was investigated by density functional theory (DFT) calculations. The simulated energy-band diagram and density of states (DOS) for the trimetallic tantalate are depicted in Fig. 15. The calculated band gap of 1.6 eV is lower than experimentally determined. In addition, the DOS diagram reveals that the valence band contains O 2p, Ta 5d, and Ce 4f orbitals, whereas the conduction band is made up of Ta 5d and Ce 4f with a small degree of O 2p orbitals. While the occupied Ce 4f orbital is disseminated rather broadly, the signal of the Ce 4f in the conduction band is strong and narrow (Fig. 15B). This is due to the contribution of occupied Ce 4f orbitals to the hybridization with Ta 5d and O 2p forming the valence band and the localized nature of the unoccupied Ce 4f. Although the valence band of $Gd_2Ti_2S_2O_5$ also consists of hybridized O 2p, S 3p, and Gd 4f orbitals, the conduction band is not made up of localized Gd 4f orbitals, but it consists of S 3p + Gd 4f and Ti 3d orbitals. This localized nature of the unoccupied Ce 4f orbital in $K_4Ce_2Ta_{10}O_{30}$ is believed to be the reason for the less effective electron transfer to the catalyst surface and thus for the lower photoactivity compared to $Gd_2Ti_2S_2O_5$.

The mixed rare earth/alkaline earth niobates $H_{1-x}La_xCa_{2-x}Nb_3O_{10}$ have been prepared by an ion-exchange reaction of $K_{1-x}La_xCa_{2-x}Nb_3O_{10}$ in hydrochloric acid, and the effect of different substitutions of La^{3+} for Ca^{2+} has been investigated. The highest photocatalytic activity was achieved for $x=0.5$ with a H_2 evolution rate of 169 cm^3 h^{-1} g^{-1} under UV light

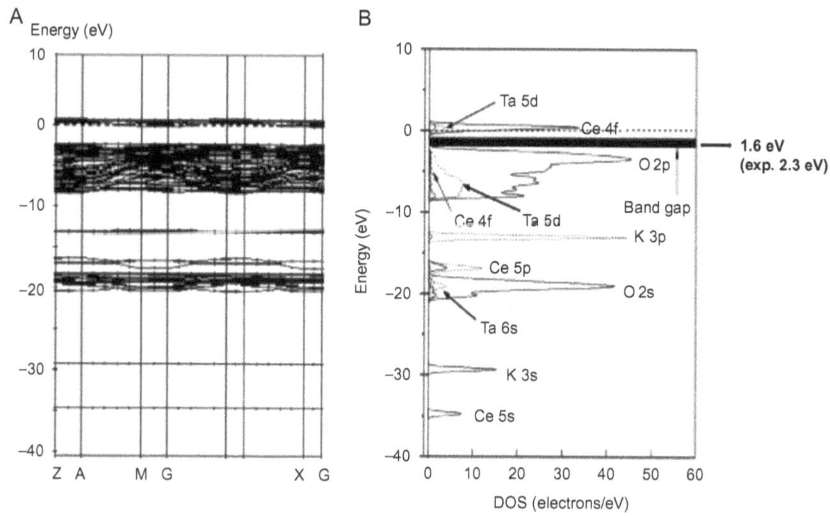

Fig. 15 (A) Energy band diagram and (B) density of states (DOS) determined by density functional theory (DFT) calculations for $K_4Ce_2Ta_{10}O_{30}$. *Reprinted from Tian, M., et al. 2006. K4Ce2M10O30 (M = Ta, Nb) as visible light-driven photocatalysts for hydrogen evolution from water decomposition. Appl. Catal., A 309, 76–84.*

irradiation (Pt loading, methanol as sacrificial agent) (Huang et al., 2008). Not only alkaline and alkaline earth but also mixed rare earth/group III and V metals are investigated. $Bi_{2-x}La_xAlNbO_7/NiO_x$, for instance, reveals a H_2 production rate of 141.4 μmol g^{-1} h^{-1} under similar conditions (Li et al., 2009).

To conclude, rare-earth-modified materials show a high photoactivity for hydrogen generation (19.8 mmol h^{-1} g^{-1}) when irradiated with UV light in the presence of cocatalysts such as NiO or Pt. The illumination with visible light, however, only yields higher evolution rates (8.4 mmol h^{-1} g^{-1}) when sacrificial agents such as lactic acid are used. The principle of UC using low energy IR radiation of the sunlight spectrum has not been successfully applied for photocatalytic water splitting until today and indicates an enormous untapped potential.

3.3 Tandem Cell Approaches

In order to improve PEC efficiency, a higher photocurrent density can be achieved by increasing the photon absorption without increasing the absorber area itself. Therefore, based on multijunction PV cell configurations, the so-called photoelectrochemical tandem cells were developed

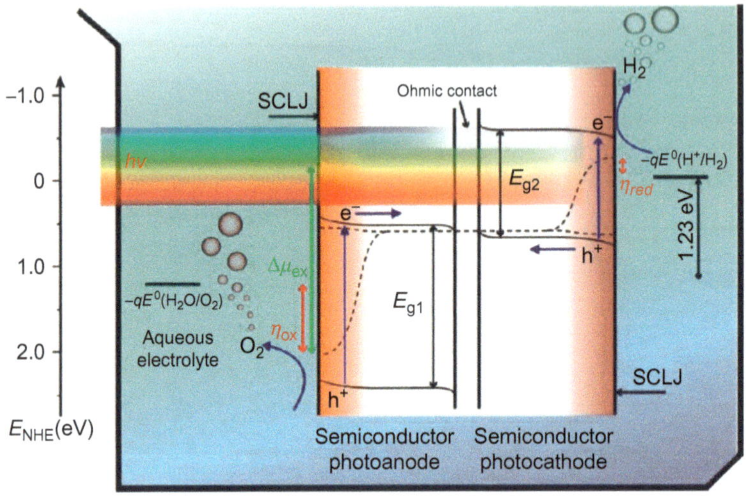

Fig. 16 Photoelectrochemical tandem cells for solar water splitting (Prévot and Sivula, 2013).

incorporating either photoanode and photocathode in a sandwich structure or a combination of a photoelectrode with an integrated photovoltaic cell (Fig. 16). Biasing of photoelectrochemical systems with external PV cells is not considered a tandem configuration per se, as the total solar absorber area is increased by the external PV system. Tandem cells can be realized as either wired, thus connecting separate electrodes via a conductive wire, or wireless, where the electrodes are connected back-to-back forming a monolith thus reducing electron losses at additional interfaces.

The challenge in all tandem devices remains the tuning of both solar absorption and photocurrent yield to match with the respective counter electrode, which limits the material selection for efficient hydrogen production. Due to the lack of accessible stable and efficient n-type and p-type lanthanoid photoelectrodes, tandem PEC cells incorporating both photoanode and photocathode have not been published so far. On the other hand most tandem PEC devices configurations consist of a photoelectrochemically active photoelectrode combined with a PV cell connected in a wireless fashion, where the PV cell provides additional bias to power the PEC active electrode at higher efficiency. The majority of tandem PEC cell configurations are based on a DSSC coupled to $BiVO_4$ photoelectrodes, with different bulk and/or surface modifications to reduce the electrodes' oxidation overpotential. Tandem cells based on rare earth materials have not been realized to the best of our knowledge.

There are a couple of reviews on rare-earth-doped materials for solar cell applications (Shalav et al., 2007; van der Ende et al., 2009; Zalas, 2014). Strümpel et al., for instance, report on the efficiency increase of silicon solar cells by using downconversion, luminescence, and upconversion mechanisms (Strümpel et al., 2007). One of the reasons for the efficiency loss in a silicon solar cell is the transmission of photons with lower energies than the silicon band gap (1.12 eV). In order to overcome this issue, UC materials such as $NaYF_4:Er^{3+}$ have been integrated (Shalav et al., 2005, 2007). However, these materials still suffer from low IR absorptions and low-conversion efficiencies resulting in EQEs of only 2.5%. Gd-doped ZnO-based dye-sensitized solar cells have been shown to enhance the energy conversion efficiency by 44.5% but still only exhibit a conversion efficiency of ~2% (Lu et al., 2011b). Not only dopants but also rare earth oxides have been investigated to increase the performance of solar cells (Corma et al., 2004; Yu et al., 2012). CeO_2 NPs, for instance, have been assembled as mirror-like scattering thin films in dye-sensitized solar cells resulting in a conversion efficiency of 8.6% (reference cell = 7.3%) (Yu et al., 2012). Nevertheless, compared to the highest published energy conversion efficiencies of single crystal (27.6%) and multicrystalline (21.3%) Si cells as well as dye-sensitized (11.9%) and perovskite (21.0%) cells, the rare-earth-modified systems lag far behind.

The efficiency of photoelectrochemical hydrogen production is determined by a combination of factors in catalyst like imperfections of crystalline structure, bulk, and surface properties, their resistance toward photocorrosion in aqueous electrolytes, and more importantly ability to drive the water splitting reaction, which makes metal oxide semiconductors as catalysts of choice (Chou et al., 2013; Shen et al., 2011; Zhang et al., 2012). Among several photophysical and photochemical considerations, the solar-to-hydrogen conversion efficiency can be enhanced by increasing the concentration of photogenerated charge carriers (e^-/h^+) together with their efficient transport and separation at the semiconductor–electrolyte junction. Consequently, a large number of pure and rare-earth-doped-metal oxides have been synthesized and studied as water oxidation catalysts (Table 3). It has been shown for a large number of compositions that the light absorption behavior and photoelectrochemical properties of commonly used large band gap semiconductors (e.g., SnO_2, ZnO, TiO_2, Fe_2O_3) can be tuned by introducing lanthanide ions (Liu et al., 2011; Uno et al., 2006). Besides altering the band gap energies, the surface activities and reaction kinetics have shown to be improved when using rare earth oxides with noble metal clusters and in conjunction with

Table 3 Rare-Earth-Containing Materials for Photoelectrochemical Water Splitting Reactions

Material	Performance	Irradiation	Cocatalyst	Sacrificial Agent	References
RE-based materials for photoelectrochemical water splitting					
TiO_2:Ce	300 μA cm^{-2} (at 1.2 V vs RHE)	Visible light	—	—	Tian et al. (2014)
TiO_2/Y_2O_3:Er,Yb-CdSe	0.8 μA cm^{-2} (at 0 V vs Ag/AgCl)	IR light	—	Na_2S/Na_2SO_3	Gonell et al. (2014)
TiO_2:Gd	5.1 mA cm^{-2} (at 1.2 V vs RHE) H_2 evolution rate: 2.4 mL h^{-1} cm^{-2}	UV light	—	—	Sudhagar et al. (2014)
WO_3:Gd	2.28 mA cm^{-2} (at 1 V vs Ag/AgCl)	AM 1.5 G	—	—	Liu et al. (2015a)
Fe_2O_3/REN	130 nA cm^{-2} (at 1.43 V vs RHE)	IR light	—	—	Zhang et al. (2012)
RE-based oxide–oxide composite materials for photoelectrochemical water splitting					
Porous Eu_2O_3 NPs	40 μA cm^{-2} (at 0.6 V vs Ag/AgCl)	AM 1.5 G	—	—	Zeng et al. (2015)
CeO_2 NRs	1.1 μA cm^{-2} (at zero bias vs Ag/AgCl)	Visible light	—	MeOH	Lu et al. (2010)
CdS NPs-CeO_2 NRs	120 μA cm^{-2} (at zero bias vs Ag/AgCl)	Visible light	—	—	You et al. (2016)

CeO_2@Ag@CdS NTAs	2.14 mA cm^{-2} (at 0.2 V vs Ag/AgCl)	Visible light	Ag	Na_2S/Na_2SO_3	Zhao et al. (2015)
Ag@CeO_2	0.015 mA cm^{-2} (at 0 V vs Ag/AgCl)	Visible light	Ag	—	Khan et al. (2014)
Mixed-metal and ternary oxide materials for photoelectrochemical water splitting					
$La_2Ti_2O_7$	H_2 evolution: 10.5 μmol/10 h	Visible light	—	—	Uno et al. (2005)
$La_2Zr_2O_7$	H_2 evolution: 9.0 μmol/10 h	Visible light	—	—	Uno et al. (2006)
$Nd_2Sn_2O_7//Fe_2O_3$	0.21 mA cm^{-2} (at 1.23 V vs RHE)	AM 1.5 G	—	—	Uno et al. (2005)
RE-based materials for photocatalytic water splitting					
TiO_2:Gd/Pt	H_2 evolution: 10.65 mmol/8 h	AM 1.5 G	Pt	MeOH	Zalas and Laniecki (2005)
CeO_x-TiO_2/Pt	H_2 evolution: 14.1 mmol h^{-1} g^{-1}	UV light	Pt	MeOH	Luo et al. (2015)
Ag–Ce/TiO_2 NTAs	H_2 evolution: 1.47 μmol h^{-1} cm^{-2}	Visible light	Ag	EtOH	Fan et al. (2015)
NiO/$NaTaO_3$:La	H_2 evolution: 19.8 mmol h^{-1} g^{-1}	UV light	NiO	—	Kudo and Kato (2000)
Pt/Y:$NaTaO_3$	H_2 evolution: 96 μmol h^{-1} g^{-1}	UV light	—	MeOH	Jana et al. (2014)

Continued

Table 3 Rare-Earth-Containing Materials for Photoelectrochemical Water Splitting Reactions—cont'd

Material	Performance	Irradiation	Cocatalyst	Sacrificial Agent	References
$Cd_{0.6}Zn_{0.4}S$:La	H_2 evolution: 1.39 mmol h^{-1} g^{-1}	AM 1.5 G without UV	—	Na_2S/Na_2SO_3	Yu et al. (2013)
S/Sm:carbon dots	H_2 evolution: 1.04 μmol h^{-1}	AM 1.5 G	—	—	Yu et al. (2013)
RE-based oxide–oxide composite materials for photocatalytic water splitting					
CeO_2 NRs	H_2 evolution: 741 μmol h^{-1} g^{-1}	AM 1.5 G	—	Na_2S/Na_2SO_3	Lu et al. (2011a)
CdS/CeO_x NWs	H_2 evolution: 473.6 μmol h^{-1} g^{-1}	Visible light	—	Na_2S/Na_2SO_3	Li et al. (2013)
CdS NP-CeO_2 NRs	H_2 evolution: 8.4 mmol h^{-1} g^{-1}	Visible light	—	Lactic acid	You et al. (2016)
CdS NW-CeO_2 NPs	H_2 evolution: 549.5 μmol h^{-1} g^{-1}	Visible light	—	Lactic acid	Zhang et al. (2015)
Rh/$(Ga_{1-x}Zn_x)(N_{1-x}O_x)$/$Ln_2O_3$	H_2 evolution: 49 μmol h^{-1} g^{-1}	Visible light	Rh	—	Yoshida et al. (2013)
Mixed-metal and ternary oxide materials for photocatalytic water splitting					
$Gd_2Ti_2S_2O_5$/Pt	H_2 evolution: 120 μmol h^{-1} g^{-1}	Visible light	Pt	Na_2S/Na_2SO_3	Ishikawa et al. (2004)
$La_2Ti_2O_7$:N/Pt	H_2 evolution: 10 μmol h^{-1} g^{-1}	Visible light	Pt	—	Meng et al. (2013)

$La_2Ti_2O_7/In_2O_3$ $La_2Ti_2O_7$	H_2 evolution: 68.14 μmol g^{-1} h^{-1} 10.6 μmol h^{-1} g^{-1}	AM 1.5 G	—	MeOH	Hu et al. (2014)
$Au/K_2La_2Ti_3O_{10}$	H_2 evolution: 841 μmol h^{-1} g^{-1}	UV light	Au	—	Tai et al. (2004)
$Rb_2La_2Ti_3O_{10}$	H_2 evolution: 869 μmol h^{-1} g^{-1}	UV light	Ni	—	Takata et al. (1998)
La_3TaO_7/NiO	H_2 evolution: 158 μmol h^{-1} g^{-1}	UV light	NiO	—	Abe et al. (2004a)
$K_4Ce_2Ta_{10}O_{30}$	H_2 evolution:	Visible light		Na_2SO_3	Tian et al. (2006)
	30 μmol h^{-1} g^{-1}		—		
	55 μmol h^{-1} g^{-1}		Pt		
$H_{1-x}La_xCa_{2-x}Nb_3O_{10}Pt$	H_2 evolution: 169 cm^3 h^{-1} g^{-1}	UV light	Pt	MeOH	Huang et al. (2008)
$Bi_{2-x}La_xAlNbO_7$	H_2 evolution:	UV light		MeOH	Li et al. (2009)
	105.3 μmol h^{-1} g^{-1}		—		
	141.4 μmol h^{-1} g^{-1}		NiO_x		

other metal oxides (Meng et al., 2013; Shi et al., 2011; Zalas and Laniecki, 2005). For example, noble metal clusters (Ag, Au, Pt) strongly adsorb on lanthanide oxide through chemical interactions driven by a surface-induced redox chemistry, whereby the partially occupied f-orbitals are additionally filled. Similarly, the presence of lanthanide oxides in oxide–oxide composites can enhance the direct diffusion of vacancies responsible for oxygen hopping in the lattice that ultimately substantiate the formation of surface peroxides suggested as the possible intermediate step in the water oxidation reaction. Despite their high catalytic activity, noble-metal-doped catalysts gradually degrade due to sinering or evaporation of nanoscopic metal clusters demaning thermally robust materials. In search of new catalytic materials, nanostructured ternary semiconducting metal oxides such as lanthanide-based perovskites (ABO_3: $LnMnO_3$, $LnCoO_3$) have received significant attention owing largely to their unique properties emerging from electronic correlations and cooperative effects, which are less prominent in monometallic binary phases. In these compounds, the electronic structure of the transition metal (stabilized by Ln counterpart) plays a crucial role in determining the oxygen evolution reaction. For instance, the ability of the singly occupied d-orbital ($t_{2g}{}^3e_g{}^1$ configuration for Mn(III)) to form σ-bonds with anion adsorbates enhances the interaction between water molecules and the catalyst surface.

4. CONCLUSIONS

The world-wide quest for efficient, clean, and sustainable energy solutions in order to address the finite nature of fossil fuels has led to tremendous efforts devoted to the conversion of solar energy into chemical fuels, for instance, by photoassisted water splitting that produces "solar hydrogen" as safe and renewable source of energy. For instance, hydrogen generated in this manner could then power fuel cells that produce virtually no polluting by-products because the hydrogen combines with oxygen to produce water again; however, the (photo)oxidation of water signifies a challenging electrochemical process involving two- and four-electron transfer uphill reactions that demand photocatalytic materials capable of accelerating chemical kinetics and mobility of photogenerated electron–hole pairs. In this context, the rare earth materials offer promising potential due to their unique electronic structure and comparable size of their ions (lanthanide contraction) enabling doped compositions and solid solutions. In addition to their thermal stability and oxygen uptake and release properties, lanthanide oxides can alter the optical absorption and surface chemistry of photoanode materials to make them more active and selective for water activation.

Among several photophysical and photochemical considerations, the solar-to-hydrogen conversion efficiency can be enhanced by increasing the concentration of photogenerated charge carriers (e^-/h^+) together with their efficient transport and separation at the semiconductor–electrolyte junction. The challenges of PEC technology are multifaceted demanding of concepts enabling integration of new photocatalysts into innovative device architectures and more importantly convincing demonstration of water-splitting processes with high efficiency, scale-up potential, and long-term operational stability at Technology Readiness Levels desired for industrial implementation.

ACKNOWLEDGMENTS

Authors are thankful to the University of Cologne for providing the infrastructural support. J.L. is thankful to Fonds der chemischen Industrie for a PhD fellowship. The financial support in the framework of the DFG priority program (SPP 1613; *"Fuels Produced Regeneratively Through Light-Driven Water Splitting: Clarification of the Elemental Processes Involved and Prospects for Implementation in Technological Concepts"*) and the Framework Program of the European Commission (FP7) that funded the European Project SOLAROGENIX (http://www.solarogenix.eu) is gratefully acknowledged.

REFERENCES

Abe, R., et al., 2004a. Photocatalytic water splitting into H2 and O2 over R3TaO7 and R3NbO7 (R = Y, Yb, Gd, La): effect of crystal structure on photocatalytic activity. J. Phys. Chem. B 108, 811–814.

Abe, R., Higashi, M., Zou, Z., Sayama, K., Abe, Y., 2004b. Photocatalytic water splitting into H2 and O2 over R2Ti2O7 (R = Y, Rare Earth) with pyrochlore structure. Chem. Lett. 33, 954–955.

Al-Shankiti, I., Al-Otaibi, F., Al-Salik, Y., Idriss, H., 2013. Solar thermal hydrogen production from water over modified CeO2 materials. Top. Catal. 56, 1129–1138.

Auzel, F., 2004. Upconversion and anti-stokes processes with f and d ions in solids. Chem. Rev. 104, 139–174.

Bradley, D.C., Ghotra, J.S., Alan Hart, F., 1973. Low co-ordination numbers in lanthanide and actinide compounds. Part I. The preparation and characterization of tris{bis(trimethylsilyl)-amido}lanthanides. J. Chem. Soc. Dalton Trans. 10, 1021–1023.

Bunzli, J.-C.G., Piguet, C., 2005. Taking advantage of luminescent lanthanide ions. Chem. Soc. Rev. 34, 1048–1077.

Chou, J.-C., Lin, S.-A., Lee, C.-Y., Gan, J.-Y., 2013. Effect of bulk doping and surface-trapped states on water splitting with hematite photoanodes. J. Mater. Chem. A 1, 5908.

Corma, A., Atienzar, P., Garcia, H., Chane-Ching, J.-Y., 2004. Hierarchically mesostructured doped CeO2 with potential for solar-cell use. Nat. Mater. 3, 394–397.

Cotton, S., 2006. Lanthanide and Actinide Chemistry. John Wiley & Sons, UK, pp. 1–7.

Dey, S., Naidu, B.S., Rao, C.N.R., 2015. Ln0.5A0.5MnO3 (Ln = Lanthanide, A = Ca, Sr) perovskites exhibiting remarkable performance in the thermochemical generation of CO and H2 from CO2 and H2O. Chem. A Eur. J. 21, 7077–7081.

Fan, X., et al., 2015. High-efficiency photoelectrocatalytic hydrogen generation enabled by Ag deposited and Ce doped TiO2 nanotube arrays. Ceram. Int. 41, 5107–5116.

Gonell, F., et al., 2014. Photon up-conversion with lanthanide-doped oxide particles for solar H2 generation. J. Phys. Chem. C 118, 11279–11284.

He, H., Andersson, D.A., Allred, D.D., Rector, K.D., 2013. Determination of the insulation gap of uranium oxides by spectroscopic ellipsometry and density functional theory. J. Phys. Chem. C 117, 16540–16551.

Hebbink, G.A., Stouwdam, J.W., Reinhoudt, D.N., van Veggel, F.C.J.M., 2002. Lanthanide(III)-doped nanoparticles that emit in the near-infrared. Adv. Mater. 14, 1147–1150.

Heer, S., Lehmann, O., Haase, M., Güdel, H.-U., 2003. Blue, green, and red upconversion emission from lanthanide-doped LuPO4 and YbPO4 nanocrystals in a transparent colloidal solution. Angew. Chem. Int. Ed. 42, 3179–3182.

Hu, S., Chi, B., Pu, J., Jian, L., 2014. Novel heterojunction photocatalysts based on lanthanum titanate nanosheets and indium oxide nanoparticles with enhanced photocatalytic hydrogen production activity. J. Mater. Chem. A 2, 19260–19267.

Huang, Y., et al., 2008. Synthesis and photochemical properties of La-doped HCa2Nb3O10. Int. J. Hydrogen Energy 33, 6432–6438.

Idriss, H., 2010. Surface reactions of uranium oxide powder, thin films and single crystals. Surf. Sci. Rep. 65, 67–109.

Ishikawa, A., et al., 2004. Oxysulfides Ln2Ti2S2O5 as stable photocatalysts for water oxidation and reduction under visible-light irradiation. J. Phys. Chem. B 108, 2637–2642.

Jamil, A., Schläfer, J., Gönüllü, Y., Lepcha, A., Mathur, S., 2016. Precursor-derived rare earth metal pyrochlores: Nd2Sn2O7 nanofibers and thin films as efficient photoabsorbers. Cryst. Growth Des. 16, 5260–5267.

Jana, P., et al., 2014. Photocatalytic hydrogen production in the water/methanol system using Pt/RE:NaTaO3 (RE = Y, La, Ce, Yb) catalysts. Int. J. Hydrogen Energy 39, 5283–5290.

Jha, A.R., 2014. Rare Earth Materials: Properties and Applications. Taylor & Francis, New York, USA.

Jia, A., Su, Z., Lou, L.-L., Liu, S., 2010. Synthesis and characterization of highly-active nickel and lanthanum co-doped SrTiO3. Solid State Sci. 12, 1140–1145.

Jiang, H., Rinke, P., Scheffler, M., 2012. Electronic properties of lanthanide oxides from the GW perspective. Phys. Rev. B 86, 125115–125128.

Kato, H., Asakura, K., Kudo, A., 2003. Highly efficient water splitting into H2 and O2 over lanthanum-doped NaTaO3 photocatalysts with high crystallinity and surface nanostructure. J. Am. Ceram. Soc. 125, 3082–3089.

Kato, Y., et al., 2011. Deep-sea mud in the Pacific Ocean as a potential resource for rare-earth elements. Nat. Geosci. 4, 535–539.

Khan, M.M., et al., 2014. Electrochemically active biofilm assisted synthesis of Ag@CeO2 nanocomposites for antimicrobial activity, photocatalysis and photoelectrodes. J. Colloid Interface Sci. 431, 255–263.

Kudo, A., Kato, H., 2000. Effect of lanthanide-doping into NaTaO3 photocatalysts for efficient water splitting. Chem. Phys. Lett. 331, 373–377.

Kudo, A., Okutomi, H., Kato, H., 2000. Photocatalytic water splitting into H2 and O2 over K2LnTa5O15 powder. Chem. Lett. 29, 1212–1213. http://dx.doi.org/10.1246/cl.2000.1212.

Kundu, S., et al., 2012. Exploring the structural and electronic properties of Pt/ceria-modified TiO2 and its photocatalytic activity for water splitting under visible light. J. Phys. Chem. C 116, 14062–14070.

Le Gal, A., Abanades, S., 2012. Dopant incorporation in ceria for enhanced water-splitting activity during solar thermochemical hydrogen generation. J. Phys. Chem. C 116, 13516–13523.

Le Gal, A., Abanades, S., Bion, N., Le Mercier, T., Harlé, V., 2013. Reactivity of doped ceria-based mixed oxides for solar thermochemical hydrogen generation via two-step water-splitting cycles. Energy Fuel 27, 6068–6078.
Lehmann, B., 2014. Economic geology of rare-earth elements in 2014: a global perspective. Eur. Geol. 37, 21–24.
Li, Y., Chen, G., Zhang, H., Li, Z., 2009. Electronic structure and photocatalytic water splitting of lanthanum-doped Bi2AlNbO7. Mater. Res. Bull. 44, 741–746.
Li, W., et al., 2013. CdS/CeOx heterostructured nanowires for photocatalytic hydrogen production. J. Mater. Chem. 1, 4190–4193.
Liu, G., et al., 2011. Visible-light-driven photocatalysts: (La/Bi + N)-codoped NaNbO3 by first principles. J. Appl. Phys. 109, 063103.
Liu, Y., et al., 2015a. Enhancement of the photoelectrochemical performance of WO3 vertical arrays film for solar water splitting by gadolinium doping. J. Phys. Chem. C 119, 14834–14842.
Liu, R., Liang, F., Zhou, W., Yang, Y., Zhu, Z., 2015b. Calcium-doped lanthanum nickelate layered perovskite and nickel oxide nano-hybrid for highly efficient water oxidation. Nano Energy 12, 115–122.
Lu, X., et al., 2010. Facile synthesis of free-standing CeO2 nanorods for photoelectrochemical applications. Chem. Commun. 46, 7721–7723.
Lu, X., et al., 2011a. Redox cycles promoting photocatalytic hydrogen evolution of CeO2 nanorods. J. Mater. Chem. 21, 5569–5572.
Lu, L., Li, R., Peng, T., Fan, K., Dai, K., 2011b. Effects of rare earth ion modifications on the photoelectrochemical properties of ZnO-based dye-sensitized solar cells. Renew. Energy 36, 3386–3393.
Luo, S., et al., 2015. Hierarchical heterogeneity at the CeOx–TiO2 interface: electronic and geometric structural influence on the photocatalytic activity of oxide on oxide nanostructures. J. Phys. Chem. B 119, 2669–2679.
Méndez-Ramos, J., Ruiz-Morales, J.C., Acosta-Mora, P., del-Castillo, J., Yanes, A.C., 2013. Rare-earth doped nano-glass-ceramics for extending spectral response of water-splitting semiconductor electrodes by high intense UV-blue up-conversion: turning the sun into blue. J. Power Sources 238, 313–317.
Meng, F., et al., 2013. Photocatalytic generation of hydrogen with visible-light nitrogen-doped lanthanum titanium oxides. Catal. Today 199, 48–52.
Prévot, M.S., Sivula, K., 2013. Photoelectrochemical tandem cells for solar water splitting. J. Phys. Chem. C 117, 17879–17893.
Primo, A., Marino, T., Corma, A., Molinari, R., García, H., 2011. Efficient visible-light photocatalytic water splitting by minute amounts of gold supported on nanoparticulate CeO2 obtained by a biopolymer templating method. J. Am. Ceram. Soc. 133, 6930–6933.
Prokofiev, A.V., Shelykh, A.I., Melekh, B.T., 1996. Periodicity in the band gap variation of Ln2X3 (X = O, S, Se) in the lanthanide series. J. Alloys Compd. 242, 41–44.
Riedel, E., Janiak, C., 2007. Anorganische Chemie. Walter de Gruyter, Berlin, Germany.
Rivera, V.A.G., Ferri, F.A., Marega Jr, E., 2012. Localized surface plasmon resonances: noble metal nanoparticle interaction with rare-earth ions. In: Kim, K.Y. (Ed.), Plasmonics—Principles and Applications InTech, Croatia. http://dx.doi.org/10.5772/50753.
Sastri, V.S., Bünzli, J.-C., Rao, V.R., Rayudu, G.V.S., Perumareddi, J.R., 2003. Modern Aspects of Rare Earths and Their Complexes. Elsevier, Netherlands, pp. 1–72.
Shalav, A., Richards, B.S., Trupke, T., Krämer, K.W., Güdel, H.U., 2005. Application of NaYF4:Er3+ up-converting phosphors for enhanced near-infrared silicon solar cell response. Appl. Phys. Lett. 86, 013505.

Shalav, A., Richards, B.S., Green, M.A., 2007. Luminescent layers for enhanced silicon solar cell performance: up-conversion. Sol. Energy Mater. Sol. Cells 91, 829–842.

Shannon, R., 1976. Revised effective ionic radii and systematic studies of interatomic distances in halides and chalcogenides. Acta Crystllogr. Sect. A: Found. Crystallogr. 32, 751–767.

Shen, S., Shi, J., Guo, P., Guo, L., 2011. Visible-light-driven photocatalytic water splitting on nanostructured semiconducting materials. Int. J. Nanotechnol. 8, 523.

Shi, Z., Zhang, X., Yao, S., 2011. Synthesis and photocatalytic properties of lanthanum doped anatase TiO2 coated Fe3O4 composites. Rare Metals 30, 252–257.

Strümpel, C., et al., 2007. Modifying the solar spectrum to enhance silicon solar cell efficiency—an overview of available materials. Sol. Energy Mater. Sol. Cells 91, 238–249.

Sudhagar, P., Devadoss, A., Nakata, K., Terashima, C., Fujishima, A., 2014. Enhanced photoelectrocatalytic water splitting at hierarchical Gd3+:TiO2 nanostructures through amplifying light reception and surface states passivation. J. Electrochem. Soc. 162, H108–H114.

Tai, Y.-W., Chen, J.-S., Yang, C.-C., Wan, B.-Z., 2004. Preparation of nano-gold on K2La2Ti3O10 for producing hydrogen from photo-catalytic water splitting. Catal. Today 97, 95–101.

Takata, T., Tanaka, A., Hara, M., Kondo, J.N., Domen, K., 1998. Recent progress of photocatalysts for overall water splitting. Catal. Today 44, 17–26.

Tian, M., et al., 2006. K4Ce2M10O30 (M = Ta, Nb) as visible light-driven photocatalysts for hydrogen evolution from water decomposition. Appl. Catal., A 309, 76–84.

Tian, M., Wang, H., Sun, D., Peng, W., Tao, W., 2014. Visible light driven nanocrystal anatase TiO2 doped by Ce from sol–gel method and its photoelectrochemical water splitting properties. Int. J. Hydrogen Energy 39, 13448–13453.

Uno, M., Kosuga, A., Okui, M., Horisaka, K., Yamanaka, S., 2005. Photoelectrochemical study of lanthanide titanium oxides, Ln2Ti2O7 (Ln = La, Sm, and Gd). J. Alloys Compd. 400, 270–275.

Uno, M., et al., 2006. Photoelectrochemical study of lanthanide zirconium oxides, Ln2Zr2O7 (Ln = La, Ce, Nd and Sm). J. Alloys Compd. 420, 291–297.

van der Ende, B.M., Aarts, L., Meijerink, A., 2009. Lanthanide ions as spectral converters for solar cells. Phys. Chem. Chem. Phys. 11, 11081–11095.

Wang, J., Yin, S., Komatsu, M., Sato, T., 2005. Lanthanum and nitrogen co-doped SrTiO3 powders as visible light sensitive photocatalyst. J. Eur. Ceram. Soc. 25, 3207–3212.

Weber, A.S., Grady, A.M., Koodali, R.T., 2012. Lanthanide modified semiconductor photocatalysts. Cat. Sci. Technol. 2, 683–693.

Wu, L., et al., 2015. Facile synthesis of Ag@CeO2 core–shell plasmonic photocatalysts with enhanced visible-light photocatalytic performance. J. Hazard. Mater. 300, 93–103.

Wu, X., et al., 2017. Control strategy on two −/four-electron pathway of water splitting by multidoped carbon based catalysts. ACS Catal. 7, 1637–1645.

Xie, S., et al., 2015. Efficient and stable photoelectrochemical water oxidation by ZnO photoanode coupled with Eu2O3 as novel oxygen evolution catalyst. J. Power Sources 297, 9–15.

Xu, H., et al., 2009. Synthesis, characterization and photocatalytic activities of rare earth-loaded BiVO4 catalysts. Appl. Surf. Sci. 256, 597–602.

Yoshida, M., Maeda, K., Lu, D., Kubota, J., Domen, K., 2013. Lanthanoid oxide layers on rhodium-loaded (Ga1–xZnx)(N1–xOx) photocatalyst as a modifier for overall water splitting under visible-light irradiation. J. Phys. Chem. C 117, 14000–14006.

You, D., Pan, B., Jiang, F., Zhou, Y., Su, W., 2016. CdS nanoparticles/CeO2 nanorods composite with high-efficiency visible-light-driven photocatalytic activity. Appl. Surf. Sci. 363, 154–160.

Yu, H., et al., 2012. Cubic CeO_2 nanoparticles as mirror-like scattering layers for efficient light harvesting in dye-sensitized solar cells. Chem. Commun. 48, 7386–7388.
Yu, Y.-G., et al., 2013. Doping La into the depletion layer of the $Cd_{0.6}Zn_{0.4}S$ photocatalyst for efficient H_2 evolution. Chem. Commun. (Camb.) 49, 10142–10144.
Zalas, M., 2014. Gadolinium-modified titanium oxide materials for photoenergy applications: a review. J. Rare Earths 32, 487–495.
Zalas, M., Laniecki, M., 2005. Photocatalytic hydrogen generation over lanthanides-doped titania. Sol. Energy Mater. Sol. Cells 89, 287–296.
Zeng, C.-H., et al., 2015. Synthesis of porous europium oxide particles for photoelectrochemical water splitting. Electrochim. Acta 165, 396–401.
Zhang, A., Zhang, J., 2010. Effects of europium doping on the photocatalytic behavior of $BiVO_4$. J. Hazard. Mater. 173, 265–272.
Zhang, F.X., et al., 2007. Structural change of layered perovskite $La_2Ti_2O_7$ at high pressures. J. Solid State Chem. 180, 571–576.
Zhang, M., et al., 2012. Improving Hematite's solar water splitting efficiency by incorporating rare-earth upconversion nanomaterials. J. Phys. Chem. Lett. 3, 3188–3192.
Zhang, C., et al., 2014. Facile electrochemical synthesis of CeO_2 hierarchical nanorods and nanowires with excellent photocatalytic activities. New J. Chem. 38, 2581–2586.
Zhang, X., Zhang, N., Xu, Y.-J., Tang, Z.-R., 2015. One-dimensional CdS nanowires-CeO_2 nanoparticles composites with boosted photocatalytic activity. New J. Chem. 39, 6756–6764.
Zhao, M., Li, H., Shen, X., Ji, Z., Xu, K., 2015. Facile electrochemical synthesis of CeO_2@Ag@CdS nanotube arrays with enhanced photoelectrochemical water splitting performance. Dalton Trans. 44, 19935–19941.
Zhou, T., Hu, J., Li, J., 2011. Er^{3+} doped bismuth molybdate nanosheets with exposed {0 1 0} facets and enhanced photocatalytic performance. Appl Catal B 110, 221–230.

PART II

Nanostructured Photocatalysts and Photoelectrodes

CHAPTER SIX

Artificial Photosynthesis on III-Nitride Nanowire Arrays

Sheng Chu*, Xianghua Kong*, Srinivas Vanka*, Hong Guo*, Zetian Mi*,†,1

*McGill University, Montreal, QC, Canada
†University of Michigan, Ann Arbor, MI, United States
[1]Corresponding author: e-mail address: ztmi@umich.edu

Contents

Semiconductors and Semimetals, Volume 97
ISSN 0080-8784
http://dx.doi.org/10.1016/bs.semsem.2017.02.004

1. INTRODUCTION

The development of a clean, cheap, renewable, and abundant energy source to replace our reliance on fossil fuels is a great scientific and technological challenge. Among various renewable energy options, solar energy is the largest and considered to be the ultimate solution for the Terawatt challenge faced by human society (Lewis and Nocera, 2006). The amount of solar energy intercepted by the earth in 1 h can satisfy global energy consumption for an entire year. To date, solar photovoltaic cells are used worldwide to convert solar energy into electricity on a large scale. However, because of the diurnal cycle and the intermittency of the terrestrial solar resource, the major issue remains in the lack of efficient and cost-effective energy storage for large-scale applications. Artificial photosynthesis, which mimics the natural photosynthesis process, has been considered as a potentially viable technology for solar energy storage (Kim et al., 2015; Styring, 2012; Tachibana et al., 2012). In such a process, the solar energy is converted and stored in the form of energy-rich solar fuels (e.g., hydrogen, methanol) via chemical reaction such as water splitting and CO_2 reduction. Solar fuels could be used around the clock, as the chemical energy stored during the daytime can be converted to electrical or mechanical power whenever and wherever needed.

In the form of artificial photosynthesis, both photocatalytic and photoelectrochemical (PEC) routes have been extensively studied. Semiconductor material, as an artificial leaf, plays an essential role in the transduction of energy from sunlight to chemical fuels. The semiconductor captures solar energy to generate energetic electron–hole pairs with sufficient potentials to accomplish the chemical reactions of water splitting and CO_2 reduction. Over the past four decades, the development has primarily focused on oxides consisting of metal cations with d^0 and d^{10} configurations (i.e., Ti^{4+}, Zr^{4+}, Nb^{5+}, Ta^{5+}, W^{6+}, Ga^{3+}, Ge^{4+}, In^{3+}, Sn^{4+}, and Sb^{5+}) (Kudo and Miseki, 2009; Maeda and Domen, 2007). Despite their inherent chemical stability, such metal oxide photocatalysts usually possess highly positive valence band (+3.0 V vs NHE or higher) formed by O_{2p} orbitals. Considering the bottom of the conduction band needs to be sufficiently negative for the water or CO_2 reduction potentials (+0 V vs NHE or lower if taking into account the large overpotential in the case of CO_2 reduction), the photocatalytic materials are only mostly active under UV irradiation (~4% in solar spectrum) due to their large bandgap. In parallel, some nonoxide photocatalysts (e.g., cadmium sulfide) are widely studied for their appropriate band levels for water splitting and CO_2 reduction under visible light (~43% in solar spectrum). However, they are generally unstable against photocorrosion in the absence of a sacrificial electron donor (Abe, 2010). In this regard, the development of stable and highly efficient photo(electro)catalyst for artificial photosynthesis under visible light is still a grand challenge.

Recently, III-nitride materials, e.g., GaN, InN, and their ternary alloys, have drawn considerate attention due to their unique electrical, optical, and structural properties (Kibria and Mi, 2016). Due to the more negative potential of the N_{2p} orbital compared to the O_{2p} orbital, III-nitride semiconductors have a narrow bandgap that can absorb the visible light, while still straddle the redox potentials of water splitting and CO_2 reduction (Fig. 1). For example, InGaN is the only known semiconductor material whose conduction and valence band edges straddle water redox potentials under deep visible and even possibly near-infrared light irradiation (Moses and Van de Walle, 2010). Moreover, in contrast to conventional III–V compounds, wherein the chemical bonds are mostly covalent, the chemical bonds in III-nitrides such as GaN are strongly ionic, resulting in extreme stability under harsh photo(electro)chemical reaction conditions (Jung et al., 2008; Luo et al., 2008; Moustakas, 2013). Additionally, III-nitrides have high absorption coefficient and large charge carrier mobility, leading to excellent photon absorption and charge carrier extraction for efficient solar water splitting and CO_2 reduction (Bhuiyan et al., 2003;

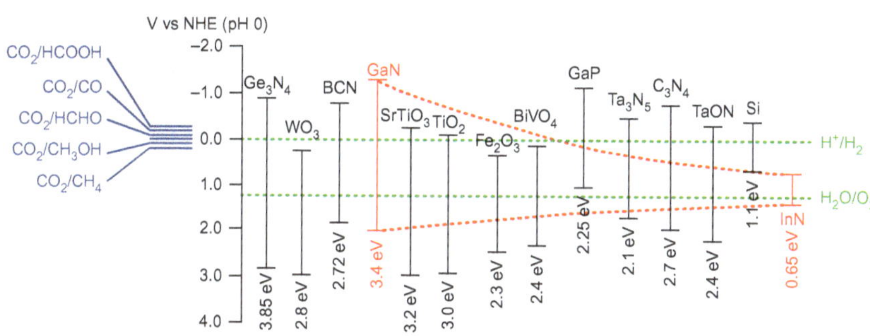

Fig. 1 Band edge positions of commonly reported nitride photocatalysts. The oxidation and reduction potentials of water are also shown (*green dotted lines*). The *red dotted line* represents the band edge positions of $In_xGa_{1-x}N$ with x increasing from *left* to *right* (0–1). The reduction potentials of CO_2 to various value added products are also shown. *Reproduced from Kibria, M.G., Mi, Z., 2016. Artificial photosynthesis using metal/nonmetal-nitride semiconductors: current status, prospects, and challenges. J. Mater. Chem. A 4 (8), 2801–2820 with permission from The Royal Society of Chemistry.*

Wu, 2009; Yotsuhashi et al., 2011). On the other hand, one-dimensional (1D) nanostructures, such as nanowire arrays, are attractive owing to their large surface-to-volume ratio, high crystal quality that are free of dislocations or defects, and enhanced light absorption via light trapping and anti-reflection effects (Hwang et al., 2012; Jung et al., 2008; Liu et al., 2014).

In this chapter, we will focus on the recent progress in artificial photosynthesis on III-nitride nanowire arrays. First, the photocatalytic properties of III-nitrides for water splitting and CO_2 reduction are discussed for a better understanding of the fundamental mechanisms. Then great emphasis is paid to the design and development of efficient III-nitride materials (i.e., GaN and InGaN) for water splitting and CO_2 reduction under visible light, either in the form of photocatalytic or PEC approach. Finally, conclusions and future prospects of III-nitride materials for artificial photosynthesis are presented.

2. PHOTOCATALYTIC PROPERTIES OF III-NITRIDES

Light harvesting, charge carrier separation and transfer, and surface electrochemical reaction are three main steps involved in a photocatalytic process (Fig. 2). The initial step involves the light absorption and the generation of electron–hole pairs. When the energy of incident light is larger than that of a bandgap of a semiconductor, excited electrons and holes are created in the conduction and valence bands, respectively. InGaN, with the incorporation of different In contents, possesses a direct and tunable bandgap from 3.4 to 0.65 eV across the ultraviolet, visible, and near-infrared spectrum, therefore, can cover nearly the entire solar spectrum (César et al., 2011; Moses and Van de Walle, 2010). In addition, the light trapping is enhanced in nanowire arrays due to the nanowire geometry (Liu et al., 2014). The second step consists of charge separation and migration of photogenerated carriers. Charge recombination (including bulk and surface recombination) and separation/migration processes are two competitive processes inside the semiconductor photocatalyst that largely affect the efficiency of the photocatalytic reaction. Carrier lifetime, diffusion length, crystallinity, and particle size are all important factors that strongly affect this step. It is worth mentioning that nearly defect-free and high crystalline quality III-nitride nanowires can be grown by molecular beam epitaxy, with the recombination process of charge carriers minimized in the bulk photocatalyst (Kibria and Mi, 2016). Moreover, the charge carrier separation and extraction in III-nitride nanowire is largely dominated by diffusion

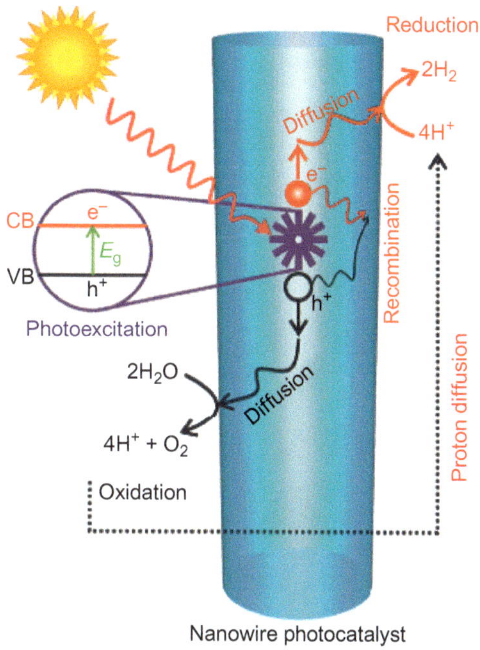

Fig. 2 Schematic illustration of the main process steps in water splitting, including photoexcitation, carrier generation, diffusion, recombination, water oxidation, proton diffusion, and reduction reaction on the surface of nanowire photocatalysts. *Reproduced from Kibria, M.G., Mi, Z., 2016. Artificial photosynthesis using metal/nonmetal-nitride semiconductors: current status, prospects, and challenges. J. Mater. Chem. A 4 (8), 2801–2820 with permission from The Royal Society of Chemistry.*

under nearly flat band conditions. Take InGaN/GaN nanowires, for example, nearly 90% of the charge carriers can readily diffuse to the nanowire surface, as the diffusion lengths of photoexcited charge carriers (~200–300 nm) are much larger than the lateral dimensions (<100 nm) of nanowire photocatalyst (Zhang et al., 2014). The final step involves the surface chemical reaction via the energetic charge carriers. To promote this step, proper cocatalysts are usually loaded on the light harvesting semiconductor to serve as the reaction sites and catalyze the reactions. This is because the photogenerated electrons and holes are likely to recombine if there are no suitable active sites available on the surface. Therefore, it is important to load with suitable cocatalyst to accelerate the slow surface reaction (10^{-6}–1 s) to compete with the fast recombination process of charge carriers (on the order of 10^{-9}–10^{-6} s) (Fan et al., 2013; Tang et al., 2008; Yang et al., 2013).

In the following, we discuss the photocatalytic properties of III-nitrides from three detailed aspects: water oxidation, proton reduction, and CO_2 adsorption and deformation.

2.1 Water Oxidation

The overall water splitting consists of two half-reactions, i.e., water oxidation and proton reduction. The water oxidation is essentially the first step, as described by Eq. (1):

$$2H_2O \rightarrow O_2 + 4e^- + 4H^+ \quad E^\circ = +0.82\ \text{V vs NHE at pH 7} \tag{1}$$

For unassisted photocatalytic water oxidation, the valence band maximum of semiconductor should be positioned at more positive potential than +0.82 V vs NHE (pH 7). Water oxidation is often the bottleneck for the overall water splitting reaction as it requires multiple intermediate steps which involve four-electron transfer and reaction (Tang et al., 2008; Yang et al., 2014). Kinetically, water oxidation competes with fast electron–hole bulk/surface recombination, surface O_2 adsorption, and self-oxidation of the photocatalyst. Recently, by using ab initio molecular dynamics simulation study, it was revealed that the nonpolar GaN surfaces were very reactive for spontaneous dissociation of ~83% of the water molecules, and the photogenerated holes had sufficient standard free-energy to drive the water oxidation (Shen et al., 2009, 2010). Moreover, it was found that the near-surface band structure played a critical role in electron/hole extraction. In the case of *p*-type III-nitrides, the presence of downward band bending can accelerate proton reduction at the nanowire surface, as shown in Fig. 3A. However, at the same time, it raises the barrier for hole-diffusion and impedes the hole transfer for water oxidation. Recently, we reported that with increasing Mg incorporation, the downward band bending in the near-surface region can be precisely tuned (Fig. 3B), indicating the transport of both photogenerated carriers can be balanced by optimizing Mg doping level (Chowdhury et al., 2015; Kibria et al., 2014, 2015). As a consequence, water oxidation, the limiting half-reaction for overall water splitting as well as the primary reaction required for H_2 production, is enhanced due to accelerated hole transport. It is worth mentioning that the electronic structures of semiconductors (such as band alignment) can be accurately calculated even with doping concentration as low as 0.1% (Michaud-Rioux et al., 2016).

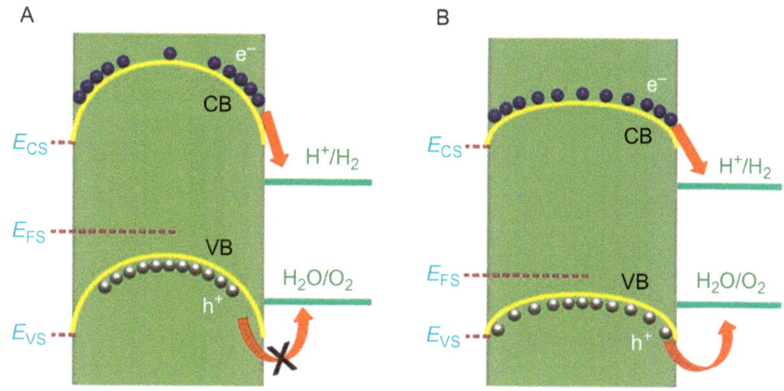

Fig. 3 (A) Illustration of the presence of severe downward band bending that hinders hole transport toward the surface and (b) reduced downward band bending with optimum Mg dopant incorporation to enhance hole transport for oxidation. E_{CS}, E_{VS}, and E_{FS} denote the conduction band minimum, valence band maximum, and Fermi level at the surface, respectively (Chowdhury et al., 2015). Used in accordance with the Creative Commons Attribution (CC BY) license, AIP Publishing LLC.

2.2 Proton Reduction

Next, protons (H^+) released in water oxidation are reduced by the photogenerated electrons, as described by Eq. (2):

$$4H^+ + 4e^- \rightarrow 2H_2 \quad E^o = -0.41\,V \text{ vs NHE at pH } 7 \tag{2}$$

For unassisted photocatalytic proton reduction, the conduction band minimum of semiconductor should be positioned at more negative potential than −0.41 V vs NHE at pH 7. Theoretically, bandgap for water splitting is at least 1.23 eV corresponding to a light wavelength of ~1000 nm. In practice, however, in order to provide sufficient kinetic overpotentials to overcome entropic and other parasitic losses, the optimum bandgap should be in the range of 1.5–2.5 eV. On the GaN nonpolar surface, effective free-energy barrier for proton diffusion was very small, thus leading to enhanced migration of protons from water oxidation sites to proton reduction sites and finally improving H_2 evolution efficiency (Wang et al., 2012).

2.3 CO_2 Adsorption and Deformation

CO_2 is a linear molecule with a high thermodynamic stability. The thermodynamic potentials for various CO_2 reduction products (pH 7 vs NHE) are given in Eqs. (3)–(8) (Li et al., 2016):

$$CO_2 + e^- \rightarrow CO_2^{\bullet -} \quad E^o = -1.90\,V \tag{3}$$

$$CO_2 + 2H^+ + 2e^- \rightarrow HCOOH \quad E^o = -0.61\,V \tag{4}$$

$$CO_2 + 2H^+ + 2e^- \rightarrow CO + H_2O \quad E^o = -0.53\,V \tag{5}$$

$$CO_2 + 4H^+ + 4e^- \rightarrow HCHO + H_2O \quad E^o = -0.48\,V \tag{6}$$

$$CO_2 + 6H^+ + 6e^- \rightarrow CH_3OH + H_2O \quad E^o = -0.38\,V \tag{7}$$

$$CO_2 + 8H^+ + 8e^- \rightarrow CH_4 + 2H_2O \quad E^o = -0.24\,V \tag{8}$$

Single electron reduction of CO_2 to $CO_2^{\bullet -}$ radical (Eq. 3) is not feasible as it requires a high negative thermodynamically unfavorable potential. Thus, proton-coupled multielectron steps for CO_2 reduction are generally more favorable from the thermodynamic point of view. The adsorption and deformation of CO_2 molecule on the photocatalyst surface is the first and foremost step. It has been well recognized that on conventional metal oxide surfaces, the adsorption of CO_2 is dominated by defect sites (oxygen vacancies) (Funk et al., 2006; Indrakanti et al., 2009; Liu et al., 2012; Markovits et al., 1996; Qiu et al., 2015; Rodriguez et al., 2012; Sorescu et al., 2011; Takahashi et al., 2002; Thompson et al., 2003), which inevitably leads to carrier loss, instability-related issues, and surface deactivation due to the competing O_2 adsorption. Recently, AlOtaibi et al. revealed that CO_2 molecules can be spontaneously adsorbed and deformed on clean nonpolar surfaces of wurtzite Ga(In)N using density functional theory calculations (AlOtaibi et al., 2016). It was further discovered that, with the incorporation of Mg dopant, owing to the enhanced adsorption and deformation of CO_2 molecules on the surface of the nanowires and the reduced surface potential barrier, the photoactivity for CO_2 reduction was enhanced by nearly 50-fold. The most energetically stable CO_2 adsorption configuration on the GaN($10\bar{1}0$) surface is shown in Fig. 4A (top view) and B (side view). Intriguingly, the preferable adsorption site for CO_2 molecule is the surface Ga–N dimer. The strong interaction between CO_2 molecule and the photocatalyst results in originally linear CO_2 molecule's heavy deformation with an O–C–O angle (θ in Fig. 4A) of 128.27 degree, thus forming a tridentate carbonate species. Meanwhile, two C–O bonds are so weakened that both bond lengths are enlarged. In addition, the adsorption energy (E_{ad}) and deformation energy ($E_{def}^{CO_2}$) are −1.76 and 2.55 eV, respectively. Here the adsorption energy represents the net energy gained upon adsorption. The deformation energy, which denotes the energy increase from a linear CO_2 molecule to a buckled one, describes the degree of activation of the surface CO_2 molecule. Differential charge density in Fig. 4C and

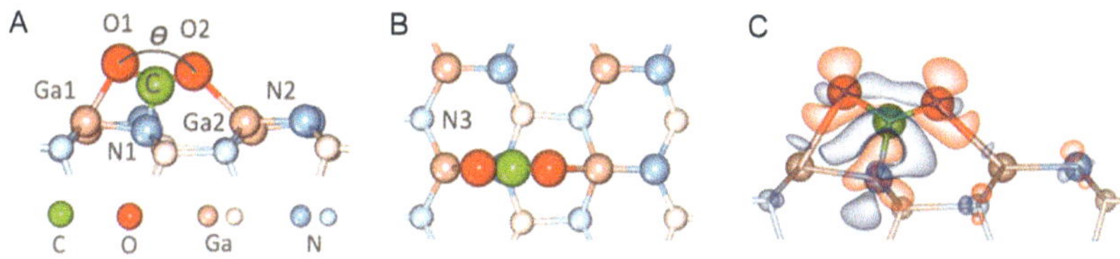

Fig. 4 Calculated structural and electronic properties of the GaN *m*-plane and CO_2 adsorption. (A and B) *Side* and *top view*, respectively, of the most stable fully relaxed configuration of CO_2 adsorbed on GaN *m*-plane. (C) Differential charge density of the adsorbed CO_2 molecule. The *pink* suggests electron charge gain, while the *light blue* indicates electron charge reduction. Isosurface contours of electron density differences are drawn at ± 0.005 e/Å^3. *Reprinted with permission from AlOtaibi, B., Kong, X., Vanka, S., Woo, S.Y., Pofelski, A., Oudjedi, F., Fan, S., Kibria, M.G., Botton G.A., Ji, W., Guo, H., Mi, Z., 2016. Photochemical carbon dioxide reduction on Mg-doped Ga(In)N nanowire arrays under visible light irradiation. ACS Energy Lett. 1, 246–252. Copyright 2016, American Chemical Society.*

Bader charge analysis before and after the adsorption illustrate the bonding mechanism between CO_2 and the photocatalyst surface qualitatively and quantitatively. Obviously, for the CO_2 molecule itself, substantial charge reduction was found around the C–O bonds, which was in accordance with the significantly weakened and elongated C–O bonds. In short, the charge transfer activates the CO_2 molecule upon the adsorption on pristine GaN surfaces. The more charge transfer between CO_2 molecule and photocatalyst surface, the stronger is the activation of CO_2, and the higher is the rate of CO_2 reduction. For Mg-doped GaN($10\bar{1}0$) surface, more surface-mediated charge, namely, 0.479e ($e = 1.60 \times 10^{-19}$ C), than the quantity of a clean surface (0.395e), is transferred to the CO_2 molecule, thus enhanced CO_2 reduction activities. Meanwhile, with respect to Ge-doped surface, electron charge of 0.398e is increased on CO_2 molecule, thus do not offer any clear advantage, compared to the undoped surface.

3. PHOTOCHEMICAL WATER SPLITTING

Photochemical water splitting offers a simple and low cost approach to produce H_2 and O_2 from water using renewable solar energy. The key for practical water splitting application is to develop efficient and stable photocatalyst, which generally consists of semiconductor and cocatalyst nanoparticles. Over the past four decades, various inorganic and organic systems have been developed as photocatalysts for water splitting. Although many photocatalytic materials have been studied, most of them are only active in the UV region or capable of drive the half reactions of water splitting in the presence of sacrificial reagent under visible light (Chen et al., 2010; Kudo and Miseki, 2009). To date, there are very few available photocatalytic materials that can achieve overall water splitting under visible light, which still suffer from low visible light activity and/or poor stability (Takata et al., 2016).

Recently, research on III-nitride photocatalysts (e.g., GaN and InGaN) for water splitting has drawn significant attention due to their suitable band structure and distinctive stability. Turner et al. reported that GaN thin film, either *p*- or *n*-type, exhibited band-edge potentials suitable for overall water splitting (Beach et al., 2003). Maeda et al. confirmed that GaN powder could meet the thermodynamic requirements for overall water splitting and function as a stable photocatalyst under UV irradiation, with $Rh_{2-y}Cr_yO_3$ nanoparticles as a cocatalyst (Maeda et al., 2007). However, the photocatalytic activity of GaN powder was moderate (H_2 production rate: $63\,\mu mol\,{g_{cat}}^{-1}\,h^{-1}$), probably because of inefficient light absorption, low crystalline quality, and small surface area. On the other hand, the bandgap of GaN, with the incorporation of In, can be tuned to encompass nearly the entire solar spectrum (César et al., 2011; Moses and Van de Walle, 2010). In addition, one-dimensional (1D) nanostructures, such as nanowires, provide a promising platform for its large lateral surface, high crystallinity, and enhanced light absorption (Hwang et al., 2012; Jung et al., 2008; Liu et al., 2014).

In this section, we show that overall water splitting can be achieved using wafer-level III-nitride nanowire arrays (e.g., GaN and InGaN) under UV and visible light irradiation. Additionally, by tuning the surface Fermi level with controlled Mg doping, we demonstrate that the efficiency can be significantly enhanced. Finally, we discuss the atomic-scale origin of the long-term stability and high performance of III-nitride nanostructures for photochemical water splitting, using the combination of experimental and theoretical approaches. This section includes the design, synthesis, and characterization of III-nitride photocatalysts, highlighting their great potential for efficient and stable solar water splitting.

3.1 Multiband InGaN/GaN Nanowire Arrays for Overall Pure Water Splitting

We have demonstrated, for the first time, that efficient overall water splitting can be achieved on wafer-level GaN nanowire arrays grown by molecular beam epitaxy (Wang et al., 2011). In this study, core–shell Rh/Cr_2O_3 nanoparticles were incorporated as cocatalysts. As shown in Fig. 5A, steady and nearly stoichiometric evolution of H_2 and O_2 gases was observed during 18 h of operation (H_2 production rate: $\sim 4\,\text{mmol}\,g_{cat}^{-1}\,h^{-1}$), with the photocatalytic activity nearly two orders of magnitudes higher than the previously reported GaN particulate sample. This work establishes the potential of III-nitride nanowire arrays as viable photocatalysts for achieving stable and efficient solar water splitting to produce clean H_2 fuels.

To utilize the abundant visible-light, multiband InGaN/GaN nanowire heterostructure was developed for overall water splitting under visible-light irradiation (Kibria et al., 2013). InGaN ternary segments and quantum dots were incorporated into the GaN nanowires. With the use of Rh/Cr_2O_3 cocatalysts, InGaN/GaN exhibited stable and efficient activity for overall water splitting under wavelengths up to ~560 nm, as shown

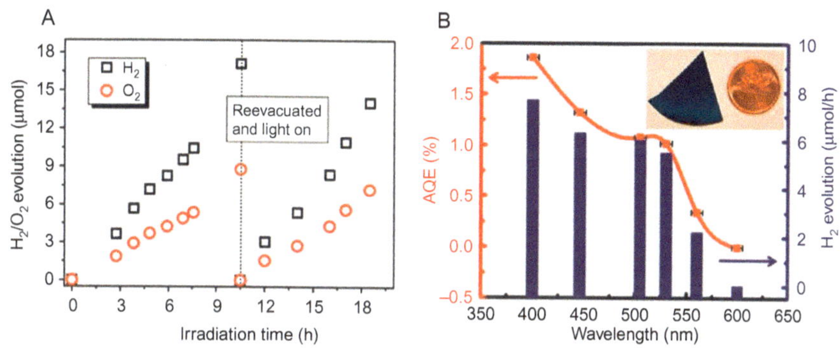

Fig. 5 (A) Experimental overall photocatalytic water splitting on Rh/Cr_2O_3 core–shell nanostructure deposited GaN nanowires under a 300 W full arc xenon lamp irradiation for a duration of 18 h. No degradation of the photocatalytic activity was observed during the course of two cycles of reaction. (B) Apparent quantum efficiency (AQE) (*red line* and *solid square symbol, left axis*) and H_2 evolution rate (*blue columns, right axis*) as a function of wavelength of the incident light. The horizontal error bars represent the full-width-half-maximum of the optical filters. The H_2 evolution rate was derived from ~2 h of overall water splitting under each optical filter. The *solid red line* is a guide to the eye. The *inset* shows a typical nanowire sample used for overall water splitting. *Panel (A) reprinted with permission from Wang, D.F., Pierre, A., Kibria, M.G., Cui, K., Han, X.G., Bevan, K.H., Guo, H., Paradis, S., Hakima, A.R., Mi, Z., 2011. Wafer-level photocatalytic water splitting on GaN nanowire arrays grown by molecular beam epitaxy. Nano Lett. 11 (6), 2353–2357. Copyright 2011, American Chemical Society. Panel (B) reprinted with permission from Kibria, M.G., Nguyen, H.P.T., Cui, K., Zhao, S., Liu, D., Guo, H., Trudeau, M.L., Paradis, S., Hakima, A.R., Mi, Z., 2013. One-step overall water splitting under visible light using multiband InGaN/GaN nanowire heterostructures. ACS Nano 7 (9), 7886–7893. Copyright 2013, American Chemical Society.*

in Fig. 5B. The H_2 production rate was calculated to be 92 mmol g_{cat}^{-1} h^{-1} for overall water splitting, which was nearly an order of magnitude higher than the above-mentioned GaN-based photocatalyst. At ~440–450 nm wavelengths, the internal quantum efficiency was estimated to be ~13%. In addition, the InGaN/GaN nanowire photocatalyst showed stable photocatalytic activity during the course of ~18 h of operation. Moreover, detailed structural and surface analysis further confirmed the high stability of the nanowire during water splitting reaction. This work demonstrates the use of III-nitrides as efficient and stable photocatalysts for photochemical water splitting under visible light.

3.2 Tuning the Surface Fermi Level for Efficient Overall Water Splitting

The surface Fermi level and band bending play a critical role in heterogeneous photocatalysis and the photochemical processes at the surface. While the concept of band bending is widespread in solid-state physics, the detailed control and tuning of band bending in the field of photochemistry or photocatalysis has remained largely unexplored (Zhang and Yates, 2012). It is well known that the upward (or downward) band bending exists for *n* (or *p*-type) semiconductors. Such near-surface band bending is necessary for driving efficient charge carrier separation in a PEC system, wherein oxidation and reduction reactions occur on different electrode surfaces. However, for photochemical water splitting, the presence of any surface band bending creates an additional energy barrier for charge carrier transport, either electron (in the case of *n*-type semiconductors) or hole (in the case of *p*-type semiconductors) diffusion toward the semiconductor–water interface.

Recently, by precisely engineering the surface charge properties, Kibria et al. demonstrated that the efficiency of overall water splitting was enhanced by nearly two orders of magnitude in both GaN and InGaN nanowires (Kibria et al., 2014, 2015). The near-surface band bending of GaN and InGaN was precisely tuned through well-controlled Mg-dopant incorporation. It was found that, with increasing Mg dopant incorporation in GaN nanowires, the near-surface band bending, i.e., $E_F - E_V$ can be continually tuned from 2.6 to 0.5 eV. This indicated the near-surface region can be transformed from *n*-type to weakly *p*-type, providing a viable approach to control the surface electronic properties and charge carrier transfer at the semiconductor–liquid interface. The internal quantum efficiency and hydrogen evolution rate on *p*-type GaN nanowires reached ~51% and

$\sim 4.0\,\text{mol}\,g_{cat}^{-1}\,h^{-1}$, respectively, which was two orders of magnitude higher than undoped GaN, as shown in Fig. 6A. In addition, the nanowires remained very stable after over 50 mol gas (hydrogen and oxygen) was produced during the course of ~22 h of reaction, with a turnover number exceeding 10,000. In order to achieve high efficient overall water splitting under visible light, the effect of *p*-type Mg doping in enhancing the efficiency was further demonstrated for InGaN nanowires. As shown in Fig. 6B, the high rate overall water splitting was measured under wavelengths up to nearly 500 nm, corresponding to the band absorption edge of $In_{0.2}Ga_{0.8}N$ nanowires. An apparent quantum efficiency of ~12.3% was estimated for overall water splitting under visible light illumination (400–475 nm), which was two times higher than the previous world record value of ~6% around 420 nm (Maeda et al., 2008, 2010). In addition, a solar-to-hydrogen conversion efficiency of ~1.8% was achieved under

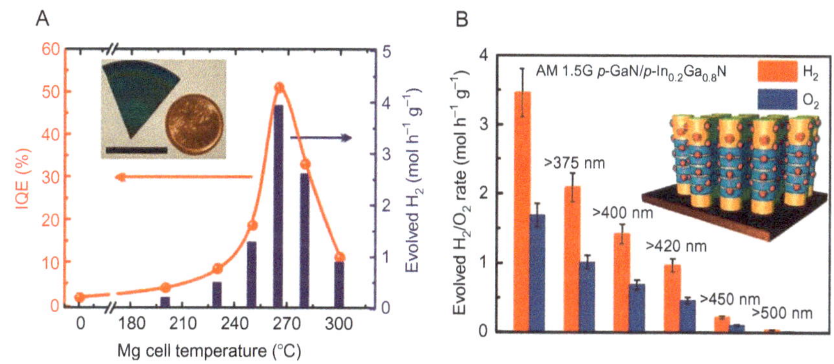

Fig. 6 (A) H_2 evolution rate and internal quantum efficiency (IQE) for different GaN:Mg samples in overall pure water splitting. Reactions were performed using ~0.387 mg GaN nanowire catalyst in pure water under illumination of a xenon lamp (300 W). The sample size is shown in the *inset*. Scale bar 2 cm. (B) H_2 and O_2 evolution rates in overall water splitting with AM1.5G filter, and with different long-pass filters. Visible light activity is clearly demonstrated. The *inset* shows a schematic of core/shell Rh/Cr_2O_3 nanoparticle decorated double-band *p*-GaN/*p*-$In_{0.2}Ga_{0.8}N$ nanowire photocatalyst on Si substrate. The error bar is defined by the s.d. *Panel (A) reprinted with permission from Kibria, M.G., Zhao, S., Chowdhury, F.A., Wang, Q., Nguyen, H.P.T., Trudeau, M.L., Guo, H., Mi, Z., 2014. Tuning the surface Fermi level on* p-type *gallium nitride nanowires for efficient overall water splitting. Nat. Commun. 5, 3825. Copyright 2014, Macmillan Publishers Ltd: Nature Communications. Panel (B) reprinted with permission from Kibria, M.G., Chowdhury, F.A., Zhao, S., AlOtaibi, B., Trudeau, M.L., Guo, H., Mi, Z., 2015. Visible light-driven efficient overall water splitting using* p-type *metal-nitride nanowire arrays. Nat. Commun. 6, 6797. Copyright 2015, Macmillan Publishers Ltd: Nature Communications.*

concentrated sunlight, which is among the highest values reported to date (Kibria and Mi, 2016).

3.3 Atomic-Scale Origin of the Long-Term Stability and High Performance of Ga(In)N Nanowires

To gain insight into the long-term stability and high performance of III-nitride nanowire arrays for photochemical overall water splitting as described earlier, the structural properties and electronic structure of crystalline GaN nanowire arrays were recently investigated in detail at the atomic level, both experimentally and theoretically (Kibria et al., 2016). It was found that GaN nanowires, when grown under nitrogen-rich conditions using molecular beam epitaxy, exhibited N-termination not only for their top surfaces but also for their side faces. The orientation of the N- and Ga-centered bond orbitals at the surface of the nanowires was determined by polarization-dependent XAS at the N 1s and Ga 2p edges (Kibria et al., 2016). The polarization dependence of the Ga 2p and N 1s XAS spectra shows that the Ga valence orbitals are oriented twice as well as the N orbitals. This is consistent with a model of the nanowire surfaces, where N atoms with distorted bond orbitals passivate the surface and charge it negatively, while bulk-like Ga atoms lie underneath. This observation is consistent with a N-termination of the nanowires not only for their top surfaces but also for their side faces, schematically illustrated in Fig. 7. The presence of N-rich side faces is also supported by high-resolution scanning transmission electron microscopy (STEM) studies with high-resolution electron energy loss spectroscopy (EELS) line profile analysis and elemental imaging. Ab initio materials modeling further revealed that the N-terminated wurtzite GaN *m*-plane surfaces are indeed energetically more stable. The key requirements for achieving efficient and stable water splitting on *p*-GaN nanowire-arrays are summarized in Fig. 7, which include *p*-type doping, engineered optimum surface band bending, and N-termination of the top (polar) and side (nonpolar) surfaces by using a Ga seeding layer and N-rich growth conditions, respectively. Such a highly specific configuration leads to negatively charged surfaces which, together with the optimized Mg-dopant incorporation, reduce the energy barrier for injecting holes into the electrolyte and generating H_2 from water oxidation reaction. Furthermore, the N-terminated surfaces protect the GaN nanowires against attack by the electrolyte (oxidation and photocorrosion) and thus, the

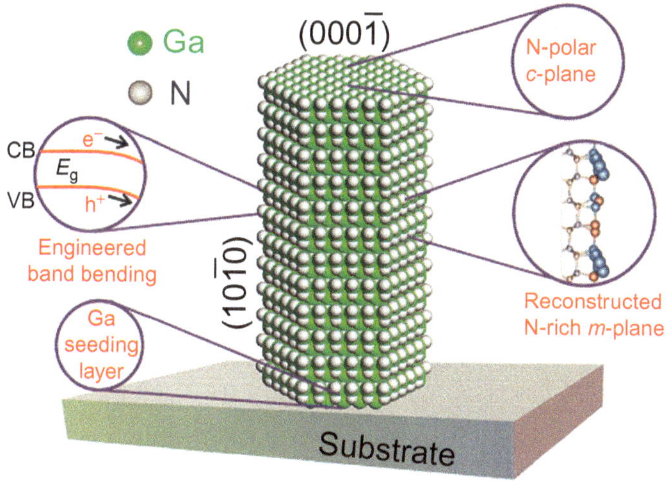

Fig. 7 Schematic of the key atomic and electronic structure requirements for long-term stability and high performance of *p*-GaN nanowire arrays for photocatalytic overall solar water splitting in pure water. (1) Engineered optimum surface band bending with controlled *p*-type doping for efficient carrier extraction, (2) N-polar *c*-plane and associated reverse polarization for enhanced carrier separation, and (3) reconstructed N-rich *m*-plane to function as a passivation layer against photocorrosion. *Reprinted with permission from Kibria, M. G., Qiao, R.M., Yang, W.L., Boukahil, I., Kong, X.H., Chowdhury, F.A., Trudeau, M.L., Ji, W., Guo, H., Himpsel, F.J., Vayssieres, L., Mi, Z. 2016. Atomic-scale origin of long-term stability and high performance of* p-GaN *nanowire arrays for photocatalytic overall pure water splitting. Adv. Mater. 28 (38), 8388–8397. Copyright 2016, John Wiley & Sons.*

engineered optimum surface band bending can be maintained during the photocatalytic overall solar water splitting reaction. The N-terminated top surfaces also provide the necessary reverse polarization of the wurtzite structure crucial for optimized photogenerated charge separation and transport in *p*-GaN for efficient overall solar water splitting in pure water and without any sacrificial agents. For comparison, conventional GaN materials are grown by metalorganic chemical vapor deposition (Kuykendall et al., 2003; Nakamura, 1991) and exhibit predominantly Ga-terminated surfaces, which, together with the presence of large densities of defects and dislocations, explains the commonly reported low efficiency and poor stability for water splitting. Such an atomic-level understanding of the key requirements for efficient and stable water splitting sheds light on the development of efficient and stable water splitting material system without using any extra protection/passivation layers (Hu et al., 2014; Kenney et al., 2013).

4. PHOTOELECTROCHEMICAL WATER SPLITTING

PEC water splitting offers a potentially efficient and affordable way to produce renewable H_2 fuels. Compared to the photovoltaic electrolysis, PEC cell combines the harvesting of solar energy with electrolysis of water into a simple and fully integrated unit. Upon light illumination on semiconducting materials, photogenerated electrons and holes are collected at the cathodic and anodic electrodes, respectively, and then transferred to electrocatalytic sites to conduct the respective hydrogen and oxygen evolution reactions. Since the demonstration of PEC water splitting concept in 1972 (Fujishima and Honda, 1972), metal oxide semiconductors have been extensively investigated as the photoelectrodes because they are relatively stable under harsh PEC conditions. However, the efficiency remains very low for practical application, mainly due to the intrinsically wide bandgap and/or low carrier mobility (e.g., TiO_2, Fe_2O_3) (Li et al., 2013). At the same time, PEC cells composed of III–V group semiconductors have been demonstrated with high solar-to-hydrogen conversion efficiency (~12.4%), but shows very poor stability (Khaselev and Turner, 1998). Recently, III-nitride semiconductors, e.g., InGaN, have been considered as promising materials for PEC water splitting, which has tunable bandgap across nearly the entire solar spectrum, high carrier mobility, and excellent chemical stability (Benton et al., 2013; Ebaid et al., 2015; Fujii et al., 2005; Hwang et al., 2012; Kamimura et al., 2013; Li et al., 2008; Luo et al., 2008; Tao et al., 2016; Kibria et al., 2016). In this section, we discuss the recent development of InGaN-based photoanode, photocathode, and dual-photoelectrodes for PEC water splitting application.

4.1 InGaN Nanowire Photoanodes

Previously reported InGaN photoelectrodes generally exhibit very low photocurrent densities (sub-mA/cm^2 under AM1.5G one sun illumination), due to the presence of extensive defects, dislocations, and indium phase separation (Chen et al., 2008; Li et al., 2008). Due to the large lattice mismatch (~11%) between InN and GaN, it is highly challenging to grow In-rich InGaN nanowires without the formation of In-rich nanoclusters, which act as charge carrier recombination center and degrade the device performance. Recently, Fan et al. developed In-rich InGaN nanowires with significantly suppressed phase separation (Fan et al., 2016b). Illustrated in Fig. 8A, the indium content map (normalized to the thickness in projection)

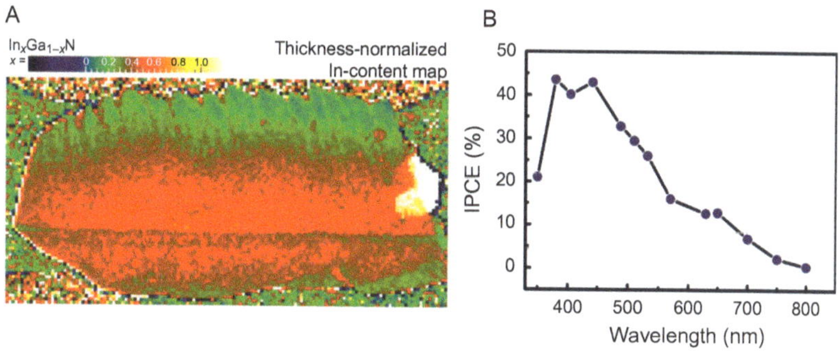

Fig. 8 (A) The thickness-projected indium-content map derived using STEM–EELS spectrum imaging, displayed in pseudocolor, indicative of relatively uniform distribution of indium in the nanowire (~50% in the bulk, ~30% at the near-surface region). (B) IPCE of InGaN nanowires photoanode (Fan et al., 2016b). Used in accordance with the Creative Commons Attribution (CC BY) license, AIP Publishing LLC.

displayed showed the average indium composition was ~50% within the bulk of the nanowire. The resulting $In_{0.5}Ga_{0.5}N$ nanowire photoanode exhibited a photocurrent density of 7.3 mA/cm^2 at an applied bias of 1.2 V vs NHE under AM1.5G one sun illumination in 1 M HBr. The incident photon-to-current efficiency (IPCE) of this InGaN nanowire photoanode is >10% at 650 nm (Fig. 8B), which is not possible for most oxide semiconductor photoelectrodes due to their large bandgap.

In order to enhance the long-term stability of InGaN nanowire photoanode, AlOtaibi et al. reported vertically aligned InGaN/GaN core/shell nanowire arrays, as shown in Fig. 9A (AlOtaibi et al., 2013). The thin GaN shell could protect the InGaN segment from photocorrosion and oxidation without adversely affecting charge carrier transport. As shown in Fig. 9B, the InGaN/GaN nanowires showed high stability for a period of 10 h. Further studies of the structural properties for the nanowire before and after the reaction confirmed that the morphology and chemical structure were unchanged during the PEC operation.

4.2 Monolithically Integrated InGaN Nanowire/Si Double-Band Photoanode

To achieve high efficiency, unassisted PEC water splitting, it is essential to develop a double-junction structure with matched bandgap. An ideal dual-band photoelectrode consists of a top and bottom junction with energy bandgap values of ~1.75 and 1.13 eV, respectively, which can provide sufficient photovoltage to sustain solar-to-hydrogen conversion with efficiency

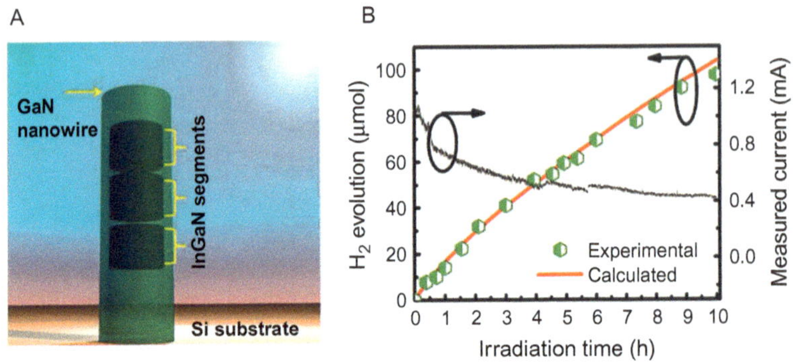

Fig. 9 (A) Schematic of InGaN/GaN core/shell nanowire structures. (B) Steady H_2 evolution in 1 M HBr at 0.2 V vs the counter electrode using a long-pass filter (>375 nm) in the two-electrode configuration. The H_2 evolution is also calculated from the measured current (*solid red curve*). *Reprinted with permission from AlOtaibi, B., Nguyen, H.P., Zhao, S., Kibria, M.G., Fan, S., Mi, Z., 2013. Highly stable photoelectrochemical water splitting and hydrogen generation using a double-band InGaN/GaN core/shell nanowire photoanode. Nano Lett. 13 (9), 4356–4361. Copyright 2013, American Chemical Society.*

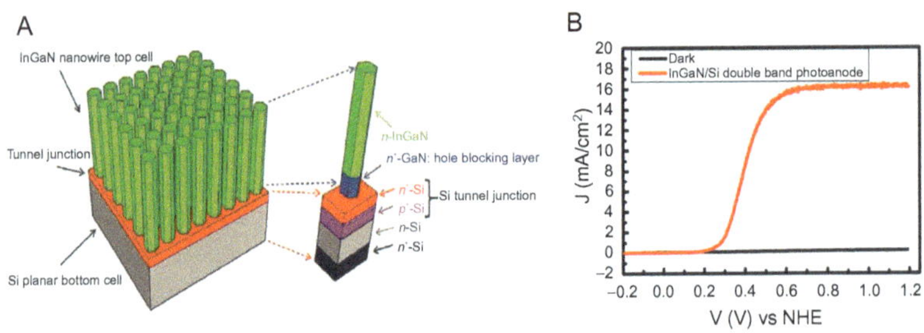

Fig. 10 (A) Schematic of the monolithically integrated InGaN nanowire/Si double-band photoanode. (B) Linear sweep voltammetry scan of InGaN nanowire/Si double-band photoanode under the illumination of AM1.5G one sun. The *black curve* represents the dark current density. The scan rate is 20 mV s^{-1}. *Adapted with permission from Fan, S., Shih, I., Mi, Z., 2016. A monolithically integrated InGaN nanowire/Si tandem photoanode approaching the ideal bandgap configuration of 1.75/1.13 eV. Adv. Energy Mater. 7, 1600952. Copyright 2016, John Wiley & Sons.*

up to ~27% under AM1.5G one sun illumination (Hu et al., 2013). Fan et al. reported the first demonstration of an InGaN/Si double-band photoanode, with the nearly ideal bandgap configuration of 1.75/1.13 eV for maximum solar-to-hydrogen conversion (Fan et al., 2016a). The top cell consisted of *n*-type InGaN nanowire arrays with bandgap of 1.75 eV, which were monolithically integrated with the underlying *n*-type Si solar cell substrate through n^{++}/p^{++} Si tunnel junction, as illustrated in Fig. 10A. Due to the presence of an upward band bending for *n*-type InGaN nanowires, the photogenerated holes could easily diffuse to the nanowire/electrolyte interface to drive water oxidation reaction. Photogenerated electrons, on the other hand, migrated toward the InGaN/Si heterointerface and recombined with photogenerated holes from the underlying Si solar cell through the p^{++}/n^{++} Si tunnel junction. In order to avoid the migration of photogenerated holes in InGaN nanowires toward the Si tunnel junction, an n^{+}-GaN nanowire template was first formed on the Si tunnel junction as a hole blocking layer. Under AM1.5G one sun illumination, the saturated photocurrent density reached 16.3 mA/cm^2 in 1 M HBr, as shown in Fig. 10B. The maximum applied bias photon-to-current efficiency (ABPE) of this InGaN nanowire/Si tandem photoanode was 8.3% at 0.5 V vs NHE in HBr. In addition, the photoanode showed relatively stable photocurrent for 5 h under AM1.5G one sun illumination in HBr.

4.3 Monolithically Integrated InGaN/GaN/Si Adaptive Tunnel Junction Photocathode

In addition to photoanodes, it is also important to develop highly photoactive photocathode materials for water reduction to couple with photoanodes to provide unassisted solar water splitting (Huang et al., 2015). Recently, Fan et al. developed an adaptive double-junction photocathode by integrating InGaN/GaN nanowire arrays with a planar Si solar cell wafer (Fan et al., 2015), which allowed charge carriers with different overpotentials to participate proton reduction reactions simultaneously, thereby reducing chemical loss without strict current matching requirement. The device, as schematically shown in Fig. 11A, consisted of planar

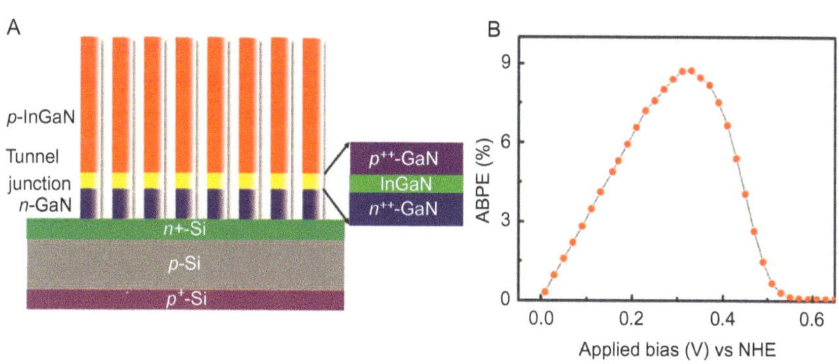

Fig. 11 (A) Schematic and (B) ABPE of the photocathode formed by InGaN tunnel junction nanowires on n^{+}-p Si substrate. *Reprinted with permission from Fan, S., AlOtaibi, B., Woo, S.Y., Wang, Y.J., Botton, G.A., Mi, Z., 2015. High efficiency solar-to-hydrogen conversion on a monolithically integrated InGaN/GaN/Si adaptive tunnel junction photocathode. Nano Lett. 15 (4), 2721–2726. Copyright 2015, American Chemical Society.*

n^+-p Si solar cell wafer, 150 nm n-GaN nanowire segment, and 600 nm p-$In_xGa_{1-x}N$ ($x=0.25$, E_g ~2.39 eV) segment along the vertical direction. The n-GaN and p-InGaN segments were connected by an n^{++}-GaN/InGaN/p^{++}-GaN polarization enhanced tunnel junction, which enabled the transport of photoexcited holes from the p-InGaN to the n-GaN within each single nanowire. Due to the relatively small offset between the Si and GaN conduction band edges and the heavy n-type doping, photoexcited electrons of the underlying Si solar cell can readily inject into the n-GaN nanowire segment. Some portion of the injected electrons can drive proton reduction on n-GaN surfaces with the rest recombining with holes from the p-InGaN in the adaptive tunnel junction. The downward surface band bending of p-InGaN facilitated the flow of electrons toward the electrolyte. This device exhibited a saturated photocurrent density of −40.6 mA/cm^2 at $V_{NHE}=0.26$ V in 1 M HBr under 1.3 sun of AM1.5G illumination. This saturated current density corresponded to an average IPCE of 72.3% over the solar spectrum (from 280 to 1100 nm). As shown in Fig. 11B, the maximum ABPE was estimated to be 8.7% at 0.33 V vs NHE.

4.4 InGaN Dual Photoelectrodes

One of the most promising PEC cell configurations to achieve high efficiency water splitting without external applied electric bias is a tandem of spatially separated photocathodes and photoanodes, which drive the reduction and oxidation of water, respectively (Walter et al., 2010). Such a two-photon dual-electrode system (Nozik, 1977) can be implemented with two semiconductors connected back-to-back in tandem, forming the top and bottom photoelectrode, which is analogous to the Z-scheme in natural photosynthesis. In this scheme, minority carriers are driven to the semiconductor/liquid junction to perform oxidation/reduction reactions, while majority carriers recombine at the photocathode/anode interface. Despite their promises, previously reported two-photon tandem photoelectrodes generally exhibit very poor performance (<0.1%), which has been largely limited by the incompatibility between the two photoelectrode materials.

Recently, AlOtaibi et al. demonstrated dual-photoelectrode devices consisting of an n-GaN nanowire photoanode and a p-InGaN nanowire photocathode can break the efficiency bottleneck of conventional two-photon tandem devices by splitting the solar spectrum spatially and spectrally, as shown in Fig. 12A (AlOtaibi et al., 2015a). It was demonstrated that

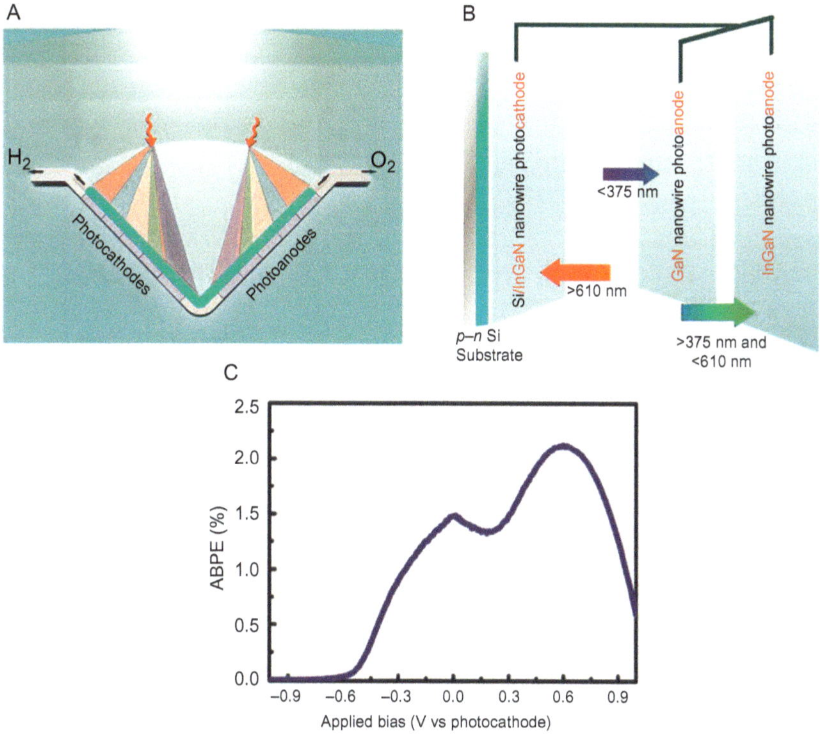

Fig. 12 (A) Schematic of the general concept of a dual-photoelectrode system under parallel illumination with the incident sunlight split spatially and spectrally on the photoanode and photocathode. Each photoanode (or photocathode) may consist of several parallel-connected anodes (or cathodes), each of which is illuminated with a portion of the solar spectrum in commensurate with its energy bandgap and light absorption capacity. While the current matching between the photoanode and photocathode is required, the current-matching for the parallel-connected photoanodes (or photocathodes) is not required, due to the parallel connection. (B) Schematic of the GaN and InGaN nanowire photoanodes, which are connected in parallel and paired with the Si/InGaN photocathode. The incident sunlight is split spectrally and spatially among the photoelectrodes. (C) The power conversion efficiency of the dual-photoelectrode device as a function of applied bias under AM1.5G 1 sun illumination. The maximum power conversion efficiency is ~2%, which occurs at ~0.6 V vs the photocathode. *Reprinted with permission from AlOtaibi, B., Fan, S., Vanka, S., Kibria, M.G., Mi, Z., 2015. A metal-nitride nanowire dual-photoelectrode device for unassisted solar-to-hydrogen conversion under parallel illumination. Nano Lett. 15 (10), 6821–6828.*

a dual-photoelectrode system consisting of GaN/InGaN nanowire photoanodes and a Si/InGaN photocathode could perform unassisted solar-to-hydrogen conversion without any bias. As shown in Fig. 12B, the GaN nanowire photoanode, InGaN photoanode, and Si/InGaN photocathode can efficiently use the solar spectrum with wavelengths >375, 375–610, and >610 nm, respectively. Under parallel illumination, a power conversion efficiency of 2% was achieved under AM1.5G one sun illumination (Fig. 12C), which was more than one order of magnitude higher than the individual photoelectrodes.

5. PHOTOCATALYTIC AND PHOTOELECTROCHEMICAL CO_2 REDUCTION

CO_2 is known as the major greenhouse gas contributing to global warming. On the other hand, CO_2 can be considered as a useful C1 feedstock for fuel or chemical synthesis due to its abundance and availability (Liu et al., 2015). However, the challenge is that CO_2 is an extremely stable molecule, and, consequently, the reduction of CO_2 requires significant energy demands and/or extremely reactive reagents using conventional approaches. Alternatively, photocatalytic and PEC routes, by mimicking the natural photosynthesis process, convert CO_2 into valuable chemical products (e.g., CO, HCOOH, CH_3OH, and CH_4) at ambient conditions with the use of sunlight as the only energy input (Kumar et al., 2012; White et al., 2015). This provides one of the best solutions to reduce our dependence on conventional fossil fuels, mitigate CO_2 emissions, and create a sustainable carbon economy. Semiconductor materials play a crucial role in the development of the photocatalytic and PEC CO_2 reduction. To date, the most widely studied materials are semiconducting oxides such as TiO_2 (Dhakshinamoorthy et al., 2012; Habisreutinger et al., 2013). Despite spectacular progress that has been made in the past decades, the systems still suffer from several fundamental efficiency bottlenecks for practical application, including weak photon absorption, limited charge separation and transport, and poor CO_2 adsorption and activation. In this regard, the design and development of extremely "talented" photocatalytic material capable of efficient CO_2 reduction under visible light is gaining increasing attention. In this section, we discuss the potential of group III-nitride semiconductors (e.g., GaN and InGaN) for solar-powered CO_2 reduction into valuable energy-rich carbonaceous products, including both photocatalytic and PEC approaches. In addition, the important role of cocatalyst in driving the selective CO_2 conversion into targeted fuels or chemicals is also presented.

5.1 CO_2 Reduction into CH_4 and CO Using GaN Nanowires

GaN is a promising semiconductor for CO_2 conversion as its conduction and valence-band edges straddle the potentials of CO_2 reduction and water oxidation in a large range of pH solutions (Kocha et al., 1995). Compared to most metal oxides, the conduction band edge of GaN located at a more negative position (1.2 V vs NHE at pH 0), which renders a large potential for the reduction of highly stable CO_2 molecule. Although GaN has been intensively studied for application in solar-to-hydrogen generation from water splitting as mentioned in the earlier sections, the first demonstration of GaN nanowire arrays for direct photochemical reduction of CO_2 into CH_4 and CO was not reported until 2015 (AlOtaibi et al., 2015b). It was found that the bare GaN nanowires showed higher photoactivity for CO production over CH_4, with an evolution rate of 47 and 1.3 $\mu mol\, g_{cat}^{-1}\, h^{-1}$ in 24 h, respectively. With the decoration of Rh/Cr_2O_3 core/shell or Pt cocatalyst on GaN nanowires (Fig. 13), the product selectivity for CH_4 was greatly enhanced. It was observed that the introduction of Rh/Cr_2O_3 cocatalyst increased the CH_4 production rate

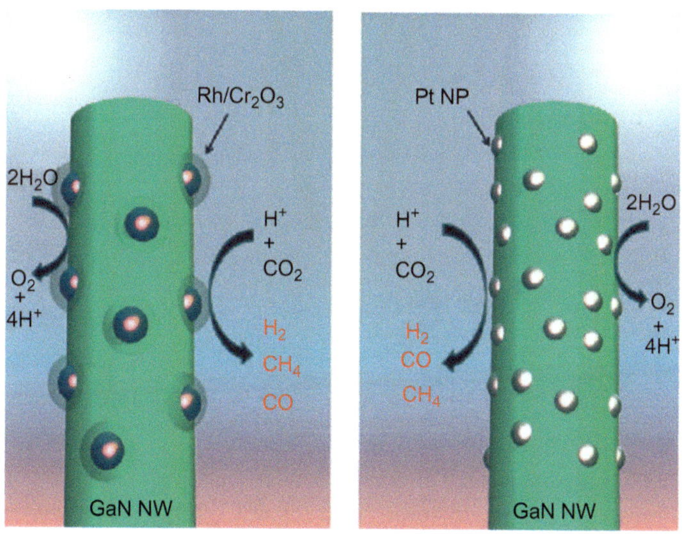

Fig. 13 Schematic of the photoreduction processes of CO_2 on Rh/Cr_2O_3 and Pt-decorated GaN nanowires. *Reprinted with permission from AlOtaibi, B., Fan, S., Wang, D.F., Ye, J.H., Mi, Z., 2015. Wafer-level artificial photosynthesis for CO_2 reduction into CH_4 and CO using GaN nanowires. ACS Catal. 5 (9), 5342–5348. Copyright 2015, American Chemical Society.*

to 3.5 μmol g_{cat}^{-1} h^{-1}, while suppressing CO production by nearly an order of magnitude with a rate of 5 μmol g_{cat}^{-1} h^{-1}. Further studies by promoting Pt nanoparticles on GaN nanowire surfaces showed that the CH_4 production rate could be further enhanced and reached 14.8 μmol g_{cat}^{-1} h^{-1}, which was an order of magnitude greater than that measured on bare GaN nanowire arrays. The CO production rate, on the other hand, was not significantly changed by adding Pt cocatalyst. Obviously, the presence of Pt and Rh/Cr_2O_3 cocatalysts on GaN greatly increased the selectivity for CH_4 formation over CO. In addition, it is worth mentioning that nearly constant CO_2 reduction rate was observed in all the above-mentioned GaN-based photocatalytic system over 24 h, indicating superior stability of GaN nanowires for CO_2 photoreduction. This work highlights the potential use of III-nitride nanowires as highly efficient photocatalysts for the direct CO_2 photoreduction into valuable chemical products and the crucial role of cocatalyst in tuning the product selectivity to achieve target chemicals/fuels.

5.2 CO_2 Reduction on InGaN/GaN Nanowires Under Visible Light

The bandgap of GaN can be tuned from 3.4 to 0.65 eV by alloying with indium, offering the unique opportunity to absorb nearly the entire solar spectrum (Moses and Van de Walle, 2010). Moreover, the energy band edges of InGaN with an In composition up to ~50%, corresponding to a bandgap of ~1.7 eV, can still straddle the CO_2 reduction and water oxidation potentials. Very recently, using Pt-decorated InGaN/GaN nanowire arrays as the photocatalyst (Fig. 14A), a high CO_2 conversion rate of ~0.5 mmol g_{cat}^{-1} h^{-1} into CH_3OH product was achieved under visible light irradiation in the presence of H_2 reagent, as shown in Fig. 14B (AlOtaibi et al., 2016). Other carbonaceous products including CO and CH_4 were also detected in the system, with a production rate of 0.15 and 0.25 mmol g_{cat}^{-1} h^{-1}, respectively. Moreover, the fundamental understanding of the interaction between CO_2 molecules and InGaN/GaN photocatalyst surfaces was elucidated using ab initio calculations. It was found that CO_2 could be spontaneously adsorbed and activated on the clean nonpolar surfaces of wurtzite III-nitride, which was not possible for most oxide semiconductors. Additionally, the surface properties of the nanowire can be finely tuned by rational dopant incorporation. For

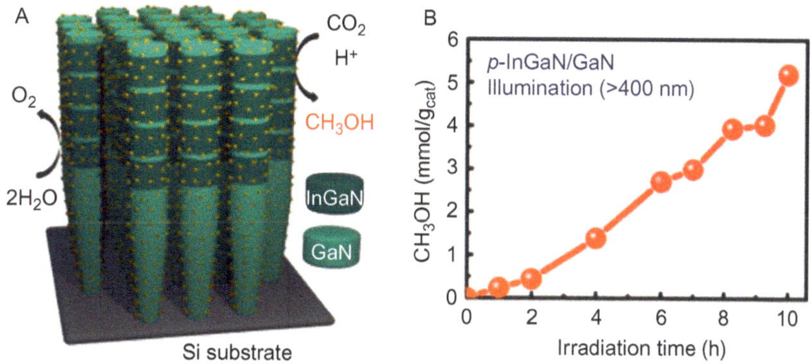

Fig. 14 (A) Schematic of a *p*-InGaN/GaN nanowire photocatalyst decorated with Pt nanoparticles. (B) CH_3OH evolution over Pt-decorated *p*-InGaN/GaN nanowires as a function of time under visible light illumination (>400 nm). *Reprinted with permission from AlOtaibi, B., Kong, X., Vanka, S., Woo, S.Y., Pofelski, A., Oudjedi, F., Fan, S., Kibria, M.G., Botton G.A., Ji, W., Guo, H., Mi, Z. 2016. Photochemical carbon dioxide reduction on Mg-doped Ga(In)N nanowire arrays under visible light irradiation. ACS Energy Lett. 1, 246–252. Copyright 2016, American Chemical Society.*

example, it was discovered that, with the incorporation of Mg dopant, the photocatalytic activity of CO_2 reduction was improved by nearly 50-fold due to the reduced surface potential barrier and the enhanced adsorption and deformation of CO_2 molecule.

5.3 Selective CO_2 Reduction to CH_4 Using a Monolithically Integrated GaN/Si Photocathode

Compared to photocatalysis, a PEC system offers the advantages of efficient charge carrier separation and collection of the oxidation and reduction products at the anodic and cathodic electrode, respectively. Recently, we demonstrated highly selective CH_4 production from CO_2 reduction using a monolithically integrated GaN/Si photocathode modified with Cu cocatalyst, as shown in Fig. 15A (Wang et al., 2016). GaN nanowires were integrated with *p–n* junction Si to form a monolithic device, which combined the advantage of strong light harvesting of Si and efficient electron extraction effect of GaN nanowire arrays. With the incorporation of copper as the cocatalyst, the device exhibited a Faradaic efficiency of 19% for the eight electrons reduction to CH_4 at −0.73 V vs RHE, which was more than 30 times higher than that for the two electrons reduction to CO (~0.6%), as shown in Fig. 15B. The remaining balance of photocurrent drove the

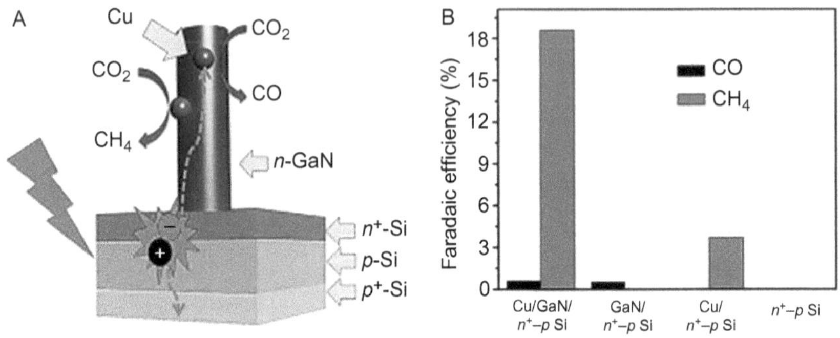

Fig. 15 (A) Schematic of the Cu/GaN/n^+-*p* Si photocathode. (B) The measured Faradaic efficiencies of products for different samples at −0.73 V vs RHE in 0.5 M $KHCO_3$ (pH 8). *Reprinted with permission from Wang, Y., Fan, S., AlOtaibi, B., Wang, Y., Li, L., and Mi, Z., 2016. A monolithically integrated gallium nitride nanowire/silicon solar cell photocathode for selective carbon dioxide reduction to methane. Chemistry 22 (26), 8809–8813. Copyright 2016, John Wiley & Sons.*

reduction of proton to form H_2. It is worth noting that Cu cocatalyst played a crucial role in CH_4 generation, as the control experiment without Cu did not detect any CH_4 but with only a small amount of CO (0.5%). In addition, GaN nanowires also contributed greatly to the high selectivity for CH_4, as the Faradic efficiency of CH_4 generation was only 3.7% over the photocathode without GaN.

5.4 Tunable Syngas Production

In contrast to the challenging six or eight electrons reduction of CO_2 to CH_3OH or CH_4 with high overpotential and poor selectivity, the two electrons reduction to CO is kinetically more feasible and a viable option for practical application (Vesborg and Seger, 2016). Moreover, CO is an important bulk chemical to form syngas (synthesis gas, CO + H_2 mixtures), which is a critical C_1 feedstock to produce various kinds of liquid fuels, such as methanol and hydrocarbons including synthetic jet, kerosene, and diesel fuels (Bell, 1981; Waugh, 1992).

Very recently, we demonstrated solar-powered syngas production in an aqueous PEC cell using a rationally designed photocathode derived from GaN nanowires, as shown in Fig. 16A (Chu et al., 2016). GaN nanowires played an important role in extracting the electrons photogenerated from the underlying *p–n* junction Si. Moreover, GaN formed a unique junction with ZnO to facilitate the transport of charge carriers. Due to the collective effects including strong light harvesting of *p–n* junction Si, efficient

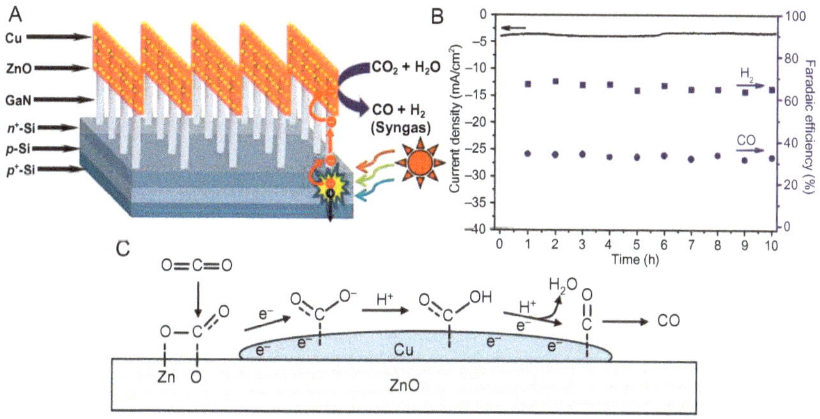

Fig. 16 (A) Schematic of Cu–ZnO/GaN/n^+-p Si photocathode. (B) Chronoamperometry data and Faradaic efficiencies for CO and H_2 of Cu–ZnO/GaN/n^+-p Si photocathode at −0.23 V vs RHE in 0.5 M $KHCO_3$ (pH 8). (C) Proposed reaction mechanism for the photoreduction of CO_2 to produce CO on Cu–ZnO dual cocatalysts. *Reprinted with permission from Chu, S., Fan, S., Wang, Y., Rossouw, D., Wang, Y., Botton, G. A. and Mi, Z., 2016. Tunable syngas production from CO_2 and H_2O in an aqueous photoelectrochemical cell. Angew. Chem. Int. Ed. 55 (46), 14260–14264. Copyright 2016, John Wiley & Sons.*

electron extraction of GaN nanowire arrays, and fast surface reaction kinetics of Cu–ZnO cocatalysts, a high Faradaic efficiency of 70% for CO was achieved at underpotential of 180 mV, which is the lowest potential ever reported in an aqueous PEC cell. Importantly, the device showed superior stability during 10 h operation, as shown in Fig. 16B. The CO/H_2 ratio in the products was constant around 1:2 with a total unity Faradaic efficiency, which is a desirable mixture that can be directly fed into the commercial reactor for methanol synthesis or Fischer–Tropsch hydrocarbon formation without further adjustment (Kang et al., 2014). In addition, a benchmark turnover number of 1330 was attained after 10 h of photoelectrolysis, significantly out-performing the state-of-the-art values reported for selective CO_2 reduction toward CO (Rosser et al., 2016; Takeda et al., 2016). Interestingly, it was found that the synergy of Cu and ZnO cocatalyst accounted for the high-performance syngas production. The reaction model of CO_2 reduction to form CO on Cu–ZnO dual cocatalysts was shown in Fig. 16C. CO_2 molecule was first adsorbed and activated on ZnO to form a destabilized bent configuration. Then a series of reaction steps involving the consecutive transfer of protons and electrons were occurred on Cu to produce CO. The combination of Cu and ZnO offered both adsorption and reaction sites with complementary chemical properties that lead to special reaction channels not seen in Cu or

ZnO alone. By exploiting a systematic study of different Cu/Zn ratios, it was found that the highest product selectivity for CO was obtained at a composition with a balanced reaction of CO_2 adsorption on ZnO and proton-coupled electron transfer on Cu. This work opens up opportunities to develop carbon-neutral syngas production from the reduction of CO_2 and H_2O using renewable solar energy, which is an important step toward solar refinery economy of the future.

6. CONCLUSIONS AND FUTURE PROSPECTS

In summary, we have provided an overview of the current status of III-nitride nanowire arrays for artificial photosynthesis, including both solar water splitting and CO_2 reduction. Although still in its early stage, III-nitride semiconductor has already demonstrated extraordinary potential to address some of the fundamental challenges of artificial photosynthesis, including low quantum efficiency, poor visible light activity, and poor stability. The research efforts on III-nitride nanowires have seen greater success in the past several years, to some extent, than what has been achieved on conventional metal oxide-based materials for solar-fuel conversion over the past four decades. For instance, it was shown that hydrogen production rate of few molars per gram photocatalyst per hour and an apparent quantum efficiency of ~12.3% were achieved for overall water splitting under visible light illumination (400–475 nm), which was not previously possible. Moreover, it is worth mentioning that III-nitride materials are well established in the semiconductor industry including solid-state lighting and power electronics, which enables the cost-effective and scalable production of solar fuels.

Although great progress has been made to date, the fascinating properties of III-nitride on artificial photosynthesis are still yet to be fully explored. An optimum artificial photosynthesis system needs to simultaneously enhance the light harvesting, charge carrier separation and transfer, and surface redox reactions. It is expected that the effective coupling of excellent photon absorption and charge carrier transport properties of III-nitride nanowire arrays with efficient cocatalyst can boost the system efficiency. The development of advanced electrocatalyst with high turnover rates and low overpotentials for water splitting and CO_2 reduction, and the engineering of interface between the cocatalyst and III-nitride semiconductor with minimum loss will maximize the overall performance, which may eventually make artificial photosynthesis devices a reality.

REFERENCES

Abe, R., 2010. Recent progress on photocatalytic and photoelectrochemical water splitting under visible light irradiation. J. Photochem. Photobiol. C Photchem. Rev. 11 (4), 179–209.

AlOtaibi, B., Nguyen, H.P., Zhao, S., Kibria, M.G., Fan, S., Mi, Z., 2013. Highly stable photoelectrochemical water splitting and hydrogen generation using a double-band InGaN/GaN core/shell nanowire photoanode. Nano Lett. 13 (9), 4356–4361.

AlOtaibi, B., Fan, S., Vanka, S., Kibria, M.G., Mi, Z., 2015a. A metal-nitride nanowire dual-photoelectrode device for unassisted solar-to-hydrogen conversion under parallel illumination. Nano Lett. 15 (10), 6821–6828.

AlOtaibi, B., Fan, S., Wang, D.F., Ye, J.H., Mi, Z., 2015b. Wafer-level artificial photosynthesis for CO_2 reduction into CH_4 and CO using GaN nanowires. ACS Catal. 5 (9), 5342–5348.

AlOtaibi, B., Kong, X., Vanka, S., Woo, S.Y., Pofelski, A., Oudjedi, F., Fan, S., Kibria, M.G., Botton, G.A., Ji, W., Guo, H., Mi, Z., 2016. Photochemical carbon dioxide reduction on Mg-doped Ga(In)N nanowire arrays under visible light irradiation. ACS Energy Lett. 1, 246–252.

Beach, J.D., Collins, R.T., Turner, J.A., 2003. Band-edge potentials of *n*-type and *p*-type GaN. J. Electrochem. Soc. 150 (7), A899–A904.

Bell, A.T., 1981. Catalytic synthesis of hydrocarbons over group-VIII metals—a discussion of the reaction-mechanism. Catal. Rev. Sci. Eng. 23 (1–2), 203–232.

Benton, J., Bai, J., Wang, T., 2013. Significantly enhanced performance of an InGaN/GaN nanostructure based photo-electrode for solar power hydrogen generation. Appl. Phys. Lett. 103, 133904.

Bhuiyan, A.G., Hashimoto, A., Yamamoto, A., 2003. Indium nitride (InN): a review on growth, characterization, and properties. J. Appl. Phys. 94 (5), 2779–2808.

César, M., Ke, Y., Ji, W., Guo, H., Mi, Z., 2011. Band gap of $In_xGa_{1-x}N$: a first principles analysis. Appl. Phys. Lett. 98 (20), 2107.

Chen, X., Matthews, K.D., Hao, D., Schaff, W.J., Eastman, L.F., 2008. Growth, fabrication, and characterization of InGaN solar cells. Phys. Status Solidi A 205 (5), 1103–1105.

Chen, X.B., Shen, S.H., Guo, L.J., Mao, S.S., 2010. Semiconductor-based photocatalytic hydrogen generation. Chem. Rev. 110 (11), 6503–6570.

Chowdhury, F.A., Mi, Z., Kibria, M.G., Trudeau, M.L., 2015. Group III-nitride nanowire structures for photocatalytic hydrogen evolution under visible light irradiation. APL Mater. 3, 104408.

Chu, S., Fan, S., Wang, Y., Rossouw, D., Wang, Y., Botton, G.A., Mi, Z., 2016. Tunable syngas production from CO_2 and H_2O in an aqueous photoelectrochemical cell. Angew. Chem. Int. Ed. 55 (46), 14260–14264.

Dhakshinamoorthy, A., Navalon, S., Corma, A., Garcia, H., 2012. Photocatalytic CO_2 reduction by TiO_2 and related titanium containing solids. Energ. Environ. Sci. 5 (11), 9217–9233.

Ebaid, M., Kang, J.H., Lim, S.H., Ha, J.S., Lee, J.K., Cho, Y.H., Ryu, S.W., 2015. Enhanced solar hydrogen generation of high density, high aspect ratio, coaxial InGaN/GaN multi-quantum well nanowires. Nano Energy 12, 215–223.

Fan, W.Q., Zhang, Q.H., Wang, Y., 2013. Semiconductor-based nanocomposites for photocatalytic H_2 production and CO_2 conversion. Phys. Chem. Chem. Phys. 15 (8), 2632–2649.

Fan, S., AlOtaibi, B., Woo, S.Y., Wang, Y.J., Botton, G.A., Mi, Z., 2015. High efficiency solar-to-hydrogen conversion on a monolithically integrated InGaN/GaN/Si adaptive tunnel junction photocathode. Nano Lett. 15 (4), 2721–2726.

Fan, S., Shih, I., Mi, Z., 2016a. A monolithically integrated InGaN nanowire/Si tandem photoanode approaching the ideal bandgap configuration of 1.75/1.13 eV. Adv. Energy Mater. 7, 1600952.

Fan, S., Woo, S.Y., Vanka, S., Botton, G.A., Mi, Z., 2016b. An $In_{0.5}Ga_{0.5}N$ nanowire photoanode for harvesting deep visible light photons. APL Mater. 4, 076106.

Fujii, K., Kusakabe, K., Ohkawa, K., 2005. Photoelectrochemical properties of InGaN for H_2 generation from aqueous water. Jpn. J. Appl. Phys. 1 44 (10), 7433–7435.

Fujishima, A., Honda, K., 1972. Electrochemical photolysis of water at a semiconductor electrode. Nature 238 (5358), 37–38.

Funk, S., Hokkanen, B., Johnson, E., Burghaus, U., 2006. Effect of oxygen vacancy sites on CO_2 adsorption dynamics: the case of rutile (1×1)-TiO_2 (110). Chem. Phys. Lett. 422 (4), 461–465.

Habisreutinger, S.N., Schmidt-Mende, L., Stolarczyk, J.K., 2013. Photocatalytic reduction of CO_2 on TiO_2 and other semiconductors. Angew. Chem. Int. Ed. 52 (29), 7372–7408.

Hu, S., Xiang, C.X., Haussener, S., Berger, A.D., Lewis, N.S., 2013. An analysis of the optimal band gaps of light absorbers in integrated tandem photoelectrochemical water-splitting systems. Energ. Environ. Sci. 6 (10), 2984–2993.

Hu, S., Shaner, M.R., Beardslee, J., Lichterman, A.M., Brunschwig, B.S., Lewis, N.S., 2014. Amorphous TiO_2 coatings stabilize Si, GaAs, and GaP photoanodes for efficient water oxidation. Science 344 (6187), 1005–1009.

Huang, Q., Ye, Z., Xiao, X.D., 2015. Recent progress in photocathodes for hydrogen evolution. J. Mater. Chem. A 3 (31), 15824–15837.

Hwang, Y.J., Wu, C.H., Hahn, C., Jeong, H.E., Yang, P.D., 2012. Si/InGaN core/shell hierarchical nanowire arrays and their photoelectrochemical properties. Nano Lett. 12 (3), 1678–1682.

Indrakanti, V.P., Schobert, H.H., Kubicki, J.D., 2009. Quantum mechanical modeling of CO_2 interactions with irradiated stoichiometric and oxygen-deficient anatase TiO_2 surfaces: implications for the photocatalytic reduction of CO_2. Energy Fuel 23 (10), 5247–5256.

Jung, H.S., Hong, Y.J., Li, Y., Cho, J., Kim, Y.J., Yi, G.C., 2008. Photocatalysis using GaN nanowires. ACS Nano 2 (4), 637–642.

Kamimura, J., Bogdanoff, P., Lahnemann, J., Hauswald, C., Geelhaar, L., Fiechter, S., Riechert, H., 2013. Photoelectrochemical properties of (In, Ga)N nanowires for water splitting investigated by in situ electrochemical mass spectroscopy. J. Am. Chem. Soc. 135 (28), 10242–10245.

Kang, P., Chen, Z.F., Nayak, A., Zhang, S., Meyer, T.J., 2014. Single catalyst electrocatalytic reduction of CO_2 in water to H_2 + CO syngas mixtures with water oxidation to O_2. Energ. Environ. Sci. 7 (12), 4007–4012.

Kenney, M.J., Gong, M., Li, Y.G., Wu, J.Z., Feng, J., Lanza, M., Dai, H.J., 2013. High-performance silicon photoanodes passivated with ultrathin nickel films for water oxidation. Science 342 (6160), 836–840.

Khaselev, O., Turner, J.A., 1998. A monolithic photovoltaic-photoelectrochemical device for hydrogen production via water splitting. Science 280 (5362), 425–427.

Kibria, M.G., Mi, Z., 2016. Artificial photosynthesis using metal/nonmetal-nitride semiconductors: current status, prospects, and challenges. J. Mater. Chem. A 4 (8), 2801–2820.

Kibria, M.G., Nguyen, H.P.T., Cui, K., Zhao, S., Liu, D., Guo, H., Trudeau, M.L., Paradis, S., Hakima, A.R., Mi, Z., 2013. One-step overall water splitting under visible light using multiband InGaN/GaN nanowire heterostructures. ACS Nano 7 (9), 7886–7893.

Kibria, M.G., Zhao, S., Chowdhury, F.A., Wang, Q., Nguyen, H.P.T., Trudeau, M.L., Guo, H., Mi, Z., 2014. Tuning the surface Fermi level on *p*-type gallium nitride nanowires for efficient overall water splitting. Nat. Commun. 5, 3825.

Kibria, M.G., Chowdhury, F.A., Zhao, S., AlOtaibi, B., Trudeau, M.L., Guo, H., Mi, Z., 2015. Visible light-driven efficient overall water splitting using *p*-type metal-nitride nanowire arrays. Nat. Commun. 6, 6797.

Kibria, M.G., Qiao, R.M., Yang, W.L., Boukahil, I., Kong, X.H., Chowdhury, F.A., Trudeau, M.L., Ji, W., Guo, H., Himpsel, F.J., Vayssieres, L., Mi, Z., 2016. Atomic-scale origin of long-term stability and high performance of *p*-GaN nanowire arrays for photocatalytic overall pure water splitting. Adv. Mater. 28 (38), 8388–8397.

Kim, D., Sakimoto, K.K., Hong, D.C., Yang, P.D., 2015. Artificial photosynthesis for sustainable fuel and chemical production. Angew. Chem. Int. Ed. 54 (11), 3259–3266.

Kocha, S.S., Peterson, M.W., Arent, D.J., Redwing, J.M., Tischler, M.A., Turner, J.A., 1995. Electrochemical investigation of the gallium nitride-aqueous electrolyte interface. J. Electrochem. Soc. 142 (12), L238–L240.

Kudo, A., Miseki, Y., 2009. Heterogeneous photocatalyst materials for water splitting. Chem. Soc. Rev. 38 (1), 253–278.

Kumar, B., Llorente, M., Froehlich, J., Dang, T., Sathrum, A., Kubiak, C.P., 2012. Photochemical and photoelectrochemical reduction of CO_2. Annu. Rev. Phys. Chem. 63, 541–569.

Kuykendall, T., Pauzauskie, P., Lee, S.K., Zhang, Y.F., Goldberger, J., Yang, P.D., 2003. Metalorganic chemical vapor deposition route to GaN nanowires with triangular cross sections. Nano Lett. 3 (8), 1063–1066.

Lewis, N.S., Nocera, D.G., 2006. Powering the planet: chemical challenges in solar energy utilization. Proc. Natl. Acad. Sci. U.S.A. 103 (43), 15729–15735.

Li, J., Lin, J., Jiang, H., 2008. Direct hydrogen gas generation by using InGaN epilayers as working electrodes. Appl. Phys. Lett. 93 (16), 162107.

Li, Z., Luo, W., Zhang, M., Feng, J., Zou, Z., 2013. Photoelectrochemical cells for solar hydrogen production: current state of promising photoelectrodes, methods to improve their properties, and outlook. Energ. Environ. Sci. 6 (2), 347–370.

Li, K., Peng, B., Peng, T., 2016. Recent advances in heterogeneous photocatalytic CO_2 conversion to solar fuels. ACS Catal. 6 (11), 7485–7527.

Liu, L., Zhao, H., Andino, J.M., Li, Y., 2012. Photocatalytic CO_2 reduction with H_2O on TiO_2 nanocrystals: comparison of anatase, rutile, and brookite polymorphs and exploration of surface chemistry. ACS Catal. 2 (8), 1817–1828.

Liu, C., Dasgupta, N.P., Yang, P.D., 2014. Semiconductor nanowires for artificial photosynthesis. Chem. Mater. 26 (1), 415–422.

Liu, Q., Wu, L.P., Jackstell, R., Beller, M., 2015. Using carbon dioxide as a building block in organic synthesis. Nat. Commun. 6, 5933.

Luo, W.J., Liu, B., Li, Z.S., Xie, Z.L., Chen, D.J., Zou, Z.G., Zhang, R., 2008. Stable response to visible light of InGaN photoelectrodes. Appl. Phys. Lett. 92, 262110.

Maeda, K., Domen, K., 2007. New non-oxide photocatalysts designed for overall water splitting under visible light. J. Phys. Chem. C 111 (22), 7851–7861.

Maeda, K., Teramura, K., Saito, N., Inoue, Y., Domen, K., 2007. Photocatalytic overall water splitting on gallium nitride powder. Bull. Chem. Soc. Jpn. 80 (5), 1004–1010.

Maeda, K., Teramura, K., Domen, K., 2008. Effect of post-calcination on photocatalytic activity of $(Ga_{1-x}Zn_x)(N_{1-x}O_x)$ solid solution for overall water splitting under visible light. J. Catal. 254 (2), 198–204.

Maeda, K., Higashi, M., Lu, D.L., Abe, R., Domen, K., 2010. Efficient nonsacrificial water splitting through two-step photoexcitation by visible light using a modified oxynitride as a hydrogen evolution photocatalyst. J. Am. Chem. Soc. 132 (16), 5858–5868.

Markovits, A., Fahmi, A., Minot, C., 1996. A theoretical study of CO_2 adsorption on TiO_2. J. Mol. Struct. Theochem 371, 219–235.

Michaud-Rioux, V., Zhang, L., Guo, H., 2016. RESCU: a real space electronic structure method. J. Comput. Phys. 307, 593–613.

Moses, P.G., Van de Walle, C.G., 2010. Band bowing and band alignment in InGaN alloys. Appl. Phys. Lett. 96, 021908.

Moustakas, T.D., 2013. The role of extended defects on the performance of optoelectronic devices in nitride semiconductors. Phys. Status Solidi A 210 (1), 169–174.
Nakamura, S., 1991. GaN growth using GaN buffer layer. Jpn. J. Appl. Phys. 2 Lett. 30 (10A), L1705–L1707.
Nozik, A.J., 1977. Photochemical diodes. Appl. Phys. Lett. 30 (11), 567.
Qiu, J., Zeng, G., Ha, M.-A., Ge, M., Lin, Y., Hettick, M., Hou, B., Alexandrova, A.N., Javey, A., Cronin, S.B., 2015. Artificial photosynthesis on TiO_2-passivated InP nanopillars. Nano Lett. 15 (9), 6177–6181.
Rodriguez, M.M., Peng, X., Liu, L., Li, Y., Andino, J.M., 2012. A density functional theory and experimental study of CO_2 interaction with brookite TiO_2. J. Phys. Chem. C 116 (37), 19755–19764.
Rosser, T.E., Windle, C.D., Reisner, E., 2016. Electrocatalytic and solar-driven CO_2 reduction to CO with a molecular manganese catalyst immobilized on mesoporous TiO_2. Angew. Chem. Int. Ed. 55 (26), 7388–7392.
Shen, X., Allen, P.B., Hybertsen, M.S., Muckerman, J.T., 2009. Water adsorption on the GaN (1010) nonpolar surface. J. Phys. Chem. C 113 (9), 3365–3368.
Shen, X., Small, Y.A., Wang, J., Allen, P.B., Fernandez-Serra, M.V., Hybertsen, M.S., Muckerman, J.T., 2010. Photocatalytic water oxidation at the GaN $(10\bar{1}0)$ – water interface. J. Phys. Chem. C 114 (32), 13695–13704.
Sorescu, D.C., Lee, J., Al-Saidi, W.A., Jordan, K.D., 2011. CO_2 adsorption on TiO_2 (110) rutile: insight from dispersion-corrected density functional theory calculations and scanning tunneling microscopy experiments. J. Chem. Phys. 134 (10), 104707.
Styring, S., 2012. Artificial photosynthesis for solar fuels. Faraday Discuss. 155, 357–376.
Tachibana, Y., Vayssieres, L., Durrant, J.R., 2012. Artificial photosynthesis for solar water-splitting. Nat. Photonics 6 (8), 511–518.
Takahashi, H., Yuki, K., Nitta, T., 2002. Chemical modification of rutile TiO_2 (110) surface by ab initio calculations for the purpose of CO_2 adsorption. Fluid Phase Equilib. 194, 153–160.
Takata, T., Pan, C.S., Domen, K., 2016. Design and development of oxynitride photocatalysts for overall water splitting under visible light irradiation. Chem. Rev. 3 (1), 31–37.
Takeda, H., Ohashi, K., Sekine, A., Ishitani, O., 2016. Photocatalytic CO_2 reduction using Cu(I) photosensitizers with a Fe(II) catalyst. J. Am. Chem. Soc. 138 (13), 4354–4357.
Tang, J., Durrant, J.R., Klug, D.R., 2008. Mechanism of photocatalytic water splitting in TiO_2. Reaction of water with photoholes, importance of charge carrier dynamics, and evidence for four-hole chemistry. J. Am. Chem. Soc. 130 (42), 13885–13891.
Tao, T., Zhi, T., Liu, B., Li, M.X., Zhuang, Z., Dai, J.P., Li, Y., Jiang, F.L., Luo, W.J., Xie, Z.L., Chen, D.J., Chen, P., Li, Z.S., Zou, Z.G., Zhang, R., Zheng, Y.D., 2016. Significant improvements in InGaN/GaN nano-photoelectrodes for hydrogen generation by structure and polarization optimization. Sci. Rep. 6, 20218.
Thompson, T.L., Diwald, O., Yates, J.T., 2003. CO_2 as a probe for monitoring the surface defects on TiO_2 (110) temperature-programmed desorption. J. Phys. Chem. B 107 (42), 11700–11704.
Vesborg, P.C.K., Seger, B., 2016. Performance limits of photoelectrochemical CO_2 reduction based on known electrocatalysts and the case for two-electron reduction products. Chem. Mater. 28 (24), 8844–8850.
Walter, M.G., Warren, E.L., McKone, J.R., Boettcher, S.W., Mi, Q., Santori, E.A., Lewis, N.S., 2010. Solar water splitting cells. Chem. Rev. 110 (11), 6446–6473.
Wang, D.F., Pierre, A., Kibria, M.G., Cui, K., Han, X.G., Bevan, K.H., Guo, H., Paradis, S., Hakima, A.R., Mi, Z., 2011. Wafer-level photocatalytic water splitting on GaN nanowire arrays grown by molecular beam epitaxy. Nano Lett. 11 (6), 2353–2357.

Wang, J., Pedroza, L.S., Poissier, A., Fernández-Serra, M., 2012. Water dissociation at the GaN ($10\bar{1}0$) surface: structure, dynamics and surface acidity. J. Phys. Chem. C 116 (27), 14382–14389.
Wang, Y., Fan, S., AlOtaibi, B., Wang, Y., Li, L., Mi, Z., 2016. A monolithically integrated gallium nitride nanowire/silicon solar cell photocathode for selective carbon dioxide reduction to methane. Chemistry 22 (26), 8809–8813.
Waugh, K.C., 1992. Methanol synthesis. Catal. Today 15 (1), 51–75.
White, J.L., Baruch, M.F., Pander, J.E., Hu, Y., Fortmeyer, I.C., Park, J.E., Zhang, T., Liao, K., Gu, J., Yan, Y., Shaw, T.W., Abelev, E., Bocarsly, A.B., 2015. Light-driven heterogeneous reduction of carbon dioxide: photocatalysts and photoelectrodes. Chem. Rev. 115 (23), 12888–12935.
Wu, J.Q., 2009. When group-III nitrides go infrared: new properties and perspectives. J. Appl. Phys. 106, 011101.
Yang, J.H., Wang, D.G., Han, H.X., Li, C., 2013. Roles of cocatalysts in photocatalysis and photoelectrocatalysis. Acc. Chem. Res. 46 (8), 1900–1909.
Yang, L., Zhou, H., Fan, T., Zhang, D., 2014. Semiconductor photocatalysts for water oxidation: current status and challenges. Phys. Chem. Chem. Phys. 16 (15), 6810–6826.
Yotsuhashi, S., Deguchi, M., Zenitani, Y., Hinogami, R., Hashiba, H., Yamada, Y., Ohkawa, K., 2011. Photo-induced CO_2 reduction with GaN electrode in aqueous system. Appl. Phys. Express 4, 117101.
Zhang, Z., Yates, J.T., 2012. Band bending in semiconductors: chemical and physical consequences at surfaces and interfaces. Chem. Rev. 112 (10), 5520–5551.
Zhang, S., Connie, A.T., Laleyan, D.A., Nguyen, H.P.T., Wang, Q., Song, J., Shih, I., Mi, Z., 2014. On the carrier injection efficiency and thermal property of InGaN/GaN axial nanowire light emitting diodes. IEEE J. Quantum Electron. 50 (6), 483–490.

CHAPTER SEVEN

Light-Induced Water Splitting Using Layered Metal Oxides and Nanosheets

Takayoshi Oshima*, Miharu Eguchi†, Kazuhiko Maeda*,[1]
*School of Science, Tokyo Institute of Technology, Tokyo, Japan
†Eelectronic Functional Materials Group, Polymer Materials Unit, National Institute for Materials Science, Tsukuba, Japan
[1]Corresponding author: e-mail address: maedak@chem.titech.ac.jp

Contents

1. INTRODUCTION

Energy problem including the depletion of fossil fuels and the resulting global warming is one of the biggest concerns that human being confronts. Because of the background, utilization of renewable energy, especially solar

Semiconductors and Semimetals, Volume 97
ISSN 0080-8784
http://dx.doi.org/10.1016/bs.semsem.2017.02.001

energy, has attracted much interest in recent years. Semiconductor photocatalysis has been studied extensively (Abe, 2010; Kudo and Miseki, 2009; Maeda, 2011, 2013) because of the potential to convert solar energy into chemical energy such as hydrogen, which is easy to store and carry unlike electricity. An ultimate goal for the semiconductor photocatalysis is to split pure water into H_2 and O_2. However, a satisfactory system has not been devised so far.

In photocatalytic reaction on an illuminated semiconductor, first, light with energy greater than the band gap of the semiconductor excites electrons in the valence band to the conduction band, remaining holes in the valence band (Fig. 1). Next, the electrons and holes migrate to the surface of the semiconductor, and finally oxidation and reduction reactions take place by holes and electrons, respectively. On the other hand, electrons and holes are recombined on the way to the surface, which is the major factor that reduces the photocatalytic activity. Hence, suppression of the recombination is a key to enhance activity.

There are several strategies to enhance the photocatalytic activity. To increase the crystallinity and surface area is one of them. High crystallinity reduces structural defects, which act as recombination sites, in a semiconductor (Domen et al., 1986). Larger surface area provides more reaction sites, thereby facilitating surface redox reactions. However, it is difficult for a "bulk-type" photocatalyst to meet both high crystallinity and surface area for the following reason (Domen et al., 1986). In general, high

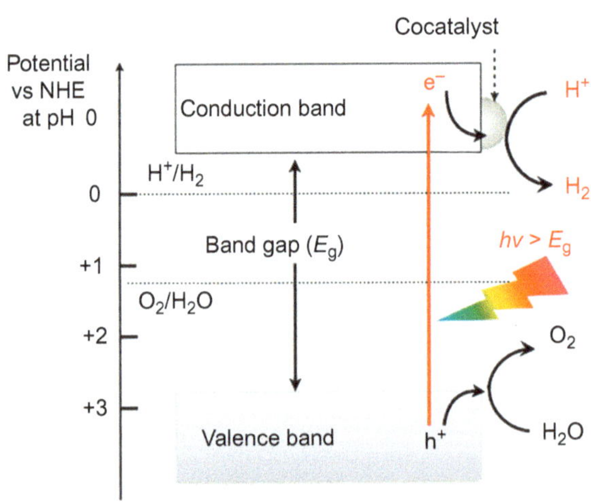

Fig. 1 Basic principle of overall water splitting using a semiconductor photocatalyst.

calcination temperature is required to synthesize a highly crystallized photocatalyst, whereby specific surface area becomes smaller by crystal growth at high temperature. On the other hand, although calcination temperature should be low in order to increase specific surface area, it also lowers crystallinity. Another way is to decrease the particle size of semiconductor (Maeda et al., 2009a). The particle size of a semiconductor photocatalyst ranges typically from several hundred nanometers to a few hundred micrometers. Longer distance for the carriers to reach surface increases the probability of recombination, resulting in lower photocatalytic performance. Although shortening particle size is advantageous for carriers to reach the surface, it is difficult to materialize it while keeping high crystallinity.

Some of semiconductor photocatalysts have layered structure and undergo exfoliation by reaction with a bulky organic base such as tetra (*n*-butyl)ammonium hydroxide (TBA^+OH^-), producing unilamellar single-crystalline colloidal sheets having a thickness of 1–2 nm, which are called "nanosheets" (Allen et al., 2010; Ebina et al., 2005, 2012; Hata et al., 2008; Ida and Ishihara, 2014; Ma et al., 2008; Sabio et al., 2010; Sarahan et al., 2008; Waller et al., 2012). The nanosheet is negatively charged, and stabilized by the bulky ammonium cation. The ammonium cations are easily exchanged into other metal cations, producing restacked nanosheet aggregates. The restacked nanosheet is usually used for photocatalytic application, as it is an easy-to-handle form. Nanosheet is believed to have several advantageous features compared to a bulk-type photocatalyst. Because nanosheet is single-crystalline, both high crystallinity and larger surface area are realized. The nanosized thickness is also beneficial for migration of photogenerated charge carriers to the surface, which reduces the possibility of recombination, eventually leading to higher activity. Finally, nanosheet can be regarded as an attractive building block to construct photocatalytic assemblies because of their high surface area and the wide variety of compositions available. The structural flexibility of nanosheets also allows one to construct multicomponent photosystems that incorporate electron donors, acceptors, catalytic nanoparticles, or photon antenna molecules (Ebina et al., 2005; Hata et al., 2008; Ma et al., 2008; Sabio et al., 2010).

In this chapter, we describe the preparation and photocatalytic properties of nanosheets. First, synthesis and characterization of perovskite-type nanosheets are summarized (Maeda et al., 2009b, 2014). Second, the photocatalytic properties of the nanosheet are discussed with regard to the lateral size and composition using half reactions of H_2 and O_2 evolution (Maeda and Eguchi, 2016; Maeda et al., 2009b, 2014; Oshima et al., 2014, 2016a), which are conventionally used as test reactions of overall water splitting (Kudo and Miseki, 2009). Overall water splitting is demonstrated with the use of cluster-sized Pt, which was synthesized in the interlayer space of restacked nanosheets (Oshima et al., 2015, 2016b). A hybrid system using metal oxide nanosheets and a redox photosensitizer to realize "dye-sensitized" H_2 evolution with visible light is also presented (Maeda et al., 2008, 2009c, 2015).

2. PREPARATION AND CHARACTERIZATION OF NANOSHEET PHOTOCATALYST WITH PEROVSKITE STRUCTURE

Nanosheets of perovskite metal oxide have been studied as photocatalysts for many years. The B site metal cation is surrounded by oxygen atoms to form octahedral units, while the A site is occupied by another metal cation to compensate charge balance of the whole crystal. Nanosheets employed as photocatalysts are typically composed of transition metal cations such as Ti^{4+}, Nb^{5+}, or Ta^{5+} having d^0-electronic configuration (Allen et al., 2010; Ebina et al., 2005, 2012; Hata et al., 2008; Ida and Ishihara, 2014; Ma et al., 2008; Sabio et al., 2010; Sarahan et al., 2008; Waller et al., 2012). The conduction band of these metal oxide nanosheets consists of empty d^0 orbital of such metals. Because the potential of d orbital of each element is different, substituting B site cation in the perovskite oxide nanosheet allows one to control the conduction band potential (E_{CB}), as demonstrated later. Although the A site cation is not directly related to the formation of conduction band, it controls the distortion of O–B–O bonding angle, which also affects the E_{CB}. In order to investigate the impact of A and B site cation on photocatalytic activity, Dion–Jacobson-type three-layer perovskite nanosheets ($A_2B_3O_{10}^-$; A = Ca, Sr; B = Nb, Ta) were synthesized by the polymerized complex (PC) method (Maeda et al., 2014). In addition, possible size effects of nanosheet were examined (Maeda et al., 2009b).

The procedure of nanosheet preparation is illustrated in Fig. 2. Layered metal oxides $KCa_{2-x}Sr_xNb_{3-y}Ta_yO_{10}$, which is the parent compound of nanosheets, were synthesized. The obtained layered materials were stirred in acidic solution to exchange interlayer K^+ cations into H^+. The protonated metal oxides were reacted with TBA^+OH^-, where the bulky TBA^+ cation was intercalated and exfoliated the layer, finally yielding nanosheet suspension. The nanosheet was restacked by proton or alkali metal cation.

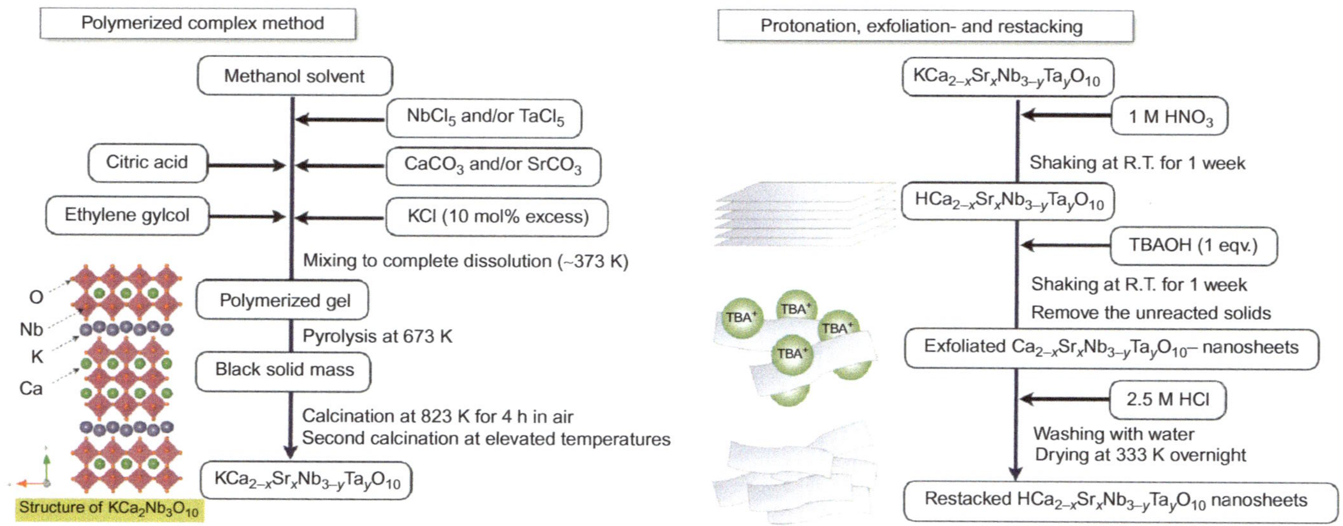

Fig. 2 Preparation schemes of layered metal oxides by the polymerized complex method and of restacked nanosheets.

2.1 Size Control of Nanosheet

The size of nanosheet can be controlled by changing the final calcination temperature in the PC method, as exemplified by $KCa_2Nb_3O_{10}$ (Maeda et al., 2009b). Powder X-ray diffraction (XRD) indicated that the diffraction peaks of the synthesized lamellar $KCa_2Nb_3O_{10}$ became stronger and narrower as the calcination temperature was increased from 1173 to 1473 K. Fig. 3A shows TEM images of exfoliated $A_2B_3O_{10}^-$ $Ca_2Nb_3O_{10}^-$ nanosheets stabilized by TBA^+, which were obtained from $KCa_2Nb_3O_{10}$ calcined at different temperatures. The lateral dimensions of the nanosheets tend to increase with raising calcination temperature. This trend is consistent with the result of XRD analysis. The TEM images clearly show that the lateral dimensions of $Ca_2Nb_3O_{10}^-$ nanosheets can be controlled by changing the calcination temperature of the parent $KCa_2Nb_3O_{10}$ powder.

The prepared nanosheets were restacked by H^+ for use as a photocatalyst. Fig. 3B shows XRD patterns of restacked $HCa_2Nb_3O_{10}$ nanosheets, along with the data for layered $HCa_2Nb_3O_{10}$ made by acid exchange of $KCa_2Nb_3O_{10}$ (calcined at 1473 K) for a comparison. The XRD patterns of the restacked nanosheets exhibit broader and weaker (001) diffraction peaks

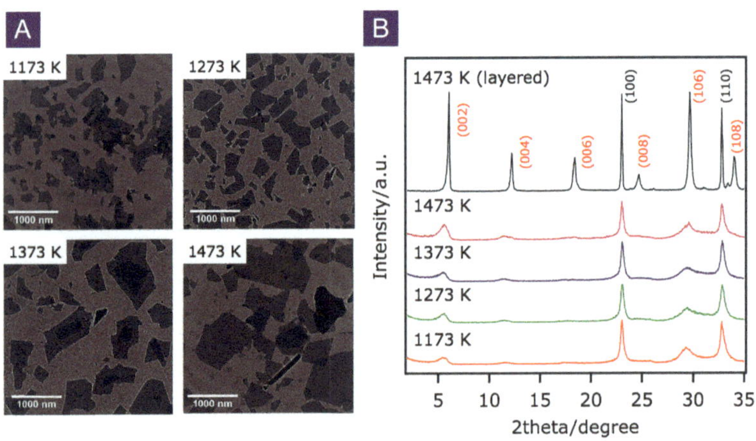

Fig. 3 (A) TEM images of exfoliated $TBA_xH_{1-x}Ca_2Nb_3O_{10}$ derived from layered $KCa_2Nb_3O_{10}$ calcined at different temperatures. (B) XRD patterns of restacked $HCa_2Nb_3O_{10}$ nanosheets derived from layered $KCa_2Nb_3O_{10}$ calcined at different temperatures. Data for layered $HCa_2Nb_3O_{10}$ (1473 K) are also given for comparison. *Reproduced from Maeda, K., Eguchi, M., Youngblood, W.J., Mallouk, T.E., 2009. Calcium niobate nanosheets prepared by the polymerized complex method as catalytic materials for photochemical hydrogen evolution. Chem. Mater. 21, 3611–3617 with permission.*

than the corresponding (002) peak in layered $HCa_2Nb_3O_{10}$. The significant reduction of intensity of the (00*l*) ($l \geq 2$) peaks compared to the parent solids indicates a much less ordered lamellar structure in the restacked materials. Strong (100) and (110) diffraction peaks corresponding to in-plane lattice directions are preserved but broadened. This indicates that the in-plane crystallinity of the $Ca_2Nb_3O_{10}^-$ sheets is preserved to some extent after the exfoliation–restacking procedure. Unlike the corresponding layered $KCa_2Nb_3O_{10}$ and $HCa_2Nb_3O_{10}$ samples, however, no distinct change with respect to calcination temperature could be identified. The specific surface area of the restacked nanosheets tended to decrease with increasing calcination temperature (72, 58, 53, and 42 $m^2\ g^{-1}$ for 1173, 1273, 1373, and 1473 K, respectively). This trend is consistent with the increase in the lateral dimensions of the nanosheets with increasing calcination temperature.

2.2 Composition and Physicochemical Properties of Nanosheets

In a similar manner, a series of substituted $Ca_2Nb_3O_{10}^-$ nanosheets were also synthesized. $HCa_{2-x}Sr_xNb_3O_{10}$ and $HCa_2Nb_{3-y}Ta_yO_{10}$ restacked nanosheets exhibited diffraction patterns similar to either $HSr_2Nb_3O_{10}$ or $HCa_2Nb_3O_{10}$ in the range of $x=0$–2 (Fig. 4A). The shape of the

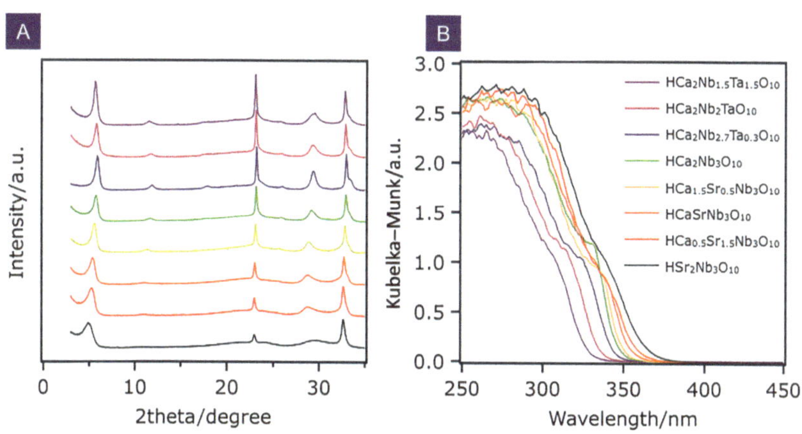

Fig. 4 (A) XRD patterns and (B) UV–visible diffuse reflectance spectra of restacked $HCa_{2-x}Sr_xNb_3O_{10}$ and $HCa_2Nb_{3-y}Ta_yO_{10}$ nanosheets. *Reproduced from Maeda, K., Eguchi, M., Oshima, T., 2014. Perovskite oxide nanosheets with tunable band-edge potentials and high photocatalytic hydrogen-evolution activity. Angew. Chem. Int. Ed. 53, 13164–13168 with permission.*

diffraction pattern underwent a gradual change from $HCa_2Nb_3O_{10}$ to $HSr_2Nb_3O_{10}$ as the Sr content in $HCa_{2-x}Sr_xNb_3O_{10}$ increases, suggesting the successful replacement of Ca into Sr. $HCa_2Nb_{3-y}Ta_yO_{10}$ also demonstrated single-phase diffraction patterns until Nb was replaced by Ta up to $y=1.5$, suggesting successful preparation of $HCa_2Nb_{3-y}Ta_yO_{10}$. However, further substitution of Nb caused the production of some impurity phases, and it was not achieved to synthesize $HCa_2Nb_{3-y}Ta_yO_{10}$ with $y>1.5$. Successful substitution of Ca and Nb for Sr and Ta, respectively, was confirmed by energy dispersive X-ray spectroscopy in the entire compositional range examined.

The substitution of A and B site cations is expected to affect band structure, reflecting the optical light absorption properties of the materials. Fig. 4B shows UV–visible diffuse reflectance spectra for the restacked samples. All samples exhibit absorption bands at around 330–370 nm. The position of the absorption edge shifted to longer wavelengths with increasing the Sr content. The band-gap energy can be estimated from the absorption edges, and it is decreased from 3.59 (for $x=0$) to 3.40 eV (for $x=2$). The band-gap energy of layered perovskite materials consisting of NbO_6 octahedral units is affected by several factors including the distortion of NbO_6 units, the thickness of perovskite layers, and the electronegativity of elements on the octahedral and 12-coordinate sites (Miseki et al., 2009). In this case, it appears that the distortion of NbO_6 units in the perovskite layers is the main factor determining the band-gap energies. As Sr^{2+} is located at the A site, the overlap of the Nb4d orbital is greater than in the case of Ca^{2+} because Sr^{2+} has a larger ionic radius (1.58 Å) compared to Ca^{2+} (1.48 Å), and thus the distortion of the NbO_6 octahedral units is relaxed. The larger overlap of Nb4d orbitals stabilizes the conduction band energy, with a positive change in energy. On the other hand, increasing the Ta content resulted in a blue-shift of the absorption edge. The estimated band gap from the absorption edge was 3.75 eV for $y=1.5$. The E_{CB} of Ta^{5+}-based oxides is located at more negative position than that of a Nb^{5+} analogue, because Ta5d orbitals have higher potential than Nb4d orbitals do (Kudo and Miseki, 2009; Maeda, 2013). The observed blue-shift should therefore result from the negative shift of E_{CB}. These results clearly demonstrate that perovskite nanosheets, which have controlled optical light absorption properties, can be obtained by changing the composition. The E_{CB} values of the restacked $HCa_{2-x}Sr_xNb_{3-y}Ta_yO_{10}$ nanosheets are summarized in Table 1.

Table 1 Band Gaps, Conduction Band Potentials, and Hydrogen Evolution Quantum Yields of Restacked $HCa_{2-x}Sr_xNb_3O_{10}$ and $HCa_2Nb_{3-y}Ta_yO_{10}$ Nanosheets

Ideal Composition	E_g/eV	E_{CB}/V vs NHE at pH 0	Apparent Quantum Yield at 300 nm[a]/%
$HCa_2Nb_3O_{10}$	3.59	−0.57	51 ± 4
$HCa_{1.5}Sr_{0.5}Nb_3O_{10}$	3.55	−0.55	48 ± 3
$HCaSrNb_3O_{10}$	3.50	−0.52	44 ± 5
$HCa_{0.5}Sr_{1.5}Nb_3O_{10}$	3.46	−0.50	39 ± 1
$HSr_2Nb_3O_{10}$	3.40	−0.47	39 ± 3
$HCa_2Nb_{2.7}Ta_{0.3}O_{10}$	3.65	−0.60	71 ± 5
$HCa_2Nb_2TaO_{10}$	3.70	−0.62	78 ± 2
$HCa_2Nb_{1.5}Ta_{1.5}O_{10}$	3.80	−0.67	65 ± 7

[a]Reaction conditions under visible light: catalyst, 25 mg (0.5 wt% Pt-loaded); aqueous solution containing 10 vol% methanol (100 mL); light source, xenon lamp (300 W) with a band-pass filter.

It is also possible to alter interlayer cations of restacked nanosheets, by simply using various acids, bases, and salts that contain different cations. As the result, nanosheets having different morphological features and different degrees of interlayer hydration could be synthesized. However, the interlayer cations did not largely affect the light absorption property (Maeda and Eguchi, 2016). For example, the absorption edge positions of $Ca_2Nb_2TaO_{10}^-$ nanosheets reacted with different restacking agents were almost similar to each other. This would be reasonable because photoexcitation, from the valence band formed by oxygen 2p orbitals to the conduction band formed by the empty Nb4d and Ta5d orbitals, occurs in the two-dimensional nanosheet.

3. H_2 EVOLUTION ACTIVITY

3.1 Effect of Nanosheet Size

As described earlier, $Ca_2Nb_3O_{10}^-$ nanosheets having different lateral sizes were synthesized by changing the calcination temperature of the parent $KCa_2Nb_3O_{10}$. First, photocatalytic H_2 evolution activity was investigated with respect to the nanosheet size using methanol as a sacrificial electron donor. Fig. 5 shows the rate of H_2 evolution from aqueous methanol

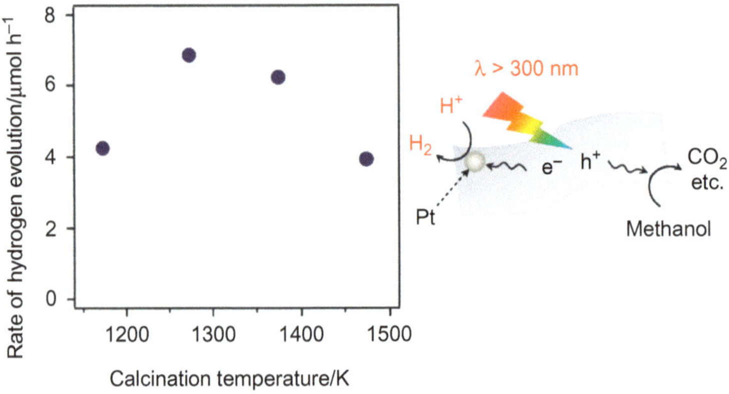

Fig. 5 Dependence of the rate of H_2 evolution from an aqueous methanol solution using 0.3 wt% Pt-loaded restacked $HCa_2Nb_3O_{10}$ nanosheets under band gap irradiation ($\lambda > 300$ nm) on the calcination temperature of $KCa_2Nb_3O_{10}$. Reaction conditions: catalyst, 5.0 mg; aqueous methanol solution (1 M, 2.0 mL); light source, xenon lamp (300 W). *Reproduced from Maeda, K., Eguchi, M., Youngblood, W.J., Mallouk, T.E., 2009. Calcium niobate nanosheets prepared by the polymerized complex method as catalytic materials for photochemical hydrogen evolution. Chem. Mater. 21, 3611–3617 with permission.*

solution over Pt-loaded restacked $HCa_2Nb_3O_{10}$ nanosheets under band gap irradiation ($\lambda > 300$ nm) as a function of the calcination temperature of $KCa_2Nb_3O_{10}$. Here Pt nanoparticles were deposited on the external surface of the restacked nanosheets as cocatalysts for H_2 evolution. All samples showed H_2 evolution activity, and no reaction took place in the dark. Interestingly, these photocatalysts show a volcano-like trend in H_2 evolution rate vs calcination temperature; namely, the activity increased and reached a maximum at 1273 K, and then lessened. The behavior can be explained in terms of crystallinity and specific surface area. The carriers generated by band gap excitation have to migrate in two-dimensional $Ca_2Nb_3O_{10}^-$ sheet without recombination so that electrons and holes react with proton and methanol, respectively. It is known that crystallinity of a semiconductor photocatalyst strongly affects the recombination of photogenerated carriers (Domen et al., 1986). In particular, poorly crystallized metal oxide, which is calcined at lower temperatures and contains structural imperfections, shows lower photocatalytic activity. Therefore, the rise in activity from 1173 to 1273 K should be ascribed to a consequence of the improved crystallinity of $HCa_2Nb_3O_{10}$ nanosheets. On the other hand, smaller specific surface area means smaller number of reaction sites, slowing down surface redox reactions. The specific surface area of the restacked $HCa_2Nb_3O_{10}$ nanosheets became smaller at elevated temperatures, which caused the growth of lateral dimensions of $Ca_2Nb_3O_{10}^-$ nanosheets (Fig. 3A). Hence, the activity-drop observed at 1373 and 1473 K, despite of higher crystallinity of nanosheet, appears to result from the loss of specific surface area.

3.2 Effect of Composition

The E_{CB} of three-layer perovskite nanosheet can be precisely tuned by altering the composition. It is expected that H_2 evolution activity is influenced by the E_{CB}, which is one of the most important factors determining the driving force for H_2 evolution. Next, substitution effect of the A/B sites in $Ca_2Nb_3O_{10}^-$ nanosheet on H_2 evolution performance was examined. As listed in Table 1, the apparent quantum yields (AQYs) of H_2 evolution on these materials were very high, giving 40%–80% at 300 nm. The performance tends to decrease with increasing the Sr content, presumably owing to a decrease in the driving force of conduction band electrons for proton reduction. On the other hand, increasing the Ta content in $HCa_2Nb_{3-y}Ta_yO_{10}$ enhanced the activity up to $y=1$, then decreasing slightly. The rise of the E_{CB} in $HCa_2Nb_{3-y}Ta_yO_{10}$ with increasing the Ta content appears to be favorable for the reduction of protons to give H_2. At the same time, however, this would lead to inferior light-harvesting (Fig. 4B), which contributes to slower H_2 evolution. The most active material, $HCa_2Nb_2TaO_{10}$, generated H_2 without noticeable degradation giving a maximum AQY of 80%. To our knowledge, the approximately 80% AQY recorded for the $HCa_2Nb_2TaO_{10}$ nanosheet is the highest value among nanosheet-based photocatalysts reported to date. The nanosized thickness and high surface area of perovskite oxide nanosheets are both beneficial in terms of prompt migration of the photogenerated charge carriers to the surface. In addition to this structural effect, the precise control of E_{CB} that maximizes the reactivity of the conduction band electrons should be responsible for the high photocatalytic activity of $HCa_2Nb_2TaO_{10}$.

Restacked $Ca_2Nb_2TaO_{10}^-$ nanosheets were further studied with respect to the structural features as photocatalysts for H_2 evolution from an aqueous methanol solution. As mentioned earlier, the structural features and the degree of interlayer hydration of the restacked material depended on the restacking agent employed. The highest photocatalytic activity was obtained for the restacked nanosheets using HCl as the restacking agent. Results of structural characterizations and photocatalytic reactions suggested that the high activity resulted from interlayer protonation (not hydration), which is favorable for the oxidation reaction.

4. O_2 EVOLUTION ACTIVITY

4.1 Exfoliation-Induced Water Oxidation Activity

In order to realize artificial photosynthesis, water should be used as an electron source. Although water oxidation is an important reaction in this sense, it is a challenging reaction because the reaction involves four-electron process, which is kinetically very slow. Metal oxide semiconductors are promising photocatalysts for water oxidation due to the enough potential of valence band for water oxidation. Early works focusing on water oxidation over metal oxides have shown that high crystallinity of a semiconductor is prerequisite for efficient water oxidation. Thus, the single-crystalline texture of nanosheet is advantageous for water oxidation.

Water oxidation into O_2 was conducted using restacked $KCa_2Nb_3O_{10}$ nanosheet and IO_3^- as a reversible electron accepter. The energy change of the total reaction is positive, so-called artificial photosynthesis-type reaction. As shown in Fig. 6, layered $KCa_2Nb_3O_{10}$ exhibited no activity. Interestingly, however, O_2 evolution was clearly observable using restacked $KCa_2Nb_3O_{10}$ nanosheets. The amount of evolved O_2 increased over time, indicating that water oxidation and reduction of IO_3^- both took place on the surface of the calcium niobate nanosheets. Analysis by ion chromatography showed that the amount of I^- produced in the liquid phase (as a result of reduction of IO_3^-) corresponded to two-thirds of the O_2 production, consistent with the amount expected from the stoichiometry. Without $NaIO_3$, no reaction occurred as well. The induced activity of water oxidation might be simply ascribed to difference in surface area because of restacking. To check the possibility, different amount of layered and restacked $KCa_2Nb_3O_{10}$ was used for photocatalytic reaction to make the total surface area similar. However, the water oxidation rate recorded using 500 mg of layered $KCa_2Nb_3O_{10}$ ($\approx$0 μmol h^{-1}) was much lower than that achieved by 20 mg of the restacked one ($\approx$0.5 μmol h^{-1}), even though the total surface areas of both catalysts were similar ($\sim$0.5 m^2). These results suggest that exfoliation and restacking process is important to induce nonsacrificial water photooxidation activity of $KCa_2Nb_3O_{10}$. It should be noted that the absorption edge of restacked $KCa_2Nb_3O_{10}$ was blue-shifted compared to layered $KCa_2Nb_3O_{10}$ one, presumably due to quantum size effects caused by nanosized thickness of the nanosheet. The induced activity of water oxidation may be attributed to the enlarged band gap with an improved drive force for surface redox reactions due to the quantum size effects.

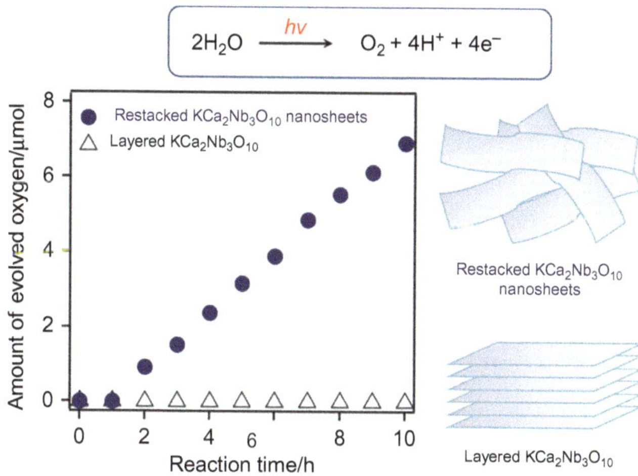

Fig. 6 Time courses of water oxidation using $KCa_2Nb_3O_{10}$-based materials under UV irradiation ($\lambda > 300$ nm). Reaction conditions: catalyst, 50 mg; 10 mM aqueous $NaIO_3$ solution, 100 mL; light source, xenon lamp (300 W). *Reproduced from Oshima, T., Ishitani, O., Maeda, K., 2014. Non-sacrificial water photo-oxidation activity of lamellar calcium niobate induced by exfoliation, Adv. Mater. Interfaces 1, 1400131.*

Aiming at enhancing the water oxidation activity, ruthenium oxide (RuO_x) was deposited on $KCa_2Nb_3O_{10}$ as a cocatalyst by an impregnation method. Deposition of 0.1 wt% RuO_x dramatically promoted water oxidation (27.3 μmol h^{-1}). Although RuO_x loading was also effective for enhancing water oxidation by layered $KCa_2Nb_3O_{10}$ (1.6 μmol h^{-1}), the promotional effect was much more pronounced for restacked $KCa_2Nb_3O_{10}$ than for the layered one, further supporting the idea that the exfoliation–restacking process is essential to inducing water oxidation activity of $KCa_2Nb_3O_{10}$.

Further investigation of cocatalyst effect shed light on the reaction mechanism for the water oxidation reaction. Table 2 summarizes the rates of O_2 evolution from aqueous $NaIO_3$ solutions on $KCa_2Nb_3O_{10}$ restacked nanosheets further modified with various metal oxides. The deposition of cobalt oxide and manganese oxide was found to reduce the O_2 evolution activity compared to the rate obtained from the bare nanosheets. By contrast, both rhodium oxide and iridium oxide facilitated O_2 evolution. Among the metal oxides assessed, the deposition of RuO_x led to the largest increase in activity. It was reported that iridium oxide and RuO_x promoted the reduction of IO_3^- (Higashi et al., 2008; Maeda et al., 2011) and that rhodium oxide could also catalyze water reduction (Hata et al., 2008; Ma

Table 2 O_2 Evolution Rate Over Various Metal Oxides (0.1 wt%) Loaded on Restacked $KCa_2Nb_3O_{10}$ Nanosheets[a]

Metal Species	O_2 Evolution Rate/μmol h^{-1}
bare	0.8
Mn	0.2
Co	0.5
Rh	1.9
Ir	5.5
Ru	27.3

[a]Reaction condition: catalysts: 50 mg; reaction solution: 10 mM $NaIO_3$ aqueous solution (100 mL); light source: 300 W xenon lamp ($\lambda > 300$ nm).

et al., 2008). It thus appears that the O_2 evolution activity of restacked $KCa_2Nb_3O_{10}$ was enhanced when a reduction reaction promoter was deposited. The reduction of IO_3^- involves a six-electron process that is kinetically unfavorable in the absence of a catalyst. The overpotentials of electrochemical reduction of IO_3^- obtained with rhodium oxide and iridium oxide were lower than that observed using cobalt oxide, and RuO_x showed the lowest overpotential (Suzuki et al., 2015). These results indicate that the reduction of IO_3^- is the rate-determining step in the oxidation reaction; thus, acceleration of the reduction of IO_3^- should lead to increased water oxidation activity.

4.2 Effect of Physicochemical Properties

As discussed earlier, H_2 evolution activity was strongly affected by the physicochemical properties of nanosheet (e.g., size, composition, etc.). The structural impact on O_2 evolution was also examined in a similar manner. Regarding the size effect, $Ca_2Nb_3O_{10}^-$ nanosheet was synthesized at various calcination temperatures, and the resulting restacked nanosheets were tested for the reaction of O_2 evolution with the aid of RuO_x as a cocatalyst. While lower calcination temperatures, which led to smaller lateral dimensions, were found to increase the specific surface area of restacked $HCa_2Nb_3O_{10}$ nanosheets (Fig. 3A), no significant correlation between the calcination temperature and specific surface area of restacked $KCa_2Nb_3O_{10}$ nanosheets was observed. This might be due to the heating process (623 K in air) in the impregnation method, which promote the reconstruction of the layered structure and may reduce any differences in specific surface area. Interestingly, O_2 evolution rate did not depend on the calcination temperature; all of the prepared materials produced O_2 at a rate of 20–25 μmol h^{-1}. These results indicate that the crystallinity of the nanosheet has little effect on the photocatalytic activity for water oxidation unlike H_2 evolution. It is thought that larger nanosheets will be more highly crystalline, and in principle this should have a positive effect on the activity. Nevertheless, no distinct correlation between the calcination temperature and the water oxidation activity area was observed. One possible explanation for the lack of correlation between the lamellar size and the activity of the nanosheets is that the reduction of IO_3^- is the rate-determining step in the reaction, as discussed in Section 4.1.

Effects of the composition of nanosheets upon photocatalytic O_2 evolution was also studied using $KCa_{2-x}Sr_xNb_3O_{10}$ restacked nanosheets. The E_{CB} for restacked $KCa_{2-x}Sr_xNb_3O_{10}$ was positively shifted with an increase in the Sr content, as discussed earlier. Based on the assumption that the valence band maximum for a metal oxide having a d^0 or d^{10} electronic configuration is approximately 3 V vs normal hydrogen electrode (NHE) at pH 0 (Scaife, 1980), the E_{CB} of $KCa_2Nb_3O_{10}$, $KCaSrNb_3O_{10}$, and $KSr_2Nb_3O_{10}$ nanosheets were calculated to be −0.5, −0.4, and −0.3 V, respectively, suggesting that restacked $KCa_2Nb_3O_{10}$ nanosheet is more advantageous with regard to the reduction of IO_3^- (+1.08 V vs NHE at pH 0) than restacked $KCaSrNb_3O_{10}$ and $KSr_2Nb_3O_{10}$ in terms of the driving force for the reaction. However, an opposite trend in the photocatalytic activities of these materials was observed. The O_2 evolution activity was found to increase with increasing the Sr concentration in $KCa_{2-x}Sr_xNb_3O_{10}$. Because the driving force for IO_3^- reduction is sufficiently large compared to the differences in the conduction band energies for these materials (0.1−0.3 V), these differences in E_{CB} evidently have little impact on the activity for water oxidation. It is also noted that the activity differed depending on the composition, although the specific surface areas of the restacked $KSr_2Nb_3O_{10}$, $KCaSrNb_3O_{10}$, and $KCa_2Nb_3O_{10}$ nanosheets were almost identical (~30 $m^2\ g^{-1}$). A possible explanation for the observed order of activity is that the activity is governed simply by the light absorption characteristics of the material. That is, the $KSr_2Nb_3O_{10}$ nanosheets, having the smallest band-gap energy among the tested materials, can more efficiently absorb the incident photon flux and thus generate a greater number of photoexited carriers. The more efficient utilization of incident photons will lead to an increase in the number of photogenerated charge carriers, which might facilitate electron transfer from the nanosheets to the RuO_x, resulting in higher activity. This result, therefore, suggests that electron transfer from the nanosheets to the RuO_x particles is the rate-determining step.

5. OVERALL WATER SPLITTING

5.1 A New Interlayer Modification Method

As demonstrated earlier, metal oxide nanosheets are promising photocatalysts for both H_2 and O_2 evolution from water. However, only a few reports of overall water splitting using nanosheet photocatalysts have been reported so far. There are several layered metal oxides that can utilize their interlayer space for water splitting (Kudo et al., 1989; Takata et al., 1997). For example, $K_4Nb_6O_{17}$ possesses two kinds of interlayer spaces, called interlayer I and II, which exhibit different ion exchange characters (Kudo et al., 1989). The ion exchange character allows us to intercalate Ni oxide nanoparticles as cocatalysts for H_2 evolution only in the interlayer I, leading to the separation of reduction/oxidation sites; namely, interlayer I, where Ni cocatalyst exists, works as a reduction site, while interlayer II functions as oxidation site. Therefore, it appears that effective utilization of interlayer space is a key to achieve overall water splitting, and that interlayer modification with a suitable cocatalyst is important for efficient water splitting. For nanosheet photocatalysts, several attempts were done to incorporate metal or metal oxide nanoparticles in the interlayer nanospace (Ebina et al., 2005; Hata et al., 2007, 2008; Ma et al., 2008). Nevertheless, an appropriate method of interlayer modification for overall water splitting had not been developed so far.

Very recently, an effective interlayer modification method, which utilizes electrostatic interaction between the negatively charged $Ca_2Nb_3O_{10}^-$ nanosheet and a cationic precursor of cocatalyst, was developed (Oshima et al., 2015). Fig. 7 depicts the preparation procedure. First, an aqueous solution containing a cationic Pt complex (e.g., $[Pt(NH_3)_4]Cl_2 \cdot H_2O$) was added into the nanosheet suspension, followed by stirring for 24 h to allow Pt precursor to adsorb onto the surface of the nanosheets. This method is referred to as the adsorption method. Then, KOH was added to precipitate the colloidal nanosheets, followed by drying in an oven at 343 K overnight and heating under a H_2 stream (20 mL min^{-1}) at 473 K to reduce cationic Pt species.

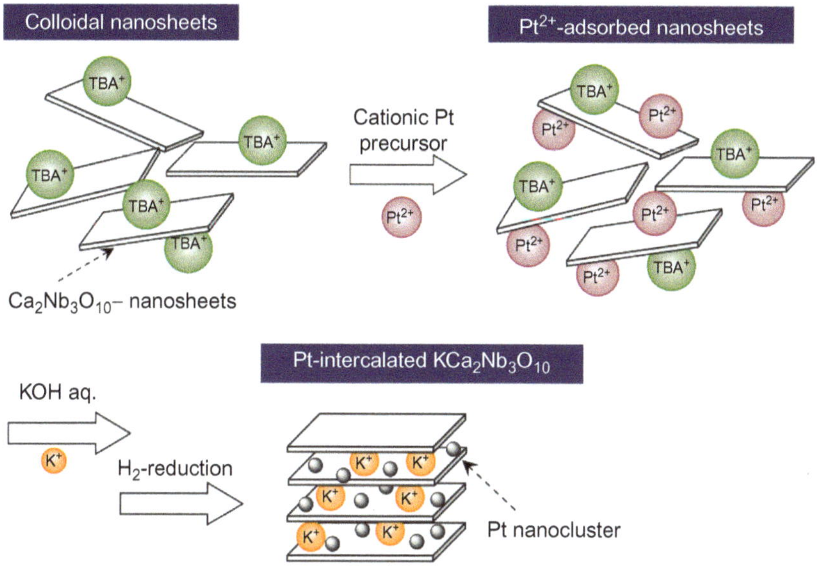

Fig. 7 Preparation of Pt-intercalated $KCa_2Nb_3O_{10}$ nanosheets through electrostatic attraction between a cationic complex and anionic nanosheets [TBA$^+$ indicates tetra (*n*-butyl)ammonium cation]. *Reproduced from Oshima, T., Lu, D., Ishitani, O., Maeda, K., 2015. Intercalation of highly dispersed metal nanoclusters into a layered metal oxide for photocatalytic overall water splitting. Angew. Chem. Int. Ed. 54, 2698–2702 with permission.*

XRD, UV–visible diffuse reflectance spectroscopy, XPS, and TEM observation revealed that Pt nanoclusters of <1 nm in size were intercalated in the interlayer nanospace of restacked $KCa_2Nb_3O_{10}$ nanosheets, although some of the deposited Pt were located on the external surface of the material (Fig. 8). On the other hand, deposition of Pt was not successful, when an anionic Pt precursor such as $H_2[PtCl_6]\cdot 6H_2O$ was employed. These results indicate that electrostatic interaction plays an important role in the adsorption method. For comparison, Pt was loaded on the pre-restacked $KCa_2Nb_3O_{10}$ nanosheets by an impregnation method using $[Pt(NH_3)_4]Cl_2\cdot H_2O$ as a precursor, followed by thermal H_2 reduction. In this case, Pt was deposited exclusively on the external surface not in the interlayers, with the average particle size of 1.8 nm (Fig. 8), which is larger than that observed in the adsorption method. The successful deposition of smaller Pt in the adsorption method is presumably because the Pt species sandwiched by $Ca_2Nb_3O_{10}^-$ nanosheets in the adsorption samples were less susceptible to aggregation upon thermal treatment due to steric hindrance. In addition, the intercalated, highly dispersed Pt nanoclusters underwent

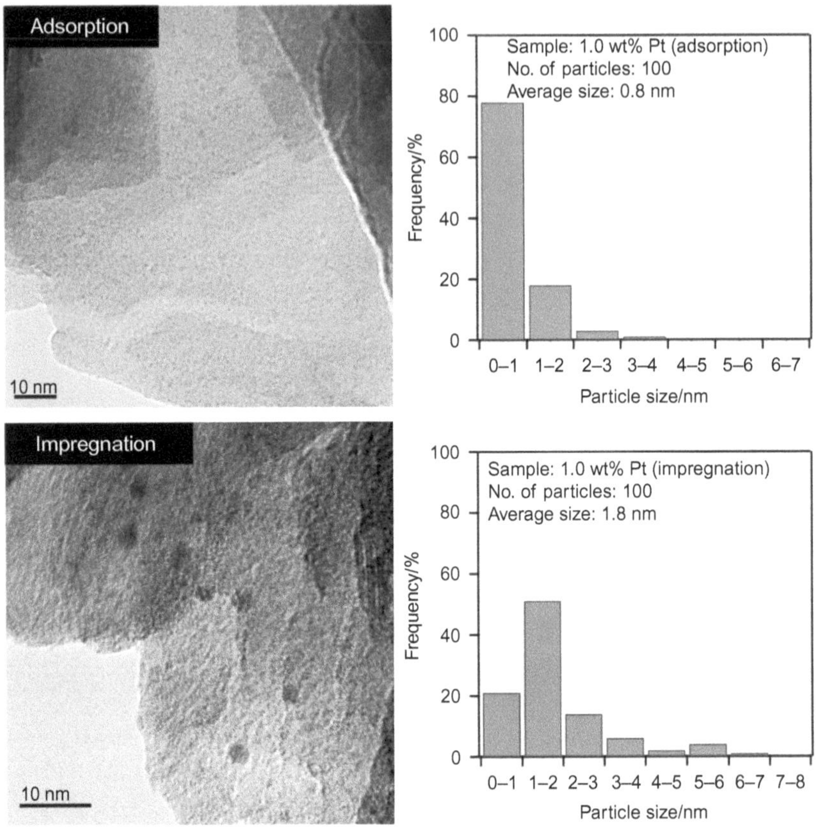

Fig. 8 Typical TEM images and corresponding particle size distributions of restacked $KCa_2Nb_3O_{10}$ nanosheets modified with 1.0 wt% Pt using an adsorption and impregnation method. *Reproduced from Oshima, T., Lu, D., Ishitani, O., Maeda, K., 2015. Intercalation of highly dispersed metal nanoclusters into a layered metal oxide for photocatalytic overall water splitting. Angew. Chem. Int. Ed. 54, 2698–2702 with permission.*

electronic interactions with the nanosheet surface, as indicated by XPS. This is favorable for photoredox catalysis, as the efficiency of photocatalytic reactions can be improved upon intimate contact between a semiconductor and a cocatalyst. It is noteworthy that this method based on simple electrostatic interaction is applicable to various metal species. For example, it is possible to intercalate highly dispersed Pd, Rh, and Ru on restacked $KCa_2Nb_3O_{10}$ nanosheet.

The as-prepared $Pt/KCa_2Nb_3O_{10}$ nanosheets were tested as photocatalysts for overall water splitting under UV and visible irradiation

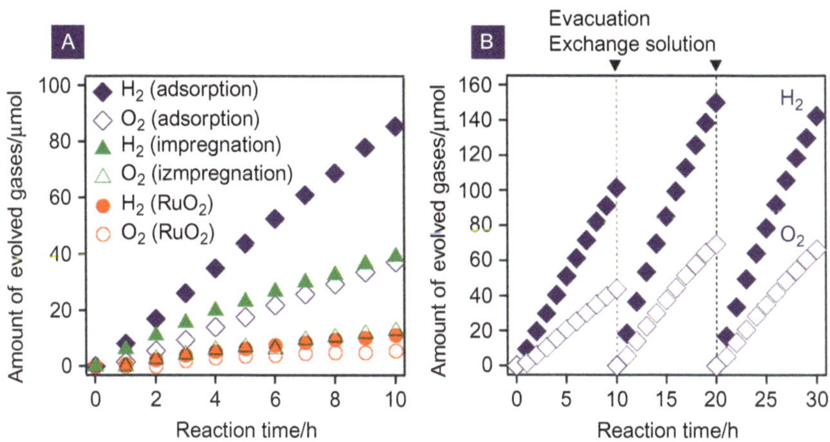

Fig. 9 (A) Time course of overall water splitting using $KCa_2Nb_3O_{10}$-based materials under UV irradiation ($\lambda > 300$ nm). (B) Recycling test using Pt(0.75 wt% Pt loaded by the adsorption method)/restacked $KCa_2Nb_3O_{10}$ nanosheets. The reaction was repeated after intermittent evacuation of the gas phase and exchange of reacted solution with fresh solution. Reaction conditions: catalyst, 50 mg; reactant solution, 10 mM aqueous NaI solution, 100 mL; light source, xenon lamp (300 W). *Reproduced from Oshima, T., Lu, D., Ishitani, O., Maeda, K., 2015. Intercalation of highly dispersed metal nanoclusters into a layered metal oxide for photocatalytic overall water splitting. Angew. Chem. Int. Ed. 54, 2698–2702 with permission.*

($\lambda > 300$ nm). Although unmodified restacked $KCa_2Nb_3O_{10}$ nanosheets produced only a small amount of H_2, Pt/$KCa_2Nb_3O_{10}$ restacked nanosheet prepared by both adsorption and impregnation methods exhibited simultaneous H_2 and O_2 evolution, as shown in Fig. 9. Here a dilute aqueous NaI solution was used as the reactant solution in order to minimize the negative impact of H_2/O_2 recombination on Pt (Abe et al., 2003). The rates of H_2 and O_2 evolution by the adsorption sample were greater than those of the impregnation sample, most likely because of the smaller size of the Pt deposits (Fig. 8). The stability of Pt-intercalated $KCa_2Nb_3O_{10}$ was tested by over 30 h of reaction. As shown in Fig. 9, the rates of H_2 and O_2 evolution were improved by a factor of about 1.5 after the first run, then exhibited stable performance without significant loss of activity. The total amount of H_2 and O_2 produced in 30 h of reaction was 573 μmol, exceeding that of Pt/$KCa_2Nb_3O_{10}$ (ca. 90 μmol), which clearly confirms the catalytic cycle of the reaction. The AQY in the steady-state was estimated to be about 3% at 300 nm. As reported by Ebina et al., restacked $KCa_2Nb_3O_{10}$ nanosheets achieved overall water splitting when RuO_x was intercalated into the interlayer (Ebina et al., 2005). However, the adsorption sample showed much higher activity than RuO_x-deposited one did. To our knowledge, the water splitting activity of the adsorption sample is the highest among nanosheet-based photocatalysts ever reported.

On another note, the effect of cocatalyst size on the photocatalytic water splitting performance had not been examined at sizes smaller than 1 nm because of the lack of an effective preparation method and a suitable photocatalyst. Thus, the results of this study revealed the superior ability of nanoclusters with diameters smaller than 1 nm to promote the water-splitting reaction for the first time.

5.2 Effect of Preparation Conditions

As demonstrated earlier, Pt nanoclusters working as efficient cocatalysts were successfully deposited on restacked $KCa_2Nb_3O_{10}$ nanosheets, achieving overall water splitting with the highest performance among nanosheet-based photocatalysts, certainly due to the small size of the deposited Pt species. In the deposition method, Pt precursor was reduced by thermal H_2 treatment, and in general, the reduction temperature largely affects the size of cocatalyst as well as photocatalytic performance. Hence, the effect of H_2 reduction temperature on the physicochemical characters of the deposited Pt and photocatalytic performance was examined (Oshima et al., 2016b).

Nanoparticulate Pt was deposited on restacked $KCa_2Nb_3O_{10}$ nanosheets in a similar manner but at different H_2 reduction temperatures. Although the difference in the Pt size distribution was observed, Pt nanoparticles having <2 nm size were mainly deposited at temperatures below 673 K, due to the suppression of Pt aggregation that originated from steric hindrance of the interlayer nanospace. In particular, most of the Pt species were smaller than 1 nm when reduced at 473 K. As the reduction temperature was increased, Pt was excluded from the interlayer to the external surface of restacked $KCa_2Nb_3O_{10}$ nanosheets because of the reconstruction of the original layered structure, which accompanied with a change in the valence state of Pt species from oxide to metal. Physicochemical analyses highlighted that the activity of Pt/$KCa_2Nb_3O_{10}$ restacked nanosheets for overall water splitting was dependent on the feature size and the valence state of Pt. In particular, the intercalated, electron-deficient Pt species realized at 473 K were advantageous in terms of suppressing undesirable backward reactions such as reduction of O_2 and/or H_2/O_2 recombination.

6. DYE-SENSITIZED H_2 EVOLUTION UNDER VISIBLE LIGHT

6.1 Utilization of the Negatively Charged Nanosheet Surface for Sensitized H_2 Evolution

Metal oxide nanosheets are shown to work as good photocatalysts for water splitting. However, they in principle work under UV light (<400 nm) because of their large band gaps. For efficient solar energy conversion, it is necessary to utilize visible light, which occupies approximately half of the solar spectrum. One of the strategies to induce visible light response of a wide-gap metal oxide is to modify a metal oxide semiconductor with a dye molecule that can absorb visible light (Bae and Choi, 2006; Furlong et al., 1986). The basic principle is essentially the same as dye-sensitized solar cells. As shown in Fig. 10, a dye molecule (redox photosensitizer) anchored on the semiconductor surface undergoes photoexcitation upon visible light, then injecting an electron into the conduction band of a semiconductor and producing one-electron oxidized species of the sensitizer. The injected electron in the semiconductor is consumed at a catalytic site such as Pt, while the oxidized sensitizer is regenerated by accepting an electron supplied from a suitable electron donor.

The primary requirement to realize efficient dye-sensitized H_2 evolution is intimate attachment of sensitizer molecules onto the surface of a metal oxide, because the lifetime of the excited state of a photosensitizer is typically too short to allow for diffusion of the molecule to the surface (Furlong et al., 1986). Ruthenium(II) trisdiimine ($Ru(bpy)_3^{2+}$, bpy = 2,2′-bipyridine; abbreviated as $\mathbf{Ru^{2+}}$, see Fig. 11) and its derivatives have been studied extensively as visible light sensitizers for photocatalytic H_2 evolution. Such Ru(II) polypyridyl sensitizers need to be anchored to the surface of a metal oxide through the bipyridyl ligands, which are functionalized typically with carboxylic acid or phosphonic acid groups, for efficient electron injection (Bae and Choi, 2006). For metal oxide nanosheets that have negatively charged surface, it is expected that $\mathbf{Ru^{2+}}$ can be immobilized on the nanosheet surface based on electrostatic attraction even without functionalization of the bpy ligand. Also, the single-crystalline texture of nanosheets as well as their high surface area appears advantageous over bulk-type oxide semiconductors such as TiO_2.

On the basis of these backgrounds, scrolled niobate nanosheets were prepared by exfoliation of lamellar $K_4Nb_6O_{17}$ to test for sensitized H_2

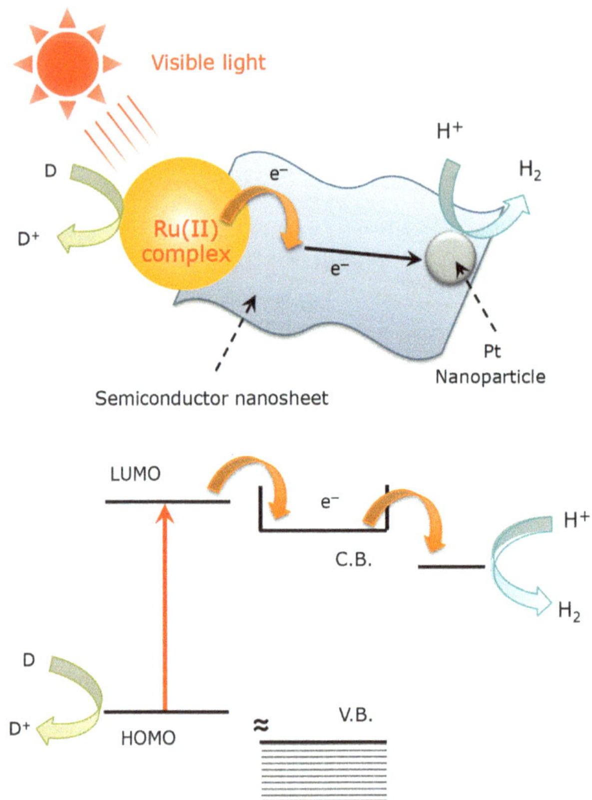

Fig. 10 Schematic illustration of H_2 evolution from water using a wide-gap semiconductor and a Ru(II) complex sensitizer under visible light. *Reproduced from Maeda, K., Sahara, G., Eguchi, M., Ishitani, O. 2015. Hybrids of a ruthenium(II) polypyridyl complex and a metal oxide nanosheet for dye-sensitized hydrogen evolution with visible light: effects of the energy structure on photocatalytic activity. ACS Catal. 5, 1700–1707 with permission.*

evolution. Because the $Nb_6O_{17}{}^{4-}$ sheets do not have mirror symmetry (i.e., the individual sheet is asymmetric), there is an intrinsic tension in the sheet, which leads to spontaneous scrolling in order to release the strain energy. The exfoliated, scrolled $Nb_6O_{17}{}^{4-}$ sheets were reassembled by HCl. TEM images of the as-prepared material, represented as NS-$H_4Nb_6O_{17}$, are shown in Fig. 12. The diameters of NS-$H_4Nb_6O_{17}$ ranges about 30 nm, with lengths of several hundred nanometers. The specific surface area of NS-$H_4Nb_6O_{17}$ was $250-300\ m^2\ g^{-1}$, much larger than that of the proton-exchanged parent solid prior to exfoliation (ca. $1-2\ m^2\ g^{-1}$). It was also confirmed by zeta-potential measurement that the surface of NS-$H_4Nb_6O_{17}$ is negatively charged in the pH range of 3–11.

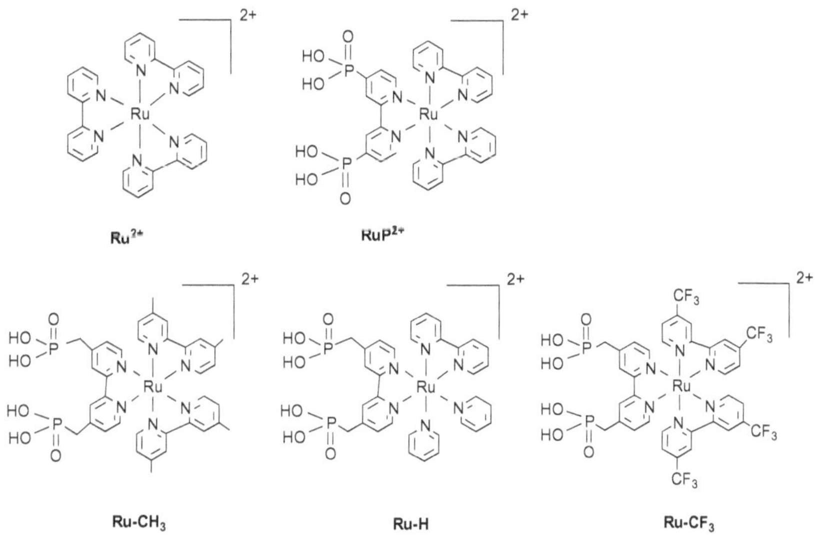

Fig. 11 Structures of **Ru^{2+}**, **RuP^{2+}**, **Ru-CH$_3$**, **Ru-H**, and **Ru-CF$_3$**.

Fig. 12 TEM images of the as-prepared NS-$K_4Nb_6O_{17}$. *Reproduced from Maeda, K., Eguchi, M., Youngblood, W.J., Mallouk, T.E. 2008. Niobium oxide nanoscrolls as building blocks for dye-sensitized hydrogen production from water under visible light irradiation. Chem. Mater. 20, 6770–6778 with permission.*

NS-$H_4Nb_6O_{17}$ was then applied as a building block for dye-sensitized H_2 evolution under visible light using ethylenediaminetetraacetic acid (EDTA) as an electron donor and Pt nanoparticles as catalysts for H_2 evolution. As expected, the negatively charged NS-$H_4Nb_6O_{17}$ binds cationic **Ru^{2+}**, allowing for electron injection from the excited state of the sensitizer and subsequent H_2 evolution without any chemical bond linkage between

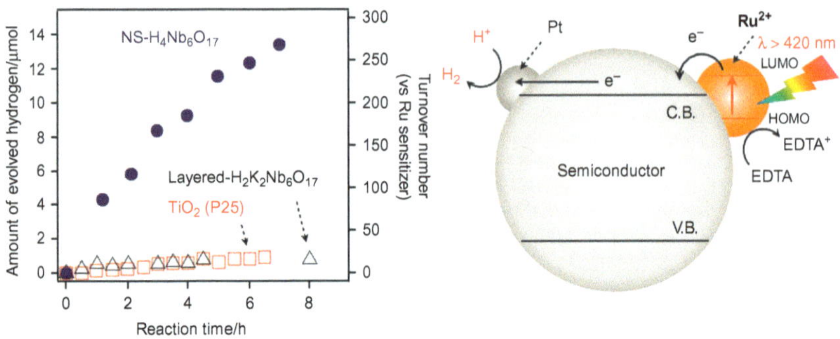

Fig. 13 Time courses of H_2 evolution from 0.1 wt% Pt-loaded semiconductors sensitized by Ru^{2+} with visible light ($\lambda > 420$ nm). Reaction conditions: catalyst, 5.0 mg; aqueous solution (2.0 mL) containing 10 mM EDTA and 50 µM Ru^{2+}; light source, xenon lamp (300 W) with a cutoff filter. *Reproduced from Maeda, K., Eguchi, M., Youngblood, W.J., Mallouk, T.E. 2008. Niobium oxide nanoscrolls as building blocks for dye-sensitized hydrogen production from water under visible light irradiation. Chem. Mater. 20, 6770–6778 with permission.*

the sensitizer and the nanoscroll surface. As shown in Fig. 13, the rate of visible light H_2 evolution in the nanoscroll-based system was 10 times higher than those of similarly sensitized P25 titania and the parent layered material, due primarily to the superior ability to bind **Ru^{2+}**. It was also found that restacked $HCa_2Nb_3O_{10}$ nanosheets achieved a similar functionality for the dye-sensitized H_2 evolution.

Choi et al. have examined the effects of anchoring groups on sensitized H_2 evolution using Ru(II) trisdiimine complexes. $Ru(bpy)_2(4,4'\text{-}(PO_3H_2)_2bpy)^{2+}$ (abbreviated as **RuP^{2+}**, see Fig. 11) was the most effective among the Ru-based sensitizers examined (Bae and Choi, 2006). It is thus of interest to compare electrostatic vs covalent anchoring of **Ru^{2+}** and **RuP^{2+}** in the nanosheet-based dye-sensitized H_2 evolution system. **RuP^{2+}**, which was anchored by a covalent linkage to the nanoscroll surface, functioned more efficiently than the electrostatically bound **Ru^{2+}** (Fig. 14), showing an AQY of ca. 25% at 450 ± 20 nm. The MLCT state luminescence of Ru(II) trisdiimine-type complexes is quenched in the presence of semiconductor particles as a result of electron injection into the conduction band. Luminescence spectroscopy was thus employed to study quenching of the MLCT excited state by NS-$H_4Nb_6O_{17}$ for **RuP^{2+}** and **Ru^{2+}**. As shown in Fig. 15, the luminescence intensity of the sensitizer decreased upon addition of NS-$H_4Nb_6O_{17}$ to the sensitizer solution in each case. The observed decrease in luminescence intensity is

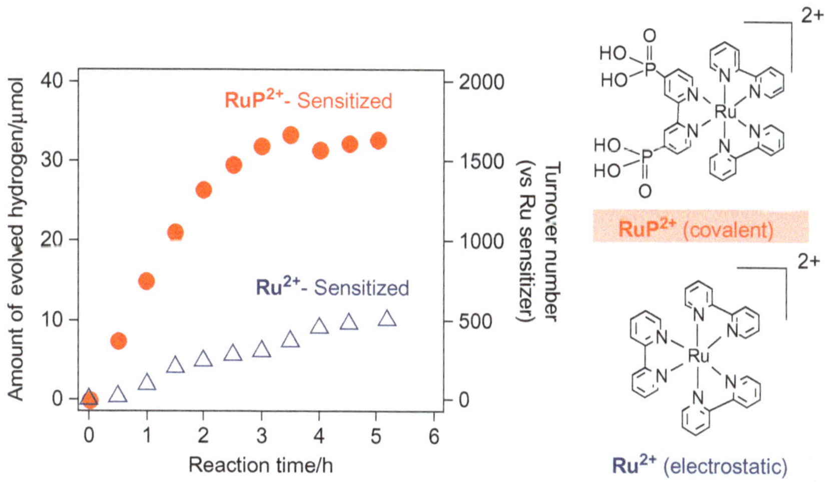

Fig. 14 Time courses of H_2 evolution from 0.3 wt% Pt-loaded NS-$H_4Nb_6O_{17}$ sensitized by **RuP^{2+}** or **Ru^{2+}** (8.0 μmol g^{-1}) with visible light ($\lambda > 420$ nm). Reaction conditions: catalyst, 5.0 mg; aqueous EDTA solution (0.01 M, 2.0 mL); light source, xenon lamp (300 W) with a cutoff filter. *Reproduced from Maeda, K., Eguchi, M., Youngblood, W.J., Mallouk, T.E. 2008. Niobium oxide nanoscrolls as building blocks for dye-sensitized hydrogen production from water under visible light irradiation. Chem. Mater. 20, 6770–6778 and Maeda, K., Eguchi, M., Lee, S.-H.A., Youngblood, W.J., Hata, H., Mallouk, T.E. 2009. Photocatalytic hydrogen evolution from hexaniobate nanoscrolls and calcium niobate nanosheets sensitized by ruthenium(II) bipyridyl complexes. J. Phys. Chem. C. 113, 7962–7969 with permission.*

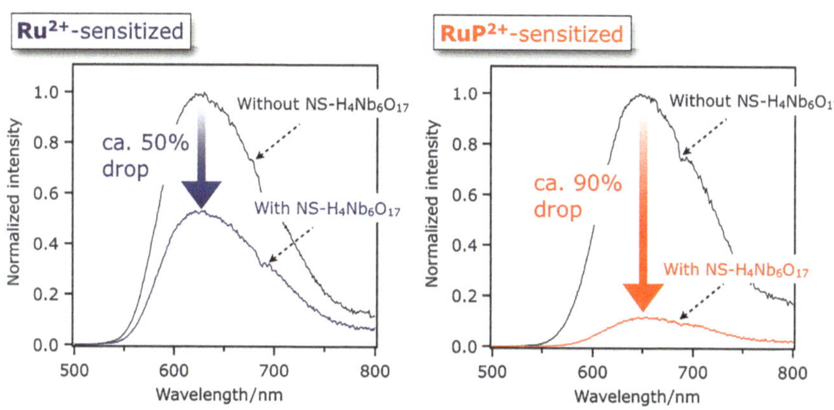

Fig. 15 Luminescence spectra for 0.3 wt% Pt-loaded NS-$H_4Nb_6O_{17}$ sensitized by **RuP^{2+}** or **Ru^{2+}** with 450 nm excitation. The spectra were acquired at room temperature under Ar atmosphere using aqueous EDTA solution (0.01 M, 2.0 mL) containing 0.3 wt% Pt-loaded NS-$H_4Nb_6O_{17}$ (5.0 mg) adsorbed with **RuP^{2+}** or **Ru^{2+}** (8.0 μmol g^{-1}). *Reproduced from Maeda, K., Eguchi, M., Lee, S.-H.A., Youngblood, W.J., Hata, H., Mallouk, T.E. 2009. Photocatalytic hydrogen evolution from hexaniobate nanoscrolls and calcium niobate nanosheets sensitized by ruthenium(II) bipyridyl complexes. J. Phys. Chem. C. 113, 7962–7969 with permission.*

mainly attributable to excited-state quenching via electron transfer to NS-$H_4Nb_6O_{17}$. The efficiency of quenching was higher for **RuP^{2+}** (ca. 90%) than for **Ru^{2+}** (ca. 50%). More efficient quenching in the **RuP^{2+}**-based system implies stronger electronic coupling between **RuP^{2+}** and NS-$H_4Nb_6O_{17}$ than obtained by electrostatic binding of **Ru^{2+}**. A primary requirement for efficient H_2 production by dye sensitization is strong adsorption of sensitizer molecules onto the surface of the metal oxide, because the dye excited state is typically too short lived to allow for diffusion of the molecule to the surface. The higher quenching efficiency observed in the **RuP^{2+}**-based system is consistent with its substantially higher rate of H_2 production.

6.2 Effect of Excited-State Potential of Ru(II) Sensitizers

It is one of the most advantageous features for metal complexes that HOMO and LUMO potentials can be easily controlled by introducing electron withdrawing or donating moiety. Electron injection from a Ru photosensitizer to the conduction band of nanosheet has a strong impact on H_2 evolution efficiency, as demonstrated by the comparative experiments using **Ru^{2+}** and **RuP^{2+}**. Thus, Ru photosensitizers with different LUMO potentials were applied to the H_2 evolution system using restacked $HCa_2Nb_3O_{10}$ nanosheets. Three different Ru(II) complexes (i.e., **Ru-CH_3**, **Ru-H**, and **Ru-CF_3**) were employed as visible light photosensitizers, and their structures are shown in Fig. 11. As the result, Pt/$HCa_2Nb_3O_{10}$ nanosheets modified with **Ru-CH_3** or **Ru-H** produced H_2 catalytically upon visible light (Fig. 16A). In contrast, the use of **Ru-CF_3** yielded a very low level of H_2 evolution.

As discussed earlier, electron injection efficiency from sensitizer to semiconductor has a large impact on H_2 evolution performance. Relative electron injection efficiencies were assessed by means of a photoelectrochemical technique using porous electrodes of $Ca_2Nb_3O_{10}^-$ nanosheet aggregates modified with a Ru(II) sensitizer. The photoelectrochemical measurements were conducted under intermittent visible light ($\lambda > 420$ nm) in an aqueous Na_2SO_4 solution (0.1 M) containing 10 mM EDTA. The $KCa_2Nb_3O_{10}$ electrode alone (without Ru) exhibited little photocurrent due to the large band gap (3.59 eV) that is unable to absorb visible light. However, electrodes that were sensitized by **Ru-CH_3** or **Ru-H** generated anodic photocurrent upon visible light, indicating that electron injection occurred from the excited-state Ru(II) sensitizers to the conduction

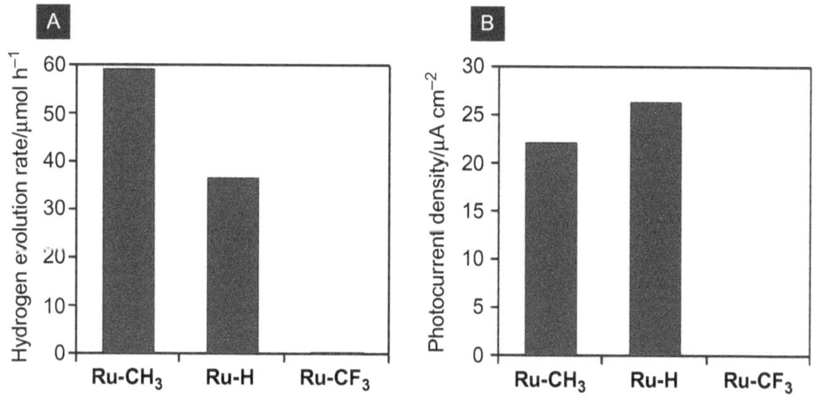

Fig. 16 (A) H_2 evolution rate on 0.5 wt% Pt-loaded $HCa_2Nb_3O_{10}$ nanosheets sensitized by different Ru(II) complexes (8.0 μmol g^{-1}) under visible light ($\lambda = 420$ nm). Reaction conditions: catalyst, 25 mg; aqueous solution containing 10 mM EDTA (100 mL); light source, xenon lamp (300 W) with a band-pass filter. (B) Photocurrent density (at −0.55 V vs Ag/AgCl) recorded using $KCa_2Nb_3O_{10}$ nanosheet electrodes sensitized by different Ru(II) complexes in aqueous 0.1 M Na_2SO_4 solution containing 10 mM EDTA under visible light ($\lambda > 420$ nm). *Reproduced from Maeda, K., Sahara, G., Eguchi, M., Ishitani, O. 2015. Hybrids of a ruthenium(II) polypyridyl complex and a metal oxide nanosheet for dye-sensitized hydrogen evolution with visible light: effects of the energy structure on photocatalytic activity. ACS Catal. 5, 1700–1707 with permission.*

band of $KCa_2Nb_3O_{10}$. The photocurrent generated at −0.55 V vs Ag/AgCl was summarized in Fig. 16B. Electrodes sensitized by **Ru-CH$_3$** or **Ru-H** exhibited similar performance, while the **Ru-CF$_3$** electrode did not show appreciable photocurrent even at more positive potential. Obviously, electrodes modified with "active" Ru complexes generated appreciable photocurrent, although the photocurrent generation was not proportional to the H_2 evolution activity.

The oxidation potential of the ^{3}MLCT excited state of the three Ru(II) complexes (E_{ox}*) shifted positively in the order of **Ru-CH$_3$**, **Ru-H**, and **Ru-CF$_3$**. The lowest performance of the **Ru-CF$_3$** systems (both in photocatalysis and photoelectrochemistry) could be explained in terms of the E_{ox}* value (E_{ox}* = −0.67 V vs Ag/AgNO$_3$), which is much more positive than those of **Ru-CH$_3$** and **Ru-H** (−1.18 and −1.08 V, respectively). The conduction band potential of $HCa_2Nb_3O_{10}$ nanosheets was estimated to be −1.13 V (vs Ag/AgNO$_3$), which is more positive than or nearly close to the E_{ox}* values of **Ru-CH$_3$** and **Ru-H**, allowing for electron injection from the ^{3}MLCT excited state of these sensitizers to restacked $HCa_2Nb_3O_{10}$ nanosheets and subsequent H_2 evolution.

6.3 Effect of Physicochemical Properties of Nanosheets

As indicated by the experiments using different Ru(II) sensitizers that have different E_{ox}* values, electron injection efficiency is definitely an important factor that determines the performance of dye-sensitized H_2 evolution. The efficiency of electron injection can be controlled by changing the conduction band potential of a semiconductor as well. Restacked nanosheets of $HCa_{2-x}Sr_xNb_3O_{10}$ and $HCa_2Nb_{3-y}Ta_yO_{10}$ having different conduction band potentials (as discussed earlier) were thus employed as a component of dye-sensitized H_2 evolution with a **Ru-CH₃** sensitizer. As shown in Fig. 17A, the sensitized $HCa_{2-x}Sr_xNb_3O_{10}$ nanosheets exhibited activity for visible-light H_2 evolution, but the H_2 evolution rate was decreased with increasing the concentration of Sr in $HCa_{2-x}Sr_xNb_3O_{10}$. Similarly, substitution of Nb in $HCa_2Nb_3O_{10}$ for Ta tended to yield an activity drop. Consequently, $HCa_2Nb_3O_{10}$ achieved the highest performance as the building block for H_2 evolution sensitized by **Ru-CH₃** with visible light among perovskite nanosheets examined.

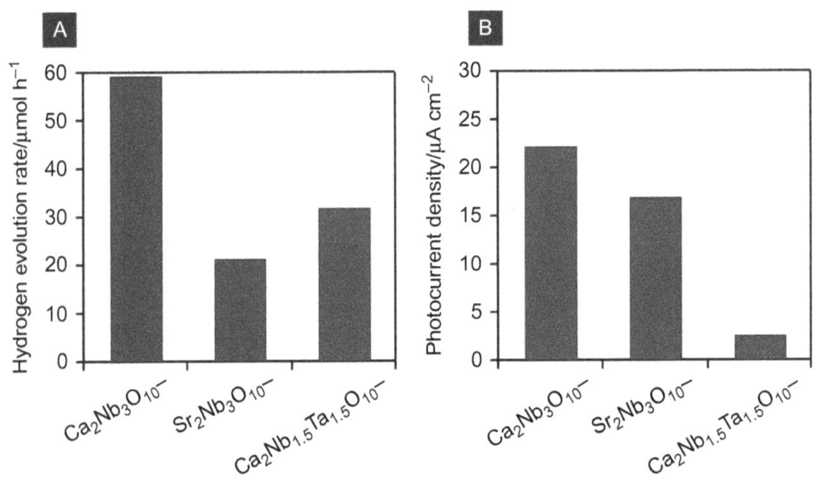

Fig. 17 (A) H_2 evolution rate on 0.5 wt% Pt-loaded nanosheets sensitized by **Ru-CH₃** (8.0 µmol g^{-1}) under visible light ($\lambda = 420$ nm). Reaction conditions: catalyst, 25 mg; aqueous solution containing 10 mM EDTA (100 mL); light source, xenon lamp (300 W) with a band-pass filter. (B) Photocurrent density (at −0.55 V vs Ag/AgCl) recorded using various nanosheet electrodes sensitized by **Ru-CH₃** in aqueous 0.1 M Na_2SO_4 solution containing 10 mM EDTA under visible light ($\lambda > 420$ nm). *Reproduced from Maeda, K., Sahara, G., Eguchi, M., Ishitani, O. 2015. Hybrids of a ruthenium(II) polypyridyl complex and a metal oxide nanosheet for dye-sensitized hydrogen evolution with visible light: effects of the energy structure on photocatalytic activity. ACS Catal. 5, 1700–1707 with permission.*

To elucidate the reason for the activity change with respect to nanosheet composition, similar photoelectrochemical measurements were conducted, which showed that the Sr-substituted material exhibited a similar level of photocurrent (~20 μA cm^{-2}) to each other (Fig. 17B). On the other hand, substitution of Nb for Ta resulted in a significant drop in photocurrent. The E_{CB} value of $HCa_{2-x}Sr_xNb_3O_{10}$ nanosheets shifts positively with an increase in the Sr content while that of $HCa_2Nb_{3-y}Ta_yO_{10}$ nanosheets shifts negatively with increasing the Ta concentration. The decrease in activity with an increase in the Ta content in $HCa_2Nb_{3-y}Ta_yO_{10}$ can be explained in terms of the reduced driving force for the electron injection, as observed in the photoelectrochemical measurements. Interestingly, however, the negatively shifted E_{CB} of Sr-rich $HCa_{2-x}Sr_xNb_3O_{10}$ has no positive impact on the electron injection event and the resulting H_2 evolution activity, in comparison to $HCa_2Nb_3O_{10}$. A plausible explanation for the activity drop observed in Sr-rich $HCa_{2-x}Sr_xNb_3O_{10}$ nanosheets is that the ability of conduction band electrons to reduce protons into H_2 is lowered due to the positive shift of the E_{CB} value. This idea could also explain the difference in H_2 evolution activity of $HCa_{2-x}Sr_xNb_3O_{10}$ nanosheets under band gap irradiation, as described in the previous section.

7. CONCLUSION AND FUTURE PERSPECTIVE

Photocatalytic properties of nanosheet was highlighted in this chapter. Nanosheet has enough potential as photocatalyst with many fascinating characters. For H_2 evolution using methanol as an electron donor, very high quantum efficiency was achieved by tuning conduction band potential. For O_2 evolution using IO_3^- as a reversible electron accepter, interestingly, O_2 evolution activity was induced by exfoliation of layered $KCa_2Nb_3O_{10}$. In addition, Pt nanocluster was deposited not only on the surface but also in the interlayer space of restacked $KCa_2Nb_3O_{10}$ nanosheet using simple electrostatic interaction between nanosheet and cationic Pt precursor, followed by thermal H_2 treatment. Surprisingly, small Pt particle was maintained even at higher reduction temperature, suggesting the usefulness of interlayer space for synthesizing small cocatalyst. The Pt-loaded $KCa_2Nb_3O_{10}$ demonstrated water splitting, and the activity was the highest among nanosheet-based photocatalyst. Finally, nanosheet was combined with Ru dye molecule as visible light absorbing moiety and visible light-driven H_2 evolution were achieved using EDTA as an electron donor. The performance was much higher than TiO_2 systems, suggesting the advantageous character of nanosheet as electron mediator. However, utilization of visible light, especially for nonsacrificial system remains a challenging topic for nanosheet. As bulk-type oxynitrides with perovskite structure, such as $ATaON_2$ (A = Ca, Sr, Ba) and $LaTiO_2N$, have been employed as visible-light-responsive photocatalysts (Maeda, 2011), development of new layered perovskite oxynitride materials that can utilize visible light as well as exfoliation to produce nanosheets would be of another interest. This is currently under investigation in our laboratory.

REFERENCES

Abe, R., 2010. Recent progress on photocatalytic and photoelectrochemical water splitting under visible light irradiation. J. Photochem. Photobiol. C Photchem. Rev. 11, 179–209.

Abe, R., Sayama, K., Arakawa, H., 2003. Significant effect of iodide addition on water splitting into H_2 and O_2 over Pt-loaded TiO_2 photocatalyst: suppression of backward reaction. Chem. Phys. Lett. 371, 360–364.

Allen, M.R., Thibert, A., Sabio, E.M., Browning, N.D., Larsen, D.S., Osterloh, F.E., 2010. Evolution of physical and photocatalytic properties in the layered titanates $A_2Ti_4O_9$(a = K, H) and in nanosheets derived by chemical exfoliation. Chem. Mater. 22, 1220–1228.

Bae, E., Choi, W., 2006. Effect of the anchoring group (carboxylate vs phosphonate) in Ru-complex-sensitized TiO_2 on hydrogen production under visible light. J. Phys. Chem. B 110, 14792–14799.

Domen, K., Kudo, A., Onishi, T., 1986. Mechanism of photocatalytic decomposition of water into H_2 and O_2 over NiO-$SrTiO_3$. J. Catal. 102, 92–98.

Ebina, Y., Sakai, N., Sasaki, T., 2005. Photocatalyst of lamellar aggregates of RuO_x-loaded perovskite nanosheets for overall water splitting. J. Phys. Chem. B 109, 17212–17216.

Ebina, Y., Akatsuka, K., Fukuda, K., Sasaki, T., 2012. Synthesis and in situ X-ray diffraction characterization of two-dimensional perovskite-type oxide colloids with a controlled molecular thickness. Chem. Mater. 24, 4201–4208.

Furlong, D.N., Wells, D., Sasse, H.F.W., 1986. Colloidal semiconductors in systems for the sacrificial photolysis of water: sensitization of TiO_2 by adsorption of ruthenium complexes. J. Phys. Chem. 90, 1107–1115.

Hata, H., Kudo, S., Kobayashi, Y., Mallouk, T.E., 2007. Intercalation of well-dispersed gold nanoparticles into layered oxide nanosheets through intercalation of a polyamine. J. Am. Chem. Soc. 129, 3064–3065.

Hata, H., Kobayashi, Y., Bojan, V., Youngblood, W.J., Mallouk, T.E., 2008. Direct deposition of trivalent rhodium hydroxide nanoparticles onto a semiconducting layered calcium niobate for photocatalytic hydrogen evolution. Nano Lett. 8, 794–799.

Higashi, M., Abe, R., Ishikawa, A., Takata, T., Ohtani, B., Domen, K., 2008. Z-scheme overall water splitting on modified–TaON photocatalysts under visible light ($\lambda < 500$ nm). Chem. Lett. 37, 138–139.

Ida, S., Ishihara, T., 2014. Recent progress in two-dimensional oxide photocatalysts for water splitting. J. Phys. Chem. Lett. 5, 2533–2542.

Kudo, A., Miseki, Y., 2009. Heterogeneous photocatalyst materials for water splitting. Chem. Soc. Rev. 38, 253–278.

Kudo, A., Sayama, K., Tanaka, A., Asakura, K., Domen, K., Maruya, K., Onishi, T., 1989. Nickel-loaded $K_4Nb_6O_{17}$ photocatalyst in the decomposition of H_2O into H_2 and O_2: structure and reaction mechanism. J. Catal. 120, 337–352.

Ma, R., Kobayashi, Y., Youngblood, W.J., Mallouk, T.E., 2008. Potassium niobate nanoscrolls incorporating rhodium hydroxide nanoparticles for photocatalytic hydrogen evolution. J. Mater. Chem. 18, 5982–5985.

Maeda, K., 2011. Photocatalytic water splitting using semiconductor particles: history and recent developments. J. Photochem. Photobiol. C Photchem. Rev. 12, 237–268.

Maeda, K., 2013. Z-scheme water splitting using two different semiconductor photocatalysts. ACS Catal. 3, 1486–1503.

Maeda, K., Eguchi, M., 2016. Structural effects of two-dimensional perovskite $Ca_2Nb_2TaO_{10}^-$ nanosheets for photocatalytic hydrogen evolution. Catal. Sci. Technol. 6, 1064–1069.

Maeda, K., Eguchi, M., Youngblood, W.J., Mallouk, T.E., 2008. Niobium oxide nanoscrolls as building blocks for dye-sensitized hydrogen production from water under visible light irradiation. Chem. Mater. 20, 6770–6778.

Maeda, K., Nishimura, N., Domen, K., 2009a. A precursor route to prepare tantalum (V) nitride nanoparticles with enhanced photocatalytic activity for hydrogen evolution under visible light. Appl. Catal. A. Gen. 370, 88–92.

Maeda, K., Eguchi, M., Youngblood, W.J., Mallouk, T.E., 2009b. Calcium niobate nanosheets prepared by the polymerized complex method as catalytic materials for photochemical hydrogen evolution. Chem. Mater. 21, 3611–3617.

Maeda, K., Eguchi, M., Lee, S.-H.A., Youngblood, W.J., Hata, H., Mallouk, T.E., 2009c. Photocatalytic hydrogen evolution from hexaniobate nanoscrolls and calcium niobate nanosheets sensitized by ruthenium(II) bipyridyl complexes. J. Phys. Chem. C 113, 7962–7969.

Maeda, K., Abe, R., Domen, K., 2011. Role and function of ruthenium species as promoters with TaON-based photocatalysts for oxygen evolution in two-step water splitting under visible light. J. Phys. Chem. C 115, 3057–3064.

Maeda, K., Eguchi, M., Oshima, T., 2014. Perovskite oxide nanosheets with tunable band-edge potentials and high photocatalytic hydrogen-evolution activity. Angew. Chem. Int. Ed. 53, 13164–13168.

Maeda, K., Sahara, G., Eguchi, M., Ishitani, O., 2015. Hybrids of a ruthenium(II) polypyridyl complex and a metal oxide nanosheet for dye-sensitized hydrogen evolution with visible light: effects of the energy structure on photocatalytic activity. ACS Catal. 5, 1700–1707.

Miseki, Y., Kato, H., Kudo, A., 2009. Water splitting into H_2 and O_2 over niobate and titanate photocatalysts with (111) plane-type layered perovskite structure. Energy Environ. Sci. 2, 306–314.

Oshima, T., Ishitani, O., Maeda, K., 2014. Non-sacrificial water photo-oxidation activity of lamellar calcium niobate induced by exfoliation. Adv. Mater. Interfaces 1, 1400131.

Oshima, T., Lu, D., Ishitani, O., Maeda, K., 2015. Intercalation of highly dispersed metal nanoclusters into a layered metal oxide for photocatalytic overall water splitting. Angew. Chem. Int. Ed. 54, 2698–2702.

Oshima, T., Eguchi, M., Maeda, K., 2016a. Photocatalytic water oxidation over metal oxide nanosheets having a three-layer perovskite structure. ChemSusChem 9, 396–402.

Oshima, T., Lu, D., Maeda, K., 2016b. Preparation of Pt-intercalated $KCa_2Nb_3O_{10}$ nanosheets and their photocatalytic activity for overall water splitting. ChemNanoMat 2, 748–755.

Sabio, E.M., Chi, M., Browning, N.D., Osterloh, F.E., 2010. Charge separation in a niobate nanosheet photocatalyst studied with photochemical labeling. Langmuir 26, 7254–7261.

Sarahan, M.C., Carroll, E.C., Allen, M., Larsen, D.S., Browning, N.D., Osterloh, F.E., 2008. $K_4Nb_6O_{17}$-derived photocatalysts for hydrogen evolution from water: nanoscrolls versus nanosheets. J. Solid State Chem. 181, 1678–1683.

Scaife, D.E., 1980. Oxide semiconductors in photoelectrochemical conversion of solar energy. Sol. Energy 25, 41–54.

Suzuki, H., Tomita, O., Higashi, M., Abe, R., 2015. Two-step photocatalytic water splitting into H_2 and O_2 using layered metal oxide $KCa_2Nb_3O_{10}$ and its derivatives as O_2-evolving photocatalysts with IO_3^-/I^- or Fe^{3+}/Fe^{2+} redox mediator. Catal. Sci. Technol. 5, 2640–2648.

Takata, T., Shinohara, K., Tanaka, A., Hara, M., Kondo, J.N., Domen, K., 1997. A highly active photocatalyst for overall water splitting with a hydrated layered perovskite structure. J. Photochem. Photobiol. A Chem. 106, 45–49.

Waller, M.R., Townsend, T.K., Zhao, J., Sabio, E.M., Chamousis, R.L., Browning, N.D., Osterloh, F.E., 2012. Single-crystal tungsten oxide nanosheets: photochemical water oxidation in the quantum confinement regime. Chem. Mater. 24, 698–704.

CHAPTER EIGHT

Nanostructured Photoelectrodes via Template-Assisted Fabrication

Rowena Yew, Siva Krishna Karuturi[1], Hark Hoe Tan, Chennupati Jagadish
Department of Electronic Materials Engineering, Research School of Physics and Engineering, The Australian National University, Canberra, ACT, Australia
[1]Corresponding author: e-mail address: Siva.Karuturi@anu.edu.au

Contents

Photoelectrochemical (PEC) cell is considered an attractive means of harvesting solar energy and producing storable chemical fuels. A photoelectrode is critical to the light harvesting and conversion efficiency of a PEC cell, wherein incident photons are absorbed, electron–hole pairs are generated, and subsequent charge transfer and catalytic conversion occur. Particularly, photoelectrodes with nanostructured morphology contribute greatly toward the performance of PEC solar energy conversion (Gratzel, 2001; Walter et al., 2010). A nanostructured electrode possesses several favorable intrinsic characteristics for efficient light harvesting, such as accessibility to a greater surface area, direct charge transfer pathways for reduced charge recombination, and strong light-trapping ability that promotes light harvesting by confining light within the photoelectrode. Thus the research community has placed enormous efforts toward the development of nanostructured semiconductor photoelectrodes for PEC solar energy conversion applications (Chen et al., 2012; van de Krol et al., 2008).

Among various approaches to make to nanostructured materials, the template-assisted method offers a simple and versatile route to produce

Semiconductors and Semimetals, Volume 97
ISSN 0080-8784
http://dx.doi.org/10.1016/bs.semsem.2017.04.002

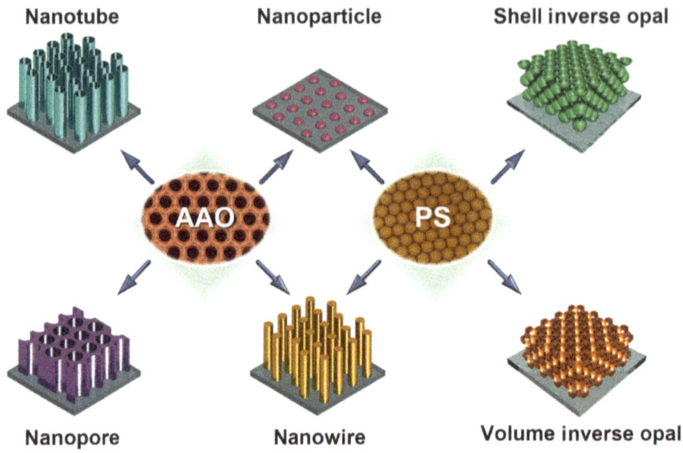

Fig. 1 Schematic illustration of template-assisted fabrication of nanostructured materials based on anodized aluminum oxide (AAO) and self-assembled opal templates. *Reproduced with permission from Zhao, H.P., et al., 2015. Template-directed construction of nanostructure arrays for highly-efficient energy storage and conversion. Nano Energy 13, 790–813. Copyright 2015 Elsevier.*

nanostructured materials of predefined morphology and optoelectronic properties. Due to their highly ordered and tunable porous structures, two types of templates, namely, self-assembled opals and anodized alumina oxide (AAO) have become widely popular as candidates for nanostructure fabrication (see Fig. 1). This chapter introduces photonic properties of opal structures; methods used to develop the templates, material infiltration, replication techniques, and summarizes the performance of as-fabricated nanostructures in photocatalytic solar energy conversion.

1. OPAL TEMPLATE-ASSISTED NANOSTRUCTURES

Due to their highly periodic arrangement, self-assembled opal structures can be considered photonic crystals. Photonic crystals are periodic dielectric structures that are designed to form the energy band structure for photons, which either allows or forbids the propagation of electromagnetic waves of certain frequency ranges, making them ideal for light-harvesting applications (Maka et al., 2003). They were first proposed by Bykov (1972), followed by Yablonovich and John in 1987. Yablonovich suggested that electron–hole radiative recombination will be severely inhibited for a three-dimensional periodic dielectric structure with an electromagnetic band gap overlapping the electronic band gap (Yablonovich, 1987), while John showed that random refractive index variation causes strong localization of photons to occur in periodic structures of high dielectric contrast (John, 1987).

In a photonic crystal with different dielectric material, light is scattered and/or diffracted, producing a band of forbidden frequencies where interference of scattered waves is destructive in all directions (see Fig. 2). Light propagation is inhibited within this region, creating a photonic band gap (PBG). The greater the refractive index contrast between the materials, the larger the PBG becomes. When light propagation is totally prohibited in a photonic crystal, a full PBG is observed. Thus, the PBG of photonic crystals is analogous to the electronic band gap in semiconductors, in that photons experience a periodic potential, but due to dielectric constants of different materials instead of the difference in potential energy of electrons (Armstrong and O'Dwyer, 2015). Photonic crystals are attractive not only for its PBG properties but also because the PBG can be tuned to specific frequencies for use in applications requiring visible and near infra-red regions of the spectrum (Maka et al., 2003).

To fully utilize the properties of photonic crystals, three-dimensional photonic crystals need to be fabricated. One of the most commonly used approaches is the formation of inverse opals through self-assembly. Inverse opals are highly ordered macroporous materials consisting of a face-centered

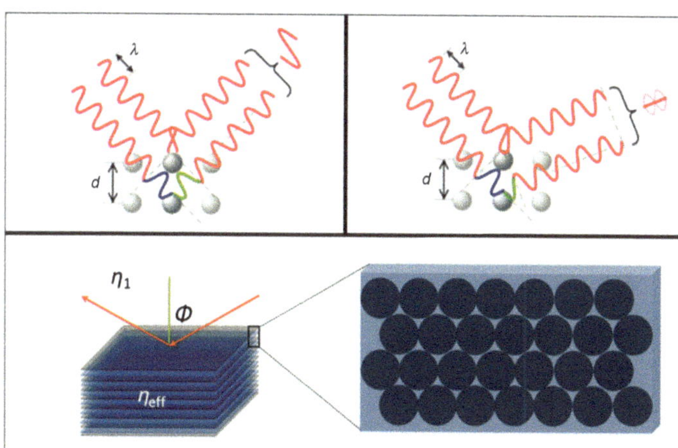

Fig. 2 Schematic illustration of how light interacts with periodic material of an effective refractive index and how light is scattered by the sphere planes. *Reproduced with permission from Armstrong, E., O'Dwyer, C., 2015. Artificial opal photonic crystals and inverse opal structures—fundamentals and applications from optics to energy storage. J. Mater. Chem. C 3(24), 6109–6143. Copyright 2015 Royal Society of Chemistry.*

cubic crystal, formed through the infiltration of a high dielectric material precursor into a sacrificial opal template, and followed by the removal of the spheres. A complete PBG can be achieved with photocatalytic material of sufficiently high refractive index. The structure and symmetry of 3D photonic crystals can be controlled through template design using artificial opal growth by colloidal self-assembly (Armstrong and O'Dwyer, 2015).

Inverse opals are ideal for photocatalysis due to their inherent structural and physical properties such as a large surface area, favorable charge transport properties, multiple scattering within the structure that improves light harvesting, and a tunable PBG that contributes positively to the absorption of light in the visible and near infra-red spectrum (Zhou et al., 2014). Another factor leading to improved photocatalytic efficiency is the "slow photon" effect. Slow photons are named as such because at the wavelengths corresponding to the edges of stop-bands, photons propagate with a strongly reduced group velocity. If the energy of slow photons overlaps with the absorbance of the material, the absorption is enhanced as a result of the increased effective optical path length (Liu et al., 2010). Chemical catalytic properties of inverse opal can be tuned by selecting an appropriate catalytically active material for templating or template infiltration. In the forthcoming sections, we will discuss some common catalytic materials such as TiO_2, Fe_2O_3, $BiVO_4$, and cadmium sulfide (CdS) used in inverse opal-based photocatalysis and the resulting catalytic performance.

1.1 Synthesis of Monodispersed Particles

Self-assembled opal templates are synthesized using silica spheres and polymer colloids, namely, polymethyl methacrylate (PMMA) and polystyrene (PS) spheres. Polymer colloid spheres are preferred over silica spheres because silica spheres are removed by wet chemical etching, whereas polymer colloid spheres can be removed easily by several methods, such as solvent removal, plasma treatment, or annealing at high temperature (Armstrong and O'Dwyer, 2015). Therefore, the choice of spheres used depends on the guest material and synthesis method. Polymer colloid spheres can be either synthesized by emulsion polymerization or purchased commercially. The most commonly employed emulsion polymerization method is carried out without the emulsifier, which is useful for the preparation of polymer colloids with narrow particle size distribution and well-characterized surface properties (Sharifi-Sanjani et al., 2004). Colloidal particles are synthesized by adding a monomer—styrene or methyl methacrylate, to distilled water and an initiator—potassium persulfate, ammonium persulfate, or azobisisobutyronitrile, which will initiate the polymerization process in a system purged with nitrogen. The system is kept at a constant temperature ranging from 50°C to 70°C for between 1 and 12h. The polymerization is then terminated by cooling the polymer colloid to room temperature and stirring it. The resulting PMMA spheres are washed twice with water by centrifugation (19,000rpm, 20min) and redispersed in water. Colloid particle sizes can be easily tuned by changing the size of the monomer used.

While this is the standard procedure for emulsion polymerization, various studies have applied modifications to their polymerization methods. Zhang et al. used comonomers potassium bicarbonate (Sigma) buffer and sodium styrene sulfonate (Aldrich), which add stability to the polymer colloids (Zhang et al., 2013). Yoon et al. used polyvinylpyrrolidone (PVP 40K) as a stabilizer for polymer colloids. Heating and reaction times for the polymer colloid suspension also varied from study to study. Stirring of the suspension is usually carried out throughout the reaction but some studies only perform vigorous stirring after polymerization has been achieved (Yoon et al., 2014).

1.2 Opal Template Assembly

Evaporation-assisted self-assembly via vertical deposition, as seen in Figs. 3 and 4, is a self-assembly method driven by lateral capillary forces, which allows for controlled and uniform deposition of the colloidal particles, by organizing them into multilayered films of highly ordered and closely packed spheres. This method is also known as the Colvin method (Jiang et al., 1999). The thickness of the film can be determined by the size of the spheres and concentration of the solution, making it the preferred method to obtain opal templates. However, the process is highly sensitive to the substrate used, environmental conditions and evaporation rate. For larger size spheres, a longer time is required to achieve well-ordered opal templates (Armstrong and O'Dwyer, 2015).

Lebrun et al. reported that substrate surface roughness is critical to the successful deposition of opals. Substrates such as glass slides, fluorine-doped tin oxide (FTO), indium-doped tin oxide, silicon, or substrates with surface roughness less than 30% of the opal diameter are preferred for opal template assembly (Lebrun et al., 2014). In addition to surface roughness, it is also important for the substrate surface to be hydrophilic to ensure uniform

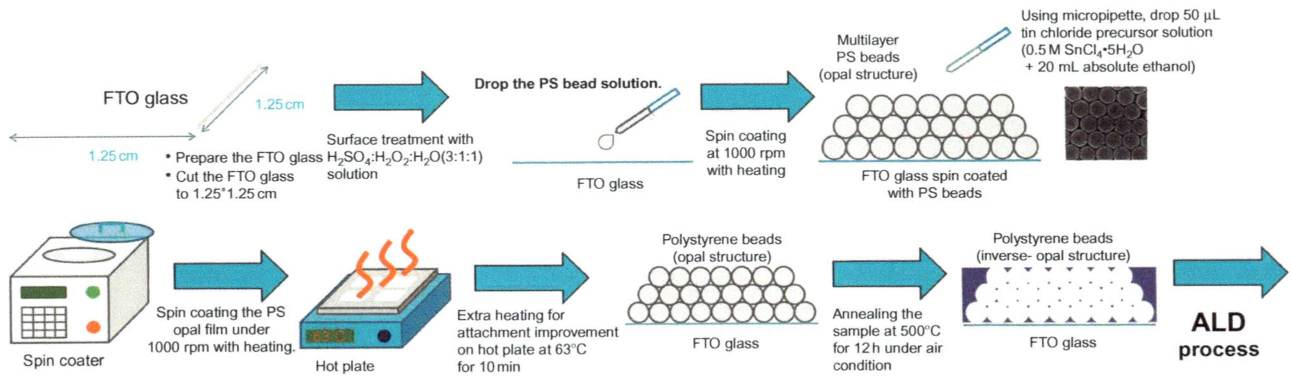

Fig. 3 Schematic illustration of opal template fabrication. *Reproduced with permission from Gun, Y., et al., 2015. Joint effects of photoactive TiO_2 and fluoride-doping on SnO_2 inverse opal nanoarchitecture for solar water splitting. ACS Appl. Mater. Interfaces 7(36), 20292–20303. Copyright 2015 American Chemical Society.*

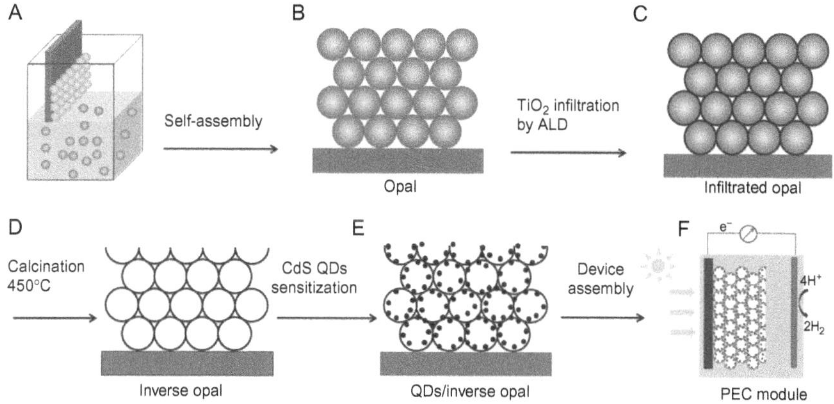

Fig. 4 Schematic illustration of inverse opal-based photoelectrode fabrication using vertical self-assembly and ALD infiltration techniques. *Reproduced with permission from Cheng, C., et al., 2012. Quantum-dot-sensitized TiO_2 inverse opals for photoelectrochemical hydrogen generation. Small 8(1), 37–42. Copyright 2012 Wiley.*

deposition of colloidal particles onto the substrate. Therefore, surface treatment of substrate is not limited to simply cleaning with ethanol, acetone, and deionized water but it is also necessary that a piranha solution of sulfuric acid (H_2SO_4) and hydrogen peroxide (H_2O_2) is used.

When preparing the aqueous solution, the ratio of water to the solution of monodispersed spheres varies from 5 to 10 wt%. The substrate is held vertically in a vial containing aqueous solution of the prepared or commercially purchased monodispersed spheres in an oven with temperature ranging from 50°C to 90°C. The aqueous solution is left to evaporate at a controlled rate, but oven temperatures cannot exceed 100°C since it would lead to the boiling of the aqueous solution. The evaporation rate is dependent on the temperature of the oven, and it should be noted that at lower rates of evaporation, the presence of moisture prevents cracking of the opal template (Zhou et al., 2014). When the aqueous solution is left to evaporate at room temperature, the evaporation rate is uncontrolled and humidity affects the quality of the opal templates. As the water evaporates and as the meniscus sweeps down the substrate, lateral capillary forces will induce ordering of the spheres on the substrate (Zhang et al., 2011).

Even though the above method is relatively straightforward, there are several limiting factors—quality of the opals is sensitive to substrate and environmental conditions, and also to the settling of the colloidal particles over time. Some studies have employed a modified method to better control the deposition of opals and to enhance the surface properties of opal templates.

Hatton et al. used an oven on a pneumatic vibration-free table as it has been noted that any vibration of the oven may cause nonuniform deposition of the opals (Hatton et al., 2010). On the other hand, Zhang et al. treated the opal templates in oxygen plasma for 90 s to make the surface hydrophilic (Zhang et al., 2013).

Dip coating is also a self-assembly method but the difference between this method and the evaporation-assisted self-assembly via vertical deposition method is that in the former, the substrate is withdrawn from the solution instead of allowing it to remain in the solution as it evaporates. Dip coating has potential for large-scale production as it provides the desired control over sample assembly. Control of the withdrawal rate, concentration of spheres, and evaporation rate, as well as the provision of several options to perfect sample quality, are advantages afforded by the dip-coating method which are required for industry applications (Armstrong and O'Dwyer, 2015). As with the evaporation-assisted self-assembly via vertical deposition method, limiting factors are the settling of the colloidal solution over time and the greater length of time required for depositing larger colloidal particles.

Kondofersky et al. employed ultraslow dip coating at the rate of $0.45\,\mathrm{mm\,min^{-1}}$ to deposit PMMA spheres onto FTO substrates. The opal films were dried in nitrogen and heated to 80°C to increase the adhesion between the PMMA spheres (Kondofersky et al., 2015). Abel et al. obtained a hexagonal closely -packed monolayer of silica microspheres deposited onto the substrate by dip coating. Substrates were immersed for 3 min and withdrawn at a rate of $120\,\mathrm{mm\,min^{-1}}$. Immersion and withdrawal were performed at 40% humidity (Abel et al., 2015).

Spin-coating colloidal particles onto a substrate is another self-assembly method that has been used to obtain opal templates. A drop of aqueous solution is spread across the substrate, which is then rotated at high speed by a spin coater, usually at greater than 1000 rpm, in order to achieve high quality and uniform films. Centripetal force together with surface tension of the solution spread particles evenly onto the substrate. Water from the solution will evaporate due to the air flow generated from rotation, leaving only colloidal particles on the substrate. The thickness of the film is determined by the number of times it is spin coated. After spin coating, samples are placed on a hot plate, to promote adhesion of the assembled spheres, as well as to remove impurities. Spin coating not only produces a highly consistent thin and uniform opal template (film), but it is also a faster process when compared to the evaporation-assisted method. However, since only a single substrate can be coated at a time, it is inefficient for large-scale production during the rotation of the substrate. Also, a large amount of the aqueous solution would be pushed off to the side and would end up being wasted.

1.3 Infiltration Methods and Template Removal

Infiltration of opal templates with catalytic material has been achieved by various methods and can be broadly classified into solution-based growth methods and physical deposition methods. Solution-growth methods include solution–gelation (sol–gel) chemistry, dip coating, spin coating, and electrochemical deposition, while physical deposition methods include atomic layer deposition (ALD), chemical vapor deposition (CVD), pulsed laser deposition (PLD), and physical vapor deposition. When multiple guest materials need to be infiltrated, a combination of infiltration techniques can be used. Fig. 4 schematically illustrates the fabrication of TiO_2/CdS inverse opal photoelectrode using the ALD technique. Not all the infiltration methods that are discussed here allow for complete infiltration of the opal template. For some methods like sol–gel chemistry, high filling fraction of opal templates can be achieved by coupling it with other methods.

The sol–gel chemistry method is the basis of solution-growth methods. Sol–gel chemistry is the preparation of inorganic polymers from solution through a transformation of liquid precursors to a solution, to a colloidal suspension of polymer particles, and finally to a network structure called a "gel"—the gelation state is a nonfluid 3D network that extends through the fluid phase (Brinker et al., 1990). Due to its simplicity and ease of fabrication, the sol–gel chemistry method is very attractive, but this method lacks the quality and precision of physical deposition methods. Voids within the opal templates are infiltrated with sol–gel precursor solution either through capillary forces by using the dip-coating method, or centrifugal force using the spin-coating method, or an electric field using the electrochemical deposition method (Galusha et al., 2008). Cracking of the opal structure that occurs during the drying of the templates is unavoidable and this is aggravated by opal defects. However, it can be mitigated by employing various assembly methods such as dip-coating, spin-coating, and electrochemical deposition (Armstrong et al., 2015). Electrochemical deposition, as seen in Fig. 5, is a potentiostatic or multipotential step method which uses a three-electrode cell- an opal-coated template substrate as the

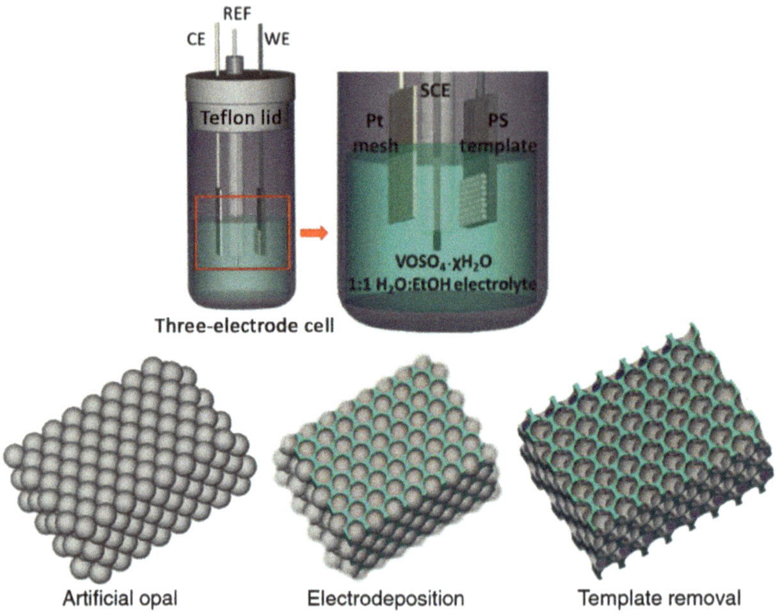

Fig. 5 Schematic diagram of a three-electrode cell used for the electrochemical deposition of vanadium oxide onto 3D opal template of PS spheres. *Reproduced with permission from Armstrong, E., et al., 2015. 3D vanadium oxide inverse opal growth by electrodeposition. J. Electrochem. Soc. 162(14), D605–D612. Copyright 2015 Journal of the Electrochemical Society.*

working electrode, platinum as the counter electrode, and a reference electrode in the precursor solution. Electrochemical deposition at room temperature occurs in a two-step electrode reaction while a constant potential is applied across the reference electrode (Fu et al., 2006; Kim et al., 2011).

Achieving both high filling fractions of the opal template and good luminescence properties of the material has proven difficult. With regards to this, CVD methods have been shown to be advantageous over solution-growth methods, since it affords a high degree of pore filling and controlled deposition. In particular, ALD has been shown to achieve close to theoretical maximum filling (Karuturi et al., 2010, 2011; Liu et al., 2011). ALD is a modified CVD growth method in which a binary synthesis reaction is split into two self-limiting surface reactions, by separate sequential exposures of the substrate to chemical precursors. It has evolved as a unique tool for nanotechnology with atomic level control of three-dimensional (3D) conformity and homogeneity. The resulting surface control and monolayer-by-monolayer growth sequence allows for the formation of uniform and defect-free films that are conformal to the substrate shape. Functional properties of resultant nanostructures, such as optical properties of photonic crystals, are critically influenced by the depth uniformity of ALD deposition. Highly precise material infiltration of the 3D colloidal crystal template can significantly improve the robustness and reliability of the template and provide greater control over functional properties such as refractive index contrast, that is required to create a PBG (Karuturi et al., 2010; Scharrer et al., 2005). Karuturi et al. in their studies of the kinetics of stop-flow ALD process for high aspect ratio template achieved infiltration close to the theoretical maximum, especially for high aspect ratio nanostructures such as opal templates. More importantly, the infiltration to bottommost layers of the opal structures is close to 96% of the maximum possible filling (Karuturi et al., 2010).

Juarez et al. employed a modified metal–organic chemical vapor deposition (MOCVD) method to infiltrate PS opal films on silicon substrates to obtain ZnO inverse opals. In the MOCVD method, after the infiltration of catalytic material, inverse opals are obtained by heating templates to 400–500°C for between 30 min and 2 h, to remove the opal template. By increasing the number of CVD cycles, 85% pore volume infiltration was achieved (Juárez et al., 2005). PLD has been used to infiltrate photocatalytic material on the opal template but it has been found that the material does not infiltrate deeply enough into the opal template structure, forming only on the surface layer of spheres, thus producing a monolayer of hollow hemispheres upon the removal of the opal template (Kim et al., 2006).

Several recent studies have employed the coassembly of polymer colloids together with the precursor solution of the intended infiltrated material, to obtain inverse opals in a single step. Ling et al. formulated a uniform mixture of monodispersed PS spheres and CdS quantum dots (QDs) dispersed in distilled water as a precursor solution, with FTO substrates inserted vertically into the mixture. Upon solvent (water) evaporation, the PS spheres and CdS QDs were coassembled on FTO substrates and the substrates were subsequently annealed at 400°C in air for 1 h to remove the opal templates, obtaining CdS inverse opals in a single step (Ling et al., 2014).

Inverse opals are photonic crystals with periodic modulation of refractive index enabled by the infiltration of material with high refractive index. Thus the removal of the opal templates increases the index contrast between the catalytic material and periodic void (air), producing the desired PBG, if not a full PBG (index contrast >2.8) (Armstrong and O'Dwyer, 2015). Removing the opal templates also creates a 3D periodic macroporous structure, which not only enhances light–matter interactions but also reduces the possibility of hole–electron recombination, a crucial step in photocatalysis (Karuturi et al., 2012a). As infiltrated catalytic materials are in the amorphous phase and show poor conductivity, calcinating the templates at high temperatures (>400°C) will convert the catalytic material from amorphous to crystalline phase and in that process, the PS spheres will also be removed (Karuturi et al., 2012b).

1.4 Photocatalytic Performance of Opal Template-Based Photoelectrodes

The early work of Fujishima and Honda (1972) on the generation of hydrogen from titanium dioxide (a photoanode) for photocatalysis has led to the extensive investigation of TiO_2 in applications for photocatalysis. TiO_2 has suitable band-edge positions (see Fig. 6) for oxygen and hydrogen evolution reactions and strong optical absorption. It is also chemically stable, highly resistant to photocorrosion, and inexpensive. Though attractive as a material for photocatalysis, TiO_2 has a large band gap of 3.2 eV, which limits light

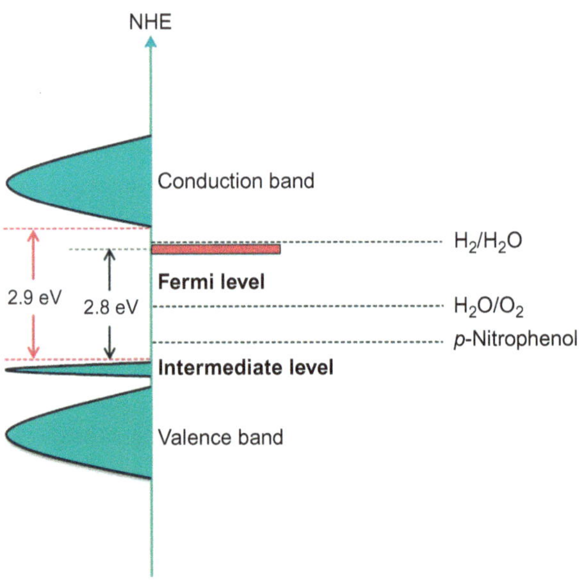

Fig. 6 Relative energy diagram of mc-TiO_2 inverse opal with respect to the reaction potentials. *Reproduced with permission from Quan, L.N., et al., 2014. Soft-template-carbonization route to highly textured mesoporous carbon-TiO_2 inverse opals for efficient photocatalytic and photoelectrochemical applications. Phys. Chem. Chem. Phys. 16(19), 9023–9030. Copyright 2014 Royal Society of Chemistry.*

absorption to the UV spectrum and it has fast electron–hole recombination due to the high density of trap states.

Liu et al. stated that titanium dioxide inverse opals exhibited better photonic properties with an absorption peak at 256 nm (Liu et al., 2010). To broaden the range of light-harvesting effects into the UV–vis range, Kim et al. balanced light-trapping effects and surface area requirements by designing a surface-textured TiO_2 inverse opal (st-TIO) structure on a few hundred nanometer scale. Gold nanoparticles were introduced to the st-TIO, which facilitated surface plasmon-enhancement of hydrogen generation, out-performing the P-25 photoelectrode by 2.58 times (Kim et al., 2013). As seen in Fig. 6, Quan et al. incorporated carbon moieties into TiO_2 mesoporous inverse opals to improve electron transport, increase visible light absorption, and decrease band gap energy, obtaining 0.087% photoelectrode efficiency for this new class of hybrid carbon-based, hierarchical inverse opal structure (Quan et al., 2014). To overcome the large band gap of TiO_2, Gun et al. created a core–shell structure between TiO_2 and fluorine-doped SnO_2, which exhibited good electron mobility and a conduction band lower than TiO_2, as seen in Figs. 7 and 8. Photocurrent density was significantly enhanced when TiO_2 was used as a shell for SnO_2 inverse opals, compared to bare SnO_2 inverse opals (Gun et al., 2015).

To overcome the limited solar-light absorption by TiO_2, sensitized TiO_2 inverse opal nanostructures with narrow-band gap semiconductor QDs such as CdS and CdSe have been investigated. Cheng et al. demonstrated

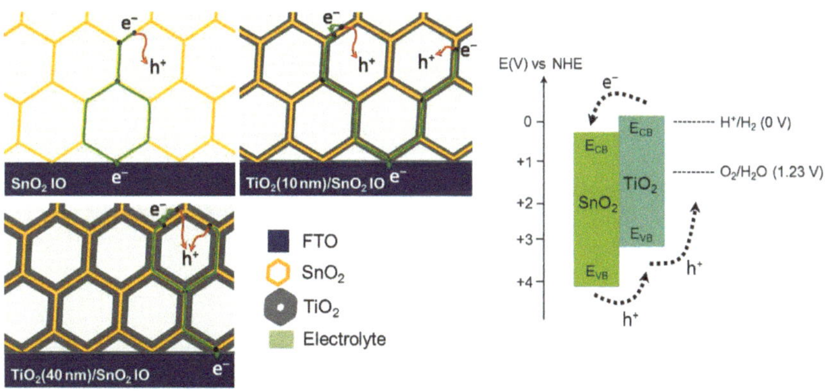

Fig. 7 Charge transfer events between SnO_2 IO and TiO_2 layer, and its energetic band diagram at the interface between SnO_2 IO and TiO_2 layer. *Reproduced with permission from Gun, Y., et al., 2015. Joint effects of photoactive TiO_2 and fluoride-doping on SnO_2 inverse opal nanoarchitecture for solar water splitting. ACS Appl. Mater. Interfaces 7(36), 20292–20303. Copyright 2015 American Chemical Society.*

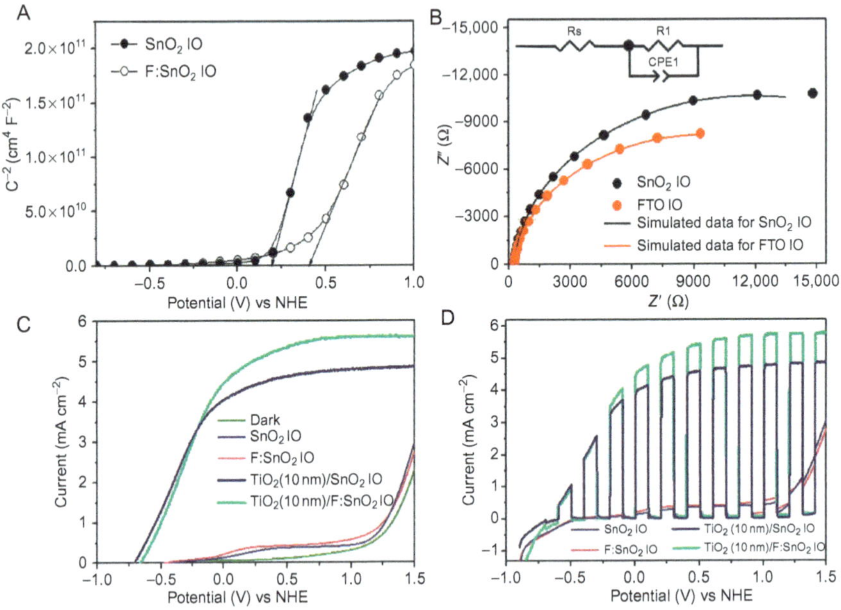

Fig. 8 (A) Mott–Schottky plot. (B) EIS spectra under 1 sun condition. (C) Linear sweep voltammetry (LSV). (D) Chopped LSV. *Reproduced with permission from Gun, Y., et al., 2015. Joint effects of photoactive TiO_2 and fluoride-doping on SnO_2 inverse opal nanoarchitecture for solar water splitting. ACS Appl. Mater. Interfaces 7(36), 20292–20303. Copyright 2015 American Chemical Society.*

nanoarchitectured electrode design (see Fig. 4) through the use of CdS QDs-sensitized, optically and electrically active TiO_2 inverse opals for PEC hydrogen production (Cheng et al., 2012). It was found that the highly ordered and percolated 3D pore structure provides efficient and fast electron-transport pathways. A photocurrent density of 4.84 mA cm^{-2} has been achieved for CdS/TiO_2 inverse opals as photoanode at 0 V vs Ag/AgCl bias under simulated solar-light illumination. This is 200 times the photocurrent realized using bare TiO_2 inverse opal photoanode. They further demonstrated that the 3D inverse opal structures can be combined with 1D ZnO nanowires to improve the specific surface area and enhance light-trapping ability (Karuturi et al., 2012a,b). Photoelectrode-based nanobushes structures (see Fig. 9) showed significantly improved light-scattering and light-harvesting ability in comparison to the conventional inverse opal-based photoelectrode.

CdS is an important optoelectronic material which has been extensively used as a photocatalyst in solar energy conversion applications for water

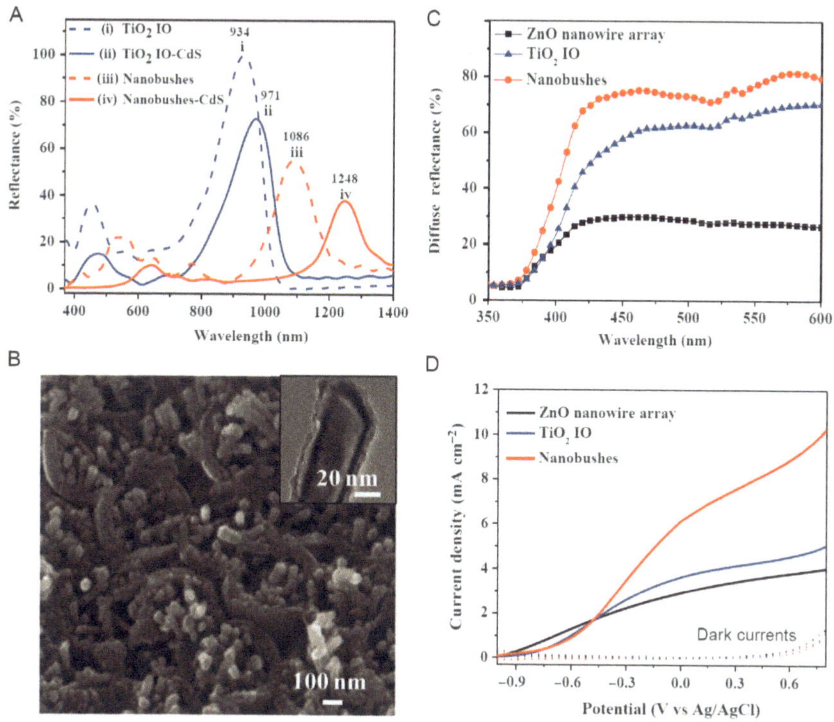

Fig. 9 CdS sensitized TiO_2/ZnO nanobushes. (A) Specular reflectance spectra showing the photonic band gap shift. (B) SEM image of the CdS sensitized TiO_2/ZnO nanobushes structure. (C) Diffuse reflectance spectra. (D) IV curves of the photoelectrodes. *Reproduced with permission from Karuturi, S.K., et al., 2012a. A novel photoanode with three-dimensionally, hierarchically ordered nanobushes for highly efficient photoelectrochemical cells. Adv. Mater. 24(30), 4157–4162. Copyright 2012 Wiley.*

splitting. Ling et al. showed that a 3D inverse opal structure of CdS exhibited high light-harvesting efficiency, a larger interface area for charge separation, higher conductivity for carrier transport and improved electrical conductivity (Ling et al., 2014). Luo et al. investigated CdSe sensitized TiO_2 inverse opal photoelectrodes to extend solar energy harvesting to the entire visible spectrum. A method of obtaining conformal CdSe films in 3D inverse opal nanostructures was proposed using a combination of ALD and ion exchange reaction processes. This photoelectrode system based on CdSe sensitized TiO_2 inverse opal nanostructures showed a photocurrent of $15.7\,mA\,cm^{-2}$ at 0 V vs Ag/AgCl, which is the highest value reported among TiO_2-based PEC cells for hydrogen generation (Luo et al., 2012).

Hematite phase iron oxide (α-Fe_2O_3) is another promising photocatalytic material of great interest due to its abundance, nontoxicity, chemical stability, the fact that its valence band edge is suitable for water oxidation, and its band gap of 2.2 eV, which allows absorption of a wide range of visible light. However, one of its major drawbacks is that it has short hole diffusion length, and does not allow light to penetrate deeply into the material; when light does penetrate, the electrons and holes are not separated far apart enough and they recombine easily. Bell et al. fabricated an α-Fe_2O_3 inverse opal film with wall thickness closer to the diffusion length to allow better charge separation and utilization at the film/electrolyte interface, leading to higher charge migration (Bell et al., 2008). To address bulk recombination at the photoanode–electrolyte interface, Zhang et al. adapted α-Fe_2O_3 inverse opals on a thin graphene interlayer film, and they reported an almost two and half times increase in incident photon-to-current efficiency (IPCE) compared to conventional α-Fe_2O_3 films on FTO substrate. This increase was attributed to the use of the modified opal structure that allowed the electrolyte to infiltrate the entire depth of the α-Fe_2O_3 photoanode, thus reaching the graphene interlayer (Zhang et al., 2013).

Using α-Fe_2O_3 inverse opal as a photoanode for photocatalysis, Shi et al. achieved the highest photocurrent density of 3.1 mA cm^{-2} at 0.5 V vs Ag/AgCl reference electrode (Shi et al., 2013a). To increase the contact surface area with electrolytes and prolong the lifetime of charge carriers, they modified α-Fe_2O_3 inverse opal structures with cobalt-phosphate (Co-Pi) to achieve a significant drop in dark current, thus improving electron transport (Shi et al., 2013b). On the other hand, Riha et al. employed an ALD deposited, α-Fe_2O_3 inverse opal scaffold, and a distributed-current-collection approach to improve the efficiency of water oxidation by threefold as compared to an optimized flat electrode (results shown in Fig. 10) (Riha et al., 2013). By coupling graphene inverse opals (GIO) with α-Fe_2O_3, Yoon et al. were able to address the short diffusion length and low absorption of α-Fe_2O_3. GIO provided 3D conducting networks for the transportation of electrons and increased the photon-trapping effect, thus greatly enhancing photocurrent density. Due to enhanced absorption, direct and fast electron transfer pathways, as well as reduction in electron–hole recombination, the photocurrent density of α-Fe_2O_3/GIO was 1.4 times higher than pristine α-Fe_2O_3 (Yoon et al., 2014). In addition, Abel et al. found that an antimony-doped tin oxide inverse opal scaffold coated with TiO_2 interlayer and Fe_2O_3 absorber, was effective in improving light-harvesting properties, showing an increase of 78% in photocurrent density as compared to planar films (Abel et al., 2015). Bismuth vanadate ($BiVO_4$) is a multinary metal oxide that has a narrow band gap of 2.4 eV and has been identified as a

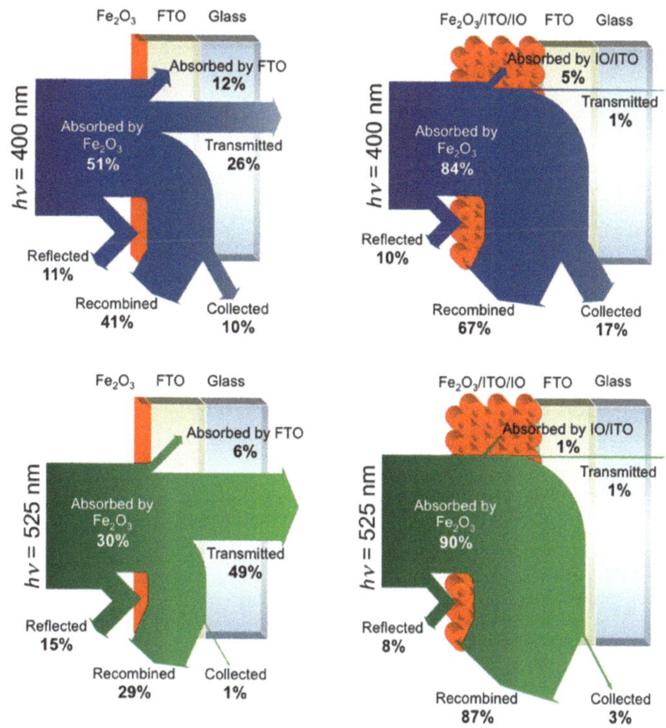

Fig. 10 Photoanode schematics of photon absorption, transmission, and conversion for flat substrates and ITO-coated Fe_2O_3 inverse opals scaffolds. *Reproduced with permission from Riha, S.C., et al., 2013. Hematite-based photo-oxidation of water using transparent distributed current collectors. ACS Appl. Mater. Interfaces 5(2), 360–367. Copyright 2013 American Chemical Society.*

promising n-type semiconductor photoanode for water splitting, but it suffers from poor electron transfer and collection. Zhang et al. fabricated an inverse opal of Al-doped ZnO together with a $BiVO_4$ photoanode to improve photogenerated electron collection. The photocurrent of $BiVO_4$/IO-AZO heterostructure shows a threefold increase compared to planar $BiVO_4$ as seen in Fig. 11 (Zhang et al., 2014).

Zou et al. analyzed the PEC performance of various ordered and disordered Mo-doped $BiVO_4$ 3D inverse opal structures to illustrate the use of nanoengineering through composition regulation and morphology innovation. Mo doping of $BiVO_4$ helps it to overcome poor electron mobility serves as electron donors. It was found that these 3D macro-mesoporous photoelectrodes were more compactly packed within a larger interstitial space, thus minimizing inner resistance of charge transport and contact

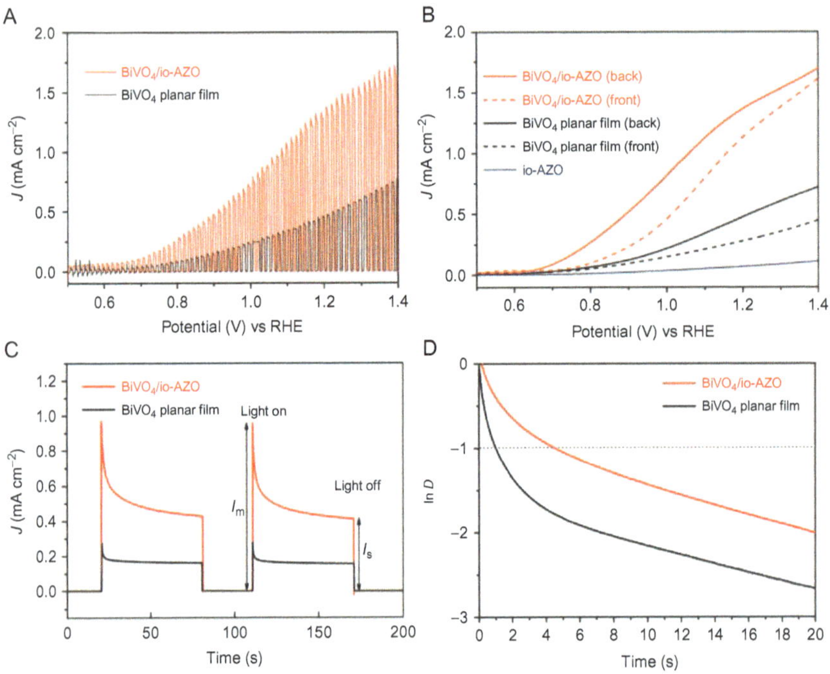

Fig. 11 (A) LSV of $BiVO_4$/io-AZO vs $BiVO_4$ planar film. (B) LSV of front and back illumination $BiVO_4$/io-AZO vs $BiVO_4$ planar film. (C) Transient photocurrent decay. (D) Transient decay times. *Reproduced with permission from Zhang, L.W., Reisner, E., Baumberg, J.J., 2014. Al-doped ZnO inverse opal networks as efficient electron collectors in $BiVO_4$ photoanodes for solar water oxidation. Energy Environ. Sci. 7(4), 1402–1408. Copyright 2014 Royal Society of Chemistry.*

resistance at the electrode/electrolyte interface. They were found to be more stable and achieved higher photocurrent densities as seen in Fig. 12 (Zhou et al., 2014).

2. AAO TEMPLATE-ASSISTED NANOSTRUCTURES IN PHOTOCATALYSIS

Since the first report of self-ordered, structured AAO materials by Masuda et al. (Masuda and Fukuda, 1995), researchers have utilized AAO templates to create a range of nanostructured materials such as nanowires, cores–shell nanowires, nanotubes, and ordered nanoparticles. AAO templates were obtained by anodizing aluminum metal under controlled anodization conditions with tunable pore dimensions and pore distributions. Pore lengths in the range of a few tenths of a nanometer to a few tenths

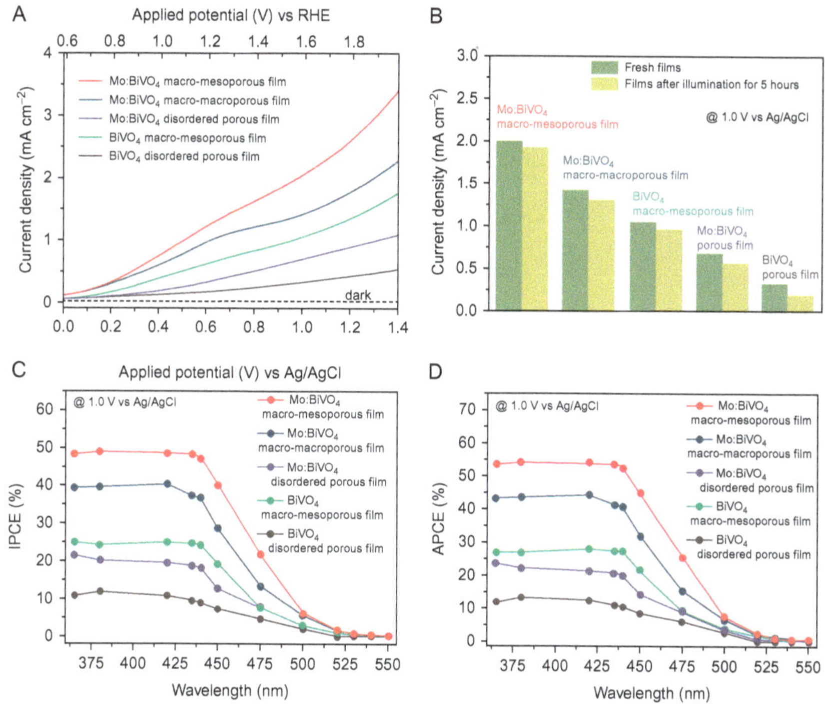

Fig. 12 (A) Photocurrent density potential curves and dark-current density potential curves of various films. (B) Photocurrent density stability of various films at 1.0V vs Ag/AgCl. (C and D) Incident photon-to-current efficiencies (IPCEs) and absorbed photon-to-current efficiencies (APCEs) of various films as a function of the wavelength at 1.0V vs Ag/AgCl as reference electrode. *Reproduced with permission from Zhou, M., et al., 2014. Photoelectrodes based upon Mo: $BiVO_4$ inverse opals for photoelectrochemical water splitting. ACS Nano 8(7), 7088–7098. Copyright 2014 American Chemical Society Nano.*

of a micrometer can be obtained by controlling the anodization time. Pore diameters and ordering could be further tuned by making use of pulse anodization and alternative mild and hard anodization methods (Lei et al., 2003; Masuda et al., 2001). Enabled by the long range ordering and tunable pore structure, an array of as-fabricated nanostructures of metals, semiconductors, carbon, and polymers using AAO templates have been investigated in various applications such as catalysis, sensors, and optoelectronics (Ji et al., 2006; Liang et al., 2007; Steinhart et al., 2002; Winkler et al., 2012).

Zero-dimensional (0D) metal nanoparticles can be assembled using AAO templates to generate surface plasmonic structures. When applied to photocatalysis, surface plasmonic structures improve light harvesting through light scattering and concentration using particle plasmons, and light trapping through plasmon polaritons (Pillai et al., 2007; Wu et al., 2011). Highly regular Au nanodots were incorporated into the TiO_2 layer by Kim et al. for improving PEC water-splitting efficiency (Kim et al., 2014). Au nanodot arrays of dot sizes 50, 63, and 83 nm with a narrow size distribution were prepared using AAO template on an indium tin oxide substrate. Titania particles of 2–3 nm in size were added to the Au nanodots using TiO_2 sols, creating a precisely controlled Au nanodot array with 110 nm of TiO_2 overlayer. The effects of Au nanodot size and TiO_2 overcoats were investigated for PEC water splitting using a three-electrode system. UV–vis measurement showed plasmon enhanced absorption for all Au nanodot arrays. Photoconversion enhancement of 25 times was found for the 50 nm Au size due to the plasmonic effect, as seen in Fig. 13.

One-dimensional (1D) nanostructures have attracted significant attention for solar energy conversion since they reduce light reflection losses and provide direct charge transfer paths (Fan et al., 2010). AAO templates are used to create highly ordered and oriented nanowires of controlled dimensions. Li et al. developed vertically aligned Ta_3N_5 nanorod arrays by nitridation of Ta_2O_5 nanorod arrays grown via a through-mask anodization method. The fabrication procedure is schematically illustrated in Fig. 14A. The robustness of the in situ-grown Ta_2O_5 nanorods enabled nitridation at a high temperature of 1000°C, resulting in good crystallinity of the Ta_3N_5 nanorods and a highly conductive interlayer between the

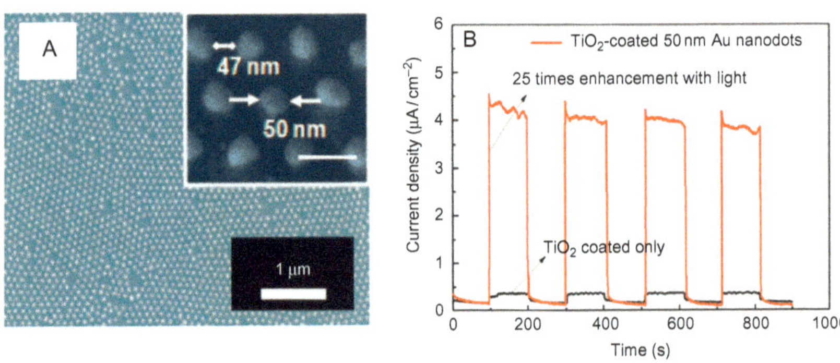

Fig. 13 (A) SEM images of 50 nm Au nanodot arrays on ITO glass substrate. (B) Photocurrent density response of the TiO_2-coated 50 nm Au nanodots arrays at an anode applied potential of 0.397 V vs RHE. *Reproduced with permission from Kim, H.J., et al., 2014. Plasmon-enhanced photoelectrochemical water splitting with size-controllable gold nanodot arrays. ACS Nano 8(10), 10756–10765. Copyright 2014 American Chemical Society Nano.*

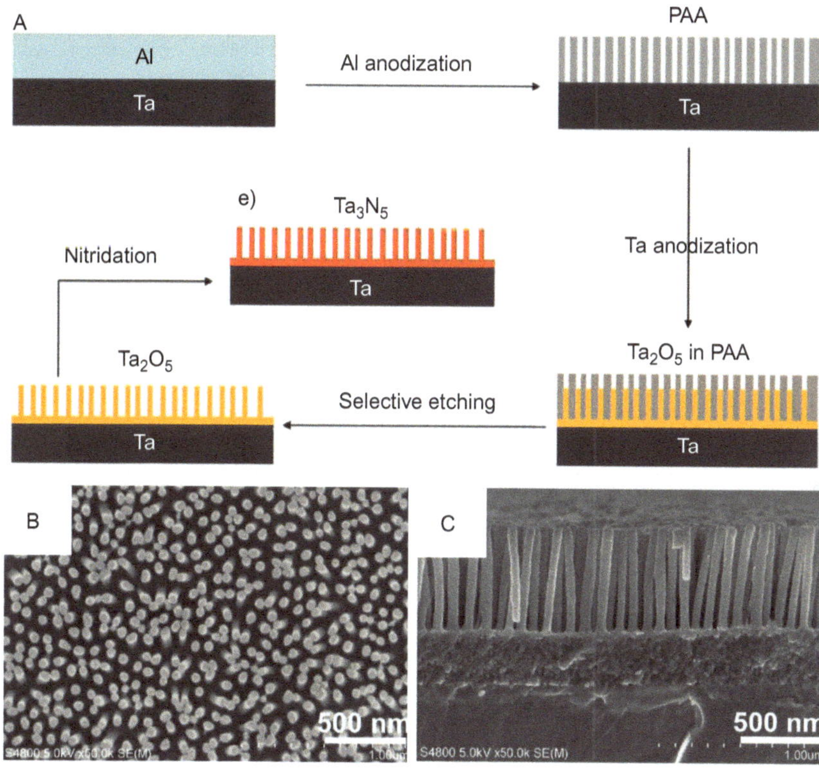

Fig. 14 (A) Schematic illustration of the fabrication procedure for vertically aligned Ta_3N_5 nanorod array using AAO template. (B) Top-view and (C) cross-sectional view SEM images of the as-fabricated Ta_3N_5 nanorod array. *Reproduced with permission from Li, Y., et al., 2013. Vertically aligned Ta3N5 nanorod arrays for solar-driven photoelectrochemical water splitting. Adv. Mater. 25(1), 125–131. Copyright 2013 Wiley.*

nanorods and the substrate (see Fig. 14B and C). A high PEC performance of 3.8 mA cm^{-2} under AM 1.5G simulated sunlight and a maximum IPCE of 41.3% was achieved by the Ta_3N_5 nanorod array photoelectrode. Photocurrent density at 1.23 V vs RHE was enhanced by a factor of 3.2 compared to that of a thin film photoelectrode (Li et al., 2013). Lee et al. prepared plasmonic photoanodes for solar water splitting by fabricating an array of Au nanorods using AAO template capped with TiO_2 overlayer. They demonstrated that 95% of the effective charge carriers derive from surface plasmon decay to hot electrons, as evidenced by fuel production efficiencies up to 20 times higher at visible compared to UV wavelengths (Lee et al., 2012). They have further demonstrated that by introducing both cobalt-based oxidation catalyst and Pt reduction catalyst into the TiO_2-capped

Au nanorod plasmonic structures, overall water-splitting performance could be further improved (Mubeen et al., 2013). An autonomous solar water-splitting device, in which charge carriers involved in the oxygen evolution and hydrogen evolution steps originate from hot electrons created by the excitation of surface plasmons in the nanostructured gold, has also been tested. Each nanorod functions as a water-splitting system without external wiring, generating 5×10^{13} H_2 molecules/cm^2/s under AM1.5 illumination, with long-term operational stability. Further, AAO templates have also been utilized to fabricate nanophotonic structures for application in solar hydrogen generation. Qiu et al. fabricated Al nanospike arrays decorated with hematite as a nanophotonic structure using a nanoimprint lithography-derived AAO template (Qiu et al., 2014). In this structure, while hematite serves as a photoactive material to generate electron–hole pairs, Al nanospikes functions as an efficient carrier collector resulting in three times higher carrier collection compared to that of a planar hematite film. Lin et al. and Tsui et al. have demonstrated nanocone arrays using a multistep anodization-etching process with controlled aspect ratios. They found that the nanocone arrays have enhanced light absorption and device performance owing to their antireflection properties (Lin et al., 2014; Tsui et al., 2014).

3. CONCLUSIONS

Over the last few years, template-assisted fabrication of nanostructures has emerged as a versatile route to produce nanostructures of defined morphology and optoelectronic properties based on a wide range of materials. In reviewing the various types of template-directed nanostructures and their fabrication methods, we are able to see that photocatalytic light-harvesting efficiency depends strongly on the structural and compositional properties of these nanostructured materials. Possessing the advantages of nanoscale dimensions, spatial architecture designs, and nanophotonic light trapping, photoelectrodes of as-fabricated nanostructured materials lead to significant improvements in solar energy conversion applications. Manipulation of the deposition methods and material components in nanostructured photoelectrodes allows one to further modify and enhance their properties to improve light harvesting, fine-tune photonic light trapping and increase the absorption of light in the visible and near infra-red spectrum. Though a wide range of template-assisted nanostructures can be constructed using self-assembled opal and AAO templates, the ability to pattern any material with any multidimensional arrangement on a large scale is limited. Further evolution in template-assisted construction methodology is needed for enhanced and multifunctional energy conversion devices and commercial applications.

ACKNOWLEDGMENTS

We acknowledge The Australian Research Council (ARC) and The Australian National Fabrication Facility (ANFF) for providing support for this work.

REFERENCES

Abel, A.J., et al., 2015. SILAR-deposited hematite films for photoelectrochemical water splitting: effects of Sn, Ti, thickness, and nanostructuring. J. Phys. Chem. C 119 (9), 4454–4465.

Armstrong, E., O'Dwyer, C., 2015. Artificial opal photonic crystals and inverse opal structures—fundamentals and applications from optics to energy storage. J. Mater. Chem. C 3 (24), 6109–6143.

Armstrong, E., et al., 2015. 3D vanadium oxide inverse opal growth by electrodeposition. J. Electrochem. Soc. 162 (14), D605–D612.

Bell, S., et al., 2008. In: Inverse opal hematite films for photocatalytic hydrogen production. 17th World Hydrogen Energy Conference.

Brinker, C.J., et al., 1990. Sol-gel thin film formation. J. Non Cryst. Solids 121 (1–3), 294–302.

Bykov, V.P., 1972. Spontaneous emission in a periodic structure. Soviet J. Exp. Theor. Phys. 35, 269.

Chen, H.M., et al., 2012. Nano-architecture and material designs for water splitting photoelectrodes. Chem. Soc. Rev. 41 (17), 5654–5671.

Cheng, C., et al., 2012. Quantum-dot-sensitized TiO_2 inverse opals for photoelectrochemical hydrogen generation. Small 8 (1), 37–42.

Fan, Z., et al., 2010. Ordered arrays of dual-diameter nanopillars for maximized optical absorption. Nano Lett. 10 (10), 3823–3827.

Fu, M., et al., 2006. ZnO nanosheets with ordered pore periodicity via colloidal crystal template assisted electrochemical deposition. Adv. Mater. 18 (8), 1001–1004.

Fujishima, A., Honda, K., 1972. Electrochemical photolysis of water at a semiconductor electrode. Nature 238 (5358), 37–38.

Galusha, J.W., et al., 2008. Optimizing sol-gel infiltration and processing methods for the fabrication of high-quality planar titania inverse opals. Chem. Mater. 20, 4925–4930.

Gratzel, M., 2001. Photoelectrochemical cells. Nature 414 (6861), 338–344.

Gun, Y., et al., 2015. Joint effects of photoactive TiO_2 and fluoride-doping on SnO_2 inverse opal nanoarchitecture for solar water splitting. ACS Appl. Mater. Interfaces 7 (36), 20292–20303.

Hatton, B., et al., 2010. Assembly of large-area, highly ordered, crack-free inverse opal films. Proc. Natl. Acad. Sci. U.S.A. 107 (23), 10354–10359.

Ji, H.X., et al., 2006. Bis(ethylenedithio)tetrathiafulvalene charge-transfer salt nanotube arrays. Adv. Mater. 18 (20), 2753–2757.

Jiang, P., et al., 1999. Single-crystal colloidal multilayers of controlled thickness. Chem. Mater. 11, 2132–2140.

John, S., 1987. Strong localization of photons in certain disordered dielectric superlattices. Phys. Rev. Lett. 58 (23), 2486–2489.

Juárez, B.H., et al., 2005. ZnO inverse opals by chemical vapor deposition. Adv. Mater. 17 (22), 2761–2765.
Karuturi, S.K., et al., 2010. Kinetics of stop-flow atomic layer deposition for high aspect ratio template filling through photonic band gap measurements. J. Phys. Chem. C 114 (35), 14843–14848.
Karuturi, S.K., et al., 2011. Gradient inverse opal photonic crystals via spatially controlled template replication of self-assembled opals. Nanoscale 3 (12), 4951–4954.
Karuturi, S.K., et al., 2012a. A novel photoanode with three-dimensionally, hierarchically ordered nanobushes for highly efficient photoelectrochemical cells. Adv. Mater. 24 (30), 4157–4162.
Karuturi, S.K., et al., 2012b. Inverse opals coupled with nanowires as photoelectrochemical anode. Nano Energy 1 (2), 322–327.
Kim, I.D., et al., 2006. Microsphere templating as means of enhancing surface activity and gas sensitivity of $CaCu_3Ti_4O_{12}$ thin films. Nano Lett. 6 (2), 193–198.
Kim, J.H., et al., 2011. Ni-NiO core-shell inverse opal electrodes for supercapacitors. Chem. Commun. (Camb.) 47 (18), 5214–5216.
Kim, K., et al., 2013. Optimization for visible light photocatalytic water splitting: gold-coated and surface-textured TiO_2 inverse opal nano-networks. Nanoscale 5 (14), 6254–6260.
Kim, H.J., et al., 2014. Plasmon-enhanced photoelectrochemical water splitting with size-controllable gold nanodot arrays. ACS Nano 8 (10), 10756–10765.
Kondofersky, I., et al., 2015. Electron collection in host-guest nanostructured hematite photoanodes for water splitting: the influence of scaffold doping density. ACS Appl. Mater. Interfaces 7 (8), 4623–4630.
Lebrun, D.M., Niklasson, G.A., Osterlund, L., 2014. Fabrication of photonic opal structures on different support materials by convective evaporation. Inera Workshop: Transition Metal Oxide Thin Films-Functional Layers in Smart Windows and Water Splitting Devices. Parallel Session of the 18th International School on Condensed Matter Physics, vol. 559, p. 012007.
Lee, J., et al., 2012. Plasmonic photoanodes for solar water splitting with visible light. Nano Lett. 12 (9), 5014–5019.
Lei, Y., et al., 2003. Ordered nanoporous nickel films and their magnetic properties. Chem. Phys. Lett. 380 (3–4), 313–318.
Li, Y., et al., 2013. Vertically aligned Ta3N5 nanorod arrays for solar-driven photoelectrochemical water splitting. Adv. Mater. 25 (1), 125–131.
Liang, C., et al., 2007. AgI/Ag heterojunction nanowires: facile electrochemical synthesis, photoluminescence, and enhanced ionic conductivity. Adv. Funct. Mater. 17 (9), 1466–1472.
Lin, Q., et al., 2014. Inverted nanocone-based thin film photovoltaics with omnidirectionally enhanced performance. ACS Nano 8 (6), 6484–6490.
Ling, T., et al., 2014. Highly conductive CdS inverse opals for photochemical solar cells. Adv. Funct. Mater. 24 (5), 707–715.
Liu, W., et al., 2010. Optimizing sol–gel infiltration for the fabrication of high-quality titania inverse opal and its photocatalytic activity. Thin Solid Films 518 (17), 4923–4927.
Liu, L., et al., 2011. TiO2 inverse-opal electrode fabricated by atomic layer deposition for dye-sensitized solar cell applications. Energ. Environ. Sci. 4 (1), 209–215.
Luo, J.S., et al., 2012. Homogeneous photosensitization of complex TiO2 nanostructures for efficient solar energy conversion. Sci. Rep. 2, 451.
Maka, T., et al., 2003. Three dimensional photonic crystals in the visible regime. Prog. Electromagn. Res. 41, 307–335.
Masuda, H., Fukuda, K., 1995. Ordered metal nanohole arrays made by a two-step replication of honeycomb structures of anodic alumina. Science 268 (5216), 1466–1468.

Masuda, H., et al., 2001. Square and triangular nanohole array architectures in anodic alumina. Adv. Mater. 13 (3), 189–192.
Mubeen, S., et al., 2013. An autonomous photosynthetic device in which all charge carriers derive from surface plasmons. Nat Nano 8 (4), 247–251.
Pillai, S., et al., Surface plasmon enhanced silicon solar cells. J. Appl. Phys., 2007. 101(9): p. 093105.
Qiu, Y., et al., 2014. Efficient photoelectrochemical water splitting with ultrathin films of hematite on three-dimensional nanophotonic structures. Nano Lett. 14 (4), 2123–2129.
Quan, L.N., et al., 2014. Soft-template-carbonization route to highly textured mesoporous carbon-TiO_2 inverse opals for efficient photocatalytic and photoelectrochemical applications. Phys. Chem. Chem. Phys. 16 (19), 9023–9030.
Riha, S.C., et al., 2013. Hematite-based photo-oxidation of water using transparent distributed current collectors. ACS Appl. Mater. Interfaces 5 (2), 360–367.
Scharrer, M., et al., Fabrication of inverted opal ZnO photonic crystals by atomic layer deposition. Appl. Phys. Lett., 2005. 86(15): p. 151113.
Sharifi-Sanjani, N., et al., 2004. Emulsifier-free emulsion polymerization of styrene. J. Appl. Polym. Sci. 94 (5), 1898–1904.
Shi, X.J., Zhang, K., Park, J.H., 2013b. Understanding the positive effects of (Co-Pi) co-catalyst modification in inverse-opal structured alpha-Fe_2O_3-based photoelectrochemical cells. Int. J. Hydrogen Energy 38 (29), 12725–12732.
Shi, X., et al., 2013a. Constructing inverse opal structured hematite photoanodes via electrochemical process and their application to photoelectrochemical water splitting. Phys. Chem. Chem. Phys. 15 (28), 11717–11722.
Steinhart, M., et al., 2002. Polymer nanotubes by wetting of ordered porous templates. Science 296 (5575), 1997.
Tsui, K.-H., et al., 2014. Low-cost, flexible, and self-cleaning 3D nanocone anti-reflection films for high-efficiency photovoltaics. Adv. Mater. 26 (18), 2805–2811.
van de Krol, R., Liang, Y., Schoonman, J., 2008. Solar hydrogen production with nanostructured metal oxides. J. Mater. Chem. 18 (20), 2311–2320.
Walter, M.G., et al., 2010. Solar water splitting cells. Chem. Rev. 110 (11), 6446–6473.
Winkler, N., et al., 2012. Large-scale highly ordered arrays of freestanding magnetic nanowires. J. Mater. Chem. 22 (32), 16627–16632.
Wu, J.-L., et al., 2011. Surface plasmonic effects of metallic nanoparticles on the performance of polymer bulk heterojunction solar cells. ACS Nano 5 (2), 959–967.
Yablonovich, E., 1987. Inhibited spontaneous emission in solid-state physics and electronics. Phys. Rev. Lett. 58, 2059–2062.
Yoon, K.Y., et al., 2014. Hematite-based photoelectrochemical water splitting supported by inverse opal structures of graphene. ACS Appl. Mater. Interfaces 6 (24), 22634–22639.
Zhang, L.W., Reisner, E., Baumberg, J.J., 2014. Al-doped ZnO inverse opal networks as efficient electron collectors in $BiVO_4$ photoanodes for solar water oxidation. Energ. Environ. Sci. 7 (4), 1402–1408.
Zhang, L., et al., 2011. Bi(2) WO(6) inverse opals: facile fabrication and efficient visible-light-driven photocatalytic and photoelectrochemical water-splitting activity. Small 7 (19), 2714–2720.
Zhang, K., et al., 2013. Inverse opal structured alpha-Fe_2O_3 on graphene thin films: enhanced photo-assisted water splitting. Nanoscale 5 (5), 1939–1944.
Zhou, M., et al., 2014. Photoelectrodes based upon Mo:BiVO4 inverse opals for photoelectrochemical water splitting. ACS Nano 8 (7), 7088–7098.

CHAPTER NINE

Nanostructured Semiconductors for Bifunctional Photocatalytic and Photoelectrochemical Energy Conversion

Songcan Wang, Jung-Ho Yun, Lianzhou Wang[1]
Nanomaterials Centre, School of Chemical Engineering and Australian Institute for Bioengineering and Nanotechnology, The University of Queensland, Brisbane, QLD, Australia
[1]Corresponding author: e-mail address: l.wang@uq.edu.au

Contents

1. INTRODUCTION

With the function of storing solar energy into chemical bonds, photocatalytic (PC) and photoelectrochemical (PEC) systems have attracted great attention due to the potential of addressing energy crisis and environmental issues (Faunce et al., 2013; Hisatomi et al., 2014; Kudo and Miseki, 2009; Lianos, 2011; Lim et al., 2009; Osterloh, 2013; Puga, 2016). Fig. 1 demonstrates the schematic of the reactions on a semiconducting photocatalyst. Under light illumination, electrons and holes are first generated inside the

Semiconductors and Semimetals, Volume 97
ISSN 0080-8784
http://dx.doi.org/10.1016/bs.semsem.2017.02.003

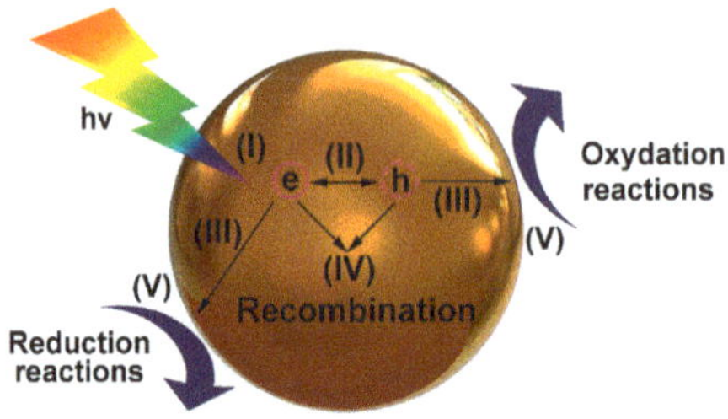

Fig. 1 Schematic of the reactions on a semiconducting photocatalyst. (I) Light absorption, (II) electron–hole pair generation, (III) charge transfer, (IV) charge recombination, and (V) redox reactions with species in the solution.

photocatalyst. Then, photoexcited electrons and holes are transferred to the surfaces of the photocatalyst for redox reactions. Due to the unstable status of photoexcited electrons and holes, recombination also occurs simultaneously, resulting in low efficiency (Luo et al., 2016). Therefore, morphology control, crystal facet engineering, and surface modification are commonly used for photocatalysts and photoelectrodes to enhance charge separation (Li et al., 2013b; Qu and Duan, 2013; Ran et al., 2014; Wang and Sasaki, 2014; Wang et al., 2014a; Yang et al., 2013). To date, most of the reported PC/PEC systems only make good use of one half reaction for hydrogen (H_2) production, carbon dioxide (CO_2) reduction, or organic pollutant degradation (Dhakshinamoorthy et al., 2012; Kim et al., 2009, 2015; Peerakiatkhajohn et al., 2016; Wang et al., 2016; Yue et al., 2015). For example, in a conventional PEC water splitting system, photoelectrodes consisting of semiconductors can convert solar energy to electron–hole pairs under light illumination. Then, hydrogen gas is generated by accepting two electrons from the photocathode, while oxygen gas is produced by donating four electrons to the photoanode. When compared to hydrogen evolution reaction (HER), oxygen evolution reaction (OER) is kinetically slow, which hinders the overall water splitting reaction, resulting in low solar-to-hydrogen (STH) efficiency (Hisatomi et al., 2014; Park et al., 2013). In addition, the production of oxygen from PEC water splitting is insignificant due to the low price. Even though the presence of hole sacrificial reagents can quickly consume holes and promote HER, it inevitably increases the cost for solar H_2 production (Lu et al., 2014). Therefore, it is more desirable to make good use of the oxidation half reaction to produce other value-added chemicals or decompose organic pollutants. In this regard, bifunctional PC/PEC systems, where useful reactions with both electrons and holes are achieved, have attracted increasing attention in recent years (Chen et al., 2012, 2015; Leng et al., 2006; Seger et al., 2012; Zong et al., 2014b).

Although there are many review articles comprehensively summarize the progresses of conventional PC/PEC systems (Acar et al., 2016; Hisatomi et al., 2014; Kang et al., 2015; Osterloh, 2013; Peerakiatkhajohn et al., 2017; Zhang et al., 2016a; Zong and Wang, 2014), there is no review about bifunctional PC/PEC systems. Therefore, it is significant to introduce some basic concepts and recent progresses of bifunctional PC/PEC systems to provide readers with better understanding of this new emerging research field. In this chapter, the configurations and working mechanisms of bifunctional PC/PEC systems will be concisely introduced. Meanwhile, recent progresses of nanostructured semiconductors for bifunctional PC/PEC systems will be summarized. Finally, the challenges and future developments of this research field will be prospected.

2. BASIC PRINCIPLES OF BIFUNCTIONAL PC/PEC SYSTEMS

2.1 Configurations and Working Mechanisms

Under light illumination, photons with energy ($h\nu$) larger than the bandgap energy (E_g) of the semiconductor can be absorbed and electrons are then excited from the valence band (VB) to the conduction band (CB), leaving the corresponding holes in the VB (Kudo and Miseki, 2009). Electrons can reduce species with the reduction potential more positive than the CB of the semiconductor, while holes can oxidize species with the oxidation potential more negative than the VB of the semiconductor, as shown in Fig. 2A. On the other hand, a bifunctional PEC cell consists of two electrodes in anodic and cathodic chambers with different electrolytes separated by an ion transport membrane. At least one electrode is composed of semiconductors and acts as the photoelectrode. Photoanodes are made of n-type semiconductors with electrons as majority carriers. Under light illumination, photoexcited holes will accumulate on the surface of the photoanode for oxidation reactions, whereas photoexcited electrons are transferred to the counter electrode via the external circuit for reduction reactions (Fig. 2B). Alternatively, p-type semiconductors can be used as photocathodes. Contrary to photoanodes, photoexcited holes are transferred to the counter electrode for oxidation reactions, while electrons will move to the surface of the photocathode for reduction reactions, as shown in Fig. 2C. In both of the above configurations, the potential of electrons/holes on the counter

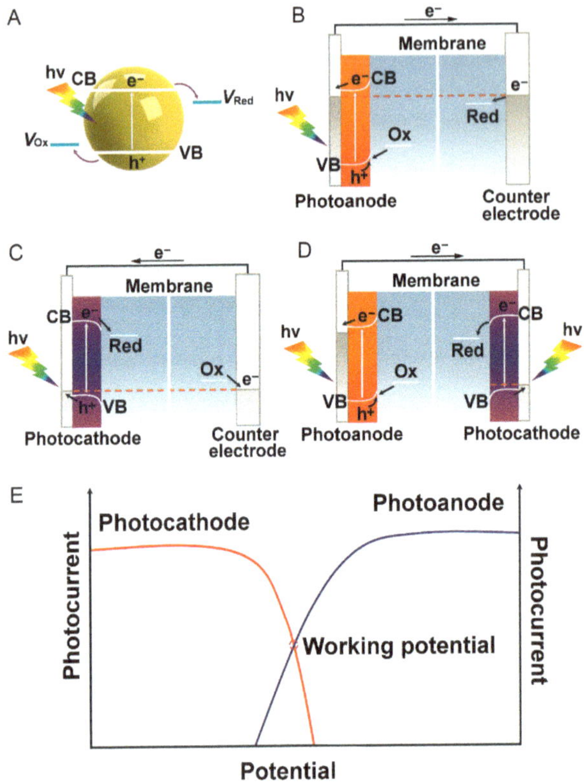

Fig. 2 (A) Energy diagram of photocatalytic reactions. Configurations of bifunctional PEC cells based on (B) a photoanode, (C) photocathode, and (D) photoanode and photocathode in tandem. (E) Working potential of the configuration in (D) based on the photocurrent vs applied potential curves of the photoanode and photocathode.

electrode conforms to the Fermi level of the photoanode/photocathode. If the potential of photoexcited electrons/holes on the counter electrode is deficient for the target reduction/oxidation reactions, an external anodic/cathodic bias can be applied to drive the redox reactions forward (Minggu et al., 2010; Zhen et al., 2016). On the other hand, a photoanode and a photocathode can be connected to form a tandem configuration (Zhang et al., 2016b). Under light illumination, the majority carriers of the photoanode (electrons) and photocathode (holes) will recombine via the external circuit, leaving the minor carriers on the surfaces of the photoanode (holes) and photocathode (electrons) for redox reactions (Fig. 2D). Theoretically, the output photocurrent and the working potential of the tandem cell are the intersection of the photocurrent–potential curves of the photoanode and photocathode, respectively (Fig. 2E). Therefore, if the connected photoanode and photocathode can deliver high photocurrent at the same potential, high solar conversion efficiency can be achieved.

2.2 Benchmarks for Determining the Performance of Bifunctional PC/PEC Systems

As the working mechanisms of bifunctional PC/PEC systems are the same as conventional PC/PEC systems (redox reactions between photoexcited electrons and holes), the benchmarks for determining the performance of a bifunctional PC/PEC system are also the same as a conventional PC/PEC system. Similar to conventional PEC cells, photocurrent vs applied potential and photocurrent vs time curves are basic parameters to determine the performance of the cell. Generally, higher photocurrent at lower applied potential and more stable photocurrent with longer time indicate higher PEC activity and higher stability of the photoelectrodes, respectively.

It should be mentioned that in an ideal case, the potential of the counter electrode in a two-electrode system is equal to the potential of the redox reaction happening on the counter electrode. However, in reality, overpotential is needed to drive the reaction take place, which depends on the properties of the counter electrode. Thus, the potential of the photoanode/photocathode measured from a two-electrode configuration is not precise. To exactly measure the potential of the working electrode against the electrolyte, a three-electrode configuration with a reference electrode is used (Hisatomi et al., 2014). Generally, Ag/AgCl electrode, saturated calomel electrode (SCE), or Hg/HgO electrode is used as the reference electrode. The measured potential of the working electrode is affected by the pH and temperature of the electrolyte. To compare the performance of the PEC cell measured in different electrolytes or with different reference electrodes, the obtained potential should be converted to the reversible hydrogen electrode (RHE). For example, the applied potential vs Ag/AgCl can be converted to RHE in accordance with the Nernst equation, as shown in Eq. (1) (Iandolo and Hellman, 2014):

$$E_{\mathrm{RHE}} = E_{\mathrm{Ag/AgCl}} + 0.059\mathrm{pH} + E^{0}_{\mathrm{Ag/AgCl}} \tag{1}$$

where E_{RHE} refers to the converted potential vs RHE. At room temperature (25°C), the value of $E^{0}_{\mathrm{Ag/AgCl}}$ (saturated KCl) is 0.197 V, and $E_{\mathrm{Ag/AgCl}}$ represents the potential vs Ag/AgCl measured experimentally.

In addition, incident photon-to-current efficiency (IPCE) is also important to measure the performance of a bifunctional PEC cell, which directly shows the percentage of incident photons at different wavelengths that are converted to current. By measuring the intensity of the incident light at different wavelengths and the corresponding photocurrent, IPCE can be calculated via Eq. (2) (Chen et al., 2010):

$$\text{IPCE}(\%) = \frac{1240J}{\lambda \times P} \times 100\% \tag{2}$$

where J refers to the photocurrent density at a certain wavelength of incident light (mA cm^{-2}), λ represents the wavelength of incident light (nm), and P is the corresponding light intensity (mW cm^{-2}).

Similar to IPCE, apparent quantum yield (AQY) is another important standard for determining the conversion ratio of incident photons. AQY can be calculated by Eq. (3) (Hisatomi et al., 2014):

$$\text{AQY} = \frac{nR}{I} \tag{3}$$

where n represents the number of electrons involved in the PC reactions, while R refers to the production rate of molecules, and I means the rate of incident photons.

Solar-to-fuel (STF) efficiency indicates the ratio of the output energy of solar fuels to the total energy of incident solar light, which is important to determine the performance of a PEC cell. Taking water splitting for instance, STH efficiency can be calculated by Eq. (4) (Chen et al., 2010):

$$\text{STH} = \frac{r_{H_2} \times \Delta G}{P \times S} \tag{4}$$

where r_{H_2} (mmol s^{-1}) represents the rate of hydrogen production, ΔG is the change in Gibbs free energy for producing one mol of H_2 (at 25°C, $\Delta G = 237$ kJ mol^{-1}), P refers to the light intensity of the incident solar light (mW cm^{-2}), and S (cm^2) is the exposed area of the photoelectrode.

If external bias is applied to a PEC cell, the input of electrical energy should also be considered when calculating the conversion efficiency. Therefore, applied bias photon-to-current efficiency (ABPE) is developed, which can be calculated by Eq. (5) (Chen et al., 2010):

$$\text{ABPE} = \frac{|J|\left|E^0_{\text{rex}} - E\right|}{P} \times 100\% \tag{5}$$

where E represents the applied potential vs RHE, E^0_{rex} is the theoretical redox potential for the target redox reactions (for water splitting, E^0_{rex} is 1.23 V vs RHE), J refers to the photocurrent density obtained at the applied potential of E, and P is the intensity of the incident solar light.

As PC systems are only powder based suspensions, all of the photocurrent related parameters cannot be used. The generation rates of the products are used to determine the activity of the photocatalyst, while time course of the target product generation under light illumination can indicate the stability. STF and AQY can also be used to determine the efficiency of PC systems.

In terms of the function for organic pollutant degradation, the kinetics of mineralization is important factors to measure the performance of the photocatalyst. Generally, the kinetics of PC degradation of organic substances can be characterized using the Langmuir-Hinshelwood model, as shown in Eq. (6) (Chong et al., 2010; Gaya and Abdullah, 2008):

$$r = -\frac{dC}{dt} = \frac{kKC}{1 + KC} \tag{6}$$

where r is the PC reaction rate, C represents the concentration of the organic substances, t refers to the reaction time under light illumination, k is the rate constant of the reaction, and K is the Langmuir adsorption constant.

When the initial concentration (C_0) of organic substances is very small (in the magnitude of mM), Eq. (6) can be simplified to an apparent first-order equation, as shown in Eq. (7) (Houas et al., 2001; Konstantinou and Albanis, 2003, 2004):

$$\ln\left(\frac{C_0}{C}\right) = kKt = k_{app}t \tag{7}$$

where k_{app} is the apparent first-order rate constant that can be obtained from the slope of the curve of $\ln(C_0/C)$ vs t.

Moreover, the organic concentrations can be expressed by the calculating chemical oxygen demand (COD) or total organic carbon to yield an in-depth understanding on the kinetics of mineralization. Theoretically, COD is the amount of O_2 that would be consumed for organic oxidation. During organic degradation, every 4 mol of electrons donated by the organic species would be accepted by 1 mol of O_2. In the bifunctional systems, organic degradation is accompanied by simultaneous H_2 production. The amount of electrons donated by the organic species is equal to the amount of electrons needed for H_2 production. Therefore, COD in a bifunctional system can be estimated with H_2 production via Eq. (8) (Wang et al., 2014c):

$$COD_{removal} = \frac{m_{H_2} M_{O_2}}{2} \tag{8}$$

where m_{H_2} is the amount of substance of H_2, and M_{O_2} is the molar mass of O_2 (32 g mol^{-1}).

3. NANOSTRUCTURED SEMICONDUCTORS FOR BIFUNCTIONAL PC/PEC SYSTEMS

Nanostructured semiconductors have been intensively studied in energy conversion field due to their unique physical, chemical, and optical properties compared to their bulk counterparts. Their advantages used in PC/PEC systems are summarized as the following (Li and Zhang, 2009):

(1) Provide more sites for the redox reactions;
(2) Increase absorption coefficient;
(3) Decrease the diffusing distance of electrons and holes to the reaction sites;
(4) Design and modify the optical and electronic properties;
(5) Reduce trapping or recombination of electrons and holes due to low surface defects;
(6) Reduce light reflection.

Therefore, the performance and efficiency of the bifunctional PC/PEC systems can be adjusted by nanotechnology. Even though numerous nanostructured semiconductors have been explored for conventional PC/PEC applications, nanomaterials utilized in bifunctional PC/PEC systems are rare (Jing et al., 2013; Wu et al., 2015; Xing et al., 2016a,b). So far, TiO_2 is still the main nanostructured semiconductors used in bifunctional PC/PEC systems. The following passage will mainly summarize the progress of TiO_2 for bifunctional PC/PEC systems. Also, other reported nanostructured semiconductors will be briefly introduced.

3.1 TiO_2

TiO_2 is the first photoanode material reported for PEC water splitting (Fujishima and Honda, 1972). Since then, TiO_2 has been intensively studied in various PC or PEC systems, because TiO_2 is stable in most chemical solutions with a broad pH range and can be easily synthesized by low-cost wet chemistry techniques, such as hydrothermal method, sol–gel process, and electrodeposition (Liu et al., 2010; Ma et al., 2014; Ochiai and Fujishima, 2012). The main drawback of TiO_2 is that it can only absorb ultraviolet (UV) light due to the large bandgap of ~3.2 eV (anatase). Strategies such as doping, sensitization, and plasmonic effect have been explored to expand the light absorption of TiO_2 from UV region to visible light region (Chen and Wang, 2014; Zhang et al., 2010).

H_2 evolution with simultaneously organic pollutant degradation in water was obtained using TiO_2 photocatalysts modified by surface fluorination and platinization (SFP) under solar light irradiation (Kim and Choi, 2010). The bifunctional performances of bare TiO_2, F-TiO_2, Pt/TiO_2, and F-TiO_2/Pt in the presence of 4-chlorophenol (4-CP) and bisphenol A (BPA) were compared, which obviously showed the superior performance of F-TiO_2/Pt. The synergic effect of SFP benefited the electron transfer process from the CB of TiO_2 to protons or water molecules for H_2 evolution, achieving the bifunctional purpose. Compared to water splitting H_2 production, this bifunctional system could be applied to purify waste water and recover energy concurrently. Similarly, with oxalic acid as water pollutant and strong reductive electron donor, PC H_2 production and spontaneous oxalic acid decomposition was achieved with Pt-TiO_2 nanoparticles as photocatalyst (Li et al., 2001). In order to simulate the practical reaction systems, the effects of common inorganic anions, pH of the solution and the loading amount of Pt on TiO_2 nanoparticles were estimated. It was found that the additives of SO_4^{2-} and $H_2PO_4^-$ significantly decreased the H_2 production rate, while NO_3^- and Cl^- showed negligible effect. In addition, Pt/TiO_2 suspensions could also achieve H_2 production with simultaneously azo-dye degradation (Patsoura et al., 2006). It was proposed that dye acted as a scavenger to consume photoexcited holes, thus increase the H_2 production rate. The concentration of dye, pH, and temperature of the solution were found to affect the PC activity of the system. With the increase of pH and temperature, the PC activity of Pt/TiO_2 suspensions also improved. The optimal loading of Pt (0.5 wt%) exhibited the maximum H_2 production rate. In this bifunctional system, azo-dye consumed photoexcited holes and was eventually mineralized to CO_2 and inorganic ions, further promoting H_2 production. Other organic pollutants such as 4-chlorophenol, urea, and urine could also be decomposed, accompanied by simultaneous H_2 production using TiO_2 photocatalysts modified with both anion adsorbates (fluoride or phosphate) and (noble) metals (Pt, Pd, Au, Ag, Cu, or Ni) under UV irradiation (Kim et al., 2012). F-TiO_2/Pt (surface fluorinated and platinized

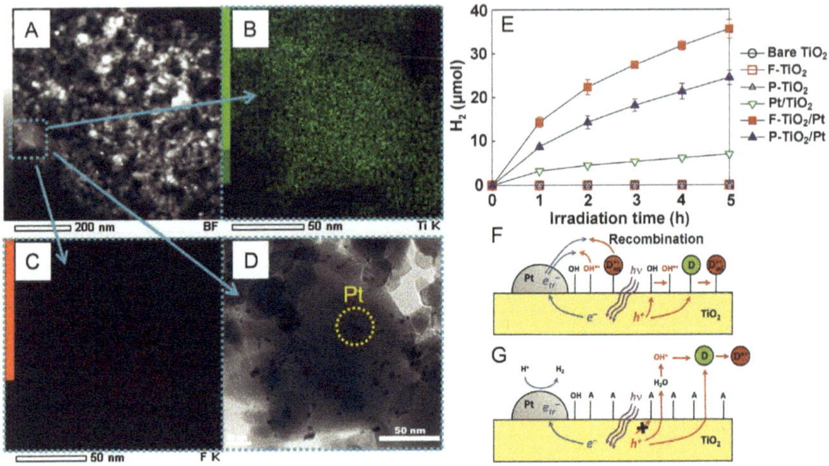

Fig. 3 TEM and EDS images of F-TiO_2/Pt: (A) Scanning TEM bright field image, EDS mapping of (B) Ti and (C) F, and (D) HRTEM image. (E) H_2 evolution in the presence of 4-CP with different photocatalysts of bare TiO_2, F-TiO_2, P-TiO_2, Pt/TiO_2, F-TiO_2/Pt, and P-TiO_2/Pt. Schematic illustration of the surface charge transfer and recombination on (F) Pt/TiO_2 and (G) F-TiO_2/Pt (or P-TiO_2/Pt). Surface absorbed fluorides/phosphates and organic substrates were denoted as A and D in the scheme, respectively. *Reproduced adapted from Kim, J., Monllor-Satoca, D., Choi, W., 2012. Simultaneous production of hydrogen with the degradation of organic pollutants using TiO_2 photocatalyst modified with dual surface components. Energy Environ. Sci. 5, 7647–7656, with permission of The Royal Society of Chemistry.*

TiO_2, as shown in Fig. 3A–D) photocatalysts worked in the acidic pH region, whereas P-TiO_2/Pt (surface phosphated and platinized) showed a consistent activity over a wide range of pH region. The synergistic effect of anions and metal modified TiO_2 improved the interfacial electron transfer and enhanced charge separation, resulting in a maximum of 20-fold increase of H_2 generation compared to only metal modified TiO_2 in the presence of 4-chlorophenol (Fig. 3E). As shown in Fig. 3F, with only Pt as cocatalyst, the overall photocatalysis was inhibited mainly due to the surface-mediated recombination between trapped charges, while with surface absorption of fluoride or phosphate, the surface OH groups were depleted and organic species were physical absorbed rather than chemical absorbed. Thus, the alternative reaction of organic substrates or water with holes took place, resulting in the production of unbound OH radicals or oxidized organic intermediates that diffused out from the surface (Fig. 3G). Compared to bare TiO_2 on which holes were trapped as surface OH radicals, the desorbing radical species on F-TiO_2 and P-TiO_2 could carry away holes, thus reducing charge recombination. Importantly, human urine could be used directly as electrolyte for H_2 evolution in this bifunctional PC system, because it could be an electron donor and pH buffer to stabilize the absorption of fluorides on the surface of TiO_2. Moreover, large amount of phosphates in human urine could achieve surface phosphatization during the reaction, which could further improve the PC activity.

TiO_2 is also commonly used as photoanodes in bifunctional PEC systems. For example, PEC water splitting and simultaneous organic pollutant decomposition were achieved using TiO_2 nanotube as photoanode under simulated solar light illumination (Wu and Zhang, 2011). The fabrication process of TiO_2 nanotube arrays (TiO_2 NTs) affected the morphology and thus exhibited different PEC performance (Fig. 4A–H). TiO_2 NTs

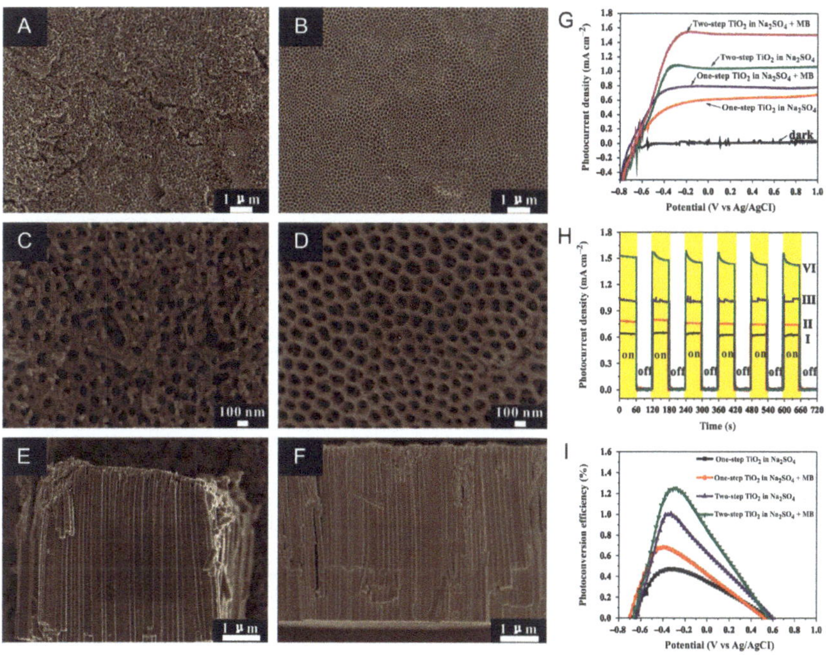

Fig. 4 SEM images of (A, C) *top* and (E) cross-sectional view of one-step TiO_2 NTs, and (B, D) *top* and (F) cross-sectional view of two-step TiO_2 NTs. (G) Photocurrent vs applied potential curves of one- and two-step TiO_2 NTs under illumination of 100 mW cm^{-2}. (H) Photocurrent vs time curves of one- and two-step TiO_2 NTs at an applied potential of 0 V. (I) Photoconversion efficiency vs applied potential curves of one- and two-step TiO_2 NTs. *Reprinted from Wu, H., Zhang, Z., 2011. Photoelectrochemical water splitting and simultaneous photoelectrocatalytic degradation of organic pollutant on highly smooth and ordered TiO_2 nanotube arrays. J. Solid State Chem. 184, 3202–3207. Copyright 2011, with permission from Elsevier.*

prepared by a two-step anodization process exhibited better performance than the counterpart prepared by a one-step process, which exhibited both highest photoconversion efficiency of 1.25% and effective photodecomposition efficiency in the presence of methylene blue (MB) as pollutant target (Fig. 4I). It was believed that the highly ordered nanostructures obtained by a two-step anodization process provided direct pathway and uniform electric field distribution for effective charge transfer/separation, as well as superior capabilities of light harvesting, resulting in enhanced performance. Carbon and nitrogen codoped TiO_2 nanotube arrays (C-N-TNTAs) were found to effectively expand the light response region for PEC H_2 production and simultaneously methyl orange (MO) degradation (Chen et al., 2015). Nitrogen doping was believed to be the main factor for reducing the band gap of TiO_2 (Fig. 5A–C). At an external applied potential of 1.0 V vs SCE under light illumination, C-N-TNTAs exhibited the best MO degradation efficiency with a rate constant of 2.3×10^{-3} s^{-1} and a maximum IPCE of 30.02% at the wavelength of 325 nm. Meanwhile, H_2 was concurrently produced in the cathodic chamber, reaching 3.2 mmol after 3 h of illumination (Fig. 5D). Under light illumination, most electrons were transferred to the cathode in this bifunctional PEC system, which maximized charge separation and thereby enhancing the photoconversion efficiency, enabling the simultaneous utilization of photoexcited holes and electrons for performing separate MO degradation and H_2 production. Multilayered BiO_x–TiO_2 photoanode composed of an under layer of TaO_x–IrO_x, a middle layer of BiO_x–SnO_2, and a top layer of BiO_x–TiO_2 deposited in a series on both sides of Ti foil was coupled with a stainless steel cathode for bifunctional PEC phenol degradation and H_2 production under UV light illumination (Park et al., 2012). With an external bias under UV light illumination, the anodic phenol oxidation and the cathodic H_2 generation rates were improved by four and three folds when compared to only light illumination and direct DC electrolysis, respectively. A heavy doping level of 25 mol% of Bi, increased the electric conductivity of TiO_2, making the BiO_x-TiO_2 layer as the key photoelectrocatalyst in this system. Moreover, Bi-doped TiO_2 anode could also be connected with a photovoltaic cell (PVC) in tandem for oxidizing a series of phenolic compounds to CO_2 with simultaneous H_2 production at the stainless steel cathode under solar light illumination (Park et al., 2008). The anodic and cathodic current efficiencies for phenolic compound oxidation and H_2 production were 3%–17% and 68%–95%, respectively. Moreover, the system exhibited energy efficiencies ranged from 30% to 70%.

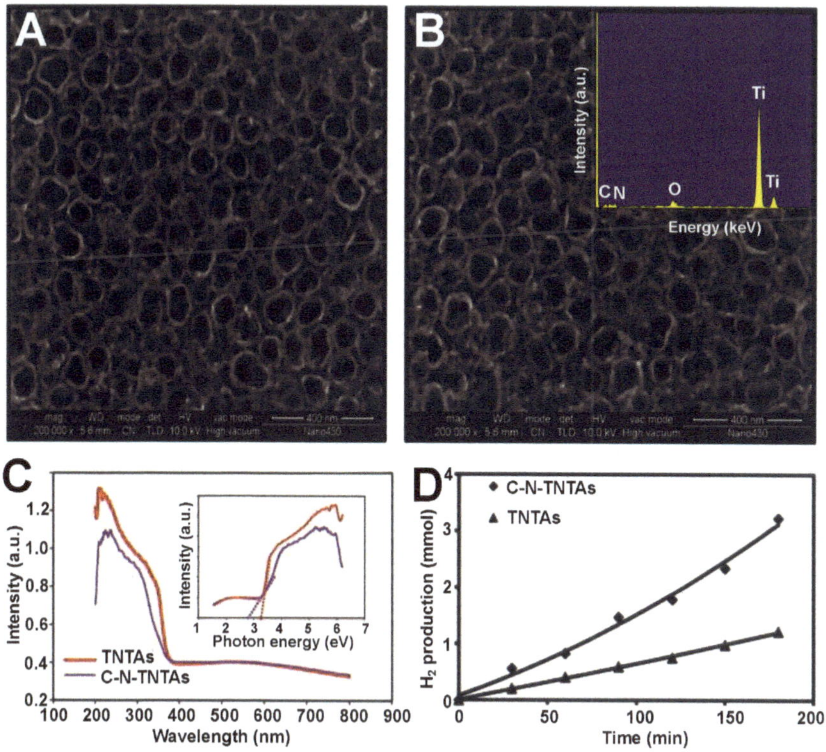

Fig. 5 FESEM images of (A) TNTAs, (B) C-N-TNTAs. (C) Their corresponding UV–vis spectra and (D) H_2 production at cathode in NaOH electrolyte (0.1 M) under 100 W low pressure Hg lamp illumination. *Inset* in (B) EDS spectrum of C-N-TNTAs and *inset* in (C) Tauc curves converted from the UV–vis absorption spectra. *Reprinted from Chen, H., Chen, K.-F., Lai, S.-W., Dang, Z., Peng, Y.-P., 2015. Photoelectrochemical oxidation of azo dye and generation of hydrogen via CN co-doped* TiO_2 *nanotube arrays. Sep. Purif. Technol. 146, 143–153. Copyright 2015, with permission from Elsevier.*

In addition to H_2 production, the reduction compartment in a bifunctional PEC cell can be used to generate other value-added chemicals. For example, bifunctional PEC decomposition of aniline and salicylic acid in the anodic chamber with simultaneous production of hydrogen peroxide (H_2O_2) in the cathodic chamber using TiO_2 as a photoanode was reported (Leng et al., 2006). They found that the method for preparing the TiO_2 photoanode and the choice of the counter electrode affected the decomposition of aniline and salicylic acid, and the production of H_2O_2, respectively. TiO_2 photoanode prepared via an optimal sol–gel process with graphite as the cathode exhibited the maximum current efficiency of 90.1% for reducing oxygen to H_2O_2. Moreover, electrons generated in the bifunctional PEC cell can also be used for CO_2 reduction. For example, nitrogen-doped TiO_2 thin film (NTTF) was employed as a photoanode coupled with a Cu counter electrode for simultaneously CO_2 reduction and MO oxidation under simulated light illumination (Peng et al., 2012). The onset potential for the bifunctional cell was approximately 1.5 V vs SCE and the maximum current density was about 0.65 mA at 2 V vs SCE. Several CO_2 reduction products such as formic acid, formaldehyde, methanol, and methane were detected in the cathodic chamber, with reaction rate constants of 6.68×10^{-7}, 2.18×10^{-4}, 8.62×10^{-4}, and $2.27 \times 10^{-4}\ s^{-1}$, respectively. In addition, the maximum faradaic efficiencies for the production of formic acid, formaldehyde, methanol, and methane were 5.01%, 1.04%, 5.41%, and 7.83%, respectively. Methanol-containing electrolyte was found to enhance CO_2 solubility, which effectively suppressed the competitive reaction of H_2 production. The bifunctional cell and the products from CO_2 reduction with different methanol concentration were shown in Fig. 6.

In a bifunctional PEC cell, the efficiency sometimes may be unusually high, which is contributed to the so-called current doubling effect (Ohno et al., 2000; Seger and Kamat, 2009). When organic substances are decomposed in the anodic chamber, intermediate radicals will form and thus injecting additional electrons into the CB of the semiconductor, causing the increase of photocurrent. Therefore, bifunctional PEC systems can also generate electricity with simultaneously H_2 production in the presence of organic species. As shown in Fig. 7A and B, a bifunctional PEC cell composed of a TiO_2 photoanode, a Nafion membrane, and a Pt cathode was used for oxidizing formic acid to generate electricity and produce H_2 (Seger et al., 2012). Under AM 1.5 G simulated solar light illumination, the cell exhibited a photocurrent density of $150\ \mu A\ cm^{-2}$ and generated $60\ \mu L\ h^{-1}\ cm^{-2}$ of H_2, corresponding to a faradaic efficiency of 88% in the presence of formic acid. It was found that temperature had a deleterious effect on the open-circuit voltage (V_{oc}), but only had a slightly positive effect on the maximum H_2 generation. Moreover, the differences in V_{oc} and polarization curves were studied with the presence of three similar single-carbon organics (Fig. 7C and D). Large voltage observed from methanol compared to formaldehyde and formic acid suggested the increased voltage was caused by the initial oxidation steps (Fig. 7E). Similarly, a two-electrode system with TiO_2 as photoanode and Pt as cathode could decompose ethanol and generate H_2 and electric current under chemical bias produced between alkaline and acidic electrolytes in the anode and cathode compartments, respectively (Antoniadou et al., 2008). Under UV light illumination

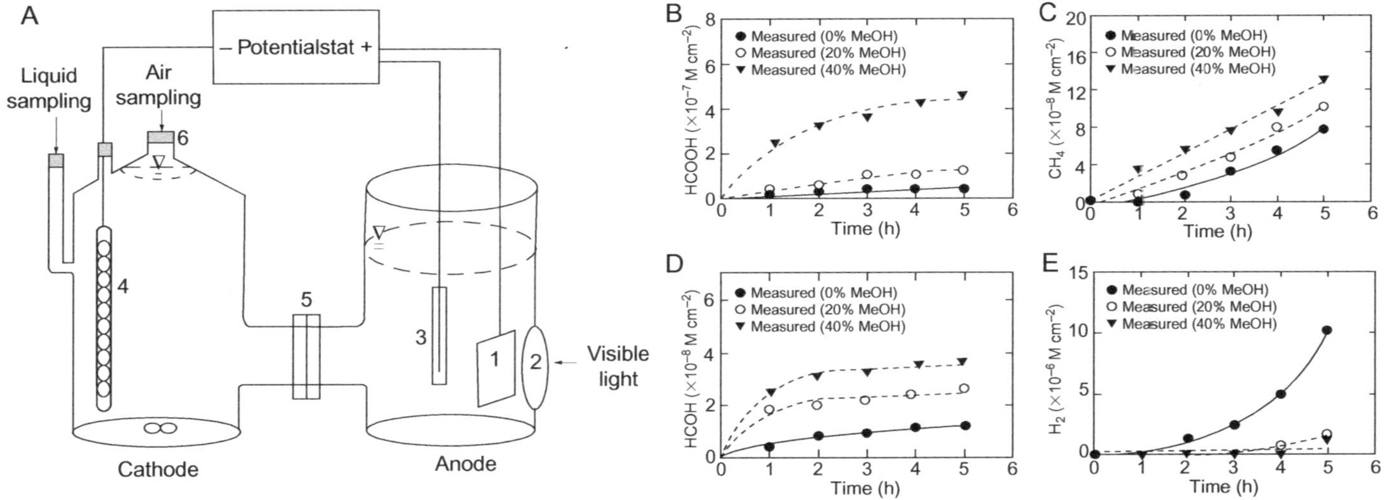

Fig. 6 (A) Schematic illustration of the bifunctional PEC cell: (1) photoanode (N-doped TiO_2 thin film), (2) quartz window, (3) reference electrode (SCE), (4) counter electrode (copper wire), (5) cation exchange membrane, and (6) septum. PEC CO_2 reduction with different methanol concentration and the corresponding products: (B) HCOOH, (C) HCOH, (D) CH_4, and (E) H_2. *Lines* are predicted based on simulated first-order model, and *dots* are experimental data. *Reprinted from Peng, Y.-P., Yeh, Y.-T., Shah, S.I., Huang, C.P., 2012. Concurrent photoelectrochemical reduction of CO_2 and oxidation of methyl orange using nitrogen-doped TiO_2. Appl. Catal. B 123–124, 414–423. Copyright 2012, with permission from Elsevier.*

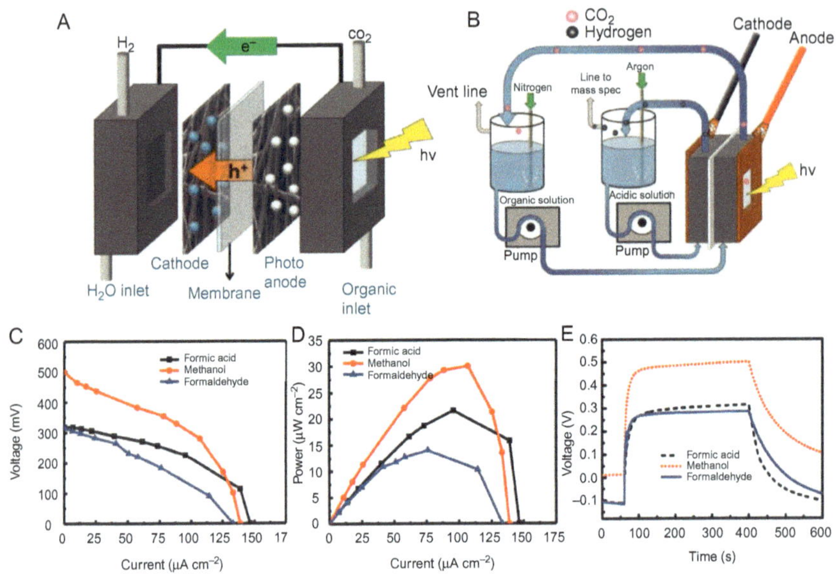

Fig. 7 (A) Schematic illustration of reactor. (B) Schematic illustration of overall H_2 generation system. (C) Polarization curves in the presence of formic acid, methanol and formaldehyde. (D) Power density curves derived from (C). (E) V_{oc} for various organics in the bifunctional PEC cell. *Reproduced adapted from Seger, B., Lu, G.Q., Wang, L.Z., 2012. Electrical power and hydrogen production from a photo-fuel cell using formic acid and other single-carbon organics. J. Mater. Chem. 22, 10709–10715, with permission of The Royal Society of Chemistry.*

($\lambda = 360$ nm, $P = 3.2$ mW cm^{-2}), with 20% (volume ratio) of ethanol, the cell exhibited a photocurrent density of 0.89 mA cm^{2}, corresponding to an IPCE of 96%.

To achieve visible light absorption, TiO_2 photoanode was functionalized with CdS and then coupled with a dark cathode made of commercial nanocrystalline TiO_2 with cast Pt nanoparticles (Antoniadou and Lianos, 2009). The system could degrade several organic substances and generate electricity. The overall efficiency of this cell was higher than the counterpart composed of pure TiO_2, while pure TiO_2 exhibited higher currents under near UV illumination. Under anaerobic conditions, H_2 evolution was observed at the dark cathode. The morphology of TiO_2 also affects the PC activity in a bifunctional PEC cell. As shown in Fig. 8A–C, short TiO_2 nanotube array (STNA) photoanode was used for electricity production and simultaneously wastewater treatment, which exhibited enhanced performance compared with TiO_2 nanoparticulate film photoanode or other long nanotube photoanode (Liu et al., 2011). The cell

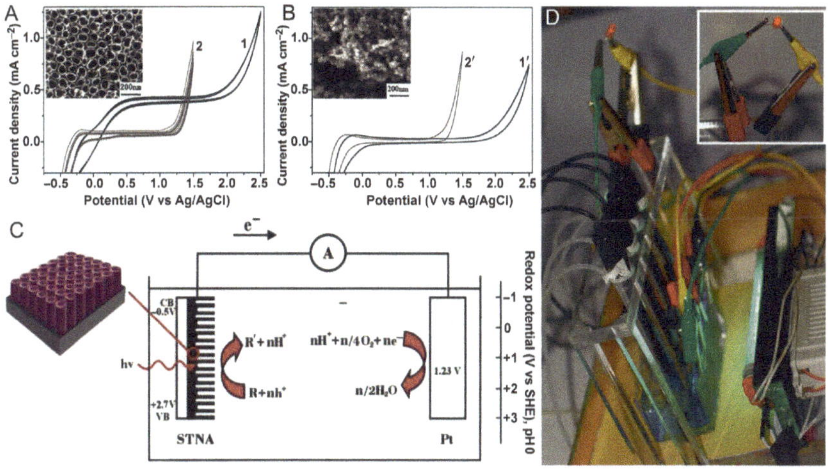

Fig. 8 Continuous cyclic voltammogram (CV) curves of STNA electrode (curve 1 and 1′) and TiO_2 nanoparticulate film electrode (curve 2 and 2′) over five cycles (A) under UV illumination and (B) in the dark, respectively. (C) Schematic illustration of working principle of the STNA-based bifunctional PEC cell. The photogenerated electrons flow from the CB of STNA to the counter electrode of Pt and the holes move toward the surface of STNA to degrade organic pollutants. (D) Four STNA-based bifunctional PEC cells in tandem to light a LED indicative lamp under UV illumination. The insets in (A) and (B) are the SEM image of typical STNA and conventional TiO_2 nanoparticulate film, respectively. *Reprinted from Liu, Y., Li, J., Zhou, B., Li, X., Chen, H., Chen, Q., Wang, Z., Li, L., Wang, J., Cai, W., 2011. Efficient electricity production and simultaneously wastewater treatment via a high-performance photocatalytic fuel cell. Water Res. 45, 3991–3998. Copyright 2011, with permission from Elsevier.*

produced the highest cell performance in acetic acid electrolyte, with a short-circuit current density of 1.42 mA cm^{-2}, open-circuit voltage of 1.48 V, and maximum power density output of 0.67 mW cm^{-2}. Four STNA-based bifunctional cells in tandem were able to light a LED indicative lamp under UV illumination (shown in Fig. 8D). In addition, the materials used in the photoanode and cathode, and the concentration and pH of electrolyte, were believed to be important factors affecting the performance of the bifunctional cell.

3.2 Other Semiconductors

Even though most of the reported bifunctional PC and PEC systems using UV responsive TiO_2 as the photocatalyst, some other visible light responsive semiconductors including CdS, $Cd_xZn_{1-x}S$, Cu_2O, WO_3, and Si have also been developed (Chen et al., 2012; Garaje et al., 2013; Wang et al., 2014c; Zong et al., 2014a,b).

3.2.1 CdS

CdS has a long charge carrier diffusion length up to the micrometer scale with a bandgap of 2.4 eV, which is beneficial for charge separation (Bhandari et al., 2013). However, CdS suffers from severe photocorrosion under light illumination (Hu et al., 2013). Therefore, CdS are commonly coupled with other semiconductors or protective layers to improve the stability (Hernandez-Alonso et al., 2009; Huang and Lu, 2015). The reported CdS used in bifunctional systems are generally in heterojunctions. For example, both powdered and immobilized Pt/CdS/TiO_2 photocatalysts were used in PC and PEC bifunctional systems for waste material degradation with simultaneously H_2 production (Daskalaki et al., 2010). $S^{2-}/SO_3{}^{2-}$ and ethanol were use as model inorganic and organic sacrificial agents/pollutants in aqueous solution. The system could achieve an AQY of 20% at 470 nm for H_2 production accompanied by decomposing inorganic or organic substances. It was found that the ratio of CdS and TiO_2 in the photocatalysts affected the PC activity for decomposing inorganic or organic substances. CdS-rich photocatalysts were more efficient for inorganic substance degradation, whereas the TiO_2-rich counterparts worked better for organic species degradation. One the other hand, CdS/$CdTiO_3$ photocatalysts (Fig. 9A and B) were synthesized by encapsulating CdS NPs on $CdTiO_3$ electrospun nanofibers for PC MB degradation and H_2 evolution from hydrolysis of ammonia–borane complex under visible light illumination (Pant et al., 2014). CdS/$CdTiO_3$ exhibited enhanced PC activity than its counterpart $CdTiO_3$ (Fig. 9C and D). The perfect recovery of activity after reaction and its steady efficiency for cyclic measurement indicated the good stability of this photocatalyst. Moreover, CdS could be surface modified by ZnS to form CdS/ZnS heterojunctions for PC H_2 production and organic degradation under visible light illumination (Wang et al., 2014c). The deposition of ZnS on the surfaces of CdS suppressed the recombination of photoexcited electron/hole pairs, resulting in faster H_2 production and enhanced stability when compared to bare CdS catalyst. By loading with Ru as the cocatalyst, the PC activity for H_2 production was increased by four times. Formic acid, methanol, and ethanol could be decomposed in this bifunctional system, in which the highest H_2 production rate of 266 mmol m^{-2} h^{-1} was obtained in formic acid solution, with an energy conversion efficiency of 3.05% under visible light illumination. Meanwhile,

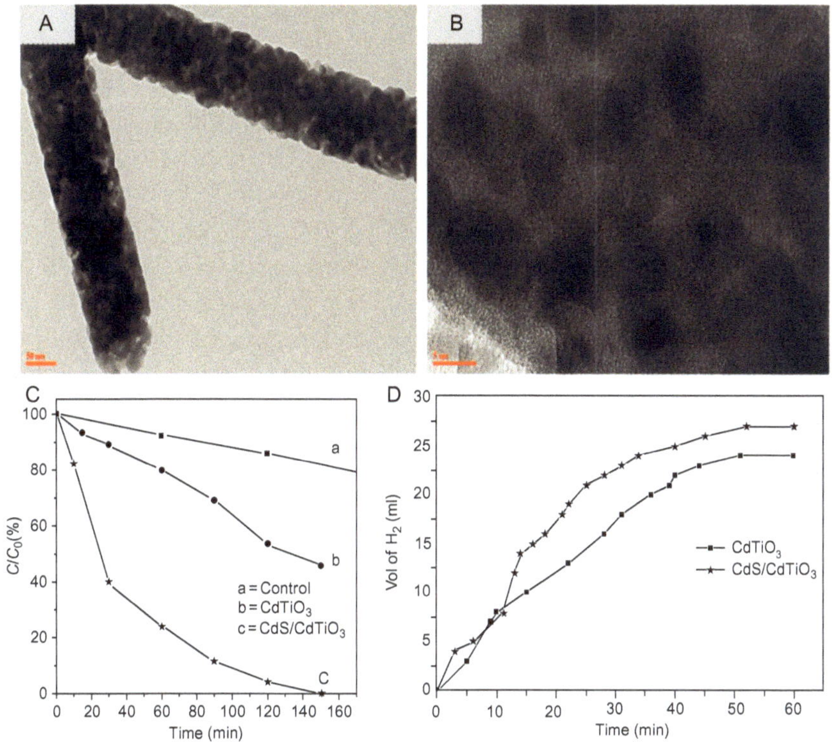

Fig. 9 (A) TEM and (B) HRTEM images of $CdS/CdTiO_3$. (C) Degradation of MB and (D) H_2 evolution from aqueous ammonia–borane with $CdS/CdTiO_3$ and $CdTiO_3$ photocatalysts, respectively. *Reprinted from Pant, B., Pant, H.R., Barakat, N.A.M., Park, M., Han, T.-H., Lim, B.H., Kim, H.-Y., 2014. Incorporation of cadmium sulfide nanoparticles on the cadmium titanate nanofibers for enhanced organic dye degradation and hydrogen release. Ceram. Int. 40, 1553–1559. Copyright 2014, with permission from Elsevier.*

the organic degradation rate in terms of the removal of COD was approximately 4272 mg COD m^{-2} h^{-1}. Ternary metal sulfides have also been explored in bifunctional systems. For example, nanostructured $Cd_xZn_{1-x}S$ ($x=0.5$–0.69) synthesized by a low-temperature solid-state method with the reactants of cadmium oxide, zinc oxide, and thiourea, was applied for H_2 production with simultaneously decomposition of H_2S under visible light illumination (Garaje et al., 2013). $Cd_{0.1}Zn_{0.9}S$ exhibited the highest H_2 production rate of 8320 μmol h^{-1} g^{-1}, which was approximately fourfold as high as that of bulk CdS (2020 μmol h^{-1} g^{-1}). The method explored in this work provided an easy, inexpensive, and pollution-free approach to prepare $Cd_xZn_{1-x}S$ NPs (sizes: 6–8 nm) with adjustable bandgaps on a large scale.

3.2.2 Cu_2O

With a narrow bandgap of ~2.0 eV, Cu_2O is one of the few metal oxides that exhibit p-type conductivity and has been intensively studied as efficient photocatalyst or photocathode for water splitting and CO_2 reduction (An et al., 2014; Morales-Guio et al., 2014; Paracchino et al., 2012a,b). However, the main drawback of Cu_2O is the self-photocorrosion in electrolyte under light illumination, resulting in poor stability (Peerakiatkhajohn et al., 2015; Zhang et al., 2013). It has been reported that the photogenerated holes are the major reason for the self-corrosion of Cu_2O (Chang et al., 2016). When used in bifunctional systems, the oxidation reactions are decomposing organic substances instead of producing O_2, which is more efficient to consume photogenerated holes. Therefore, the stability of Cu_2O is improved. For example, Cu_2O photocatalyst was reported to stoichiometrically and selectively decompose formic acid (HCOOH) into H_2 and CO_2, without the contamination of CO from the organic acid (Kakuta and Abe, 2009). Compared to water splitting, H_2 production from HCOOH degradation could prevent the oxidation of Cu_2O itself, improving the stability. In addition to suspension based PC systems, Cu_2O can also be fabricated as photocathode for bifunctional PEC systems. For example, a bifunctional PEC cell with TiO_2/Ti photoanode and Cu_2O/Cu photocathode was designed for organic pollutant degradation with simultaneous electricity generation under AM 1.5 G simulated solar light illumination without external bias, as shown in Fig. 10 (Li et al., 2013a). The bifunctional PEC cell exhibited a short-circuit current density

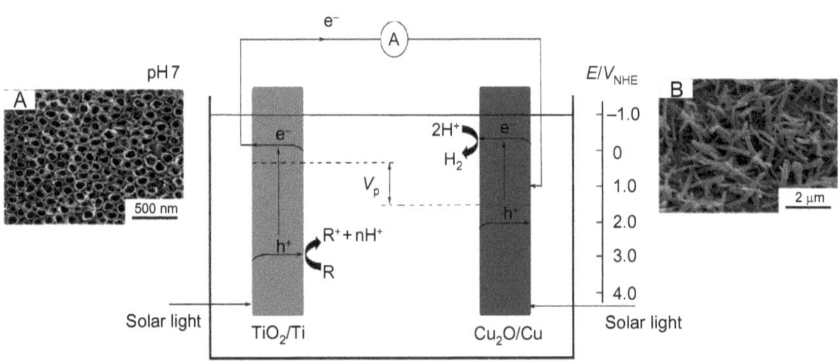

Fig. 10 Schematic illustration of working principal of TiO_2/Ti photoanode and Cu_2O/Cu photocathode-based bifunctional PEC cell. SEM images of (A) TiO_2 nanotube arrays and (B) Cu_2O nanowire arrays. *Reprinted from Li, J., Li, J., Chen, Q., Bai, J., Zhou, B., 2013a. Converting hazardous organics into clean energy using a solar responsive dual photoelectrode photocatalytic fuel cell. J. Hazard. Mater. 262, 304–310. Copyright 2013, with permission from Elsevier.*

of 0.23 mA cm^{-2}, open-circuit voltage of 0.49 V, and maximum power output of 0.36×10^{-4} W cm^{-2} in the presence of 0.05 M phenol. Moreover, the removal rates of chroma for MO, MB, and Congo red (CR) were 67%, 87%, and 63% in 8 h, respectively. Similarly, Cu_2O/Cu photocathode could also be connected with WO_3/W photoanode for organic degradation with simultaneous electricity generation (Chen et al., 2012). The interior bias was generated by the mismatched Fermi levels between the two photoelectrodes under AM 1.5 G illumination without external applied potentials. Several organic compounds such as phenol, Rhodamine B (RB), and CR were successfully degraded in this bifunctional PEC system, with the degradation rates of 58%, 63%, and 74% after 5 h of illumination, respectively.

3.2.3 Si

Si is a promising photoelectrode material for PEC applications due to its narrow bandgap (~1.12 eV), efficient light absorption, and high carrier mobility (Garnett and Yang, 2010; Peng et al., 2005; Wang et al., 2014b). However, Si is very easy to be oxidized in aqueous solution under light illumination or under anodic potential and will rapidly lose the PEC activity; thus, the surface of Si photoelectrode need to be functionalized by pinhole-free protective layers (Andoshe et al., 2016; Chen et al., 2011; Hwang et al., 2009; Lichterman et al., 2016; Noh et al., 2013; Xia et al., 2015). While most of the reported Si-based photoelectrodes are for directly splitting water into H_2 and O_2, our group innovatively introduced an indirect strategy to store solar energy in intermediate species to drive the subsequent reactions, which could achieve bifunctional solar fuel generation and value-added chemical production (Zong et al., 2014a,b). Specifically, with redox couples such as Fe^{2+}/Fe^{3+} and I^-/I_3^- to store solar energy for driving the subsequent reaction, a functionalized Si photoelectrode was connected with Pt counter electrode to achieve H_2S splitting for H_2 generation and value-added elemental sulfur production under simulated solar light illumination (Zong et al., 2014b). In the presence of I^-/I_3^- as the mediator to relay the subsequent reaction, the bifunctional PEC system exhibited excellent performance for H_2S splitting at a low external potential of 0.2 V vs RHE. Similarly, a bifunctional PEC cell composed of a functionalized Si photoelectrode and a carbon counter electrode was able to reduce O_2 to H_2O_2 and simultaneously oxidize H_2S to S with an overall efficiency of 1.1% under AM 1.5 G illumination without external bias in the presence of anthraquinone/anthrahydroquinone (AQ/AQH_2) and I^-/I_3^- redox couples, as shown in Fig. 11 (Zong et al., 2014a). The concept of

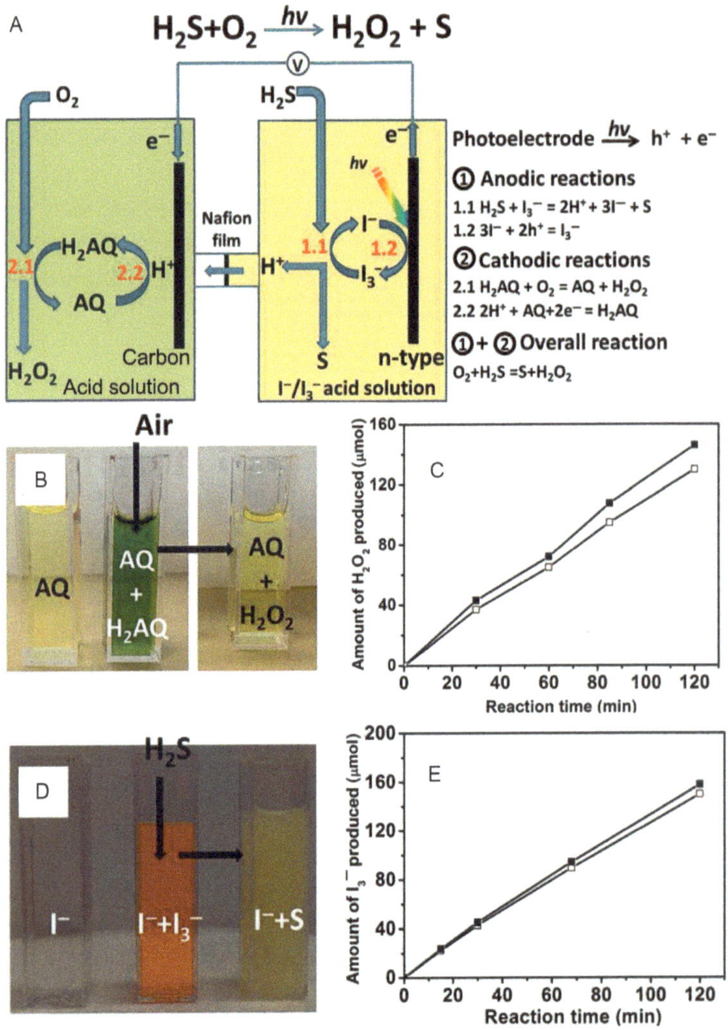

Fig. 11 (A) Schematic illustration of the bifunctional PEC cell on the selective production of H_2O_2 and S from O_2 and H_2S on n-type electrode. (B) Digital photo showing the generation of H_2O_2 after bubbling air into aqueous solution containing H_2AQ in the middle cuvette. (C) Time course of H_2O_2 production. (D) Digital photo showing the production of S after bubbling H_2S into aqueous solution containing I_3^- in the middle cuvette. (E) Time course of I_3^- production. *Reproduced adapted from Zong, X., Chen, H., Seger, B., Pedersen, T., Dargusch, M.S., Mcfarland, E.W., Li, C., Wang, L., 2014a. Selective production of hydrogen peroxide and oxidation of hydrogen sulfide in an unbiased solar photoelectrochemical cell. Energy Environ. Sci. 7, 3347–3351, with permission of The Royal Society of Chemistry.*

linking a photochemical–chemical loop through redox couples to achieve bifunctional solar fuel generation or value-added chemical production demonstrated in these two works may provide new opportunities for other challenging PEC reactions.

4. HYBRID BIFUNCTIONAL SYSTEMS

In addition to the above PC and PEC bifunctional systems, PC or PEC systems can also be combined with other systems such as PVCs, electrochemical cells, rechargeable batteries, and biosystems to achieve solar fuel production/electricity generation, and organic pollutant degradation/value-added chemical production (Canterino et al., 2009; Li et al., 2015; Zang et al., 2014; Zeng et al., 2014).

As shown in Fig. 12A, a Bi_2S_3/TiO_2 nanotube array (TNA) photoanode was connected with a Pt/Si PVC photocathode to form a hybrid system for self-biased H_2 production and electricity generation (Zeng et al., 2014). Under visible light illumination, the system exhibited an open-circuit potential of 0.766 V, generated by the mismatched Femi levels between two electrodes. The short-circuit current density, fill factor, and photoelectric conversion efficiency were 1.55 mA cm^{-2}, 0.602%, and 0.718%, respectively (Fig. 12B). Moreover, the H_2 production rate of this system was ~45.5 μmol cm^{-2} h^{-1}, which was 141 times larger than the

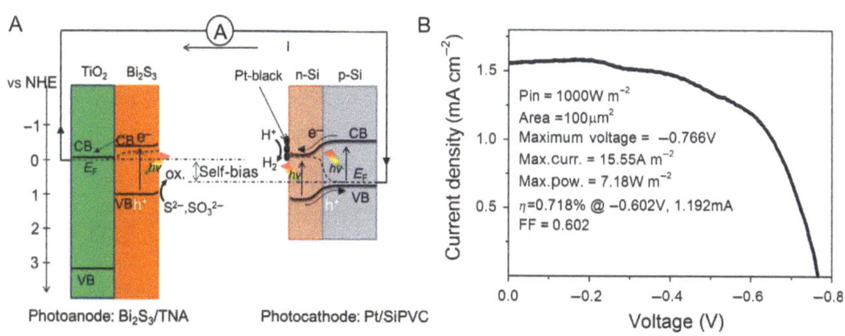

Fig. 12 (A) Energy-level diagram of the unassisted bifunctional PEC cell assembled with Bi_2S_3/TNA photoanode and Pt/SiPVC photocathode under short-circuit situation. (B) Photocurrent–voltage characteristics of the unassisted bifunctional PEC cell of Bi_2S_3/TNA-Pt/SiPVC under identical conditions. *Reprinted from Zeng, Q., Bai, J., Li, J., Li, Y., Li, X., Zhou, B., 2014. Combined nanostructured Bi_2S_3/TNA photoanode and Pt/SiPVC photocathode for efficient self-biasing photoelectrochemical hydrogen and electricity generation. Nano Energy 9, 152–160. Copyright 2014, with permission from Elsevier.*

Bi_2S_3/TNA-Pt (Pt as counter electrode) and 1.8 times larger than the Pt-Pt/Si PVC (Pt as counter electrode) systems, respectively.

Also, simultaneous organic degradation and electricity generation was achieved by combining a PC reactor (TiO_2 as photocatalyst) and an electrochemical cell (Canterino et al., 2009). The recombination system was an H-shaped cell consisting of an anodic compartment and cathodic compartment separated by a Nafion-117 cation exchange membrane. Metallic copper was used as the electrodes and cupric sulfate solution was used as the electrolyte. TiO_2 suspensions and organic substances such as formic acid, glycerol, and glucose were added in the anodic compartment. Under UV illumination, photoexcited electrons would reduce Cu^{2+} ions, while holes would oxide organic substances. The decrease of Cu^{2+} concentration in the anodic compartment caused voltage between the anode and the cathode, thereby generating electricity.

Simultaneous H_2 production and solar energy storage was achieved by introducing a Pt-modified CdS photoelectrode into a rechargeable lithium–sulfur (Li–S) battery (Li et al., 2015). Under solar light illumination, photoexcited holes could charge the Li–S battery by oxidizing S^{2-} ions to polysulfide ions, while photoexcited electrons could be used for H_2 production. Without any other energy input, a specific capacity of 792 mAh g^{-1} could be achieved by 2 h of photocharging. Meanwhile, the corresponding H_2 evolution rate was 1.02 mmol g^{-1} h^{-1}.

The combination of photoelectrodes with bioelectrodes has attracted increasing attention, because bioelectrodes are self-sustained, cost-effective, and environmental friendly when compared to abiotic materials (Du et al., 2014; Lianos, 2011; Xiao and He, 2014). For example, a hybrid bifunctional system (Bio-PEC) composed a TiO_2 nanotube array photoanode and a microbial cathode was reported for organic pollutant degradation and simultaneous electricity generation, as shown in Fig. 13 (Du et al., 2014). The system exhibited a photocurrent density of 325 μA cm^{-2} under solar simulated light illumination. The performance of a PEC cell with Pt/C as the cathode was dependent on the structure and the loading amounts of Pt/C. To achieve a similar MO degradation rate of 0.0120 min^{-1} and maximum power density of 211.32 mW m^{-2} for a Bio-PEC, the loading of Pt/C should be 50 mg. Taking the cost into account, the biocathode reported in this work was promising to replace the conventional Pt/C catalyst. Similarly, another Bio-PEC composed of a MoS_3 modified p-type Si nanowire photocathode and a microbial catalyzed bioanode was designed

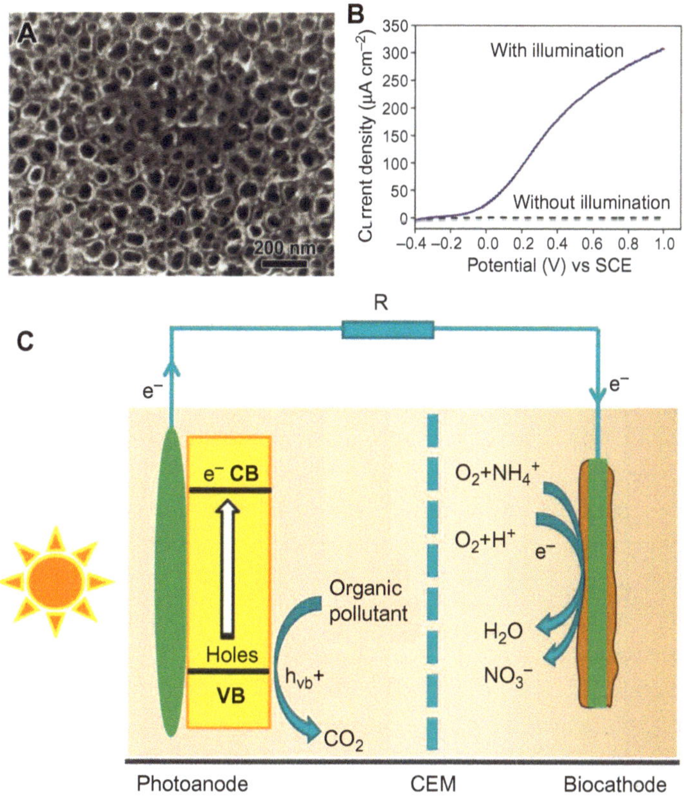

Fig. 13 (A) SEM image of TiO_2 nanotube arrays grown on the Ti substrate. (B) Photocurrent vs applied potential curve of TiO_2 electrodes in a three-electrode system with and without illumination. (C) Schematic illustration of the working principles of the bifunctional Bio-PEC cell. *Reprinted with permission from Du, Y., Feng, Y., Qu, Y., Liu, J., Ren, N., Liu, H., 2014. Electricity generation and pollutant degradation using a novel biocathode coupled photoelectrochemical cell. Environ. Sci. Technol. 48, 7634–7641. Copyright 2014 American Chemical Society.*

for simultaneous H_2 production and electricity generation (Zang et al., 2014). Under visible light illumination, microbial pollutant oxidation took place spontaneously in the bioanode, providing sufficient electrons for the photocathode reaction without external potential. The supply of electrons from the bioanode effectively reduces charge recombination, resulting in more available electrons for H_2 production. Under light illumination, the Bio-PEC delivered a maximum power density of 71 $mW\ m^{-2}$ and an average H_2 generation rate of $7.5 \pm 0.3\ \mu mol\ h^{-1}\ cm^{-2}$.

5. CONCLUSION AND OUTLOOK

PC/PEC systems have been intensively investigated for solar fuel production or pollutant degradation in the past few decades. However, most of the reported systems only make good use of the half reactions. For example, water splitting only makes good use of the photoexcited electrons for H_2 generation, while PC organic pollutant degradation only requires photoexcited holes for oxidation reactions. To better use both of the photoexcited electrons and holes, bifunctional PC/PEC systems have been developed. However, the development of bifunctional PC/PEC systems are in its infancy stage, the publications in this research field are much less than conventional PC/PEC systems. There is no doubt that bifunctional PC/PEC systems are equally significant as conventional PC/PEC systems. This new emerging field will definitely attract increasing attention. In addition to the configuration design, nanostructured semiconductors play a pivotal role in the advancement of this research field. On this ground, this chapter introduced some fundamental knowledge of bifunctional PC/PCE systems, and also provided a state-of-the-art summary on the recent progresses of nanostructured semiconductors for PC/PEC systems.

So far, the reported bifunctional systems include single PC and PEC systems, and the hybrid systems by combining PC/PEC systems with electrochemical cells, photovoltaic cells, rechargeable batteries, or biosystems, which have been concisely summarized. Although various systems have been explored, it is still challenging to achieve satisfactory efficiency for practical applications. To further improve the efficiency, adequate light absorption and efficient charge separation are required for semiconductors used in the system. Most of the reported bifunctional systems are only based on TiO_2 semiconductors. The utilization of solar light is limited in the UV region, which only accounts for ~4% of the whole solar spectrum. Therefore, it is necessary to develop more reliable visible light responsive semiconductors for bifunctional PC/PEC systems. In addition, nanostructure design and surface modification of the semiconductors are important to enhance charge separation and transfer and surface reactions. Combining the advantages of different components of the materials to form a multiple junction may have more chance to reduce charge recombination while keeping broad light absorption (Wang et al., 2017). Another challenge is to reduce the cost for fabricating the whole system. Generally, cocatalysts are needed to improve the surface kinetics of the redox reactions on the surfaces of the semiconductor to reduce energy loss. Most of the reported bifunctional systems use Pt as the cocatalyst, which is not reasonable for scale up applications. Therefore, other earth-abundant materials with competitive performance should be explored as alternative cocatalysts. The third challenge is the long-term stability. Photocatalysts are working in a tough environment, photo or chemical corrosion generally happen in PC/PEC systems. For example, even though the PC activity of CdS is high, it suffers from severe photocorrosion under light illumination. Even though the formation of heterojunctions and surface modification with protective layers can reduce self-photocorrosion to some extent, it is still challenging to overcome the intrinsic drawback of self-photocorrosion. Therefore, deeper understanding of the photocorrosion mechanism and new strategies to enhance long-term stability are still needed. Last but not least, most studies of bifunctional systems only focused on PC/PEC oxidizing one type of organic pollutant with simultaneous H_2 production or electricity generation in one simple electrolyte (e.g., NaOH, Na_2SO_4, H_2SO_4, etc.). However, in reality, there are numerous organic pollutants in wastewater. The competitive reactions between different organic species may affect the PC/PEC activity and reliability of the bifunctional system. Thus, more fundamental knowledge and more experimental work are needed to understand the reaction mechanisms of the bifunctional system toward the more efficient utilization of solar energy to address the challenging energy and environment issues.

ACKNOWLEDGMENTS

The authors would like to acknowledge the financial support from Australian Research Council through its DP and FF programs. S.W. acknowledges the support from Australian Government Research Training Program and UQ Centennial Scholarships.

REFERENCES

Acar, C., Dincer, I., Naterer, G.F., 2016. Review of photocatalytic water-splitting methods for sustainable hydrogen production. Int. J. Energy Res. 40, 1449–1473.

An, X., Li, K., Tang, J., 2014. Cu_2O/reduced graphene oxide composites for the photocatalytic conversion of CO_2. ChemSusChem 7, 1086–1093.

Andoshe, D.M., Choi, S., Shim, Y.S., Lee, S.H., Kim, Y., Moon, C.W., Kim, D.H., Lee, S.Y., Kim, T., Park, H.K., Lee, M.G., Jeon, J.M., Nam, K.T., Kim, M., Kim, J.K., Oh, J., Jang, H.W., 2016. A wafer-scale antireflective protection layer of solution-processed TiO_2 nanorods for high performance silicon-based water splitting photocathodes. J. Mater. Chem. A 4, 9477–9485.

Antoniadou, M., Lianos, P., 2009. Near ultraviolet and visible light photoelectrochemical degradation of organic substances producing electricity and hydrogen. J. Photochem. Photobiol. A 204, 69–74.

Antoniadou, M., Bouras, P., Strataki, N., Lianos, P., 2008. Hydrogen and electricity generation by photoelectrochemical decomposition of ethanol over nanocrystalline titania. Int. J. Hydrogen Energy 33, 5045–5051.

Bhandari, K.P., Roland, P.J., Mahabaduge, H., Haugen, N.O., Grice, C.R., Jeong, S., Dykstra, T., Gao, J., Ellingson, R.J., 2013. Thin film solar cells based on the heterojunction of colloidal PbS quantum dots with CdS. Sol. Energy Mater. Sol. Cells 117, 476–482.

Canterino, M., Di Somma, I., Marotta, R., Andreozzi, R., Caprio, V., 2009. Energy recovery in wastewater decontamination: simultaneous photocatalytic oxidation of an organic substrate and electricity generation. Water Res. 43, 2710–2716.

Chang, X., Wang, T., Zhang, P., Wei, Y., Zhao, J., Gong, J., 2016. Stable aqueous photoelectrochemical CO_2 reduction by a Cu_2O dark cathode with improved selectivity for carbonaceous products. Angew. Chem. 128, 8986–8991.

Chen, H., Wang, L., 2014. Nanostructure sensitization of transition metal oxides for visible-light photocatalysis. Beilstein J. Nanotechnol. 5, 696–710.

Chen, Z.B., Jaramillo, T.F., Deutsch, T.G., Kleiman-Shwarsctein, A., Forman, A.J., Gaillard, N., Garland, R., Takanabe, K., Heske, C., Sunkara, M., Mcfarland, E.W., Domen, K., Miller, E.L., Turner, J.A., Dinh, H.N., 2010. Accelerating materials development for photoelectrochemical hydrogen production: standards for methods, definitions, and reporting protocols. J. Mater. Res. 25, 3–16.

Chen, Y.W., Prange, J.D., Duhnen, S., Park, Y., Gunji, M., Chidsey, C.E., Mcintyre, P.C., 2011. Atomic layer-deposited tunnel oxide stabilizes silicon photoanodes for water oxidation. Nat. Mater. 10, 539–544.

Chen, Q., Li, J., Li, X., Huang, K., Zhou, B., Cai, W., Shangguan, W., 2012. Visible-light responsive photocatalytic fuel cell based on WO_3/W photoanode and Cu_2O/Cu photocathode for simultaneous wastewater treatment and electricity generation. Environ. Sci. Technol. 46, 11451–11458.

Chen, H., Chen, K.-F., Lai, S.-W., Dang, Z., Peng, Y.-P., 2015. Photoelectrochemical oxidation of azo dye and generation of hydrogen via CN co-doped TiO_2 nanotube arrays. Sep. Purif. Technol. 146, 143–153.

Chong, M.N., Jin, B., Chow, C.W., Saint, C., 2010. Recent developments in photocatalytic water treatment technology: a review. Water Res. 44, 2997–3027.

Daskalaki, V.M., Antoniadou, M., Li Puma, G., Kondarides, D.I., Lianos, P., 2010. Solar light-responsive Pt/CdS/TiO_2 photocatalysts for hydrogen production and simultaneous degradation of inorganic or organic sacrificial agents in wastewater. Environ. Sci. Technol. 44, 7200–7205.

Dhakshinamoorthy, A., Navalon, S., Corma, A., Garcia, H., 2012. Photocatalytic CO_2 reduction by TiO_2 and related titanium containing solids. Energy Environ. Sci. 5, 9217–9233.

Du, Y., Feng, Y., Qu, Y., Liu, J., Ren, N., Liu, H., 2014. Electricity generation and pollutant degradation using a novel biocathode coupled photoelectrochemical cell. Environ. Sci. Technol. 48, 7634–7641.

Faunce, T., Styring, S., Wasielewski, M.R., Brudvig, G.W., Rutherford, A.W., Messinger, J., Lee, A.F., Hill, C.L., Degroot, H., Fontecave, M., Macfarlane, D.R., Hankamer, B., Nocera, D.G., Tiede, D.M., Dau, H., Hillier, W., Wang, L., Amal, R., 2013. Artificial photosynthesis as a frontier technology for energy sustainability. Energy Environ. Sci. 6, 1074–1076.

Fujishima, A., Honda, K., 1972. Electrochemical photolysis of water at a semiconductor electrode. Nature 238, 37–38.

Garaje, S.N., Apte, S.K., Naik, S.D., Ambekar, J.D., Sonawane, R.S., Kulkarni, M.V., Vinu, A., Kale, B.B., 2013. Template-free synthesis of nanostructured $Cd_xZn_{1-x}S$ with tunable band structure for H_2 production and organic dye degradation using solar light. Environ. Sci. Technol. 47, 6664–6672.

Garnett, E., Yang, P., 2010. Light trapping in silicon nanowire solar cells. Nano Lett. 10, 1082–1087.

Gaya, U.I., Abdullah, A.H., 2008. Heterogeneous photocatalytic degradation of organic contaminants over titanium dioxide: a review of fundamentals, progress and problems. J. Photochem. Photobiol. C 9, 1–12.

Hernandez-Alonso, M.D., Fresno, F., Suarez, S., Coronado, J.M., 2009. Development of alternative photocatalysts to TiO_2: challenges and opportunities. Energy Environ. Sci. 2, 1231–1257.

Hisatomi, T., Kubota, J., Domen, K., 2014. Recent advances in semiconductors for photocatalytic and photoelectrochemical water splitting. Chem. Soc. Rev. 43, 7520–7535.

Houas, A., Lachheb, H., Ksibi, M., Elaloui, E., Guillard, C., Herrmann, J.M., 2001. Photocatalytic degradation pathway of methylene blue in water. Appl. Catal. B Environ. 31, 145–157.

Hu, Y., Gao, X., Yu, L., Wang, Y., Ning, J., Xu, S., Lou, X.W., 2013. Carbon-coated CdS petalous nanostructures with enhanced photostability and photocatalytic activity. Angew. Chem. Int. Ed. 52, 5636–5639.

Huang, L., Lu, J.G., 2015. Synthesis, characterizations and applications of cadmium chalcogenide nanowires: a review. J. Mater. Sci. Technol. 31, 556–572.

Hwang, Y.J., Boukai, A., Yang, P., 2009. High density n-Si/n-TiO_2 core/shell nanowire arrays with enhanced photoactivity. Nano Lett. 9, 410–415.

Iandolo, B., Hellman, A., 2014. The role of surface states in the oxygen evolution reaction on hematite. Angew. Chem. 126, 13622–13626.

Jing, L., Zhou, W., Tian, G., Fu, H., 2013. Surface tuning for oxide-based nanomaterials as efficient photocatalysts. Chem. Soc. Rev. 42, 9509–9549.

Kakuta, S., Abe, T., 2009. A novel example of molecular hydrogen generation from formic acid at visible-light-responsive photocatalyst. ACS Appl. Mater. Interfaces 1, 2707–2710.

Kang, D., Kim, T.W., Kubota, S.R., Cardiel, A.C., Cha, H.G., Choi, K.S., 2015. Electrochemical synthesis of photoelectrodes and catalysts for use in solar water splitting. Chem. Rev. 115, 12839–12887.

Kim, J., Choi, W., 2010. Hydrogen producing water treatment through solar photocatalysis. Energy Environ. Sci. 3, 1042–1045.

Kim, C., Kim, J.T., Kim, K.S., Jeong, S., Kim, H.Y., Han, Y.S., 2009. Immobilization of TiO_2 on an ITO substrate to facilitate the photoelectrochemical degradation of an organic dye pollutant. Electrochim. Acta 54, 5715–5720.

Kim, J., Monllor-Satoca, D., Choi, W., 2012. Simultaneous production of hydrogen with the degradation of organic pollutants using TiO_2 photocatalyst modified with dual surface components. Energy Environ. Sci. 5, 7647–7656.

Kim, J.H., Magesh, G., Kang, H.J., Banu, M., Kim, J.H., Lee, J., Lee, J.S., 2015. Carbonate-coordinated cobalt co-catalyzed $BiVO_4/WO_3$ composite photoanode tailored for CO_2 reduction to fuels. Nano Energy 15, 153–163.

Konstantinou, I.K., Albanis, T.A., 2003. Photocatalytic transformation of pesticides in aqueous titanium dioxide suspensions using artificial and solar light: intermediates and degradation pathways. Appl. Catal. B Environ. 42, 319–335.

Konstantinou, I.K., Albanis, T.A., 2004. TiO_2-assisted photocatalytic degradation of azo dyes in aqueous solution: kinetic and mechanistic investigations. Appl. Catal. B 49, 1–14.

Kudo, A., Miseki, Y., 2009. Heterogeneous photocatalyst materials for water splitting. Chem. Soc. Rev. 38, 253–278.

Leng, W.H., Zhu, W.C., Ni, J., Zhang, Z., Zhang, J.Q., Cao, C.N., 2006. Photoelectrocatalytic destruction of organics using TiO_2 as photoanode with simultaneous production of H_2O_2 at the cathode. Appl. Catal., A 300, 24–35.

Li, Y., Zhang, J.Z., 2009. Hydrogen generation from photoelectrochemical water splitting based on nanomaterials. Laser Photonics Rev. 4, 517–528.

Li, Y., Lu, G., Li, S., 2001. Photocatalytic hydrogen generation and decomposition of oxalic acid over platinized TiO_2. Appl. Catal., A 214, 179–185.

Li, J., Li, J., Chen, Q., Bai, J., Zhou, B., 2013a. Converting hazardous organics into clean energy using a solar responsive dual photoelectrode photocatalytic fuel cell. J. Hazard. Mater. 262, 304–310.

Li, Z., Luo, W., Zhang, M., Feng, J., Zou, Z., 2013b. Photoelectrochemical cells for solar hydrogen production: current state of promising photoelectrodes, methods to improve their properties, and outlook. Energy Environ. Sci. 6, 347–370.

Li, N., Wang, Y., Tang, D., Zhou, H., 2015. Integrating a photocatalyst into a hybrid lithium-sulfur battery for direct storage of solar energy. Angew. Chem. 127, 9403–9406.

Lianos, P., 2011. Production of electricity and hydrogen by photocatalytic degradation of organic wastes in a photoelectrochemical cell: the concept of the photofuel cell: a review of a re-emerging research field. J. Hazard. Mater. 185, 575–590.

Lichterman, M.F., Sun, K., Hu, S., Zhou, X.H., Mcdowell, M.T., Shaner, M.R., Richter, M.H., Crumlin, E.J., Carim, A.I., Saadi, F.H., Brunschwig, B.S., Lewis, N.S., 2016. Protection of inorganic semiconductors for sustained, efficient photoelectrochemical water oxidation. Catal. Today 262, 11–23.

Lim, M., Zhou, Y., Wang, L., Rudolph, V., Lu, G.Q., 2009. Development and potential of new generation photocatalytic systems for air pollution abatement: an overview. Asia-Pac. J. Chem. Eng. 4, 387–402.

Liu, G., Wang, L., Yang, H.G., Cheng, H.-M., Lu, G.Q., 2010. Titania-based photocatalysts—crystal growth, doping and heterostructuring. J. Mater. Chem. 20, 831–843.

Liu, Y., Li, J., Zhou, B., Li, X., Chen, H., Chen, Q., Wang, Z., Li, L., Wang, J., Cai, W., 2011. Efficient electricity production and simultaneously wastewater treatment via a high-performance photocatalytic fuel cell. Water Res. 45, 3991–3998.

Lu, X., Xie, S., Yang, H., Tong, Y., Ji, H., 2014. Photoelectrochemical hydrogen production from biomass derivatives and water. Chem. Soc. Rev. 43, 7581–7593.

Luo, B., Liu, G., Wang, L., 2016. Recent advances in 2D materials for photocatalysis. Nanoscale 8, 6904–6920.

Ma, Y., Wang, X., Jia, Y., Chen, X., Han, H., Li, C., 2014. Titanium dioxide-based nanomaterials for photocatalytic fuel generations. Chem. Rev. 114, 9987–10043.

Minggu, L.J., Wan Daud, W.R., Kassim, M.B., 2010. An overview of photocells and photoreactors for photoelectrochemical water splitting. Int. J. Hydrogen Energy 35, 5233–5244.

Morales-Guio, C.G., Tilley, S.D., Vrubel, H., Gratzel, M., Hu, X., 2014. Hydrogen evolution from a copper(I) oxide photocathode coated with an amorphous molybdenum sulphide catalyst. Nat. Commun. 5, 3059.

Noh, S.Y., Sun, K., Choi, C., Niu, M., Yang, M., Xu, K., Jin, S., Wang, D., 2013. Branched TiO_2/Si nanostructures for enhanced photoelectrochemical water splitting. Nano Energy 2, 351–360.

Ochiai, T., Fujishima, A., 2012. Photoelectrochemical properties of TiO_2 photocatalyst and its applications for environmental purification. J. Photochem. Photobiol., C 13, 247–262.

Ohno, T., Izumi, S., Fujihara, K., Masaki, Y., Matsumura, M., 2000. Vanishing of current-doubling effect in photooxidation of 2-propanol on TiO_2 in solutions containing Fe(III) ions. J. Phys. Chem. B 104, 6801–6803.

Osterloh, F.E., 2013. Inorganic nanostructures for photoelectrochemical and photocatalytic water splitting. Chem. Soc. Rev. 42, 2294–2320.

Pant, B., Pant, H.R., Barakat, N.A.M., Park, M., Han, T.-H., Lim, B.H., Kim, H.-Y., 2014. Incorporation of cadmium sulfide nanoparticles on the cadmium titanate nanofibers for enhanced organic dye degradation and hydrogen release. Ceram. Int. 40, 1553–1559.

Paracchino, A., Brauer, J.C., Moser, J.-E., Thimsen, E., Graetzel, M., 2012a. Synthesis and characterization of high-photoactivity electrodeposited Cu_2O solar absorber by photoelectrochemistry and ultrafast spectroscopy. J. Phys. Chem. C 116, 7341–7350.

Paracchino, A., Mathews, N., Hisatomi, T., Stefik, M., Tilley, S.D., Grätzel, M., 2012b. Ultrathin films on copper(I) oxide water splitting photocathodes: a study on performance and stability. Energy Environ. Sci. 5, 8673–8681.

Park, H., Vecitis, C.D., Hoffmann, M.R., 2008. Solar-powered electrochemical oxidation of organic compounds coupled with the cathodic production of molecular hydrogen. J. Phys. Chem. A 112, 7616–7626.

Park, H., Bak, A., Ahn, Y.Y., Choi, J., Hoffmannn, M.R., 2012. Photoelectrochemical performance of multi-layered BiO_x-TiO_2/Ti electrodes for degradation of phenol and production of molecular hydrogen in water. J. Hazard. Mater. 211–212, 47–54.

Park, Y., Mcdonald, K.J., Choi, K.S., 2013. Progress in bismuth vanadate photoanodes for use in solar water oxidation. Chem. Soc. Rev. 42, 2321–2337.

Patsoura, A., Kondarides, D.I., Verykios, X.E., 2006. Enhancement of photoinduced hydrogen production from irradiated Pt/TiO_2 suspensions with simultaneous degradation of azo-dyes. Appl. Catal. B 64, 171–179.

Peerakiatkhajohn, P., Butburee, T., Yun, J.-H., Chen, H., Richards, R.M., Wang, L., 2015. A hybrid photoelectrode with plasmonic Au@TiO_2 nanoparticles for enhanced photoelectrochemical water splitting. J. Mater. Chem. A 3, 20127–20133.

Peerakiatkhajohn, P., Yun, J.H., Chen, H., Lyu, M., Butburee, T., Wang, L., 2016. Stable hematite nanosheet photoanodes for enhanced photoelectrochemical water splitting. Adv. Mater. 28, 6405–6410.

Peerakiatkhajohn, P., Yun, J.-H., Wang, S., Wang, L., 2017. Review of recent progress in unassisted photoelectrochemical water splitting: from material modification to configuration design. J. Photonics Energy 7, 012006.

Peng, K., Xu, Y., Wu, Y., Yan, Y., Lee, S.T., Zhu, J., 2005. Aligned single-crystalline Si nanowire arrays for photovoltaic applications. Small 1, 1062–1067.

Peng, Y.-P., Yeh, Y.-T., Shah, S.I., Huang, C.P., 2012. Concurrent photoelectrochemical reduction of CO_2 and oxidation of methyl orange using nitrogen-doped TiO_2. Appl. Catal. B 123–124, 414–423.

Puga, A.V., 2016. Photocatalytic production of hydrogen from biomass-derived feedstocks. Coord. Chem. Rev. 315, 1–66.

Qu, Y., Duan, X., 2013. Progress, challenge and perspective of heterogeneous photocatalysts. Chem. Soc. Rev. 42, 2568–2580.

Ran, J., Zhang, J., Yu, J., Jaroniec, M., Qiao, S.Z., 2014. Earth-abundant cocatalysts for semiconductor-based photocatalytic water splitting. Chem. Soc. Rev. 43, 7787–7812.

Seger, B., Kamat, P.V., 2009. Fuel cell geared in reverse: photocatalytic hydrogen production using a TiO_2/nafion/Pt membrane assembly with no applied bias. J. Phys. Chem. C 113, 18946–18952.

Seger, B., Lu, G.Q., Wang, L.Z., 2012. Electrical power and hydrogen production from a photo-fuel cell using formic acid and other single-carbon organics. J. Mater. Chem. 22, 10709–10715.

Wang, L., Sasaki, T., 2014. Titanium oxide nanosheets: graphene analogues with versatile functionalities. Chem. Rev. 114, 9455–9486.

Wang, H., Zhang, L., Chen, Z., Hu, J., Li, S., Wang, Z., Liu, J., Wang, X., 2014a. Semiconductor heterojunction photocatalysts: design, construction, and photocatalytic performances. Chem. Soc. Rev. 43, 5234–5244.

Wang, X., Peng, K.Q., Hu, Y., Zhang, F.Q., Hu, B., Li, L., Wang, M., Meng, X.M., Lee, S.T., 2014b. Silicon/hematite core/shell nanowire array decorated with gold nanoparticles for unbiased solar water oxidation. Nano Lett. 14, 18–23.

Wang, X., Peng, W.-C., Li, X.-Y., 2014c. Photocatalytic hydrogen generation with simultaneous organic degradation by composite CdS–ZnS nanoparticles under visible light. Int. J. Hydrogen Energy 39, 13454–13461.

Wang, S.C., Chen, H.J., Gao, G.P., Butburee, T., Lyu, M.Q., Thaweesak, S., Yun, J.H., Du, A.J., Liu, G., Wang, L.Z., 2016. Synergistic crystal facet engineering and structural control of WO_3 films exhibiting unprecedented photoelectrochemical performance. Nano Energy 24, 94–102.

Wang, S., Yun, J.-H., Luo, B., Butburee, T., Peerakiatkhajohn, P., Thaweesak, S., Xiao, M., Wang, L., 2017. Recent progress on visible light responsive heterojunctions for photocatalytic applications. J. Mater. Sci. Technol. 33, 1–22.

Wu, H., Zhang, Z., 2011. Photoelectrochemical water splitting and simultaneous photoelectrocatalytic degradation of organic pollutant on highly smooth and ordered TiO_2 nanotube arrays. J. Solid State Chem. 184, 3202–3207.

Wu, W., Changzhong, J., Roy, V.A., 2015. Recent progress in magnetic iron oxide-semiconductor composite nanomaterials as promising photocatalysts. Nanoscale 7, 38–58.

Xia, Z.M., Zhou, X.M., Li, J., Qu, Y.Q., 2015. Protection strategy for improved catalytic stability of silicon photoanodes for water oxidation. Sci. Bull. 60, 1395–1402.

Xiao, L., He, Z., 2014. Applications and perspectives of phototrophic microorganisms for electricity generation from organic compounds in microbial fuel cells. Renew. Sustain. Energy Rev. 37, 550–559.

Xing, Z., Zong, X., Butburee, T., Pan, J., Bai, Y., Wang, L., 2016a. Nanohybrid materials of titania nanosheets and plasmonic gold nanoparticles for effective hydrogen evolution. Appl. Catal. A 521, 96–103.

Xing, Z., Zong, X., Zhu, Y., Chen, Z., Bai, Y., Wang, L., 2016b. A nanohybrid of CdTe@CdS nanocrystals and titania nanosheets with p–n nanojunctions for improved visible light-driven hydrogen production. Catal. Today 264, 229–235.

Yang, J., Wang, D., Han, H., Li, C., 2013. Roles of cocatalysts in photocatalysis and photoelectrocatalysis. Acc. Chem. Res. 46, 1900–1909.

Yue, D., Qian, X., Zhao, Y., 2015. Photocatalytic remediation of ionic pollutant. Sci. Bull. 60, 1791–1806.

Zang, G.-L., Sheng, G.-P., Shi, C., Wang, Y.-K., Li, W.-W., Yu, H.-Q., 2014. A bio-photoelectrochemical cell with a MoS3-modified silicon nanowire photocathode for hydrogen and electricity production. Energy Environ. Sci. 7, 3033–3039.

Zeng, Q., Bai, J., Li, J., Li, Y., Li, X., Zhou, B., 2014. Combined nanostructured Bi_2S_3/TNA photoanode and Pt/SiPVC photocathode for efficient self-biasing photoelectrochemical hydrogen and electricity generation. Nano Energy 9, 152–160.

Zhang, J., Wu, Y., Xing, M., Leghari, S.A.K., Sajjad, S., 2010. Development of modified N doped TiO_2 photocatalyst with metals, nonmetals and metal oxides. Energy Environ. Sci. 3, 715–726.

Zhang, Z., Dua, R., Zhang, L., Zhu, H., Zhang, H., Wang, P., 2013. Carbon-layer-protected cuprous oxide nanowire arrays for efficient water reduction. ACS Nano 7, 1709–1717.

Zhang, G., Liu, G., Wang, L., Irvine, J.T., 2016a. Inorganic perovskite photocatalysts for solar energy utilization. Chem. Soc. Rev. 45, 5951–5984.

Zhang, K., Ma, M., Li, P., Wang, D.H., Park, J.H., 2016b. Water splitting progress in tandem devices: moving photolysis beyond electrolysis. Adv. Energy Mater. 6, 1600602.

Zhen, C., Chen, R.Z., Wang, L.Z., Liu, G., Cheng, H.M., 2016. Tantalum (oxy)nitride based photoanodes for solar-driven water oxidation. J. Mater. Chem. A 4, 2783–2800.

Zong, X., Wang, L.Z., 2014. Ion-exchangeable semiconductor materials for visible light-induced photocatalysis. J. Photochem. Photobiol. C 18, 32–49.

Zong, X., Chen, H., Seger, B., Pedersen, T., Dargusch, M.S., Mcfarland, E.W., Li, C., Wang, L., 2014a. Selective production of hydrogen peroxide and oxidation of hydrogen sulfide in an unbiased solar photoelectrochemical cell. Energy Environ. Sci. 7, 3347–3351.

Zong, X., Han, J., Seger, B., Chen, H., Lu, G.M., Li, C., Wang, L., 2014b. An integrated photoelectrochemical-chemical loop for solar-driven overall splitting of hydrogen sulfide. Angew. Chem. Int. Ed. 53, 4399–4403.

CHAPTER TEN

Facet Control of Photocatalysts for Water Splitting

Jian Pan*, Gang Liu†,‡,1
*Curtin University, Perth, WA, Australia
†Shenyang National Laboratory for Materials Science, Institute of Metal Research, Chinese Academy of Sciences, Shenyang, China
‡School of Materials Science and Engineering, University of Science and Technology of China, Hefei, China
[1]Corresponding author: e-mail address: gangliu@imr.ac.cn

Contents

1. INTRODUCTION

Facet engineering has become one of the most effective modifications to nanomaterials for enhancing their performance in many applications, such as heterogeneous catalysis, gas sensing, antibacterial activity, photocatalysis, solar cells, and lithium-ion batteries, etc. This is because the reactions occurred at the surface or interface of the nanomaterials are very sensitive to the exposed surface atomic structures and the

Semiconductors and Semimetals, Volume 97
ISSN 0080-8784
http://dx.doi.org/10.1016/bs.semsem.2017.04.003

corresponding physical and chemical properties. For instance, a favorable adsorption mode of dye molecules anchored on the surface of semiconductors can increase the dye loading and accelerate the electron injection in dye-sensitized solar cells (DSSC) (Wu et al., 2011). The effective adsorption of water molecules on the surface of photocatalysts is the prerequisite for the subsequent photocatalytic reactions (Diebold, 2003). Li^+ surface insertion with a smaller energy barrier could give improved lithium storage at high rate in lithium ion batteries (Sun et al., 2010). The separation of reduction and oxidation reactive sites located on different facets is beneficial to the H_2 and/or O_2 evolution from water splitting (Mu et al., 2016).

So far, nanomaterials enclosed by well-defined facets have been widely explored. The faceted crystals have shown encouraging improvement in the field of environment and energy applications. Among them, photocatalytic water splitting is a significant process to convert solar energy into chemical energy by dissociating water into hydrogen (H_2) and oxygen (O_2). In this chapter, we mainly focus on the facet controlling of photocatalysts for water-splitting processes, including the water reduction, water oxidation, and overall water splitting.

2. FACETED PHOTOCATALYSTS FOR WATER REDUCTION

Water reduction is a key step to generate H_2 from water. A large number of semiconductor materials have been developed as photocatalysts for H_2 evolution from water splitting. To achieve the water reduction, the conduction band minimum (CBM) of the semiconductor photocatalysts should be more negative than the reduction potential of H^+/H_2. Among the photocatalysts satisfied this request, TiO_2 is one of the most investigated photocatalysts for H_2 evolution. The tailored TiO_2 with well-defined facets has also been widely studied (Liu et al., 2010c, 2011b, 2014). In this section, we mainly focus on the facet engineering of TO_2 for water reduction. And then, some other faceted photocatalysts for water reduction will be briefly introduced.

2.1 Anatase TiO_2

TiO_2 crystals with tailored facets have been one of the hottest research topics since the breakthrough in synthesizing high-quality anatase crystals with a high percentage of reactive high-energy {001} facets in 2008 (Yang et al., 2008). The extensive studies on the synthesis of faceted TiO_2, the unusual properties of the exposed facets, and the modification associated with surface have been achieved and reported (Liu et al., 2014). These faceted crystals have been actively used in nearly all research fields and have exhibited encouraging performance improvements in each application. Several reviews have comprehensively summarized the progress made in the facet engineering of TiO_2 (Fang et al., 2011; Liu et al., 2011a,b, 2014; Wen et al., 2011b). Here we classify all the studies by applications and target their performance in photocatalytic water splitting.

Anatase is the most intensely studied TiO_2 polymorph among the three natural phases (anatase, rutile, and brookite). Numerous studies have focused on the morphology control and facet engineering of the anatase crystals. The equilibrium shape of an anatase crystal according to the Wulff construction and surface energies calculation in vacuum is a slightly truncated bipyramid enclosed by more than 94% {101} and fewer 6% {001} facets (the left pane in Fig. 1) (Lazzeri et al., 2001). All the crystal shapes in the right panel of Fig. 1 show the facets have been developed in anatase crystals. The surface energies of low-index facets are in the order {111} ($1.09 J m^{-2}$) > {001} ($0.90 J m^{-2}$) > {010} ($0.53 J m^{-2}$) > {101} ($0.44 J m^{-2}$) (Lazzeri et al., 2001, 2002). The surface energy is highly related

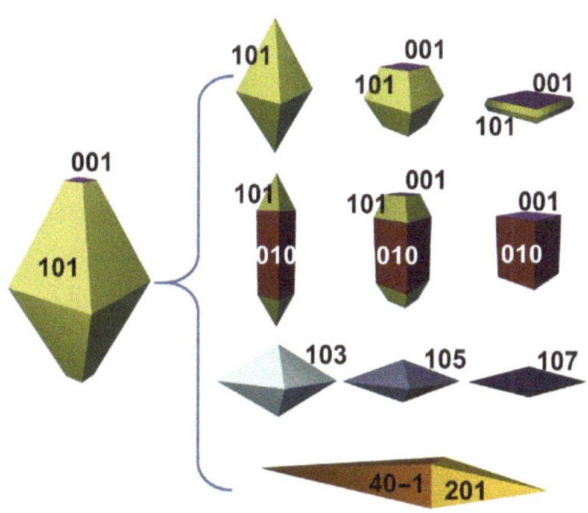

Fig. 1 Equilibrium crystal shape of anatase TiO_2 through the Wulff construction and the evolved other shapes. *Reprinted with permission from Liu, G., Yang, H.G., Pan, J., Yang, Y.Q., Lu, G.Q. & Cheng, H.M. 2014. Titanium dioxide crystals with tailored facets. Chem. Rev. 114, 9559–9612. Copyright 2014 American Chemical Society.*

to the density of undercoordinated Ti atoms. The {110} surface is the only low-index surface with fourfold coordinated Ti (Ti_{4c}) atoms; both {001} and {010} surfaces contain only fivefold coordinated Ti (Ti5c) atoms; and the most stable {101} surface has 50% Ti5c and 50% sixfold coordinated Ti (Ti6c) atoms.

So far, the production of low-index facets ({101}, {001}, {010}, and {110}) and high-index facets ({103}, {105}, {107}, {201}, {401}, {301}, and {106}) has been achieved in anatase crystals (Liu et al., 2014). Except the synthetic studies of these facets, they have been investigated in many applications. Among the application investigations, only a few studies of the low-index facets were associated with H_2 evolution from photocatalytic water splitting. Most of the rest works were focused on photodegradation (Liu et al., 2010a), lithium battery (Sun et al., 2010), DSSC (Wu et al., 2011), CO_2 reduction (Pan et al., 2011b), etc. The studies on the faceted anatase TiO_2 for photocatalytic water splitting are summarized in Table 1.

As we can see from Table 1, the faceted anatase TiO_2 exhibited enhanced performance in photocatalytic water splitting. However, the H_2 evolution rates were unpredictable with various shapes of crystals under different test conditions. Although some studies supported the trend of the higher activities of anatase crystal with higher percentage areas of {001} facet (Yang et al., 2011), some studies showed that a higher proportion of {001} did not always have a higher activity (Gordon et al., 2012; Pan et al., 2011a; Zhao et al., 2011a). These phenomena could be attributed to the following factors.

(i) Anisotropic surface electronic structures. Because of the different atomic arrangements and configuration in each surface, anisotropic surface electronic structures are logically expected. One remarkable example is that nanosized anatase crystals with 82% {101} and 18% {001} facets have a blue-shift absorption edge by around 9 nm with respect to microsized anatase crystals with 28% {101} and 72% {001} facets (Liu et al., 2010b). The origin of the bandgap difference in the two crystals is attributed to the different atomic configurations on {001} and {101} surface. Further study also found similar phenomenon by comparing microsized anatase crystals with a predominance of {001}, {101}, and {010} facets, respectively (Pan et al., 2011a).

Table 1 Anatase TiO_2 Crystals With Tailored Facets for H_2 Evolution From Photocatalytic Water Splitting

Faceted Anatase TiO_2 Crystals	Conditions	H_2 Evolution (μmol h^{-1} g^{-1})	References
Decahedral particles with {001} and {101} facets; 9.4 m^2 g^{-1}	50 vol% methanol solution with $H_2PtCl_6 \cdot 6H_2O$; 400 W Hg lamp	ca. 23,000	Amano et al. (2009)
Crystals with 4.5% {001} and 95.5% {101} facets; 134 m^2 g^{-1}	50 vol% methanol solution; 1 wt% Pt; 150 W Xe lamp with an atmospheric filter	ca. 1500 (fluorinated) ca. 2150 (defluorinated)	Gordon et al. (2012)
Crystals with 54.3% {001} and 45.7% {101} facets; 130 m^2 g^{-1}		ca. 1000 (fluorinated) ca. 1000 (defluorinated)	
Crystals with >60% {001} and <40% {101} facets; 99 m^2 g^{-1}		ca. 450 (fluorinated) ca. 250 (defluorinated)	
Hollow mesocrystals with dominant {101} facets; 42 m^2 g^{-1}	10 vol% methanol solution; 2 wt% Pt; 300 W Xe lamp	ca. 750	(Jiao et al., 2012)
Crystals with 100% {101} facet	20 vol% methanol solution; No Pt; 300 W Xe lamp	0	(Liu et al., 2013)
Crystals with 7.1% {001} and 92.9% {101} facets		ca. 110	
Crystals with 14.9% {001} and 75.1% {101} facets		ca. 180	
Crystals with 27.9% {001} and 72.1% {101} facets		ca. 120	
Crystals with 51.2% {001} and 48.8% {101} facets		ca. 10	

Continued

Table 1 Anatase TiO_2 Crystals With Tailored Facets for H_2 Evolution From Photocatalytic Water Splitting—cont'd

Faceted Anatase TiO_2 Crystals	Conditions	H_2 Evolution (μmol h^{-1} g^{-1})	References
Boron-doped sphere with {001} facet	10 vol% methanol solution; 1 wt% Pt; 300 W Xe lamp	ca. 720 (boron-free shell) ca. 200 (boron-containing shell)	Liu et al. (2012a)
Micron-sized crystals with 72% {001} facet; 3 μm	10 vol% methanol solution; 1 wt% Pt; 300 W Xe lamp	30 (fluorinated) 250 (defluorinated)	Liu et al. (2010b)
Nanosized crystals with 18% {001} facet; 21 m^2 g^{-1}		700 (fluorinated) 2050 (defluorinated)	
Anatase sheets with 56% {001} and 44% {101} facets	10 vol% methanol solution; 1 wt% Pt; 300 W Xe lamp	2332 μmol h^{-1} m^{-2} (oxygen-deficient surface) 1333 μmol h^{-1} m^{-2}	Liu et al. (2009a)
Nitrogen-doped anatase sheets with 60% {001} facet	10 vol% methanol solution; 1 wt% Pt; 300 W Xe lamp	1483 μmol h^{-1} m^{-2} (full spectrum) 81 μmol h^{-1} m^{-2} (visible spectrum)	Liu et al. (2009b)
Crystals with 40% {001} and 60% {101} facets	10 vol% methanol solution; 1 wt% Pt; 300 W Xe lamp	ca. 80 (fluorinated) ca. 160 (defluorinated)	Pan et al. (2011a)
Crystals with 24% {001} and 76% {101} facets		ca. 80 (fluorinated) ca. 220 (defluorinated)	
Crystals with 14% {001} and 33% {101} and 53% {010} facets		ca. 80 (fluorinated) ca. 380 (defluorinated)	

Table 1 Anatase TiO_2 Crystals With Tailored Facets for H_2 Evolution From Photocatalytic Water Splitting—cont'd

Faceted Anatase TiO_2 Crystals	Conditions	H_2 Evolution (μmol h^{-1} g^{-1})	References
Silicon-doped truncated-bipyramidal crystals with {001} and {201} facets	10 vol% methanol solution; 1 wt% Pt; 300 W Xe lamp	ca. 600 μmol h^{-1} m^{-2} (visible spectrum)	Wu et al. (2013)
Silicon-doped truncated-bipyramidal crystals with {001} and {101} facets		ca. 300 μmol h^{-1} m^{-2} (visible spectrum)	
Nitrogen-doped nanosheets with 67% {001} facet	25 vol% methanol solution; 0.5 wt% Pt; 350 W Xe lamp	865 (visible spectrum)	Xiang et al. (2011)
Sulfur-doped TiO_2 box with {001} facet	10 vol% methanol solution; 1 wt% Pt; 300 W Xe lamp	7054.5	Wen et al. (2011a)
TiO_2 box with {001} facet	20 vol% methanol solution; 1 wt% Pt; 300 W Xe lamp	7550	Xie et al. (2011)
5.0-nm thick nanosheets with 69% {001} facet	10 vol% methanol solution; 1 wt% Pt; 300 W Xe lamp	4335	Yang et al. (2011)
2.7-nm thick nanosheets with 82% {001} facet		7381	
1.6-nm thick nanosheets with 77% {001} facet		6958	
Nanosheets with 75% {001} facet	25 vol% methanol solution; 2 wt% Pt; 350 W Xe lamp	16,675	Yu et al. (2010)

Continued

Table 1 Anatase TiO_2 Crystals With Tailored Facets for H_2 Evolution From Photocatalytic Water Splitting—cont'd

Faceted Anatase TiO_2 Crystals	Conditions	H_2 Evolution (μmol h^{-1} g^{-1})	References
Flower-like microspheres with {010} facet; 96 m^2 g^{-1}	50 vol% methanol solution; 125 W Hg lamp	714.5 (no Pt) ca. 63,500 (1 wt% Pt)	Zhao et al. (2013b)
Nanorods with {010} facet; 27 m^2 g^{-1}		319.2 (no Pt) ca. 15,000 (1 wt% Pt)	
Microsphere consisting of decahedrons with 45% {001} facet; 67.2 m^2 g^{-1}	20 vol% methanol solution; 1 wt% Pt; 300 W Xe lamp	ca. 5900	Zheng et al. (2011a)
Crystals with {111} facets	18.5 vol% methanol solution; 0.5 wt% Pt; 300 W Xe lamp	ca. 12,000	Xu et al. (2013)
Pt-doped TiO_2 nanocrystals with 15.88% {001} facet; 124 m^2 g^{-1}	AM 1.5G	5400 (no sacrificial agent) 11,200 (20% methanol solution)	Banerjee et al. (2015)
Nanosheets with {001} and {101} facets	25 vol% methanol solution; 300 W Xe lamp	0 (no cocatalysts) 560 (1 wt% Co_3O_4) 2920 (1 wt% Pt) 5280 (1 wt% Pt and 1 wt% Co_3O_4)	Meng et al. (2016)

(ii) Anisotropic molecule adsorption. The effective adsorption of water molecule on the surface is important to the photocatalytic reactions. Most theoretical and experimental results support the molecular adsorption of water on anatase {101} surface (Selloni et al., 1998), and the dissociative adsorption of water on anatase {001} surface (Vittadini et al., 1998). The dissociative adsorption of water is also favorable on anatase {100} surface.

(iii) Synergistic effects of different facets. The different facets coexisting in one particle should not be considered as isolated surfaces, but as interactive surfaces with synergistic effects. It was found that the reduction and oxidation sites located on anatase {101} and {001} facets, respectively (Murakami et al., 2009). The spatial separation of reduction and oxidation sites will lead to the corresponding spatial separation of photoexcited charge carriers on different facets for improving the photocatalytic reactions. The very straightforward proof of the synergistic effects of different facets was the comparison of H_2 production of a group of bare anatase nanocrystals with the tunable percentage of {001} facet from 0% to 51.2% (Liu et al., 2013). As shown in Table 1, the sample with 14.9% {001} facet exhibited the highest activity, whereas the one with 100% {101} facet showed negligible activity.

2.2 Rutile TiO_2

Rutile is the thermodynamically most stable phase among the three natural phases of TiO_2. It has a wide range of applications, due to its unusual properties, such as high optical, high chemical stability, high refractive index, high dielectric constant, and excellent scattering efficiency. According to the Wulff construction with computed surface energies of the facets, Ramamoorthy et al. predicted that the shape of rutile under equilibrium conditions is a macroscopic crystal constructed with {110} (15.6 meV au^{-2}), {100} (19.6 meV au^{-2}), {011} (24.4 meV au^{-2}), and {001} (28.9 meV au^{-2}) facets (Ramamoorthy et al., 1994).

In Fig. 2, the left panel shows the equilibrium crystal shape of rutile TiO_2, while the right panel shows the possible shapes based on the former one. So far, the low-index facets ({110}, {011}, {001}, and {111}) (Liu et al., 2014) and high-index {331} facet (Truong et al., 2015) of rutile TiO_2 have been realized and applied in many applications, such as photodegradation (Aoyama et al., 2012; Cao et al., 2011; Cha et al., 2013; Zhang et al., 2015a), DSSC (Du et al., 2016; Hao et al., 2011),

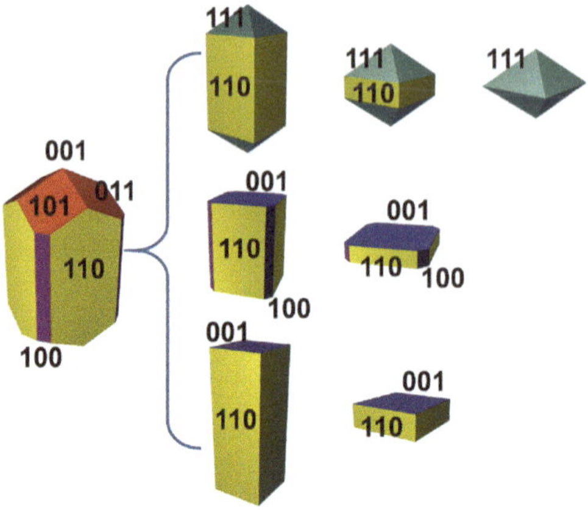

Fig. 2 Equilibrium crystal shape of rutile TiO_2 through the Wulff construction and the evolved other shapes of rutile TiO_2. *Reprinted with permission from Liu, G., Yang, H.G., Pan, J., Yang, Y.Q., Lu, G.Q., Cheng, H.M. 2014. Titanium dioxide crystals with tailored facets. Chem. Rev. 114, 9559–9612. Copyright 2014 American Chemical Society.*

lithium-ion battery (Chen and Lou, 2011; Hong et al., 2012), photocatalytic antibacterial (Sun et al., 2011), photoelectrocatalytic oxidation of water (Chen et al., 2014; Liu et al., 2012c; Zhang et al., 2013), and hydrogen evolution from photocatalytic water splitting (Bai et al., 2012; Wu et al., 2015b,c; Zhang et al., 2016b; Zhao et al., 2014). Those reports associated with faceted rutile TiO_2 crystals for H_2 evolution from photocatalytic water splitting are summarized in Table 2.

Similar to the anatase TiO_2, the rutile surface also has anisotropic molecule adsorption. For instance, a defect-free (110) surface adsorbs water molecules at low coverages in most studies (Brookes et al., 2001; Diebold, 2003; Henderson, 1996). The adsorption states of water on the rutile {100} surface are controversial. Current option tends to believe that water can dissociate on rutile {100} (Diebold, 2003), to some extent. The dissociative adsorption of water is favorable on both rutile (011) and (001) surfaces (Barnard et al., 2005; Fahmi and Minot, 1994).

Furthermore, the spatial separation of the photoexcited charge carriers between different facets also exists in rutile crystals. The early studies by Ohno et al. observed the selective photodeposition of Pt and PbO_2 particles on rutile {110} and {011} facets, respectively, indicating that rutile {110} and {011} facets provide reduction and oxidation sites, respectively (Ohno et al., 2002). The obvious separation of reduction and oxidation sites on faceted rutile crystals is attributed to photoexcited electron and hole transfer between {011} and {110} facets, which is driven by the higher electronic energy levels of {011} facet.

Table 2 Rutile TiO_2 Crystals With Tailored Facets for H_2 Evolution From Photocatalytic Water Splitting

Faceted Anatase TiO_2 Crystals	Conditions	H_2 Evolution ($\mu mol\ h^{-1}\ g^{-1}$)	References
Rutile sphere with {110} and {001} facets	10 vol% methanol solution; Cu as cocatalyst; 400 W Hg lamp	22,000	Bai et al. (2012)
Rutile nanorods with reduced {110} facet	20 vol% methanol solution; 0.5 wt% Pt; 300 W Xe lamp	ca. 6000 (full spectrum) ca. 80 (visible spectrum)	Zhao et al. (2014)
Rutile crystal with high ratio of {110}/{111} facets exposed	10 vol% methanol solution; 1 wt% Pt; 300 W Xe lamp	ca. 1	Wu et al. (2015b)
Rutile crystal with low ratio of {110}/{111} facets exposed		ca. 4	
Rutile crystal with 100% {111} facet		ca. 30	
Mesoporous rutile crystal with high ratio of {110}/{111} facets exposed		ca. 320	
Mesoporous rutile crystal with low ratio of {110}/{111} facets exposed		ca. 480	
Mesoporous rutile crystal with 100% {111} facet		ca. 800	
Boron-doped rutile microsphere with {111} and {110} facets	10 vol% methanol solution; 1 wt% Pt; 300 W Xe lamp	ca. 152 (boron-free shell) ca. 60 (boron contained shell)	Wu et al. (2015c)
Crystals with 94% {110} and 6% {111} facets	20 vol% methanol solution; 300 W Xe lamp; photodeposition (pd.) or impregnation deposition (imp.) Pt or Au	89.0 (pd. Au) 24.9 (imp. Au) 215.79 (pd. Pt) 89.05 (imp. Pt)	Zhang et al. (2016b)
Crystals with 73% {110} and 27% {111} facets		34.1 (pd. Au) 2.1 (imp. Au) 32.5 (pd. Pt) 3.0 (imp. Pt)	
Crystals with 27% {110} and 73% {111} facets		16.1 (pd. Au) 1.2 (imp. Au) 11.0 (pd. Pt) 1.9 (imp. Pt)	

At present, the investigations of faceted rutile TiO_2 are not as much as the studies of faceted anatase TiO_2. More attention should be focused on the systematic study on each rutile facet and the synergistic effects between each other. For instance, the high-energy {111} facet is the most studied one among the low-index facets. Except the studies summarized in Table 2, there are some other investigations on {111} facet (Liu et al., 2012c; Sun et al., 2011; Zhang et al., 2013, 2015a). However, some contradictive results are still under debate. Lai et al. reported that rutile crystals with nearly 100% {111} facet exhibited highest photocatalytic activity in OH radical generation, comparing to the one with majority {110} facet (Lai et al., 2013). Wu et al. also reported that rutile crystals with wholly {111} facet had highest H_2 evolution from photocatalytic water splitting, comparing to the one with a high ratio of {110}/{111} facets (Wu et al., 2015b). In contrast, the results from Zhang et al. showed that the rutile crystals with a high ratio of {110}/{111} facets showed highest H_2 evolution, no matter with photodeposition or impregnation deposition of Au or Pt nanoparticles as cocatalysts (Zhang et al., 2016b).

2.3 Brookite TiO_2

Brookite is the least investigated among the three natural phases of TiO_2, mainly due to the extreme difficulty in acquiring its pure form. The metastable brookite always appears as a minor by-product of anatase or rutile in most synthesis. However, when the particle size is in the range of 11–35 nm, it is reported that brookite is more stable than anatase (Zhang and Banfield, 2000). Moreover, there are many important progresses in the synthesis of pure brookite phase (Buonsanti et al., 2008; Kobayashi et al., 2011; Ohno et al., 2011), the morphology control of brookite TiO_2 for the application of lithium battery (Dambournet et al., 2011), photodegradation (Katsumata et al., 2012), and H_2 evolution (Lin et al., 2012).

Brookite TiO_2 is usually considered to be photocatalytically inactive. However, in recent years, the facet engineering of brookite TiO_2 greatly enhanced its performance in water reduction. Kandiel et al. synthesized high-quality brookite TiO_2 nanorods with {210}, {111}, and {−110} facet exposed via hydrothermal method (Kandiel et al., 2010). A water-soluble precursor, titanium bis(ammonium lactate) dihydroxide, was hydrothermally treated in the presence of different concentrations of urea to produce tailored TiO_2 crystals with different anatase/brookite ratios. The photoactivities of pure anatase nanoparticles, of anatase/brookite mixtures, and of pure brookite nanorods were assessed by photocatalytic water reduction. The hydrogen evolution results indicated that the photoactivities of anatase/brookite mixtures and pure brookite were higher than that of pure anatase nanoparticles, even with the lower surface area. It was demonstrated that the higher photocatalytic activity was attributed to the conduction band edge of brookite TiO_2 shifted more negative than that of anatase TiO_2.

Although brookite TiO_2 exhibited excellent photocatalytic activity, the function of the brookite facets exposed in photocatalytic water splitting was not clearly figured out. Several works have reported successful synthesis of brookite TiO_2 single crystals with specific facets. However, most of them were targeting the synthesis study (Ohno et al., 2011) or other applications (Kandiel et al., 2010; Zhao et al., 2013c). For instance, Katsumata et al. demonstrated that brookite TiO_2 nanoparticles exposed with {001} and {210} facets exhibited better performance than Degussa P25 in photodegradation, due to the noncovalent bonding oxygens and bridging oxygens on the {001} and {210} surfaces acted as active sites (Katsumata et al., 2012).

The function of brookite facets in the photocatalytic process was revealed by Li and his coworkers. They prepared impressive brookite TiO_2 single-crystal nanosheets surrounded with four {210}, two {101}, and two {201} facets by hydrothermally treating a novel in situ-generated precursor, $[Ti(C_3H_4O_3)_3]^{2-}$, with an asymmetric structure for the formation of low symmetric brookite TiO_2 (Lin et al., 2012). Selective photodeposition revealed that the {201} facets served as oxidation sites, while the remaining {210} and {101} facets acted as reduction sites. In the performance tests, faceted brookite TiO_2 exhibited much higher photocatalytic activity in photodegradation and $^{\bullet}OH$ generation, comparing to other brookite samples (as shown in Fig. 3). In their further work, the faceted brookite TiO_2 crystals with different percentage of {210} facets were investigated in photodegradation and H_2 evolution (Xu et al., 2015). The quasi-octahedral crystals with 70% {210} facet showed the highest performance in OH radical generation, photodegradation, and H_2 evolution, due to the unsaturated coordinative Ti and O atoms at the exposed surface.

Zhao et al. systematically studied the surface atomic structure and electronic structure of brookite {121} and {211} facets. In their investigation, brookite TiO_2 nanorods with dominant {121} facets and brookite TiO_2 nanosheets with dominant {211} facets were prepared and assessed by a series of characterization and tests (Zhao et al., 2015). The DFT (Density functional theory) calculation was employed to predict that the atomic ratio of Ti_{4c}:Ti_{5c}:Ti_{6c} for {121} facet is 12.5%:75%:12.5%, whereas the atomic

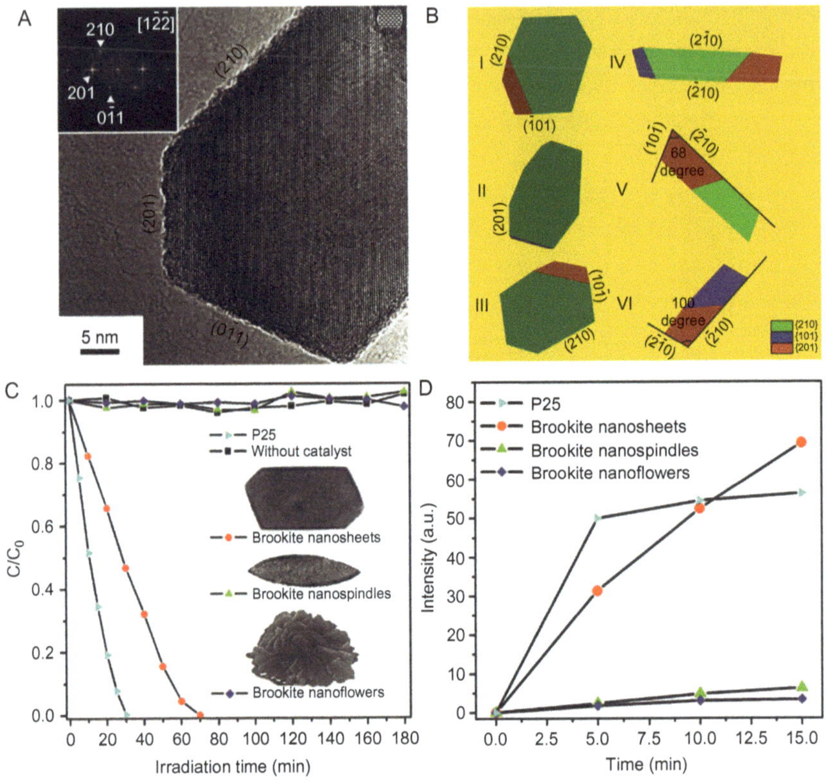

Fig. 3 (A) HRTEM images and (*insets*) corresponding FFT results of brookite nanosheets. (B) Nanosheet model in various orientations. (C) MO photocatalytic degradation in the presence of brookite nanosheets, nanoflowers, and nanospindles, Degussa P25, and no catalyst. (D) Fluorescence intensities for these samples at 425 nm using terephthalic acid as a fluorescence probe of •OH. *Reprinted with permission from Lin, H., Li, L., Zhao, M., Huang, X., Chen, X., Li, G., Yu, R. 2012. Synthesis of high-quality brookite TiO2 single-crystalline nanosheets with specific facets exposed: tuning catalysts from inert to highly reactive. J. Am. Chem. Soc. 134, 8328–8331. Copyright 2012 American Chemical Society.*

ratio of Ti_{5c}:Ti_{6c} for {211} facet is 75%:25%. The surface energies are 0.66 and 0.59 J m^{-2} for the {121} and {211} facets, respectively. The analysis of the surface electronic structure indicated that the {121} facet has lower VB potential and {211} facet has higher CB potential. For the photocatalytic performance, the nanorod with dominant {121} facets exhibited better performance in photodegradation due to the undercoordinated atoms at surface and the lower VB potential, while the nanosheets with dominant {211} facets showed superior hydrogen productivity due to the higher CB potential.

2.4 Others

Cuprous oxide, Cu_2O, is a p-type oxide semiconductor with a direct bandgap of 2 eV. Theoretical light to hydrogen conversion efficiency of Cu_2O is 18% (Paracchino et al., 2011). In recent years, Cu_2O nanomaterials have drawn much attention, due to their outstanding morphology-dependent applications in many applications (Shang and Guo, 2015). Especially, the Cu_2O polyhedra enclosed by clear facets have been intensively studied, from 6 facets cubes to 74 facets crystals with low-index {100}, {111}, {110} and high-index {211}, {522}, {544} facets (Sun and Yang, 2014). For instance, Sun et al. prepared highly symmetric multifacet (8-, 14-, 26-, 50-facet) polyhedral Cu_2O microcrystals via a complex-precursor solution route, and observed facet-dependent activity in photodegradation (Sun et al., 2012). Zhang et al. reported that Cu_2O polyhedra with high-index {544} and {104} facets showed high H_2 evolution rate from photocatalytic water splitting under visible light (as shown in Fig. 4) (Zhang et al., 2014c).

Cadmium sulfide, CdS, is one of the most popular photocatalysts due to its broad absorption of visible light and suitable conduction band position for water reduction. The shape control and facet engineering of CdS have recently been investigated (Chen et al., 2015; Li et al., 2012; Wang et al.,

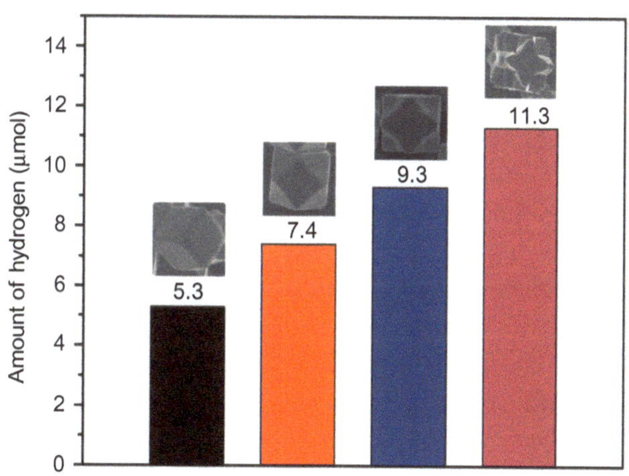

Fig. 4 The performance of photocatalytic hydrogen production under visible light irradiation over Cu_2O with different morphologies. *Reprinted with permission from Zhang, L., Shi, J., Liu, M., Jing, D., Guo, L., 2014. Photocatalytic reforming of glucose under visible light over morphology controlled Cu2O: efficient charge separation by crystal facet engineering. Chem. Commun. 50, 192–194. Copyright 2014 Royal Society of Chemistry.*

2015a; Yu et al., 2014). Among the studies, the high energy {0001} facet of CdS showed significant effects on the performance of CdS in H_2 evolution. Li et al. synthesized CdS micro/nano leaves with majority high energy {0001} facet exposed via controlled hydrothermal process (Li et al., 2012). The tailored CdS leaves showed enhance photocatalytic activity for H_2 evolution. Without any noble metal cocatalysts, CdS leaves exhibited an H_2 evolution rate of 740.9 $\mu mol\ h^{-1}$ under visible light irradiation, which was about 6.2 times higher than that of spheroidal CdS particles. It was demonstrated that the enhanced performance of CdS leaves could be attributed to the high crystallinity, the exposed high energy {0001} facet, and wider bandgap. Wang et al. prepared CdS nanocrystals with tunable shape by controlling the CdS growth along [0001] and [10 − 11] orientation via wet-chemistry synthesis (Wang et al., 2015a). Nanocone, nanofrustum, and nanoplate crystals with different ratio of {10–11}/{0001} facets were obtained. With largest {0001} facet, the nanoplates exhibited the highest photocatalytic activity for H_2 evolution, comparing to the nanocones and nanofrustums (as shown in Fig. 5).

3. FACETED PHOTOCATALYSTS FOR WATER OXIDATION

Water oxidation is also one of the key steps for both photocatalytic water splitting and photoelectrochemical (PEC) water splitting. However, its efficiency is relatively low comparing with the photocatalytic reduction of water. Tang et al. demonstrated that photocatalytic O_2 evolution from water occurs on a timescale of approximately 1 s and requiring four holes, while recombination takes place on the order of microseconds under reasonably weak laser excitation (Tang et al., 2008). Therefore, the water oxidation is considered as the bottleneck of water splitting.

The ideal photocatalysts or photoanodes should meet the following essential requirements: (1) the thermodynamically favorable valence band maximum (VBM) that is more positive than the potential of O_2 evolution; (2) a broad light absorption; 3) the efficient separation and transfer of charge carriers; (4) the sufficient active sites for O_2 evolution at the surface. Given the published results, the crystal facet plays a crucial role in the water oxidation processes.

3.1 Ag_3PO_4

Ag_3PO_4 is an ideal candidate of high-efficiency visible light photocatalysts for photocatalytic water oxidation and photocatalytic organic contaminant

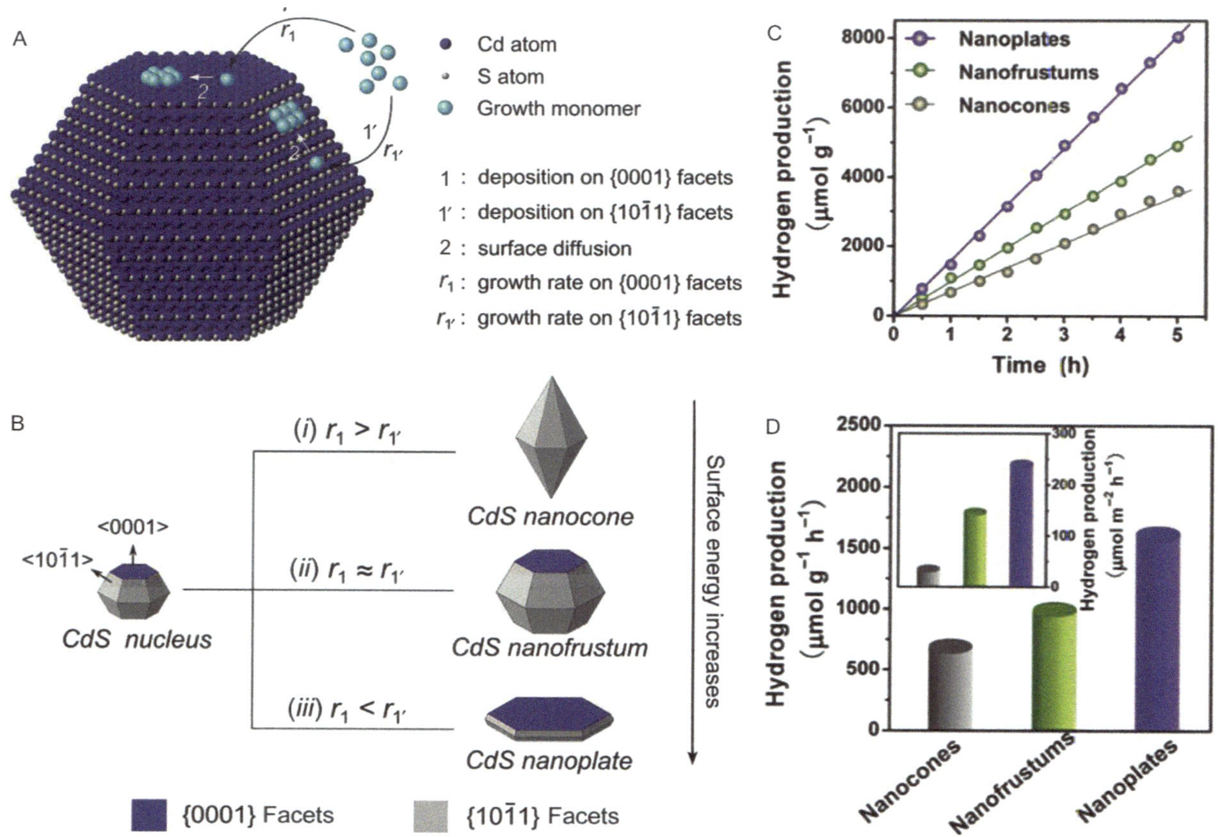

Fig. 5 Schematic illustration of (A) the major steps involved in the CdS growth process and (B) the formation of CdS nanocone, nanofrustum, and nanoplate. (C) Time courses of photocatalytic H_2 production and (D) average rate of CdS nanocones, nanofrustums, and nanoplates. *Inset* in (D) shows corresponding specific activities that were normalized with respect to their calculated surface areas. *Reprinted with permission from Wang, X.X., Liu, M.C., Zhou, Z.H., Guo, L.J. 2015. Toward facet engineering of CdS nanocrystals and their shape-dependent photocatalytic activities. J. Phys. Chem. C 119, 20555–20560. Copyright 2015 American Chemical Society.*

decomposition. Its highly positive VB position, low toxicity, and high photocatalytic ability have attracted much attention (Martin et al., 2015; Yi et al., 2010). It has been reported that Ag_3PO_4 outperformed WO_3 and $BiVO_4$ in oxygen evolution and photodegradation (Yi et al., 2010), although the photocorrosion of Ag_3PO_4 remains an issue. Many researchers have been endeavoring to enhance the performance of Ag_3PO_4 by doping (Zhang et al., 2014b), constructing heterostructures (Li et al., 2013a) and controlling the morphology of Ag_3PO_4 (Bi et al., 2011). The facet engineering of Ag_3PO_4 has been widely investigated in recent years. The encouraging results indicate that it is a promising and efficient approach for the further enhancement of the photocatalytic performance of Ag_3PO_4. In this section, we aim to summary the studies of facet engineering of Ag_3PO_4, although most of the studies targeted to the photodegradation and only a limited number of works on the water oxidation.

The crystal structure of Ag_3PO_4 is a body-centered cubic structure (BCC, space group *P4-3n* (218)) with a lattice parameter of 6.004 Å (Huang et al., 2013). Ag_3PO_4 possesses an indirect bandgap of 2.36 eV and a direct transition of 2.43 eV (Yi et al., 2010). And it has a very deep VBM, located at +2.9 V vs NHE (normal hydrogen electrode, pH 0). Similar to the most visible-light photocatalysts for oxygen evolution, its CBM locates at +0.45 V vs NHE (pH 0), which means it cannot realize the overall water splitting without external support.

The very first work on the facet-dependent photocatalytic activity of Ag_3PO_4 was reported in 2011. Bi et al. synthesized Ag_3PO_4 single crystal in rhombic dodecahedron shape with {110} facets exposed and one in cubic shape with {100} facets exposed by using CH_3COOAg or $[Ag(NH_3)_2]^+$ complexes as the silver ion precursors, respectively (Bi et al., 2011). The better performance of rhombic dodecahedrons in photodegradation suggested that the {110} fact is more reactive than {100} facet. Because the calculated surface energy of {110} facet is higher than that of {100} facets, it was speculated that {110} facets provided more active sites and enhanced the chemical adsorption of the organic dye, thus resulting in a higher photocatalytic activity. This milestone inspired more researchers to devote themselves to the synthesis and study of the faceted Ag_3PO_4. Except synthesizing rhombic dodecahedron single crystals to obtain {110} facet of Ag_3PO_4, Guo et al. (Wang et al., 2012) and Teng et al. (Wang et al., 2013) synthesized Ag_3PO_4 tetrapod single crystals with {110} facets exposed, respectively. Both as-prepared Ag_3PO_4 tetrapods exhibited enhanced performance in photodegradation.

Later, Zheng et al. developed a facile route to synthesize single-crystalline tetrahedral Ag_3PO_4 microcrystals with {111} facets exposed, which possess higher surface energy than those of {110} and {100} facets (Zheng et al., 2013). The tetrahedral Ag_3PO_4 showed a superior photocatalytic activity compared to the Ag_3PO_4 crystals with {110} and {100} facets exposed. Combining with DFT calculation, they concluded three advantages of {111} facets: the {111} facets possess highest surface energy (1.65 J m^{-2}) than that of {110} facet (1.31 J m^{-2}) and {100} facet (1.10 J m^{-2}); largest bandgap (2.736 eV) with higher conduction band position; and better separation of photogenerated electrons and holes. Beyond the most investigations on the low-index facets, the high-index facets are rarely studied. One interesting research reported by Jiao et al. is the synthesis of trisoctahedron Ag_3PO_4 microcrystals enclosed by {221} and {332} facets based on the heteroepitaxial growth procedure (Jiao et al., 2013). Au@Ag core–shell nanorods were used as starting materials. The trisoctahedron Ag_3PO_4 exhibited significantly enhanced photocatalytic activity compared with cubic Ag_3PO_4 for RhB dye photodegradation.

In a thermodynamic view, photodegradation is a downhill reaction, while water oxidation is an uphill reaction. Furthermore, water oxidation requires four holes, and photodegradation initially requires one hole. Consequently, the water oxidation is more difficult, thermodynamically and kinetically. However, Ag_3PO_4 possesses high photocatalytic activity in water oxidation. For example, Yi et al. first demonstrated that the photocatalytic water oxidation was possible on Ag_3PO_4, reporting peak quantum yields of ca. 90% at 420 nm with $AgNO_3$ as sacrificial agents (Yi et al., 2010).

At present, there are only a limited number of works report the influence on water oxidation by the exposing facets of Ag_3PO_4. Bi et al. reported that cubic Ag_3PO_4 with {100} facets exhibited higher photocurrent, stability, and conversion efficiency than common spherical Ag_3PO_4 in PEC water splitting (Bi et al., 2012). Martin et al. predicted that the {111} facet possesses the highest surface energy {1.65 J m^{-2}}, over two times higher than that of {110} (0.78 J m^{-2}) and {100} (0.67 J m^{-2}) facets by theoretical modeling (Martin et al., 2013). Then they synthesized tetrahedral Ag_3PO_4 crystals with {111} facets by using a novel kinetic control method in the absence of surfactants. In the photocatalytic water oxidation, the activity of the tetrahedral Ag_3PO_4 with {111} facet was nearly 12 times greater than cubic Ag_3PO_4 with {100} facet and rhombic dodecahedron Ag_3PO_4 with {110} facet (as shown in Fig. 6). And the internal quantum yield reached 98% at 400 nm and >80% from 365 to 500 nm. The authors concluded that the high activity of {111} facet was attributed to the higher mobility of holes and the overabundance of dangling P—O bonds as oxidation sites.

Fig. 6 SEM micrograph of Ag_3PO_4 crystals: (A) tetrahedron, (B) cubic, (C) rhombic dodecahedron. (D) Oxygen yield comparison of Ag_3PO_4 facets using a 300 W Xe light source under full arc irradiation, with $AgNO_3$ acting as an electron scavenger. *Reprinted with permission from Martin, D.J., Umezawa, N., Chen, X.W., Ye, J.H., Tang, J. W. 2013. Facet engineered Ag3PO4 for efficient water photooxidation. Energ. Environ. Sci. 6, 3380–3386. Copyright 2013 Royal Society of Chemistry.*

3.2 Bismuth vanadate

Bismuth vanadate ($BiVO_4$) is another one of the most promising photocatalysts for water oxidation and has been intensively studied in the past decade, due to its proper VBM, narrow bandgap (ca. 2.0 eV) for visible light absorption, and low cost. Its theoretical solar-to-hydrogen (STH) conversion efficiency reaches 9.2% with a maximum photocurrent of 7.5 mA cm^{-2} under standard AM1.5 solar irradiations (Prevot and Sivula, 2013). However, the actual efficiency of $BiVO_4$ is still unsatisfactory, as it suffers from three issues. 1. Low charge separation. Approximately 60%–80% of the electron–hole pairs recombine before reaching the photocatalyst surface, mainly due to the localization of the photoinduced electrons (Abdi et al., 2013; Cooper et al., 2014; Kweon and Hwang, 2013). 2. Low oxygen evolution kinetics. Although sulfite oxidation occurs extremely fast at the surface of $BiVO_4$ (McDonald and Choi, 2012), oxygen evolution rate of $BiVO_4$ is very low without cocatalysts (Li et al., 2014). 3. Low CBM. The CBM of $BiVO_4$ locates below the reversible hydrogen electrode (RHE) (Zhao et al., 2011b). Therefore, the water oxidation of $BiVO_4$ usually requires sacrificial agents in photocatalytic water splitting or an external bias in PEC water splitting.

Targeting above issues, many modifications have been conducted on $BiVO_4$-based materials in the past decade, including doping, morphology control, facet engineering, etc. Many significant and encouraging breakthroughs have been summarized in some reviews (Huang et al., 2014; Li et al., 2013c; Park et al., 2013).

$BiVO_4$ exists in three polymorphs, orthorhombic pucherite, tetragonal dreyerite, and monoclinic clinobisvanite. Among them, the monoclinic phase exhibits good thermal stability and photocatalytic activity (Park et al., 2013). Here we mainly focus on the facet engineering of the monoclinic phase $BiVO_4$.

Due to the importance of crystal facet in photocatalysis, the synthesis of crystals with highly active facets has attracted much attention. Xi et al. successfully synthesized, for the first time, well-defined m-$BiVO_4$ nanoplates with exposed {001} facets by an easy and straightforward hydrothermal route in the absence of template or organic surfactant (Xi and Ye, 2010). The exposed {001} facets of the m-$BiVO_4$ nanoplates lead to a remarkable enhancement of the visible-light photocatalytic degradation of organic molecules and photocatalytic oxidation of water for O_2 generation. Wang et al. synthesized monoclinic $BiVO_4$ crystals with preferentially exposed {040} facets were hydrothermally synthesized by using a trace amount of $TiCl_3$ as the directing agent (Wang et al., 2011). They found that the photocatalytic activity of $BiVO_4$ for oxygen evolution was proportionally correlated with the exposed surfaces of the {040} facet, and the active sites with a BiV_4 structure on the exposed {040} facet are assigned to be responsible for the high activity of O_2 evolution. Yang et al. found that the higher activity of {010} facet is attributed to the higher charge carrier mobility, easier adsorption of water, and lower overpotential for O_2 evolution, through a systematic study on water oxidation on $BiVO_4$ {010} and {011} facets by DFT calculations (Yang et al., 2013).

Recently, some investigations have demonstrated that the photoreduction and oxidation can occur preferentially at different facets, due to the

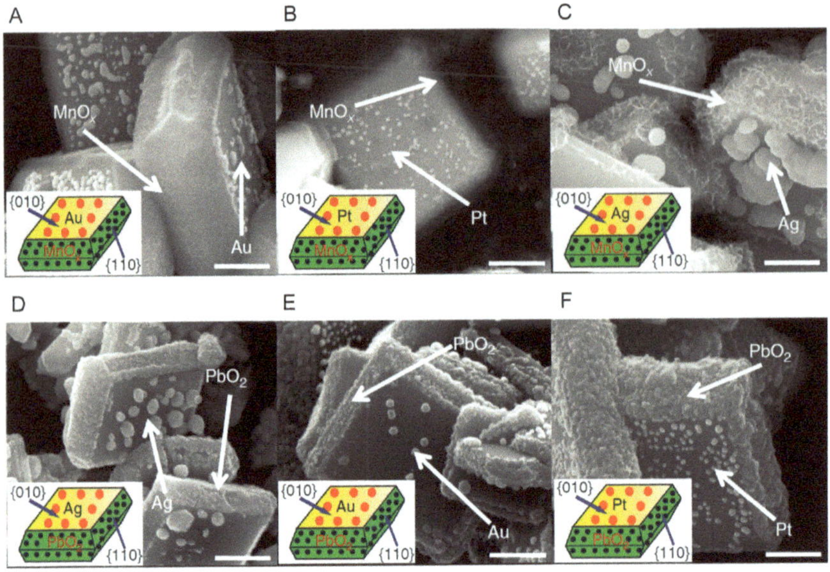

Fig. 7 SEM images of dual components photodeposited on the surface of $BiVO_4$. (A) Au/MnO_x/$BiVO_4$, (B) Pt/MnO_x/$BiVO_4$, (C) Ag/MnO_x/$BiVO_4$, (D) Ag/PbO_2/$BiVO_4$, (E) Au/PbO_2/$BiVO_4$, and (F) Pt/PbO_2/$BiVO_4$. The contents of the deposited metals/metal oxides are all 5 wt%. Scale bar, 500 nm. *Reprinted with permission from Li, C.J., Zhang, P., Lv, R., Lu, J.W., Wang, T., Wang, S.P., Wang, H.F., Gong, J.L. 2013. Selective deposition of Ag3PO4 on monoclinic BiVO4(040) for highly efficient photocatalysis. Small 9, 3951–3956. Copyright 2013 Nature Publishing Group.*

distribution of photogenerated charge carriers. Li et al. provided experimental evidence for the separation of electrons and holes between the {010} and {110} crystal facets of $BiVO_4$ by using photochemical labeling (Li et al., 2013b). It was found that Ag, Au, and Pt were all solely reduced on the {010} facets, while MnO_x and PbO_2 were oxidized on the {110} facets (as shown in Fig. 7). These results indicated that the photogenerated electrons and holes tend to accumulate on the {010} and {110} facets, respectively, leading to the reduction and oxidation reactions taking place on the {010} and {110} facets. They also found that only when the reduction/oxidation cocatalysts were selectively deposited on the corresponding reduction/oxidation reaction facets, the photocatalytic performance can be most greatly enhanced. In later work, they continued to investigate the two $BiVO_4$-based photocatalysts systems (Pt/MnO_x/$BiVO_4$ and Pt/Co_3O_4/$BiVO_4$) for oxygen evolution (Li et al., 2014). The O_2 evolution rates of $BiVO_4$ with cocatalysts deposited by different methods are summarized in Table 3. And it was found that the enhanced photocatalytic performances were due to not only the intrinsic nature of charge separation

Table 3 Photocatalytic Water Oxidation Performances of $BiVO_4$ With Cocatalysts Deposited by Different Methods

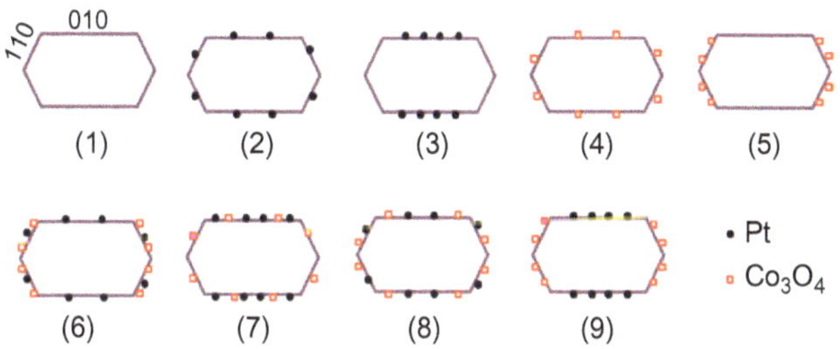

Entry	Pt Cocatalyst	Co_3O_4 Cocatalyst	The Amount of O_2 Evolution (μmol h^{-1})
1	—	—	1.0
2	Imp.	—	8.7
3	P.D.	—	16.1
4	—	Imp.	1.8
5	—	P.D.	2.5
6	Imp.	P.D.	87.0
7	P.D.	Imp.	84.2
8	Imp.	Imp.	9.6
9	P.D.	P.D.	160.3

Imp. presents impregnation method; P.D. presents photodeposition method. Reaction conditions: 0.15 g sample, 150 mL 0.02 M $NaIO_3$ aqueous solution, 300 W Xe lamp ($\lambda > 420$ nm), top irradiation, reaction time: 1 h. The contents of the deposited Pt and Co_3O_4 are optimized to be 0.5 and 0.075 wt%, respectively.

Reprinted with Permission from Li, R.G., Han, H.X., Zhang, F.X., Wang, D.G., Li, C. 2014. Highly efficient photocatalysts constructed by rational assembly of dual-cocatalysts separately on different facets of BiVO4. Energ. Environ. Sci., 7, 1369–1376. Copyright 2014 Royal Society of Chemistry.

between the {010} and {110} facets, but also the synergetic effect of dual-cocatalysts deposited on different facets of $BiVO_4$.

The new theory is also supported by the other works. The activation energies of carrier hopping and the mobility of electron/hole transport along seven low-index crystal orientations of bulk $BiVO_4$ were calculated using a small polaron model (Liu et al., 2015). The results showed that (1) the mobility of electrons is much larger than that of holes along the same crystal axis orientations, (2) the facets of (010) and (110) are highly exposed, and (3) for the exposed facets of (010) and (110), the mobility of electrons along [010] is five orders of magnitude larger than that of the [110] direction, and the mobility of holes along [110] is two orders of magnitude larger than that of the [010] direction. Furthermore, Zhu et al. employed spatially resolved surface photovoltage spectroscopy to reveal the anisotropic photoinduced charge transfer in a $BiVO_4$ single crystal (Zhu et al., 2015). The results showed that $BiVO_4$ single crystals with the different percentage of {010} facets possess the different strength of the built-in electric field between the {010} and {110} facets, which lead to highly anisotropic photoinduced hole distribution. Tan et al. also found that the plate-like $BiVO_4$ with dominant reduction functional {010} facets has superior photooxidation activities than the one with dominant oxidation functional {110} facets (Tan et al., 2016). The enlargement of {010} facets relative to {110} in $BiVO_4$ was indispensable for better suppression of electron trapping not only to improve photoreduction efficiency but also for effective photooxidation reactions.

The faceted engineering of $BiVO_4$ was also introduced into the photoanode for PEC water splitting. Kim et al. introduced {040} facets into the $BiVO_4$ photoanode, which lead to an increase in the energy conversion efficiency and oxygen evolution rate. The photocurrent density of the (040)-BVO photoanode is determined to be 0.94 mA cm^{-2} under AM 1.5 G illumination and produces 42.1% of the absorbed photon-to-current conversion efficiency at 1.23 V (vs RHE). In addition, Li et al. decorated Ag@AgCl on the {040} facets of $BiVO_4$ to construct Z-scheme for PEC water splitting (Li et al., 2015). And Li et al. used silver to modify the {040} facets of $BiVO_4$ to enhance the performance of photoanode (Li et al., 2016).

3.3 WO_3

WO_3 is a typical n-type semiconductor with a narrow indirect bandgap of 2.5–2.8 eV, which allows WO_3 capture approximately 12% of the solar spectrum and absorb visible light up to 500 nm. WO_3 becomes one of the most promising candidate for water oxidation due to its many excellent intrinsic properties, including high resistance against photocorrosion (Zheng et al., 2015), excellent chemical stability (Liu et al., 2012b), good hole mobility (10 cm^2 V^{-1} s^{-1}) (Butler, 1977), long diffusion length (150 nm), and favorable valence band position for water oxidation. However, the potential of CBM of WO_3 (0.3 V vs NHE at pH 0) is below the hydrogen redox. Therefore, WO_3 can only work as photocatalyst for oxygen evolution in the present of sacrificial agents or as photoanode under an external bias.

The crystal structure of WO_3 is a typical ABO_3 perovskite-like structure, which is constructed of WO_6 octahedra sharing corners and edges (Zheng et al., 2011b). WO_3 has five phases: tetragonal (α-WO_3), orthorhombic (β-WO_3), monoclinic I (γ-WO_3), triclinic (δ-WO_3), and monoclinic II (ε-WO_3) (Zheng et al., 2015). Among them, γ-WO_3 is the most stable phase in bulk WO_3 at room temperature. The facet engineering of WO_3 introduced in this section is mainly on γ-WO_3.

The monoclinic I WO_3 (γ-WO_3) is the most studied WO_3 for water oxidation. Many significant results have shown that the photocatalytic activity of WO_3 for oxygen evolution is highly affected by its surface atomic structures and corresponding electronic structures. Xie et al. reported a facile and new route of synthesizing γ-WO_3 crystals with different facets exposed (Xie et al., 2012). One was a quasi-cubic-like WO_3 crystal with a nearly equal percentage of {002}, {200}, and {020} facets, and the other was a rectangular sheet-like WO_3 crystal with predominant {002} facet (as shown in Fig. 8). DFT was used to predict a surface energy order of {002} ($1.56\,J\,m^{-2}$) > {020} ($1.54\,J\,m^{-2}$) > {200} ($1.43\,J\,m^{-2}$), suggesting that {200} is the most stable and {002} is the most unstable. Due to the crystal facet electronic structure effects, the quasi-cubic-like WO_3 possessed a smaller bang gap (2.71 eV) with a lower CBM. Meanwhile, the rectangular

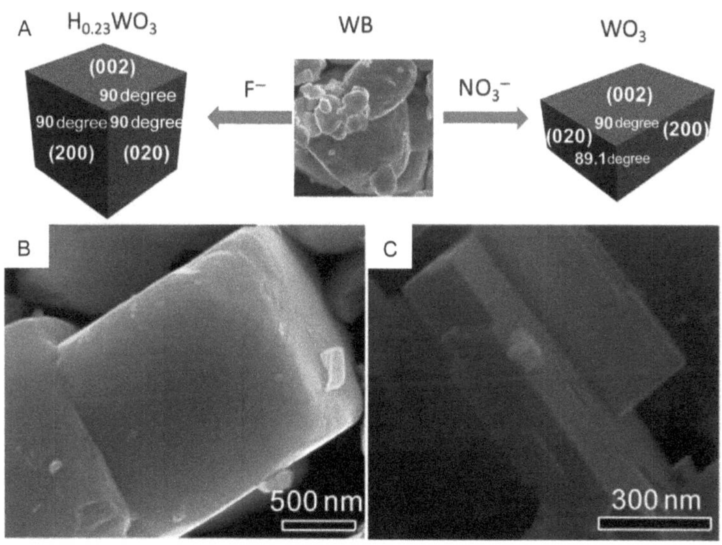

Fig. 8 (A) Schematic of the growth of a cubic $H_{0.23}WO_3$ crystal and rectangular sheet-like monoclinic WO_3 crystal from WB precursor in the presence of HF and HNO_3 aqueous solutions; SEM images of (B) quasi-cubic-like and (C) rectangular sheet-like WO_3 crystals. *Reprinted with permission from Xie, Y.P., Liu, G., Yin, L.C., Cheng, H.M. 2012. Crystal facet-dependent photocatalytic oxidation and reduction reactivity of monoclinic WO3 for solar energy conversion. J. Mater. Chem. 22, 6746–6751. Copyright 2012 Royal Society of Chemistry.*

sheet-like WO_3 had a larger bandgap (2.79 eV) with an elevated CBM. The former one showed a much higher oxygen evolution rate in the photocatalytic water oxidation, and the later one showed an ability to photoreduce CO_2/H_2O to CH_4. The photocatalytic activity studies revealed the substantial power of crystal facet electronic structure effects and broaden the photocatalytic applications of WO_3.

Another example of the facet-dependent photocatalytic activity of WO_3 for oxygen evolution was reported by Zhang et al. Monoclinic WO_3 rectangular nanoplates with dominated {100} facets was synthesized by a simple one-step hydrothermal method (Zhang et al., 2016a). Compared with normal WO_3, the {100} faceted WO_3 had a similar bandgap (2.84 eV vs 2.85 eV) with a 0.31 eV lower VBM. The theoretical calculation demonstrated that the density of states (DOS) of VBM of {100} locates at higher energy mainly because of the negative shift of O 2p orbitals. Meanwhile, the total DOS of CBM of {100} has the same shift because of the W 5d DOS. The results of photocatalytic activity tests showed that the {100} faceted WO_3 exhibited 5.1 times oxygen evolution rate as the normal WO_3 in the photocatalytic water oxidation.

Except for the photocatalytic water oxidation, γ-WO_3 has also been widely studied in PEC water splitting. The maximum theoretical STH efficiency of WO_3 is 4.8%, with a maximum photocurrent density of 3.9 mA cm^{-2} under AM 1.5 G irradiation (100 mW cm^{-2}) (Li and Wu, 2015). Many works have been reported that the photocurrent of WO_3 can be enhanced by the {002} facet. Zhang et al. synthesized orthorhombic $WO_3{\cdot}H_2O$ nanoplates with a clear shape first, then transferred the orthorhombic phase $WO_3{\cdot}H_2O$ to monoclinic phase WO_3 at high temperature. Thus, the most exposed {020} facet of the *O*-$WO_3{\cdot}H_2O$ nanoplate was transformed to the most reactive {002} facet of γ-WO_3 nanoplate. In the PEC water oxidation, the photocurrent of the γ-WO_3 photoanode reached 0.38 mA cm^{-2} at +1.0 V vs Ag/AgCl (Zheng et al., 2014). Zhang et al. synthesized in situ growth of monoclinic WO_3 nanomultilayers with preferentially exposed (002) facets via a one-step hydrothermal method (Zhang et al., 2015c). The WO_3 nanomultilayers exhibited an excellent photocurrent of 1.62 mA cm^{-2} at 1.25 V (vs Ag/AgCl) under AM1.5 G irradiation, and an IPCE of 40% at 400 nm at a bias of 0.67 V (vs Ag/AgCl). Furthermore, Qamar et al. also found that the photocurrent obtained from {002} faceted WO_3 nanocuboids was approximately twice that of spherical WO_3 under the same condition (Qamar et al., 2015).

Recently, a breakthrough to WO_3 photoanode was reported by Wang et al. (2016a). A facile hydrothermal method was employed for in situ growing WO_3 plate-like arrays on a transparent FTO substrate. WO_3 nanoplate arrays with dominant {200} and {002} facets were obtained, respectively. In the PEC water-splitting test, the photoanode with exposed {002} facets show a much higher photocurrent density (3.1 mA cm^{-2} at 1.23 V vs RHE) than that of the one with exposed {200} facets (2.1 mA cm^{-2} at 1.23 V vs RHE). With the optimized synthesis condition, a two-step hydrothermal method lead to the WO_3 photoanode exhibit extraordinary performance. Even without any cocatalysts for oxygen evolution, a remarkable photocurrent density of 3.7 mA cm^{-2} at 1.23 V vs RHE under AM 1.5 G irradiation, which is 93% of the theoretical photocurrent of WO_3 was obtained. And the highest IPCE reached 67% at 350 nm at 1.23 V vs RHE. The DFT calculations were carried out to demonstrate the facet effect on PEC water splitting. To drive the water oxidation, the {200} facet needs to overcome the whole energy change of 1.62 eV, while the {002} facet only needs to overcome 1.49 eV. Hence, the water oxidation occurred on {002} facet is easier than that on the {200} facet, contributing the better performance of {002} facets.

3.4 Others

Beyond Ag_3PO_4, $BiVO_4$, and WO_3, three of the most studied photocatalysts for water oxidation, the facet engineering research has been extended to other photocatalysts for water oxidation.

Bismuth oxychloride, BiOCl, has been synthesized in a variety of shapes and found the facet engineering is a promising strategy to improve its photocatalytic activity. Jiang et al. synthesized BiOCl nanosheets with {001} and {010} facets exposed, respectively. The nanosheets with dominant {001} facets exhibited a higher photocurrent than that of the nanosheets with dominant {010} facets, indicating the more efficient photoinduced charge separation and transfer in former one (Jiang et al., 2012). DFT calculations have further suggested that the surface energy of the {001} facet (2.42 J m^{-2}) is much higher than that of the {010} facet (0.51 J m^{-2}) (Zhao et al., 2013a). So far, several works have proved that the {001} facet of BiOCl enhanced its performance in PEC water splitting (Haider et al., 2016), CO_2 reduction (Zhang et al., 2015b), and photodegradation (Wu et al., 2015a).

Ceria, CeO_2, has also been studied for O_2 evolution. Jiang et al. synthesized CeO_2 nanocubes with isotropic {100} facet and nanorods with dominant {100} and {110} facets via a hydrothermal method (Jiang et al., 2015). In the photocatalytic performance tests, variable photocatalytic activity priorities were found in the photocatalytic oxidation of volatile organic compounds, oxygen evolution, and •OH generation, due to the different surface atomic structure and electronic structure.

$AgSbO_3$ is one of the photocatalyst for visible-light-driven water oxidation. Shi et al. first synthesized single-crystal nanosheets-based $NaSbO_3$ hierarchical cuboidal microclusters with exposed {001} facets, then performed an ion exchange on the as-prepared $NaSbO_3$ precursor to obtain $AgSbO_3$ hexagonal nanosheets with {001} facets exposed. The results showed that {001} facet enhanced the visible-light-driven oxygen evolution.

4. FACETED PHOTOCATALYSTS FOR OVERALL WATER SPLITTING

Overall water splitting is the ultimate goal of photocatalytic water splitting to generate a stoichiometric mixture of H_2 and O_2. It requires both the reduction of H^+ ions to H_2 by photogenerated electrons and the oxidation of H_2O to O_2 by holes. The surface of the photocatalysts plays a crucial role in the reaction processes. Because, even the band positions and the potentials of corresponding electrons and holes are thermodynamically sufficient for the overall water splitting, the photocatalysts could not produce H_2 and O_2 at the same time without surface active sites (Xing et al., 2012). Over the past 45 years, many photocatalysts were reported to exhibit ability for overall water splitting under the ultraviolet light irradiation (Chen et al., 2010). However, the photocatalytic overall water splitting is still facing many issues, as low efficiency, high cost due to noble metal cocatalysts, and deficient fundamental mechanistic understanding.

Among the photocatalysts for overall water splitting, perovskite-type oxides are the most common ones, which possess the general formula ABO_3. Due to their wide range of ferro-, piezo-, and pyroelectrical properties and electrooptical effects, they have been used as electronic, structure, magnetic, and refractory materials. In this section, we introduce $NaTaO_3$ and $SrTiO_3$ as photocatalysts for overall water splitting and review the studies of their facet engineering.

4.1 $NaTaO_3$

$NaTaO_3$ is one of the most studied photocatalysts that can realize overall water splitting. Back in 1998, Kato and Kudo found that alkali tantalate, $ATaO_3$ (A = Li, Na, K), with perovskite structure exhibited photocatalytic activity for water splitting into H_2 and O_2 under UV irradiation (Kato and Kudo, 1998). Later, the order of photocatalytic activity obtained was $KTaO_3 < NaTaO_3 < LiTaO_3$ (Kato and Kudo, 2003). However, NiO loaded $NaTaO_3$ exhibited highest photocatalytic activity, which led to numerous investigations devoted into the study of $NaTaO_3$ (Zhang et al., 2014a).

Although there are many research outputs of $NaTaO_3$, the most studies mainly concentrated on the effects of synthesis methods (Lan et al., 2016; Yan et al., 2009), doping (Kudo and Kato, 2000; Zhang et al., 2014a), cocatalysts loading (Kato and Kudo, 2003; Sakata et al., 2008), heterostructure (Deng et al., 2013; Xing et al., 2015), and crystal structures (Hu and Teng, 2007; Lin et al., 2006) on the performance of overall water splitting. However, some of these works partially contained the investigation of the surface of $NaTaO_3$.

In the early research stage, Kato and Kuto reported that doping with lanthanides Sr, Ba, La, Pr, Nd, Sm, Gd, Tb, and Dy improved the photocatalytic activity of $NaTaO_3$ for water splitting (Kudo and Kato, 2000). Among them, lanthanum was the most effective dopant, resulting in the maximum apparent quantum yield of the $NiO/NaTaO_3$:La was 56% at 270 nm (Kato et al., 2003). They found that ordered nanosteps at the surface of La-doped $NaTaO_3$, while the surface of nondoped $NaTaO_3$ was flat. Furthermore, the nanostep structure was also observed for other lanthanum-doped $NaTaO_3$ crystals. After charge location probe experiments, they demonstrated that the high performance of overall water splitting was attributed to the separated reaction sites at the nanostep structure, as the edges for the reduction sites and the grooves for the oxidation sites (as shown in Fig. 9).

The morphology of $NaTaO_3$ crystals affected by the dopants was due to the difference in ionic radii between dopants and Na cation. Ionic radii of La^{3+}, Ca^{2+}, Sr^{2+}, and Ba^{2+} in 12-coordination are similar to the radius of Na^+, while ionic radii of La^{3+}, Ca^{2+}, Sr^{2+}, and Ba^{2+} in 6-coordination are smaller than that of Ta^{5+}. Therefore, the distortion of the local structure

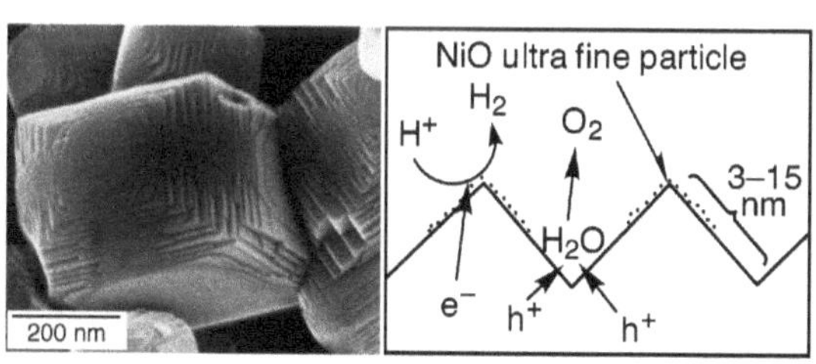

Fig. 9 SEM image of La-doped $NaTaO_3$; Mechanism of overall water splitting at the nanostep surface of $NiO/NaTaO_3$:La crystals. *Reprinted with permission from Kato, H., Asakura, K., Kudo, A. 2003. Highly efficient water splitting into H-2 and O-2 over lanthanum-doped NaTaO3 photocatalysts with high crystallinity and surface nanostructure. J. Am. Chem. Soc. 125, 3082–3089. Copyright 2003 American Chemical Society.*

caused by the Na replacement leads to the surface nanostep structure (Kudo et al., 2007). Although the authors did not identify the facets of the La-doped $NaTaO_3$, the grooves between facets played an important role in the water oxidation. The geometry of the walls on the both side of the groove would facilitate the multihole injection and the coupling between intermediate species to form O_2 (Kato et al., 2003).

The function of the flat surface of $NaTaO_3$ in the processes of water splitting was not identified. Recently, Zhang et al. investigated the {001} facets of cubic structured $NaTaO_3$ to identify their functions in the charge separation and surface reaction in overall water splitting (Zhang et al., 2016c). In charge location probe experiments, they found that no preferential deposition of noble metals (Pt, Ag) and metal oxides (MnO_2, PdO_2) on the six equivalent {001} facets. This result revealed that the photogenerated electrons and holes cannot be spatially separated among the different equivalent {001} facets of $NaTaO_3$. It also means high symmetry and exposed isotropic facets are not suitable for charge separation between facets.

According to the synthesis study, we can find that it is easy to synthesize the quasi-cubic $NaTaO_3$ crystals without any capping agents, no matter in hydrothermal reaction or in solid-state reaction. However, the studies of faceted engineering of $NaTaO_3$ are still rare. The synthesis and deep study of facets of $NaTaO_3$ are desirable to be developed for enhancing their effects on the photocatalytic activity of $NaTaO_3$.

4.2 $SrTiO_3$

$SrTiO_3$ is another popular photocatalyst for overall water splitting, which has simple cubic perovskite structure with an indirect bandgap of 3.1–3.7 eV (Grabowska, 2016). Because $SrTiO_3$ only responds to UV light, many studies were focused on modifying the band structure of $SrTiO_3$ to shift its optical absorption to the visible light region (Kanhere and Chen, 2014). For example, introducing a new mid-gap state between VB and CB by metal doping (Kato and Kudo, 2002; Yu et al., 2011), or changing the VB or CB by nonmetal doping (Ohno et al., 2005; Wang et al., 2003). Only a few investigations involved the effect of exposing facets on its photocatalytic performance.

Kato et al. used flux treatment with various molten salts on $SrTiO_3$ powders and found that the morphology of $SrTiO_3$ was highly depended on the kinds of molten salts (Kato et al., 2013). Large cubic crystals with mainly {100} facets exposed were obtained by the LiCl flux treatment; Smaller truncated cubes with {100} and {110} facets were obtained by the NaCl or $SrCl_2$ flux treatment. In contrast, $SrTiO_3$ crystals treated by the KCl flux exposed {100}, {120}, and {121} facets. In the photocatalytic performance test, it was found that all flux-treated samples exhibited higher activities than the irregular $SrTiO_3$ particles without treatment. Furthermore, the function of the exposing facets was discussed. The cubes with {100} facets and the truncated cubes with {100} and {110} facets exhibited similar photocatalytic activities, indicating {110} facets had no significantly positive effects. The KCl-treated $SrTiO_3$ with high-indexed {120} and {121} facets exhibited the highest activity. By the chemical probe test, it was found that the reduction sites located on {120} facet, whereas the oxidation sites located on {121} facet. The efficient separation of the reaction sites is responsible for the dramatic enhancement of photocatalytic activity.

The high-indexed {023} facet was also developed in recent years. Wang et al. successfully synthesized $SrTiO_3$ single crystals enclosed with {023} and {001} facets via a one-step solvothermal method (Wang et al., 2015b). It was found that the ratio of {023} and {001} facets could be adjusted by changing the concentrations of NaOH and ethanolamine during the solvothermal process. The selective photodeposition revealed that the {023} facet provided active oxidation sites, whereas the {001} facets provided reduction active sites. The performance of faceted $SrTiO_3$ in overall water splitting was optimized by tuning the ratio of {023} and {001} facets. Based on this work, Wang et al. continued to improve the performance of this faceted $SrTiO_3$ by introducing surface reconstructure on the {023} and {001} facets (Wang et al., 2016b).

In Section 4.1, it was mentioned that the high symmetry and isotropic facets of $NaTaO_3$ are not suitable for photogenerated charge separation. $SrTiO_3$ also has high-symmetry structure. However, Mu et al. found that the photogenerated charge carriers can be separated on the anisotropic facets of 18-facet $SrTiO_3$ (Mu et al., 2016). They synthesized the exposed facets of $SrTiO_3$ nanocrystals from isotropic facets (6-facet $SrTiO_3$) to anisotropic facets (18-facet $SrTiO_3$). The in situ photodeposition method was employed to reveal the locations of the reduction and oxidation catalytic sites. For the 6-facet $SrTiO_3$, both reduction and oxidation active sites appeared randomly on all facets. In contrast, the reduction and oxidation active sites separately located on the {001} and {110} facets of 18-facet $SrTiO_3$, respectively (as shown in Fig. 10). In addition, they further compared the photodeposition method and impregnation method of cocatalysts loading

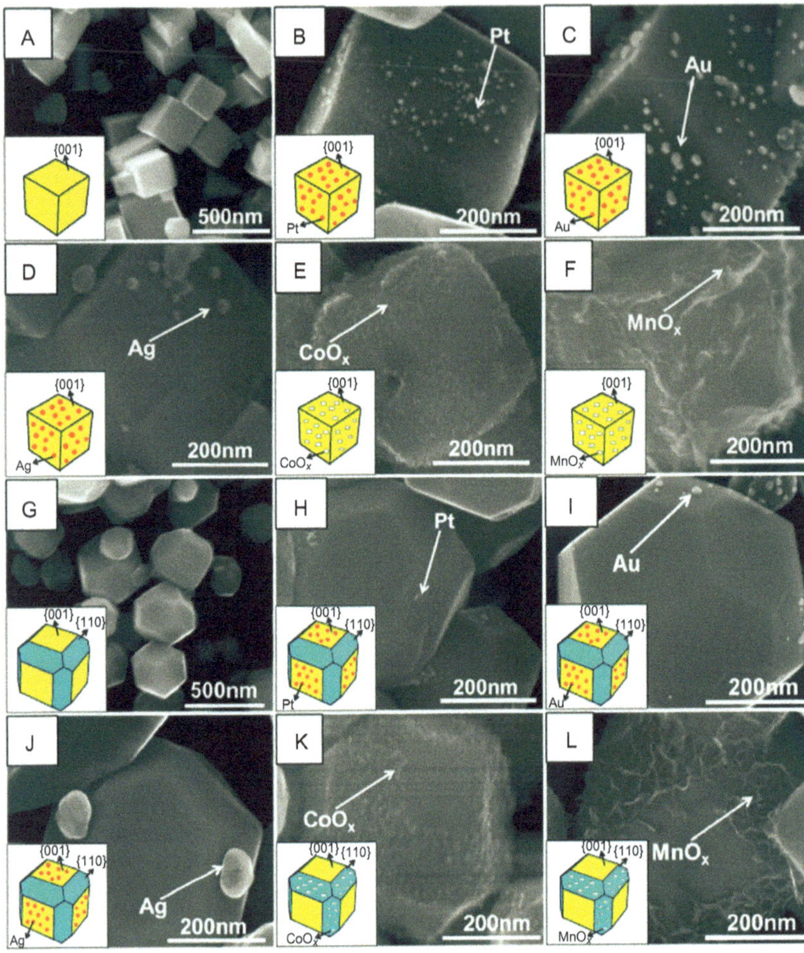

Fig. 10 The SEM images of the 6-facet and 18-facet $SrTiO_3$ nanocrystals with photodeposition of noble metals or metal oxides. (A) 6-facet $SrTiO_3$, (B) Pt/6-facet $SrTiO_3$, (C) Au/6-facet $SrTiO_3$, (D) Ag/6-facet $SrTiO_3$, (E) Co_3O_4/6-facet $SrTiO_3$, (F) MnO_x/6-facet $SrTiO_3$, (G) 18-facet $SrTiO_3$, (H) Pt/18-facet $SrTiO_3$, (I) Au/18-facet $SrTiO_3$, (J) Ag/18-facet $SrTiO_3$, (K) Co_3O_4/18-facet $SrTiO_3$, and (L) MnO_x/18-facet $SrTiO_3$. The contents of the deposited noble metal/metal oxides are all 2.0 wt%. *Reprinted with permission from Mu, L.C., Zhao, Y., Li, A.L., Wang, S.Y., Wang, Z.L., Yang, J.X., Wang, Y., Liu, T.F., Chen, R.T., Zhu, J., Fan, F.T., Li, R.G., Li, C. 2016. Enhancing charge separation on high symmetry SrTiO3 exposed with anisotropic facets for photocatalytic water splitting. Energ. Environ. Sci. 9, 2463–2469. Copyright 2016 Royal Society of Chemistry.*

on 18-facet $SrTiO_3$. The results summarized in Table 4 lead to the conclusion that the superior performance of $SrTiO_3$ with anisotropic exposed facets was attributed to the enhancement of photogenerated charge separation and spatial separation of the redox active sites for overall water splitting.

Table 4 Photocatalytic Overall Water-Splitting Performance of 18-Facet $SrTiO_3$ Nanocrystals With Cocatalysts Deposited

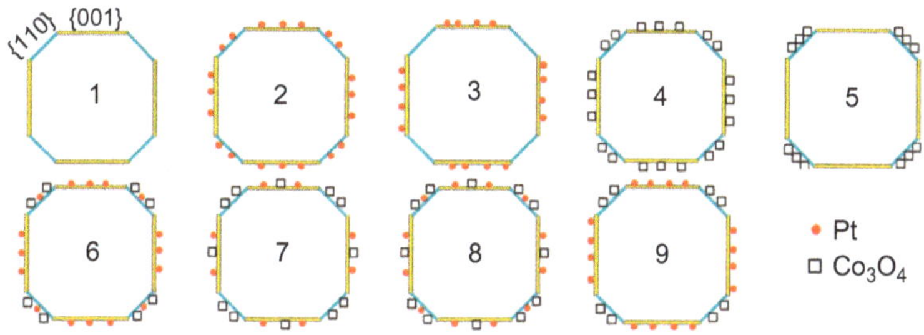

Entry	Pt	Co_3O_4	H_2	O_2	H_2/O_2
1	—	—	0	0	—
2	Imp.	—	1099	493	2.2
3	P.D.	—	2083	1143	1.8
4	—	Imp.	0	0	—
5	—	P.D.	0	0	—
6	Imp.	P.D.	749	406	1.8
7	P.D.	Imp.	1206	571	2.1
8	Imp.	Imp.	534	279	1.9
9	P.D.	P.D.	4089	2095	2.0

Imp. presents impregnation method; P.D. presents photodeposition method. Reaction time: 1 h. The contents of the deposited Pt and Co3O4 are optimized to be 0.1 and 0.01 wt%, respectively. The activities are normalized by the number of Pt active sites. Gas: mmol h^{-1} m^{-2}.

Reprinted with permission from Mu, L.C., Zhao, Y., Li, A.L., Wang, S.Y., Wang, Z.L., Yang, J.X., Wang, Y., Liu, T.F., Chen, R.T., Zhu, J., Fan, F.T., Li, R.G., Li, C. 2016. Enhancing charge separation on high symmetry SrTiO3 exposed with anisotropic facets for photocatalytic water splitting. Energ. Environ. Sci. 9, 2463–2469. Copyright 2016 Royal Society of Chemistry

5. CONCLUSION

The unprecedented progress of tailored photocatalysts with well-defined facets has been achieved in recent years, from the synthetic study of faceted crystals to the intrinsic property investigation of exposed surface, and to the overall performance in different applications.

At the initial research stage, researchers pursued to synthesize high-quality photocatalyst crystals enclosed by the high-index facets/high surface energy facets/active facets, and investigated the unique properties of each facet, and employed them in different applications. With the further development of facet engineering of photocatalysts, it was found that the different facets coexisting in one crystal should not be considered as isolated surface. Thus, synergistic effects between coexisting different facets, such as the charge transfer between facets and spatial separation of redox active sites, started to get more and more attention. Moreover, the facet controls of photocatalysts need to be adjusted in different applications for optimizing the overall performance. In other words, there is no universal structures of photocatalyst apply to all applications.

Undoubtedly, the faceted photocatalysts exhibited enhanced performance in the photocatalytic water splitting. Some photocatalysts have been comprehensively studied in the facet engineering, like TiO_2. For some other faceted photocatalysts, the studies are still in the initial stage. But they could progress quite fast, referring to the research outcomes of TiO_2. However, no matter at any stage, there are still many meaningful issues worth to be explored. For example, developing new facets with new features, heterostructure of faceted photocatalysts, combining facet engineering with other modifications to optimize photocatalytic performance, etc.

Overall, the encouraging improvements in performance and new performance features are stimulating the fast progress of the facet engineering of photocatalysts, feeding back to the rational design and fabrication of efficient material systems. Along with the knowledge of facet engineering accumulated, new material systems with outstanding performance will constantly be developed.

REFERENCES

Abdi, F.F., Firet, N., van de Krol, R., 2013. Efficient BiVO4 thin film photoanodes modified with cobalt phosphate catalyst and W-doping. ChemCatChem 5, 490–496.

Amano, F., Prieto-Mahaney, O.-O., Terada, Y., Yasumoto, T., Shibayama, T., Ohtani, B., 2009. Decahedral single-crystalline particles of anatase titanium(IV) oxide with high photocatalytic activity. Chem. Mater. 21, 2601–2603.

Aoyama, Y., Oaki, Y., Ise, R., Imai, H., 2012. Mesocrystal nanosheet of rutile TiO2 and its reaction selectivity as a photocatalyst. CrstEngComm 14, 1405–1411.

Bai, H., Liu, Z., Sun, D.D., 2012. Facile preparation of monodisperse, carbon doped single crystal rutile TiO2 nanorod spheres with a large percentage of reactive (110) facet exposure for highly efficient H-2 generation. J. Mater. Chem. 22, 18801–18807.

Banerjee, B., Amoli, V., Maurya, A., Sinha, A.K., Bhaumik, A., 2015. Green synthesis of Pt-doped TiO2 nanocrystals with exposed (001) facets and mesoscopic void space for photo-splitting of water under solar irradiation. Nanoscale 7, 10504–10512.

Barnard, A.S., Zapol, P., Curtiss, L.A., 2005. Modeling the morphology and phase stability of TiO2 nanocrystals in water. J. Chem. Theory. Comput. 1, 107–116.

Bi, Y., Ouyang, S., Umezawa, N., Cao, J., Ye, J., 2011. Facet effect of single-crystalline Ag3PO4 sub-microcrystals on photocatalytic properties. J. Am. Chem. Soc. 133, 6490–6492.

Bi, Y.P., Hu, H.Y., Ouyang, S.X., Lu, G.X., Cao, J.Y., Ye, J.H., 2012. Photocatalytic and photoelectric properties of cubic Ag3PO4 sub-microcrystals with sharp corners and edges. Chem. Commun. 48, 3748–3750.

Brookes, I.M., Muryn, C.A., Thornton, G., 2001. Imaging water dissociation on TiO2(110). Phys. Rev. Lett. 87, 266103.

Buonsanti, R., Grillo, V., Carlino, E., Giannini, C., Kipp, T., Cingolani, R., Cozzoli, P.D., 2008. Nonhydrolytic synthesis of high-quality anisotropically shaped brookite TiO(2) nanocrystals. J. Am. Chem. Soc. 130, 11223–11233.

Butler, M.A., 1977. Photoelectrolysis and physical properties of the semiconducting electrode WO_2. J. Appl. Phys. 48, 1914.

Cao, T., Li, Y., Wang, C., Shao, C., Liu, Y., 2011. One-step nonaqueous synthesis of pure phase TiO2 nanocrystals from TiCl4 in butanol and their photocatalytic properties. J. Nanomater.

Cha, S.I., Hwang, K.H., Kim, Y.H., Yun, M.J., Seo, S.H., Shin, Y.J., Moon, J.H., Lee, D.Y., 2013. Crystal splitting and enhanced photocatalytic behavior of TiO2 rutile nano-belts induced by dislocations. Nanoscale 5, 753–758.

Chen, J.S., Lou, X.W., 2011. Unusual rutile TiO2 nanosheets with exposed (001) facets. Chem. Sci. 2, 2219–2223.

Chen, X.B., Shen, S.H., Guo, L.J., Mao, S.S., 2010. Semiconductor-based photocatalytic hydrogen generation. Chem. Rev. 110, 6503–6570.

Chen, J.Y., Zhang, H.M., Liu, P.R., Wang, Y., Liu, X.L., Li, G.Y., An, T.C., Zhao, H.J., 2014. Vapor-phase hydrothermal synthesis of rutile TiO2 nanostructured film with exposed pyramid-shaped (111) surface and superiorly photoelectrocatalytic performance. J. Colloid Interface Sci. 429, 53–61.

Chen, J.Z., Wu, X.J., Yin, L.S., Li, B., Hong, X., Fan, Z.X., Chen, B., Xue, C., Zhang, H., 2015. One-pot synthesis of CdS nanocrystals hybridized with single-layer transition-metal dichalcogenide nanosheets for efficient photocatalytic hydrogen evolution. Angew. Chem. Int. Ed. 54, 1210–1214.

Cooper, J.K., Gul, S., Toma, F.M., Chen, L., Glans, P.A., Guo, J.H., Ager, J.W., Yano, J., Sharp, I.D., 2014. Electronic structure of monoclinic BiVO4. Chem. Mater. 26, 5365–5373.

Dambournet, D., Belharouak, I., Ma, J.W., Amine, K., 2011. Toward high surface area TiO2 brookite with morphology control. J. Mater. Chem. 21, 3085–3090.

Deng, Y.Y., Chen, Y.J., Chen, B.G., Ma, J.H., 2013. Preparation, characterization and photocatalytic activity of CuBi2O4/NaTaO3 coupled photocatalysts. J. Alloys Compd. 559, 116–122.

Diebold, U., 2003. The surface science of titanium dioxide. Surf. Sci. Rep. 48, 53–229.

Du, Y.E., Li, J., Liu, Y.F., Niu, X.J., Guo, F., Feng, Q., 2016. Synthesis of {110}-faceted rutile TiO2 nanocrystals from tetratitanate nanoribbons for improving dye-sensitized solar cell performance. RSC Adv. 6, 9717–9724.

Fahmi, A., Minot, C., 1994. A theoretical investigation of water-adsorption on titanium-dioxide surfaces. Surf. Sci. 304, 343–359.

Fang, W.Q., Gong, X.-Q., Yang, H.G., 2011. On the unusual properties of anatase TiO2 exposed by highly reactive facets. J. Phys. Chem. Lett. 2, 725–734.

Gordon, T.R., Cargnello, M., Paik, T., Mangolini, F., Weber, R.T., Fornasiero, P., Murray, C.B., 2012. Nonaqueous synthesis of TiO2 nanocrystals using TiF4 to engineer morphology, oxygen vacancy concentration, and photocatalytic activity. J. Am. Chem. Soc. 134, 6751–6761.

Grabowska, E., 2016. Selected perovskite oxides: characterization, preparation and photocatalytic properties—a review. Appl. Catal. B Environ. 186, 97–126.

Haider, Z., Zheng, J.Y., Kang, Y.S., 2016. Surfactant free fabrication and improved charge carrier separation induced enhanced photocatalytic activity of {001} facet exposed unique octagonal BiOCl nanosheets. Phys. Chem. Chem. Phys. 18, 19595–19604.

Hao, F., Lin, H., Zhou, C., Liu, Y., Li, J., 2011. Bifunctional single-crystalline rutile nanorod decorated heterostructural photoanodes for efficient dye-sensitized solar cells. Phys. Chem. Chem. Phys. 13, 15918–15924.

Henderson, M.A., 1996. An HREELS and TPD study of water on TiO2(110): the extent of molecular versus dissociative adsorption. Surf. Sci. 355, 151–166.

Hong, Z., Xu, Y., Liu, Y., Wei, M., 2012. Unique ordered TiO2 superstructures with tunable morphology and crystalline phase for improved lithium storage properties. Chem.-Eur. J. 18, 10753–10760.

Hu, C.C., Teng, H., 2007. Influence of structural features on the photocatalytic activity of NaTaO3 powders from different synthesis methods. Appl. Catal. A Gen. 331, 44–50.

Huang, G.F., Ma, Z.L., Huang, W.Q., Tian, Y., Jiao, C., Yang, Z.M., Wan, Z., Pan, A.L., 2013. Ag3PO4 semiconductor photocatalyst: possibilities and challenges. J. Nanomater.

Huang, Z.F., Pan, L., Zou, J.J., Zhang, X.W., Wang, L., 2014. Nanostructured bismuth vanadate-based materials for solar-energy-driven water oxidation: a review on recent progress. Nanoscale 6, 14044–14063.

Jiang, J., Zhao, K., Xiao, X.Y., Zhang, L.Z., 2012. Synthesis and facet-dependent photoreactivity of BiOCl single-crystalline nanosheets. J. Am. Chem. Soc. 134, 4473–4476.

Jiang, D., Wang, W.Z., Zhang, L., Zheng, Y.L., Wang, Z., 2015. Insights into the surface-defect dependence of photoreactivity over CeO2 nanocrystals with well-defined crystal facets. Acs Catal. 5, 4851–4858.

Jiao, W., Wang, L., Liu, G., Lu, G.Q., & Cheng, H.-M., 2012. Hollow anatase TiO2 single crystals and mesocrystals with dominant {101} facets for improved photocatalysis activity and tuned reaction preference. Acs Catal. 2, 1854–1859.

Jiao, Z.B., Zhang, Y., Yu, H.C., Lu, G.X., Ye, J.H., Bi, Y.P., 2013. Concave trisoctahedral Ag3PO4 microcrystals with high-index facets and enhanced photocatalytic properties. Chem. Commun. 49, 636–638.

Kandiel, T.A., Feldhoff, A., Robben, L., Dillert, R., Bahnemann, D.W., 2010. Tailored titanium dioxide nanomaterials: anatase nanoparticles and brookite nanorods as highly active photocatalysts. Chem. Mater. 22, 2050–2060.

Kanhere, P., Chen, Z., 2014. A review on visible light active perovskite-based photocatalysts. Molecules 19, 19995–20022.

Kato, H., Kudo, A., 1998. New tantalate photocatalysts for water decomposition into H-2 and O-2. Chem. Phys. Lett. 295, 487–492.

Kato, H., Kudo, A., 2002. Visible-light-response and photocatalytic activities of TiO2 and SrTiO3 photocatalysts codoped with antimony and chromium. J. Phys. Chem. B 106, 5029–5034.

Kato, H., Kudo, A., 2003. Photocatalytic water splitting into H-2 and O-2 over various tantalate photocatalysts. Catal. Today 78, 561–569.

Kato, H., Asakura, K., Kudo, A., 2003. Highly efficient water splitting into H-2 and O-2 over lanthanum-doped NaTaO3 photocatalysts with high crystallinity and surface nanostructure. J. Am. Chem. Soc. 125, 3082–3089.

Kato, H., Kobayashi, M., Hara, M., Kakihana, M., 2013. Fabrication of SrTiO3 exposing characteristic facets using molten salt flux and improvement of photocatalytic activity for water splitting. Cat. Sci. Technol. 3, 1733–1738.

Katsumata, K.-I., Ohno, Y., Tomita, K., Taniguchi, T., Matsushita, N., Okada, K., 2012. Synthesis of amphiphilic brookite nanoparticles with high photocatalytic performance for wide range of application. Acs Appl. Mater. Interfaces 4, 4846–4852.

Kobayashi, M., Petrykin, V., Tomita, K., Kakihana, M., 2011. Hydrothermal synthesis of brookite-type titanium dioxide with snowflake-like nanostructures using a water-soluble citratoperoxotitanate complex. J. Cryst. Growth 337, 30–37.

Kudo, A., Kato, H., 2000. Effect of lanthanide-doping into NaTaO3 photocatalysts for efficient water splitting. Chem. Phys. Lett. 331, 373–377.

Kudo, A., Niishiro, R., Iwase, A., Kato, H., 2007. Effects of doping of metal cations on morphology, activity, and visible light response of photocatalysts. Chem. Phys. 339, 104–110.

Kweon, K.E., Hwang, G.S., 2013. Surface structure and hole localization in bismuth vanadate: a first principles study. Appl. Phys. Lett. 103, 4.

Lai, Z., Peng, F., Wang, H., Yu, H., Zhang, S., Zhao, H., 2013. A new insight into regulating high energy facets of rutile TiO2. J. Mater. Chem. A 1, 4182–4185.

Lan, N.T., Phan, L.C., Hoang, L.H., Huan, B.D., Hong, L.V., Anh, T.X., Chinh, H.D., 2016. Hydrothermal synthesis, structure and photocatalytic properties of La/Bi Co-doped NaTaO3. Mater. Trans. 57, 1–4.

Lazzeri, M., Vittadini, A., Selloni, A., 2001. Structure and energetics of stoichiometric TiO2 anatase surfaces. Phys. Rev. B 63, 115409.

Lazzeri, M., Vittadini, A., Selloni, A., 2002. Structure and energetics of stoichiometric TiO2 anatase surfaces (vol 63, art no 155409, 2001). Phys. Rev. B 65, 119901.

Li, J.T., Wu, N.Q., 2015. Semiconductor-based photocatalysts and photoelectrochemical cells for solar fuel generation: a review. Cat. Sci. Technol. 5, 1360–1384.

Li, C.X., Han, L.J., Liu, R.J., Li, H.H., Zhang, S.J., Zhang, G.J., 2012. Controlled synthesis of CdS micro/nano leaves with (0001) facets exposed: enhanced photocatalytic activity toward hydrogen evolution. J. Mater. Chem. 22, 23815–23820.

Li, C.J., Zhang, P., Lv, R., Lu, J.W., Wang, T., Wang, S.P., Wang, H.F., Gong, J.L., 2013a. Selective deposition of Ag3PO4 on monoclinic BiVO4(040) for highly efficient photocatalysis. Small 9, 3951–3956.

Li, R.G., Zhang, F.X., Wang, D.G., Yang, J.X., Li, M.R., Zhu, J., Zhou, X., Han, H.X., Li, C., 2013b. Spatial separation of photogenerated electrons and holes among {010} and {110} crystal facets of BiVO4. Nat. Commun. 4, 7.

Li, Z., Luo, W., Zhang, M., Feng, J., Zou, Z., 2013c. Photoelectrochemical cells for solar hydrogen production: current state of promising photoelectrodes, methods to improve their properties, and outlook. Energ. Environ. Sci. 6, 347–370.

Li, R.G., Han, H.X., Zhang, F.X., Wang, D.G., Li, C., 2014. Highly efficient photocatalysts constructed by rational assembly of dual-cocatalysts separately on different facets of BiVO4. Energ. Environ. Sci. 7, 1369–1376.

Li, H.Y., Sun, Y.J., Cai, B., Gan, S.Y., Han, D.X., Niu, L., Wu, T.S., 2015. Hierarchically Z-scheme photocatalyst of Ag@AgCl decorated on BiVO4 (040) with enhancing photoelectrochemical and photocatalytic performance. Appl. Catal. B Environ. 170, 206–214.

Li, J.Q., Zhou, J., Hao, H.J., Zhu, Z.F., 2016. Silver-modified specific (040) facet of BiVO4 with enhanced photoelectrochemical performance. Mater. Lett. 170, 163–166.

Lin, W.H., Cheng, C., Hu, C.C., Teng, H.S., 2006. NaTaO3 photocatalysts of different crystalline structures for water splitting into H-2 and O-2. Appl. Phys. Lett. 89, 3.

Lin, H., Li, L., Zhao, M., Huang, X., Chen, X., Li, G., Yu, R., 2012. Synthesis of high-quality brookite TiO2 single-crystalline nanosheets with specific facets exposed: tuning catalysts from inert to highly reactive. J. Am. Chem. Soc. 134, 8328–8331.

Liu, G., Yang, H.G., Wang, X., Cheng, L., Lu, H., Wang, L., Lu, G.Q., Cheng, H.-M., 2009a. Enhanced photoactivity of oxygen-deficient anatase TiO2 sheets with dominant {001} facets. J. Phys. Chem. C 113, 21784–21788.

Liu, G., Yang, H.G., Wang, X., Cheng, L., Pan, J., Lu, G.Q., Cheng, H.-M., 2009b. Visible light responsive nitrogen doped anatase TiO2 sheets with dominant {001} facets derived from TiN. J. Am. Chem. Soc. 131, 12868–12869.

Liu, S., Yu, J., Jaroniec, M., 2010a. Tunable photocatalytic selectivity of hollow TiO2 microspheres composed of anatase polyhedra with exposed {001} facets. J. Am. Chem. Soc. 132, 11914–11916.

Liu, G., Sun, C., Yang, H.G., Smith, S.C., Wang, L., Lu, G.Q., Cheng, H.-M., 2010b. Nanosized anatase TiO2 single crystals for enhanced photocatalytic activity. Chem. Commun. 46, 755–757.

Liu, G., Wang, L.Z., Yang, H.G., Cheng, H.M., Lu, G.Q., 2010c. Titania-based photocatalysts-crystal growth, doping and heterostructuring. J. Mater. Chem. 20, 831–843.

Liu, S., Yu, J., Jaroniec, M., 2011a. Anatase TiO2 with dominant high-energy {001} facets: synthesis, properties, and applications. Chem. Mater. 23, 4085–4093.

Liu, G., Yu, J.C., Lu, G.Q., Cheng, H.-M., 2011b. Crystal facet engineering of semiconductor photocatalysts: motivations, advances and unique properties. Chem. Commun. 47, 6763–6783.

Liu, G., Pan, J., Yin, L., Irvine, J.T.S., Li, F., Tan, J., Wormald, P., Cheng, H.-M., 2012a. Heteroatom-modulated switching of photocatalytic hydrogen and oxygen evolution preferences of anatase TiO2 microspheres. Adv. Funct. Mater. 22, 3233–3238.

Liu, X., Wang, F.Y., Wang, Q., 2012b. Nanostructure-based WO3 photoanodes for photoelectrochemical water splitting. Phys. Chem. Chem. Phys. 14, 7894–7911.

Liu, X., Zhang, H., Yao, X., An, T., Liu, P., Wang, Y., Peng, F., Carroll, A.R., Zhao, H., 2012c. Visible light active pure rutile TiO2 photoanodes with 100% exposed pyramid-shaped (111) surfaces. Nano Res. 5, 762–769.

Liu, C., Han, X., Xie, S., Kuang, Q., Wang, X., Jin, M., Xie, Z., Zheng, L., 2013. Enhancing the photocatalytic activity of anatase TiO2 by improving the specific facet-induced spontaneous separation of photogenerated electrons and holes. Chem. Asian J. 8, 282–289.

Liu, G., Yang, H.G., Pan, J., Yang, Y.Q., Lu, G.Q., Cheng, H.M., 2014. Titanium dioxide crystals with tailored facets. Chem. Rev. 114, 9559–9612.

Liu, T.F., Zhou, X., Dupuis, M., Li, C., 2015. The nature of photogenerated charge separation among different crystal facets of BiVO4 studied by density functional theory. Phys. Chem. Chem. Phys. 17, 23503–23510.

Martin, D.J., Umezawa, N., Chen, X.W., Ye, J.H., Tang, J.W., 2013. Facet engineered Ag3PO4 for efficient water photooxidation. Energ. Environ. Sci. 6, 3380–3386.

Martin, D.J., Liu, G.G., Moniz, S.J.A., Bi, Y.P., Beale, A.M., Ye, J.H., Tang, J.W., 2015. Efficient visible driven photocatalyst, silver phosphate: performance, understanding and perspective. Chem. Soc. Rev. 44, 7808–7828.

McDonald, K.J., Choi, K.S., 2012. A new electrochemical synthesis route for a BiOI electrode and its conversion to a highly efficient porous BiVO4 photoanode for solar water oxidation. Energ. Environ. Sci. 5, 8553–8557.

Meng, A.Y., Zhang, J., Xu, D.F., Cheng, B., Yu, J.G., 2016. Enhanced photocatalytic H-2-production activity of anatase TiO2 nanosheet by selectively depositing dual-cocatalysts on {101} and {001} facets. Appl. Catal. B Environ. 198, 286–294.

Mu, L.C., Zhao, Y., Li, A.L., Wang, S.Y., Wang, Z.L., Yang, J.X., Wang, Y., Liu, T.F., Chen, R.T., Zhu, J., Fan, F.T., Li, R.G., Li, C., 2016. Enhancing charge separation on high symmetry SrTiO3 exposed with anisotropic facets for photocatalytic water splitting. Energ. Environ. Sci. 9, 2463–2469.

Murakami, N., Kurihara, Y., Tsubota, T., Ohno, T., 2009. Shape-controlled anatase titanium(IV) oxide particles prepared by hydrothermal treatment of peroxo titanic acid in the presence of polyvinyl alcohol. J. Phys. Chem. C 113, 3062–3069.

Ohno, T., Sarukawa, K., Matsumura, M., 2002. Crystal faces of rutile and anatase TiO2 particles and their roles in photocatalytic reactions. New J. Chem. 26, 1167–1170.

Ohno, T., Tsubota, T., Nakamura, Y., Sayama, K., 2005. Preparation of S, C cation-codoped SrTiO3 and its photocatalytic activity under visible light. Appl. Catal. A Gen. 288, 74–79.

Ohno, Y., Tomita, K., Komatsubara, Y., Taniguchi, T., Katsumata, K.-I., Matsushita, N., Kogure, T., Okada, K., 2011. Pseudo-cube shaped brookite (TiO2) nanocrystals synthesized by an oleate-modified hydrothermal growth method. Cryst. Growth Des. 11, 4831–4836.

Pan, J., Liu, G., Lu, G.Q., Cheng, H.-M., 2011a. On the true photoreactivity order of {001}, {010}, and {101} facets of anatase TiO2 crystals. Angew. Chem. Int. Ed. 50, 2133–2137.

Pan, J., Wu, X., Wang, L., Liu, G., Lub, G.Q., Cheng, H.-M., 2011b. Synthesis of anatase TiO2 rods with dominant reactive {010} facets for the photoreduction of CO2 to CH4 and use in dye-sensitized solar cells. Chem. Commun. 47, 8361–8363.

Paracchino, A., Laporte, V., Sivula, K., Gratzel, M., Thimsen, E., 2011. Highly active oxide photocathode for photoelectrochemical water reduction. Nat. Mater. 10, 456–461.

Park, Y., McDonald, K.J., Choi, K.S., 2013. Progress in bismuth vanadate photoanodes for use in solar water oxidation. Chem. Soc. Rev. 42, 2321–2337.

Prevot, M.S., Sivula, K., 2013. Photoelectrochemical tandem cells for solar water splitting. J. Phys. Chem. C 117, 17879–17893.

Qamar, M., Alhooshani, K., Ahmed, M.I., Azad, A.M., Merzougui, B., 2015. Amelioration in the visible-light-driven photocatalysis by <002> faceted WO3 nanocuboids. Cat. Com. 70, 21–25.

Ramamoorthy, M., Vanderbilt, D., Kingsmith, R.D., 1994. 1st-principles calculations of the energetics of stoichiometric TIO2 surfaces. Phys. Rev. B 49, 16721–16727.

Sakata, Y., Matsuda, Y., Yanagida, T., Hirata, K., Imamura, H., Teramura, K., 2008. Effect of metal ion addition in a Ni supported Ga(2)O(3) photocatalyst on the photocatalytic overall splitting of H(2)O. Catal. Lett. 125, 22–26.

Selloni, A., Vittadini, A., Gratzel, M., 1998. The adsorption of small molecules on the TiO2 anatase(101) surface by first-principles molecular dynamics. Surf. Sci. 402, 219–222.

Shang, Y., Guo, L., 2015. Facet-controlled synthetic strategy of Cu2O-based crystals for catalysis and sensing. Adv. Sci. 2, 22.

Sun, S.D., Yang, Z.M., 2014. Recent advances in tuning crystal facets of polyhedral cuprous oxide architectures. RSC Adv. 4, 3804–3822.

Sun, C.H., Yang, X.H., Chen, J.S., Li, Z., Lou, X.W., Li, C., Smith, S.C., Lu, G.Q., Yang, H.G., 2010. Higher charge/discharge rates of lithium-ions across engineered TiO2 surfaces leads to enhanced battery performance. Chem. Commun. 46, 6129–6131.

Sun, L., Qin, Y., Cao, Q., Hu, B., Huang, Z., Ye, L., Tang, X., 2011. Novel photocatalytic antibacterial activity of TiO2 microspheres exposing 100% reactive {111} facets. Chem. Commun. 47, 12628–12630.

Sun, S.D., Kong, C.C., You, H.J., Song, X.P., Ding, B.J., Yang, Z.M., 2012. Facet-selective growth of Cu-Cu2O heterogeneous architectures. CrstEngComm 14, 40–43.

Tan, H.L., Wen, X.M., Amal, R., Ng, Y.H., 2016. BiVO4 {010} and {110} relative exposure extent: governing factor of surface charge population and photocatalytic activity. J. Phys. Chem. Lett. 7, 1400–1405.

Tang, J.W., Durrant, J.R., Klug, D.R., 2008. Mechanism of photocatalytic water splitting in TiO(2). Reaction of water with photoholes, importance of charge carrier dynamics, and evidence for four-hole chemistry. J. Am. Chem. Soc. 130, 13885–13891.

Truong, Q.D., Kato, H., Kobayashi, M., Kakihana, M., 2015. Hierarchical structures of rutile exposing high-index facets. J. Cryst. Growth 418, 86–91.

Vittadini, A., Selloni, A., Rotzinger, F.P., Gratzel, M., 1998. Structure and energetics of water adsorbed at TiO2 anatase (101) and (001) surfaces. Phys. Rev. Lett. 81, 2954–2957.

Wang, J.S., Yin, S., Zhang, Q.W., Saito, F., Sato, T., 2003. Mechanochemical synthesis of SrTiO3-xFx with high visible light photocatalytic activities for nitrogen monoxide destruction. J. Mater. Chem. 13, 2348–2352.

Wang, D.E., Jiang, H.F., Zong, X., Xu, Q.A., Ma, Y., Li, G.L., Li, C., 2011. Crystal facet dependence of water oxidation on BiVO4 sheets under visible light irradiation. Chem.-Eur. J. 17, 1275–1282.

Wang, H., He, L., Wang, L.H., Hu, P.F., Guo, L., Han, X.D., Li, J.H., 2012. Facile synthesis of Ag3PO4 tetrapod microcrystals with an increased percentage of exposed {110} facets and highly efficient photocatalytic properties. CrstEngComm 14, 8342–8344.

Wang, J., Teng, F., Chen, M.D., Xu, J.J., Song, Y.Q., Zhou, X.L., 2013. Facile synthesis of novel Ag3PO4 tetrapods and the {110} facets-dominated photocatalytic activity. CrstEngComm 15, 39–42.

Wang, X.X., Liu, M.C., Zhou, Z.H., Guo, L.J., 2015a. Toward facet engineering of CdS nanocrystals and their shape-dependent photocatalytic activities. J. Phys. Chem. C 119, 20555–20560.

Wang, B., Shen, S.H., Guo, L.J., 2015b. SrTiO3 single crystals enclosed with high-indexed {023} facets and {001} facets for photocatalytic hydrogen and oxygen evolution. Appl. Catal. B Environ. 166, 320–326.

Wang, S.C., Chen, H.J., Gao, G.P., Butburee, T., Lyu, M.Q., Thaweesak, S., Yun, J.H., Du, A.J., Liu, G., Wang, L.Z., 2016a. Synergistic crystal facet engineering and structural control of WO3 films exhibiting unprecedented photoelectrochemical performance. Nano Energy 24, 94–102.

Wang, B., Shen, S.H., Guo, L.J., 2016b. Surface reconstruction of facet-functionalized SrTiO3 nanocrystals for photocatalytic hydrogen evolution. ChemCatChem 8, 798–804.

Wen, C.Z., Hu, Q.H., Guo, Y.N., Gong, X.Q., Qiao, S.Z., Yang, H.G., 2011a. From titanium oxydifluoride (TiOF2) to titania (TiO2): phase transition and non-metal doping with enhanced photocatalytic hydrogen (H-2) evolution properties. Chem. Commun. 47, 6138–6140.

Wen, C.Z., Jiang, H.B., Qiao, S.Z., Yang, H.G., Lu, G.Q., 2011b. Synthesis of high-reactive facets dominated anatase TiO2. J. Mater. Chem. 21, 7052–7061.

Wu, X., Chen, Z., Lu, G.Q., Wang, L., 2011. Nanosized anatase TiO2 single crystals with tunable exposed (001) facets for enhanced energy conversion efficiency of dye-sensitized solar cells. Adv. Funct. Mater. 21, 4167–4172.

Wu, L., Jiang, H.B., Tian, F., Chen, Z., Sun, C., Yang, H.G., 2013. Ti0.89Si0.11O2 single crystals bound by high-index {201} facets showing enhanced visible-light photocatalytic hydrogen evolution. Chem. Commun. 49, 2016–2018.

Wu, Y.H., Yuan, B., Li, M.R., Zhang, W.H., Liu, Y., Li, C., 2015a. Well-defined BiOCl colloidal ultrathin nanosheets: synthesis, characterization, and application in photocatalytic aerobic oxidation of secondary amines. Chem. Sci. 6, 1873–1878.

Wu, T.T., Kang, X.D., Kadi, M.W., Ismail, I., Liu, G., Cheng, H.M., 2015b. Enhanced photocatalytic hydrogen generation of mesoporous rutile TiO2 single crystal with wholly exposed {111} facets. Chinese J. Catal. 36, 2103–2108.

Wu, T.T., Xie, Y.P., Yin, L.C., Liu, G., Cheng, H.M., 2015c. Switching photocatalytic H-2 and O-2 generation preferences of rutile TiO2 microspheres with dominant reactive facets by boron doping. J. Phys. Chem. C 119, 84–89.

Xi, G.C., Ye, J.H., 2010. Synthesis of bismuth vanadate nanoplates with exposed {001} facets and enhanced visible-light photocatalytic properties. Chem. Commun. 46, 1893–1895.

Xiang, Q., Yu, J., Wang, W., Jaroniec, M., 2011. Nitrogen self-doped nanosized TiO2 sheets with exposed {001} facets for enhanced visible-light photocatalytic activity. Chem. Commun. 47, 6906–6908.

Xie, S., Han, X., Kuang, Q., Fu, J., Zhang, L., Xie, Z., Zheng, L., 2011. Solid state precursor strategy for synthesizing hollow TiO2 boxes with a high percentage of reactive {001} facets exposed. Chem. Commun. 47, 6722–6724.

Xie, Y.P., Liu, G., Yin, L.C., Cheng, H.M., 2012. Crystal facet-dependent photocatalytic oxidation and reduction reactivity of monoclinic WO3 for solar energy conversion. J. Mater. Chem. 22, 6746–6751.

Xing, J., Fang, W.Q., Zhao, H.J., Yang, H.G., 2012. Inorganic photocatalysts for overall water splitting. Chem. Asian J. 7, 642–657.

Xing, G.J., Tang, C.H., Zhang, B., Zhao, L.X., Su, Y.G., Wang, X.J., 2015. A highly uniform ZnO/NaTaO3 nanocomposite: Enhanced self-sensitized degradation of colored pollutants under visible light. J. Alloys Compd. 647, 287–294.

Xu, H., Reunchan, P., Ouyang, S., Tong, H., Umezawa, N., Kako, T., Ye, J., 2013. Anatase TiO2 single crystals exposed with high-reactive {111} facets toward efficient H-2 evolution. Chem. Mater. 25, 405–411.

Xu, Y.S., Lin, H.F., Li, L.P., Huang, X.S., Li, G.S., 2015. Precursor-directed synthesis of well-faceted brookite TiO2 single crystals for efficient photocatalytic performances. J. Mater. Chem. A 3, 22361–22368.

Yan, S.C., Wang, Z.Q., Li, Z.S., Zou, Z.G., 2009. Photocatalytic activities for water splitting of La-doped-NaTaO3 fabricated by microwave synthesis. Solid State Ion. 180, 1539–1542.

Yang, H.G., Sun, C.H., Qiao, S.Z., Zou, J., Liu, G., Smith, S.C., Cheng, H.M., Lu, G.Q., 2008. Anatase TiO2 single crystals with a large percentage of reactive facets. Nature 453, 638–641.

Yang, X.H., Li, Z., Liu, G., Xing, J., Sun, C., Yang, H.G., Li, C., 2011. Ultra-thin anatase TiO2 nanosheets dominated with {001} facets: thickness-controlled synthesis, growth mechanism and water-splitting properties. CrstEngComm 13, 1378–1383.

Yang, J.X., Wang, D.G., Zhou, X., Li, C., 2013. A theoretical study on the mechanism of photocatalytic oxygen evolution on BiVO4 in aqueous solution. Chem.-Eur. J. 19, 1320–1326.

Yi, Z.G., Ye, J.H., Kikugawa, N., Kako, T., Ouyang, S.X., Stuart-Williams, H., Yang, H., Cao, J.Y., Luo, W.J., Li, Z.S., Liu, Y., Withers, R.L., 2010. An orthophosphate semiconductor with photooxidation properties under visible-light irradiation. Nat. Mater. 9, 559–564.

Yu, J., Qi, L., Jaroniec, M., 2010. Hydrogen production by photocatalytic water splitting over Pt/TiO2 nanosheets with exposed (001) facets. J. Phys. Chem. C 114, 13118–13125.

Yu, H., Ouyang, S., Yan, S., Li, Z., Yu, T., Zou, Z., 2011. Sol-gel hydrothermal synthesis of visible-light-driven Cr-doped SrTiO3 for efficient hydrogen production. J. Mater. Chem. 21, 11347–11351.

Yu, J.G., Yu, Y.F., Zhou, P., Xiao, W., Cheng, B., 2014. Morphology-dependent photocatalytic H-2-production activity of CdS. Appl. Catal. B Environ. 156, 184–191.

Zhang, H.Z., Banfield, J.F., 2000. Understanding polymorphic phase transformation behavior during growth of nanocrystalline aggregates: Insights from TiO2. J. Phys. Chem. B 104, 3481–3487.

Zhang, H., Liu, X., Wang, Y., Liu, P., Cai, W., Zhu, G., Yang, H., Zhao, H., 2013. Rutile TiO2 films with 100% exposed pyramid-shaped (111) surface: photoelectron transport properties under UV and visible light irradiation. J. Mater. Chem. A 1, 2646–2652.

Zhang, P., Zhang, J.J., Gong, J.L., 2014a. Tantalum-based semiconductors for solar water splitting. Chem. Soc. Rev. 43, 4395–4422.

Zhang, S.N., Zhang, S.J., Song, L.M., 2014b. Super-high activity of Bi3+ doped Ag3PO4 and enhanced photocatalytic mechanism. Appl. Catal. B Environ. 152, 129–139.

Zhang, L., Shi, J., Liu, M., Jing, D., Guo, L., 2014c. Photocatalytic reforming of glucose under visible light over morphology controlled Cu2O: efficient charge separation by crystal facet engineering. Chem. Commun. 50, 192–194.
Zhang, J., Liu, P.L., Lu, Z.D., Xu, G.L., Wang, X.Y., Qian, L.S., Wang, H.B., Zhang, E.P., Xi, J.H., Ji, Z.G., 2015a. One-step synthesis of rutile nano-TiO2 with exposed {111} facets for high photocatalytic activity. J. Alloys Compd. 632, 133–139.
Zhang, L., Wang, W., Jiang, D., Gao, E., Sun, S., 2015b. Photoreduction of CO2 on BiOCl nanoplates with the assistance of photoinduced oxygen vacancies. Nano Res. 8, 821–831.
Zhang, J.J., Zhang, P., Wang, T., Gong, J.L., 2015c. Monoclinic WO3 nanomultilayers with preferentially exposed (002) facets for photoelectrochemical water splitting. Nano Energy 11, 189–195.
Zhang, N., Chen, C., Mei, Z.W., Liu, X.H., Qu, X.L., Li, Y.X., Li, S.Q., Qi, W.H., Zhang, Y.J., Ye, J.H., Roy, V.A.L., Ma, R.Z., 2016a. Monoclinic tungsten oxide with {100} facet orientation and tuned electronic band structure for enhanced photocatalytic oxidations. Acs Appl. Mater. Interfaces 8, 10367–10374.
Zhang, Q., Li, R.G., Li, Z., Li, A.L., Wang, S.Y., Liang, Z.X., Liao, S.J., Li, C., 2016b. The dependence of photocatalytic activity on the selective and nonselective deposition of noble metal cocatalysts on the facets of rutile TiO2. J. Catal. 337, 36–44.
Zhang, Q., Li, Z., Wang, S.Y., Li, R.G., Zhang, X.W., Liang, Z.X., Han, H.X., Liao, S.J., Li, C., 2016c. Effect of redox cocatalysts location on photocatalytic overall water splitting over cubic NaTaO3 semiconductor crystals exposed with equivalent facets. Acs Catal. 6, 2182–2191.
Zhao, X., Jin, W., Cai, J., Ye, J., Li, Z., Ma, Y., Xie, J., Qi, L., 2011a. Shape- and size-controlled synthesis of uniform anatase TiO2 nanocuboids enclosed by active {100} and {001} facets. Adv. Funct. Mater. 21, 3554–3563.
Zhao, Z.Y., Li, Z.S., Zou, Z.G., 2011b. Electronic structure and optical properties of monoclinic clinobisvanite BiVO4. Phys. Chem. Chem. Phys. 13, 4746–4753.
Zhao, K., Zhang, L.Z., Wang, J.J., Li, Q.X., He, W.W., Yin, J.J., 2013a. Surface structure-dependent molecular oxygen activation of BiOCl single-crystalline nanosheets. J. Am. Chem. Soc. 135, 15750–15753.
Zhao, J., Zou, X.-X., Su, J., Wang, P.-P., Zhou, L.-J., Li, G.-D., 2013b. Synthesis and photocatalytic activity of porous anatase TiO2 microspheres composed of {010}-faceted nanobelts. Dalton Trans 42, 4365–4368.
Zhao, M., Li, L., Lin, H., Yang, L., Li, G., 2013c. A facile strategy to fabricate large-scale uniform brookite TiO2 nanospindles with high thermal stability and superior electrical properties. Chem. Commun. 49, 7046–7048.
Zhao, Z., Tan, H.Q., Zhao, H.F., Lv, Y., Zhou, L.J., Song, Y.J., Sun, Z.C., 2014. Reduced TiO2 rutile nanorods with well-defined facets and their visible-light photocatalytic activity. Chem. Commun. 50, 2755–2757.
Zhao, M., Xu, H., Chen, H.R., Ouyang, S.X., Umezawa, N., Wang, D.F., Ye, J.H., 2015. Photocatalytic reactivity of {121} and {211} facets of brookite TiO2 crystals. J. Mater. Chem. A 3, 2331–2337.
Zheng, Z., Huang, B., Lu, J., Qin, X., Zhang, X., Dai, Y., 2011a. Hierarchical TiO2 microspheres: synergetic effect of {001} and {101} facets for enhanced photocatalytic activity. Chemistry 17, 15032–15038.
Zheng, H.D., Ou, J.Z., Strano, M.S., Kaner, R.B., Mitchell, A., Kalantar-Zadeh, K., 2011b. Nanostructured tungsten oxide—properties, synthesis, and applications. Adv. Funct. Mater. 21, 2175–2196.
Zheng, B.J., Wang, X., Liu, C., Tan, K., Xie, Z.X., Zheng, L.S., 2013. High-efficiently visible light-responsive photocatalysts: Ag3PO4 tetrahedral microcrystals with exposed {111} facets of high surface energy. J. Mater. Chem. A 1, 12635–12640.

Zheng, J.Y., Song, G., Hong, J.S., Van, T.K., Pawar, A.U., Kim, D.Y., Kim, C.W., Haider, Z., Kang, Y.S., 2014. Facile fabrication of WO3 nanoplates thin films with dominant crystal facet of (002) for water splitting. Cryst. Growth Des. 14, 6057–6066.
Zheng, J.Y., Haider, Z., Van, T.K., Pawar, A.U., Kang, M.J., Kim, C.W., Kang, Y.S., 2015. Tuning of the crystal engineering and photoelectrochemical properties of crystalline tungsten oxide for optoelectronic device applications. CrstEngComm 17, 6070–6093.
Zhu, J., Fan, F.T., Chen, R.T., An, H.Y., Feng, Z.C., Li, C., 2015. Direct imaging of highly anisotropic photogenerated charge separations on different facets of a single BiVO4 photocatalyst. Angew. Chem. Int. Ed. 54, 9111–9114.

CHAPTER ELEVEN

Black Titanium Dioxide for Photocatalysis

Yan Liu*,†, Xiaobo Chen*,[1]
*University of Missouri, Kansas City, MO, United States
†College of Environment, Sichuan Agricultural University, Chengdu, China
[1]Corresponding author: e-mail address: chenxiaobo@umkc.edu

Contents

1. INTRODUCTION

The rapid expansions of globalization, industrialization, and manufacturing have caused serious energy shortage and environmental pollution problems. Photocatalysis could be the best solution, in absorbing the sunlight to produce hydrogen (H_2) from water and degrading organic pollutants (Bai et al., 2014; Borgarello et al., 1981; Chen and Mao, 2007; Chen et al., 2012; Fujishima and Honda, 1972; Gao et al., 2015; Hoffmann et al., 1995;

Semiconductors and Semimetals, Volume 97
ISSN 0080-8784
http://dx.doi.org/10.1016/bs.semsem.2017.04.001

Jacobs et al., 2010; Lewicka et al., 2013; Linsebigler et al., 1995; Ma et al., 2014; Nakata and Fujishima, 2012; Nozik, 1975; Weir et al., 2012). Titanium dioxide (TiO_2) nanomaterials have been studied over the past few decades for this purpose. However, their most common phases: anatase and rutile, have band gaps of 3.2 eV (387 nm) and 3.0 eV (413 nm), respectively (Chen and Selloni, 2014; Dette et al., 2014). This limits their photocatalytic activity to the UV region of the solar spectrum, which accounts for less than 5% over the full solar energy. Meanwhile, the rapid recombination rate of photogenerated electron–hole pairs leads to a low quantum efficient and a poor photocatalytic activity (Schneider et al., 2014). Until now, numerous efforts have been made to tune the electronic and optical properties (Khan et al., 2002; Liu et al., 2015a; Park et al., 2006; Williams et al., 2008). For example, introducing alien atoms (metal (Chambers et al., 2001; Deng et al., 2009; Hoa, 2015; Jiao et al., 2014; Kneiß et al., 2014; Liu et al., 2013a; Matsumoto et al., 2001; Peng et al., 2012; Roy et al., 2015; Yao et al., 2010; Yu et al., 2009; Zhang et al., 2015), nonmetal (Cheng et al., 2014; Cong et al., 2011; Dong et al., 2008; Eslami et al., 2016; Hwang et al., 2015; Luo et al., 2012; Mathis, 2015; Xing et al., 2010, 2011; Zhang et al., 2010), and metal and non-metal codoping (Cong et al., 2011; Dong et al., 2008; Eslami et al., 2016; Mathis, 2015; Xing et al., 2010, 2011; Zhang et al., 2010)) or compounds (metal oxide (Bai et al., 2012; Cheng et al., 2014; Feng et al., 2015; Hwang et al., 2015; Jiang et al., 2015; Luo et al., 2012; Qiu et al., 2015; Štengl et al., 2013; Zhang et al., 2013), metal chalcogenides (AlHaddad et al., 2016; Bubenhofer et al., 2012; Gao et al., 2009; Kang et al., 2011; Li et al., 2014a; Liu et al., 2012; Qin et al., 2015; Zhang et al., 2014), semiconductor alloy chalcogenides (Li et al., 2014b, 2015a; Zhou et al., 2012), and graphene and graphene oxide (Ferrighi et al., 2016; Li et al., 2015b, 2016; Niu et al., 2014; Wang and Sasaki, 2014; Williams et al., 2008)) into the TiO_2 matrix has been used to alter the electronic structure for enhancing visible light absorption. Creating intrinsic defects (oxygen vacancies and Ti^{3+}) into the TiO_2 lattice (Kwon et al., 2015; Li et al., 2015c; Lin et al., 2015; Liu et al., 2009, 2016a; Mao et al., 2014; Qiu et al., 2015; Ren et al., 2015; Xing et al., 2014; Zhou et al., 2016) or modifying the structural factors of TiO_2 (phase, morphology, structure, and porosity) (Cao et al., 2016; Garcia et al., 2015; Liu et al., 2015b) has also been shown to affect its photochemical activity. In 2011, a black TiO_2 was reported with a narrowed band gap about 1.5 eV and substantial solar-driven photocatalytic activities (Chen et al., 2011). That discovery has triggered world-wide research interests in black TiO_2

nanomaterials. Here, we would like to summarize the synthesis, properties, and photocatalytic applications of the black TiO_2 nanomaterials.

2. SYNTHESIS

2.1 Chemical Routes

2.1.1 Hydrogenation

Here, the hydrogenation process refers to the treatment under a hydrogen-containing environment or hydrogen plasma for a certain period of time at some temperatures (Yan et al., 2016). Hydrogenation of inorganic nanomaterials can induce the following typical situations: (a) structural alteration, e.g., from crystalline phase to disordered phase, forming crystalline/disordered core/shell nanostructures; (b) partial chemical reduction, introducing lower valence state, or oxygen vacancies; (c) deep chemical reduction, forming completely reduced metallic phase; and (d) complete chemical reduction, forming a metallic phase, but the followed exposure to air induced oxidation on the surface to form a disordered layer, e.g., resulting in crystalline/disordered metal/oxide core/shell nanostructures (Yan et al., 2016).

2.1.1.1 Hydrogen Thermal Treatment

Hydrogen thermal treatment of TiO_2 nanocrystals is a simple and straightforward method to prepare black TiO_2 (Becker and Hosler, 1965; Breckenridge and Hosler, 1953; Cronemeyer, 1952, 1959; Cronemeyer and Gilleo, 1951; Henrich and Kurtz, 1981; Pan et al., 1992; Sekiya et al., 2004; Wallace et al., 2011). Hydrogenated rutile TiO_2 single crystals were reported in 1951 with a long wavelength absorption (Cronemeyer and Gilleo, 1951), and with an increased electrical conductivity in 1958 (Cronemeyer, 1959). Pale blue or dark blue TiO_2 was obtained when annealed in hydrogen (Sekiya et al., 2004). Black TiO_2 nanoparticles were obtained by treating pure white TiO_2 nanoparticles under a 20.0-bar pure H_2 atmosphere at about 200°C for 5 days (Chen et al., 2011). The white and black TiO_2 samples are shown in Fig. 1A (Chen et al., 2011). The high-resolution transmission electron microscopy images of white and black TiO_2 nanocrystals are shown in Fig. 1B and C (Chen et al., 2011). The black TiO_2 featured a well-crystallized lattice core surrounded by a disordered

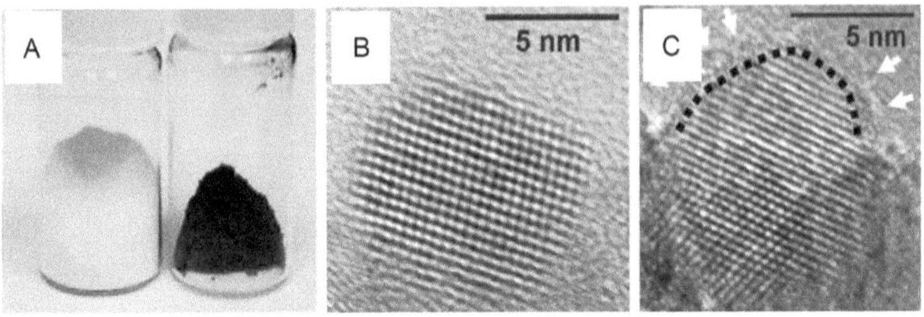

Fig. 1 (A) Pictures of white and black TiO_2 nanoparticles. HR-TEM images of (B) white and (C) black TiO_2 nanoparticles (Chen et al., 2011). *Reprinted with permission from Chen, X., Liu, L., Yu, P.Y., Mao, S.S., 2011. Increasing solar absorption for photocatalysis with black hydrogenated titanium dioxide nanocrystals. Science 331, 746–750. Copyright 2011. AAAS.*

shell structure (Chen et al., 2011), due to the possible hydrogen doping, Ti–H and O–H bond formation (Chen et al., 2011).

The color of the hydrogenated TiO_2 nanosheets could be changed from white to blue, then finally to gray, as shown in Fig. 2 (Yu et al., 2013). Hydrogenated anatase TiO_2 nanosheets were obtained under a H_2 gas flow (50 sccm) at atmospheric pressure in a quartz tube at 500–700°C for a fixed period of time (0.5–1 h) (Yu et al., 2013). The hydrogenation temperature and time have a great influence on the concentration of Ti^{3+} defects and oxygen vacancies (Yu et al., 2013). After hydrogenated under ambient hydrogen pressure at 400°C for 1 h (Li et al., 2013), the color of the TiO_2 microspheres changed to gray from white and displayed a strong light absorption from UV light toward the infrared region (Li et al., 2013).

Ordered mesoporous black TiO_2 materials were obtained by hydrogenation at 500°C for 3 h under normal pressure conditions, with a constant heating rate of 5°C min^{-1} (Fig. 3A) (Zhou et al., 2014). The thermally stable and high-surface-area mesoporous TiO_2 was synthesized by an evaporation-induced self-assembly method combined with an ethylene diamine encircling process (Zhou et al., 2014). The absorption spectra of the ordered mesoporous black TiO_2 materials can extend the optical absorption from UV light to visible and infrared light regions, and the band gap of the ordered mesoporous black TiO_2 material is narrowed to be about 2.82 eV (Zhou et al., 2014).

Black TiO_2 nanocrystals were also obtained by hydrogenating commercial P25 TiO_2 under 35 bar hydrogen atmosphere at room temperature for

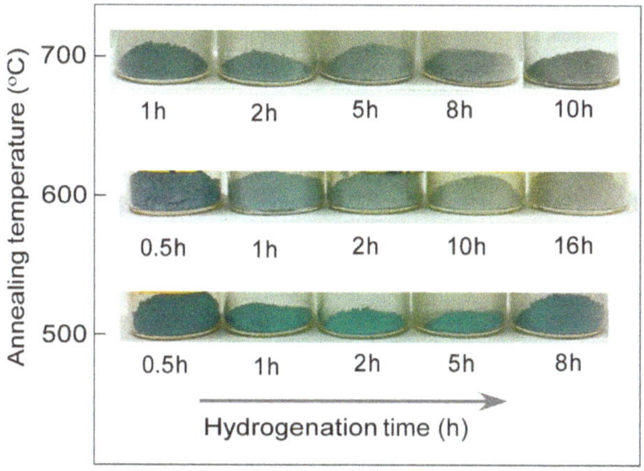

Fig. 2 Photographs of hydrogenated anatase TiO_2 nanosheets samples prepared with a H_2 gas flow at temperatures of 500–700°C (Yu et al., 2013). *Reprinted with permission from Yu, X., Kim, B., Kim, Y.K., 2013. Highly enhanced photoactivity of anatase TiO_2 nanocrystals by controlled hydrogenation-induced surface defects. ACS Catal. 3, 2479–2486. Copyright 2013. American Chemical Society.*

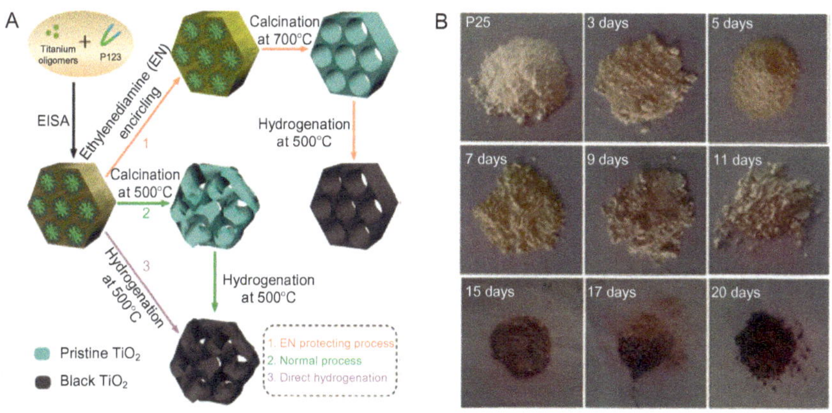

Fig. 3 (A) Schematic synthesis process for the ordered mesoporous black TiO_2 materials (Zhou et al., 2014). (B) Photographic images of P25 treated under hydrogen for 0–20 days at 35 bar hydrogen atmosphere at room temperature (Lu et al., 2014). *Panel (A) Reprinted with permission from Zhou, W., Li, W., Wang, J.-Q., Qu, Y., Yang, Y., Xie, Y., Zhang, K., Wang, L., Fu, H., Zhao, D., 2014. Ordered mesoporous black TiO_2 as highly efficient hydrogen evolution photocatalyst. J. Am. Chem. Soc. 136, 9280–9283. Copyright 2014. American Chemical Society. Panel (B) Reprinted with permission from Lu, H., Zhao, B., Pan, R., Yao, J., Qiu, J., Luo, L., Liu, Y., 2014. Safe and facile hydrogenation of commercial Degussa P25 at room temperature with enhanced photocatalytic activity. RSC Adv. 4, 1128–1132. Copyright 2014. The Royal Society of Chemistry.*

0–20 days (Lu et al., 2014). The hydrogenated TiO_2 treated for more than 15 days have dark appearance and a crystalline-disordered core–shell structure (Lu et al., 2014). The images of P25 treated under hydrogen for 0–20 days are shown in Fig. 3B (Lu et al., 2014).

The color of hydrogenated anatase TiO_2 nanotubes was changed under the following different atmospheres (i) an anatase TiO_2 nanotube layer (air); (ii) this layer converted with Ar (Ar) or H_2/Ar(H_2/Ar) to black titania under atmospheric pressure; (iii) the anatase layer converted with a high pressure H_2 treatment (20 bar, 500°C for 1 h) (HP-H_2); and (iv) a high pressure conditions but mild heating (H_2, 20 bar, 200°C for 5 days) (Liu et al., 2014). After the reduction treatments, the color of TiO_2 was changed from white to black (Ar/H_2), deep purple (Ar), light blue (H_2, 500°C for 1 h), and gray (H_2, 200°C for 5 days) (Liu et al., 2014). Apparently, the hydrogenation conditions played a critical role in the color of TiO_2.

2.1.1.2 Hydrogen Plasma

Plasma reactors have been long proposed to reduction of metal oxides due to the presence of highly active species such as atomic hydrogen in these reactors (Palmer et al., 2002). Long et al. and Degout et al. prepared reduced TiO_2 with atmospheric-pressure hydrogen plasma and carbon-reducing agents (Degout et al., 1984; Long et al., 1981). Kitamura et al. performed the reduction with Ar–H_2 (5% H_2) plasma, and the TiO_2 was slightly reduced to a mixture of Ti_2O_3 and Ti_3O_5 (Kitamura et al., 1993).

Recently, hydrogenated black TiO_2 nanoparticles were prepared with hydrogen plasma (200 W) at 500°C for 4–8 h (Wang et al., 2013a). The commercial amorphous TiO_2 was used as precursor, heated at 200°C under flowing O_2 for 1 h, then reduced in H_2 flow (Wang et al., 2013a). The black TiO_2 nanoparticles showed a significant absorption in the visible and near-infrared light (Wang et al., 2013a). Yan et al. found that the hydrogenated TiO_2 nanoparticles prepared by hydrogen plasma at 390°C for 3 h had a black color and enhanced absorption in the visible light region (Yan et al., 2013). Wang et al. studied the formation of black TiO_2 nanowire arrays (Chieh and Hsun, 2016). The pristine TiO_2 nanowire arrays were prepared by hydrothermal method (Chieh and Hsun, 2016). The black TiO_2 nanowire arrays were obtained by treating in Ar (95%)/H_2 (5%) plasma for 5 min at 25°C with the power of 200 W, constant pressure of

2 Torr (Chieh and Hsun, 2016). The hydrogenated TiO_2 nanowire arrays exhibited black color and increased absorption in the visible-light region (Chieh and Hsun, 2016).

Lepcha et al. prepared black 1D TiO_2 nanofibers by employing mild hydrogen plasma treatments, with a gas flow rate of 500 sccm, power of 15 W, and purity of 99.99% in the temperature range of 300–500°C for 3 h (Lepcha et al., 2015). The TiO_2 nanofibers were synthesized via electrospining of Ti sols, followed by annealing at 500°C for 5 h to obtain a crystalline phase (Lepcha et al., 2015). The hydrogen plasma resulted in a darkening of the initially white fibers, giving rise to black TiO_2 (Fig. 4A) (Lepcha et al., 2015). The hydrogenated black TiO_2 nanofibers had a significant absorption in the visible and near-infrared light (Fig. 4B) (Lepcha et al., 2015). Plasma treatment at higher temperatures results in a noticeably lowered reflectance (Lepcha et al., 2015).

An et al. reported that anatase/brookite bicrystalline hydrogenated TiO_2 photocatalysts prepared by hydrogen plasma extended the absorption spectra from ultraviolet to visible range (Fig. 5A) (An et al., 2016). The starting TiO_2 nanoparticles were obtained by a sol–gel method (An et al., 2016). The H_2 plasma treatment time was controlled within the 0–120 min range and the input power was 120 W (H-TiO_2 30, H-TiO_2 120), and the H_2 flow

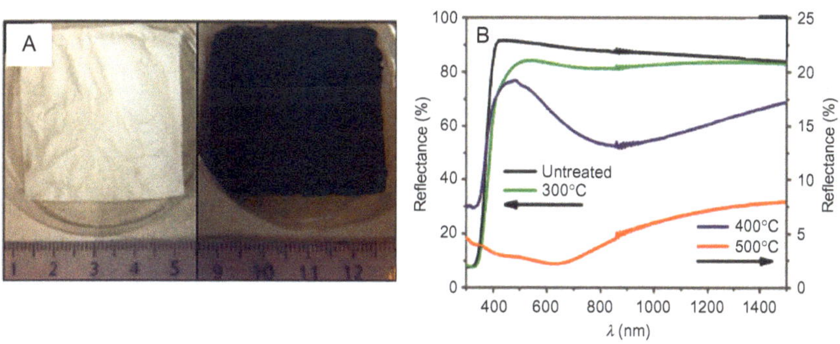

Fig. 4 (A) Optical difference between pristine TiO_2 nanofibers (*white*) and hydrogen plasma-treated fibers (*black*). (B) UV/vis/NIR reflectance spectra of the untreated and H_2 plasma-treated TiO_2 samples (Lepcha et al., 2015). *Reprinted with permission from Lepcha, A., Maccato, C., Mettenbörger, A., Andreu, T., Mayrhofer, L., Walter, M., Olthof, S., Ruoko, T.P., Klein, A., Moseler, M., Meerholz, K., Morante, J.R., Barreca, D., Mathur, S., 2015. Electrospun black titania nanofibers: influence of hydrogen plasma-induced disorder on the electronic structure and photoelectrochemical performance. J. Phys. Chem. C 119, 18835–18842. Copyright 2015. American Chemical Society.*

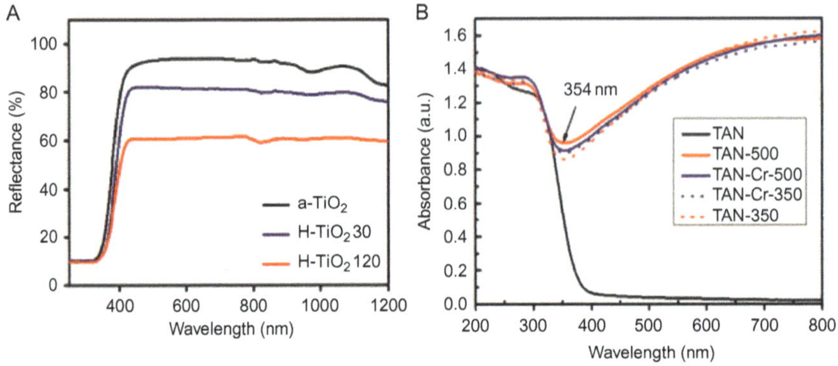

Fig. 5 (A) UV–Vis–NIR reflectance of as-synthesized TiO_2 (a-TiO_2), H-TiO_2 30, and H-TiO_2 120 (An et al., 2016). (B) UV–visible absorption spectra of the pristine TiO_2 nanotubes (TAN) and hydrogenated black TiO_2 nanotubes (TAN-500, TAN-Cr-500, TAN-Cr-350, TAN-350) (Teng et al., 2014). *Panel (A) Reprinted with permission from An, H.R., Park, S.Y., Kim, H., Lee, C.Y., Choi, S., Lee, S.C., Seo, S., Park, E.C., Oh, Y.K., Song, C.G., Won, J., Kim, Y.J., Lee, J., Lee, H.U., Lee, Y.C., 2016. Advanced nanoporous TiO_2 photocatalysts by hydrogen plasma for efficient solar-light photocatalytic application. Sci. Rep. 6, 29683. Copyright 2016. Nature. Panel (B) Reprinted with permission from Teng, F., Li, M., Gao, C., Zhang, G., Zhang, P., Wang, Y., Chen, L., Xie, E., 2014. Preparation of black TiO_2 by hydrogen plasma assisted chemical vapor deposition and its photocatalytic activity. Appl. Catal. Environ. 148–149, 339–343. Copyright 2014. Elsevier.*

rate of 50 sccm (An et al., 2016). Teng et al. prepared black TiO_2 nanotubes by hydrogen plasma in a hot filament chemical vapor deposition apparatus with hydrogen as reaction gas (Teng et al., 2014). During the treatment process, the obtained TiO_2 nanotubes were loaded in a corundum boat under the filament (Teng et al., 2014). The temperature of the filament was maintained at 2000°C, and the samples were treated for 3h at 350 and 500°C (TiO_2-350, TiO_2-500) (Teng et al., 2014). Some Cr fragments were mixed in the TiO_2 during the treatment process to study the effect of Cr(TiO_2-Cr-350, TiO_2-Cr-500) (Teng et al., 2014). All the hydrogenated TiO_2 nanotubes possessed a significant absorption in the visible and near-infrared region (Fig. 5B) (Teng et al., 2014).

2.1.2 Chemical Reduction

2.1.2.1 Aluminum Reduction

Over 50 years ago, aluminothermic reduction reaction of TiO_2 with elemental aluminum (Al) was used to produce Al-Ti alloy (Kubaschewski and Dench, 1955). Inspired by this old technique, Wang et al. prepared black TiO_2 nanoparticles using melted Al as a reductant in an evacuated two-zone vacuum furnace at 300–500°C (Wang et al., 2013b). The obtained sample had a unique crystalline core–amorphous shell structure ($TiO_2@TiO_{2-x}$), inducing significant enhancement of visible and near-infrared absorption (Wang et al., 2013b). Zhu et al. prepared black brookite TiO_2 through Al reduction at different temperatures (Zhu et al., 2013). Many oxygen vacancies and Ti^{3+} states were introduced into the distorted shell, which increased the solar energy absorption and boost the photocatalytic activity (Zhu et al., 2013).

Cui et al. prepared black TiO_2 nanotube arrays (B-TNTs) by the melted Al reduction (Cui et al., 2014). They first prepared TiO_2 nanotube arrays (TNTs) via electrochemical anodization followed by annealing at 500°C for 4h in air, and finally obtained B-TNTs with Al reduction at 500°C for 4h (Cui et al., 2014). The B-TNTs sample showed a dramatically enhanced light absorption from the visible light to the near-infrared light region (Fig. 6A), due to the additional transitions between the different energy levels of the Ti^{3+} states, oxygen vacancies, conduction band, and valence band (Cui et al., 2014). Lin et al. prepared a series of nonmetal-doped black TiO_2 by a two-step strategy (Lin et al., 2014). First, they introduced many oxygen vacancies on the amorphous surface layer surrounding the crystalline core by Al reduction (Lin et al., 2014). Then, they

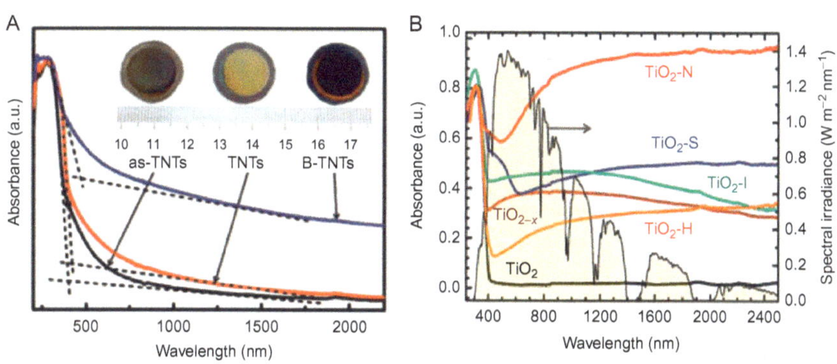

Fig. 6 (A) Absorption spectra and photographs (*inset*) of the as-TNTs, TNTs, and B-TNTs (Cui et al., 2014). (B) Diffuse reflectance spectra of TiO_{2-x} (X = H, N, S, I) and solar spectral irradiance (*right*) (Lin et al., 2014). *Panel (A) Reprinted with permission from Cui, H., Zhao, W., Yang, C., Yin, H., Lin, T., Shan, Y., Xie, Y., Gu, H., Huang, F., 2014. Black TiO_2 nanotube arrays for high-efficiency photoelectrochemical water-splitting. J. Mater. Chem. A 2, 8612–8616. Copyright 2014. The Royal Society of Chemistry. Panel (B) Reprinted with permission from Lin, C., Shin, D., Demkov, A.A., 2015. Localized states induced by an oxygen vacancy in rutile TiO_2. J. Appl. Phys. 117, 225703. Copyright 2014. The Royal Society of Chemistry.*

incorporated nonmetal element X (X=H, N, S, I) in the oxygen-deficient amorphous layers (Lin et al., 2014). All the TiO_{2-x} nanoparticles exhibited enhanced absorption in the visible light and near-infrared regions (Fig. 6B) (Lin et al., 2014).

2.1.2.2 Magnesium Reduction

Yu et al. synthesized reduced black TiO_2 nanoparticles by magnesium reduction under a 5% H_2/Ar atmosphere (Sinhamahapatra et al., 2015). Well-mixed sample of TiO_2 and magnesium powder were placed in a tube furnace and then heated at 650°C for 5 h (Sinhamahapatra et al., 2015). After the annealing treatment, the sample was stirred for 24 h in 1.0 M HCl solution, then washed and dried at 80°C (Sinhamahapatra et al., 2015). Different reduced black TiO_2 were prepared with varying the molar ratio of TiO_2 and Mg (BT-X) (Fig. 7A) (Sinhamahapatra et al., 2015). All the Mg-treated samples displayed enhanced absorbance from UV to visible and infrared region,

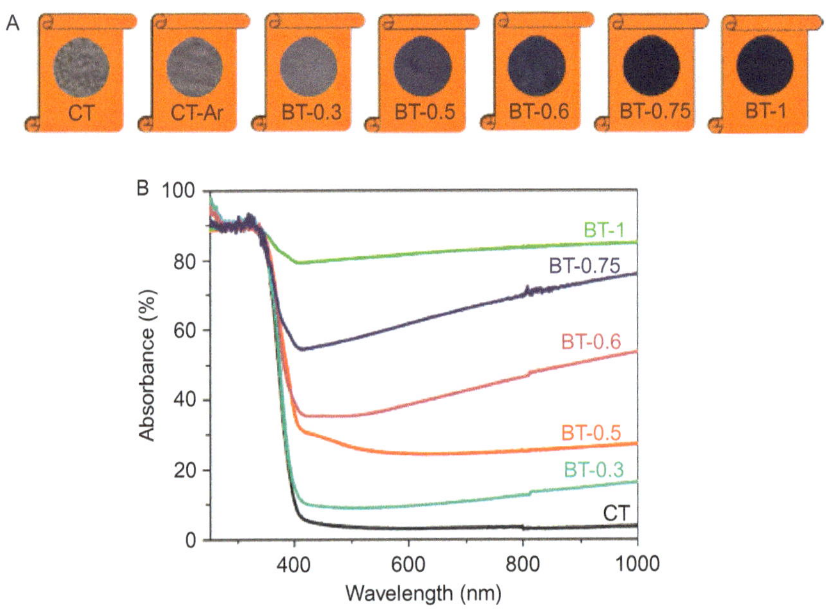

Fig. 7 (A) Photographs and (B) absorption spectra of Mg-treated TiO_2 samples. CT was commercially nano-TiO_2, CT-Ar was the sample annealed under Ar atmosphere. BT-X, X is the molar ratio of Mg with respect to TiO_2 (Sinhamahapatra et al., 2015). *Reprinted with permission from Sinhamahapatra, A., Jeon, J.P., Yu, J.S., 2015. A new approach to prepare highly active and stable black titania for visible light-assisted hydrogen production. Energy Environ. Sci. 8, 3539–3544. Copyright 2015. The Royal Society of Chemistry.*

and the absorption of light increased as the amount of Mg increases (Fig. 7B) (Sinhamahapatra et al., 2015).

2.1.2.3 $NaBH_4$ Reduction

Kang et al. prepared reduced TiO_2 nanotube arrays (TNTAs) by chemical reduction with $NaBH_4$ (Kang et al., 2013). The starting TNTAs were anodized under 80 V for 30 min in an ethylene glycol solution containing 0.3 wt% NH_4F and 2 vol% H_2O, followed by annealing at 450°C for 3 h (Kang et al., 2013). The TiO_2 extended the photocatalytic activity from the UV to visible light region (Kang et al., 2013).

Ren et al. synthesized reduced dark gray TiO_2 nanoparticles by chemical reduction with $NaBH_4$ (Ren et al., 2015). For the reduction reaction, 12 g $NaBH_4$ and 0.5 g TiO_2 were added into 60 mL water for hydrothermal reaction at 180°C for 16 h (Ren et al., 2015). The starting TiO_2 was prepared by the hydrothermal method. 5 mL of 50 wt% titanium (IV) bis(ammonium lactate)dihydroxide and 60 mL of 0.08 g L^{-1} glucose were reacted at 170°C for 8 h, followed by annealing at 500°C for 3 h (Ren et al., 2015).

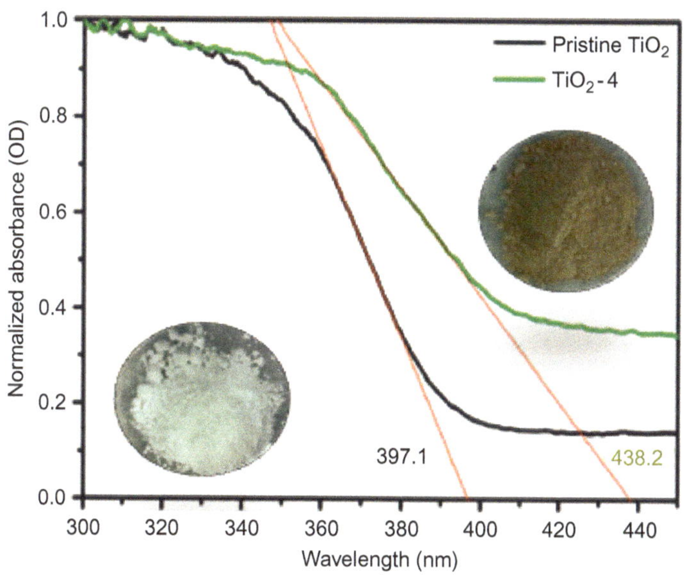

Fig. 8 UV–visible absorbance spectra and picture of pristine TiO_2 nanoparticles and $NaBH_4$-reduced TiO_2 nanoparticles (Ren et al., 2015). *Reprinted with permission from Ren, R., Wen, Z., Cui, S., Hou, Y., Guo, X., Chen, J., 2015. Controllable synthesis and tunable photocatalytic properties of Ti^{3+}-doped TiO_2. Sci. Rep. 5, 10714. Copyright 2015. Nature.*

The reduced TiO_2 nanoparticles showed a positive shift of the absorption spectra, and an absorption edge of 438.2 nm (Fig. 8) (Ren et al., 2015).

2.1.2.4 NaH Reduction

Li et al. synthesized brown TiO_2 nanoparticles by chemical reduction with NaH solution (Wang et al., 2016). Anhydrous dimethyl formamide (5.0 mL), TiO_2 (100 mg), and NaH powder (200 mg) were treated at 120–150°C in N_2 for several hours (2–4 h) (Wang et al., 2016). The color changed from white to brown after reduction by NaH (Fig. 9A) (Wang et al., 2016). The NaH-treated TiO_2 samples exhibited better optical responses than P25. Especially, the sample treated at 150°C in N_2 for 4 h displayed enhanced absorption in the visible region from 400 to 700 nm (Fig. 9B) (Wang et al., 2016).

2.1.3 Chemical Oxidation

Xin et al. prepared black brookite TiO_2 single-crystalline nanosheets by anoxidation-based hydrothermal method (Xin et al., 2016). In a typical procedure, H_2O_2 (30 mL, 30 wt%) was added in the mixed solution of TiH_2 (0.256 g) and H_2O (2 mL), and stirring for 12 h in forming a yellowish gel (Xin et al., 2016). A certain amount of NaOH (1.0 M) was added to adjust the pH to 9.0. Then $NaBH_4$ (0.4 g) was added into the mixture and

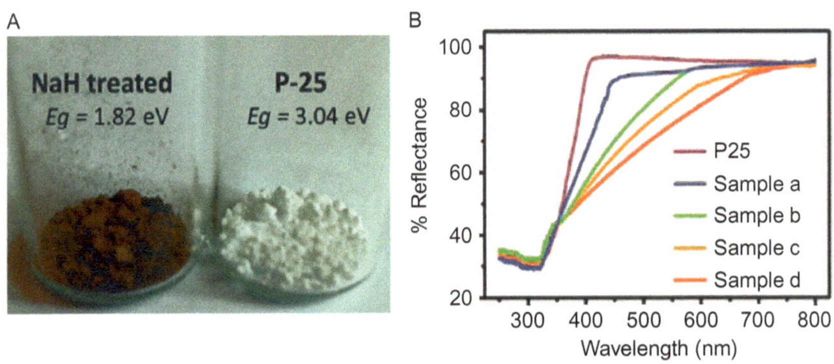

Fig. 9 (A) Photographs of NaH-treated TiO_2 (150°C in N_2 for 4 h, UV illumination) and P25 (Wang et al., 2016). (B) Reflectance spectra of P25 and NaH-treated TiO_2 sample: (a) 150°C in N_2 for 4 h, UV illumination off; (b) 120°C in N_2 for 4 h, UV illumination on; (c) 150°C in N_2 for 2 h, UV illumination on; and (d) 150°C in N_2 for 4 h, UV illumination on (Wang et al., 2016). *Reprinted with permission from Wang, M., Nie, B., Yee, K.K., Bian, H., Lee, C., Lee, H.K., Zheng, B., Lu, J., Luo, L., Li, Y.Y., 2016. Low-temperature fabrication of brown TiO_2 with enhanced photocatalytic activities under visible light. Chem. Commun. 52, 2988–2991. Copyright 2016. The Royal Society of Chemistry.*

hydrothermally treated at 180°C for 24h (Xin et al., 2016). The sample was washed with HCl (50mL, 1.0M) to eliminate the sodium boron compounds (Xin et al., 2016). After postannealing treatment, the color of the brookite changed from blue to brown at 300°C (T300), further to black at 500°C (T500), and light gray blue at 700°C (T700) (Xin et al., 2016). All the postannealing treatment nanosheets displayed enhanced absorption, and the amount of light absorbed followed this order: T500 > T300 > T700 (Fig. 10) (Xin et al., 2016).

Liu et al. synthesized blue TiO_{2-x} nanoparticles by mild hydrothermal treatment of TiH_2 in H_2O_2 aqueous solution (Liu et al., 2013b). TiH_2 powder (0.6g) was added in H_2O_2 solution and stirred for 5h at room temperature, and then the mixture was treated at 160°C for 20–27h (Liu et al., 2013b). The TiO_{2-x} nanoparticles showed a strong absorption in the UV–visible light region (Liu et al., 2013b). Pei et al. prepared gray TiO_{2-x} by hydrothermal treatment of TiO in HCl solution (Pei et al., 2013). Titanium monoxide (400mg) was reacted with HCl solution (20mL, 3M) at 160°C for 24h (Pei et al., 2013). The obtained TiO_{2-x} also exhibited enhanced visible light absorption and good stability (Pei et al., 2013).

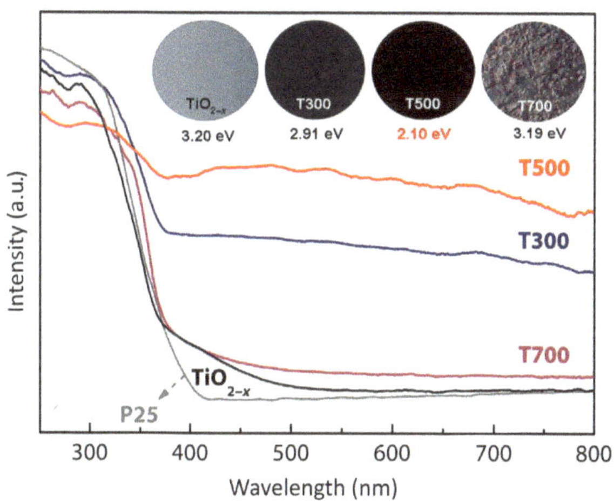

Fig. 10 UV–vis diffuse reflectance spectra of the as-prepared brookite TiO_{2-x}, T300, T500, and T700 (TiO_{2-x} was the sample without postannealing) (Xin et al., 2016). *Reprinted with permission from Xin, X., Xu, T., Wang, L., Wang, C., 2016. Ti^{3+}-self doped brookite TiO_2 single-crystalline nanosheets with high solar absorption and excellent photocatalytic CO_2 reduction. Sci. Rep. 6, 23684. Copyright 2016. Nature.*

2.2 Electrochemical

2.2.1 Electrochemical Reduction

Yang et al. prepared blue TNTAs via the combination of electrochemical anodization and reduction as illustrated in Fig. 11 (Yang et al., 2016). TNTAs film was first prepared with a Pt sheet as the counter electrode and a Ti foil as the working anode under a constant voltage at 60 V for 30 min in the ethylene glycol electrolyte solution (2 wt% H_2O and 0.3 wt % NH_4F) (Yang et al., 2016). The second anodization was performed at 30 V for 1 h on the same Ti foil, after removing the as-grown nanotube layers by ultrasonication (Yang et al., 2016), and then calcined at 450°C in air for 3 h (Yang et al., 2016). The electrochemical reduction was conducted in a 0.5 M Na_2SO_4 aqueous solution at different potentials (−1.0, −1.2, −1.4, and −1.6 V vs SCE) for 10 min (Yang et al., 2016). The self-doped TNTAs showed enhanced photoelectrochemical performance (Yang et al., 2016).

Kim et al. reported that blue and black TNTAs scan be prepared by electrochemical self-doping (Fig. 12) (Kim et al., 2016). The pristine TiO_2 nanotube arrays were prepared by anodization under 45 V for 5 h at room temperature in electrolyte-containing H_2O (2.5 wt%)/NH_4F (0.2 wt%)

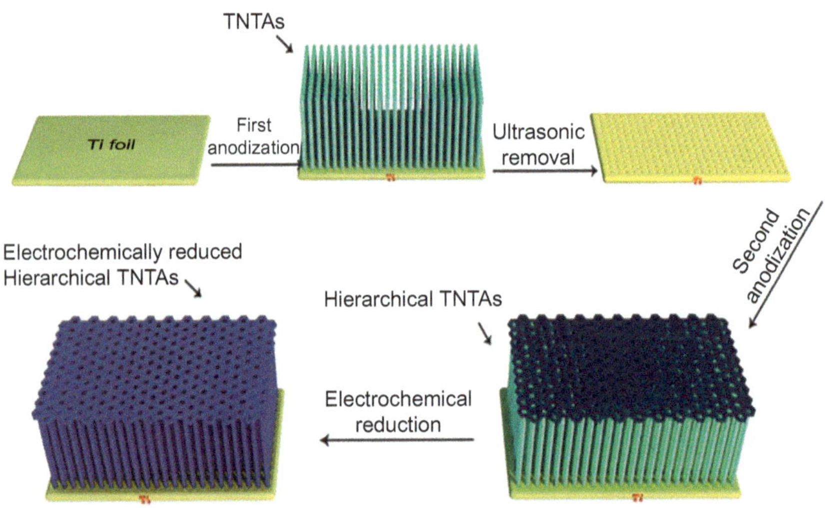

Fig. 11 Two-step anodization process for the TNTAs preparation and electrochemical reduction process to produce Ti^{3+} self-doped TNTAs (Yang et al., 2016). *Reprinted with permission from Yang, Y., Liao, J., Li, Y., Cao, X., Li, N., Wang, C., Lin, S., 2016. Electrochemically self-doped hierarchical TiO_2 nanotube arrays for enhanced visible-light photoelectrochemical performance: an experimental and computational study. RSC Adv. 6, 46871–46878. Copyright 2016. The Royal Society of Chemistry.*

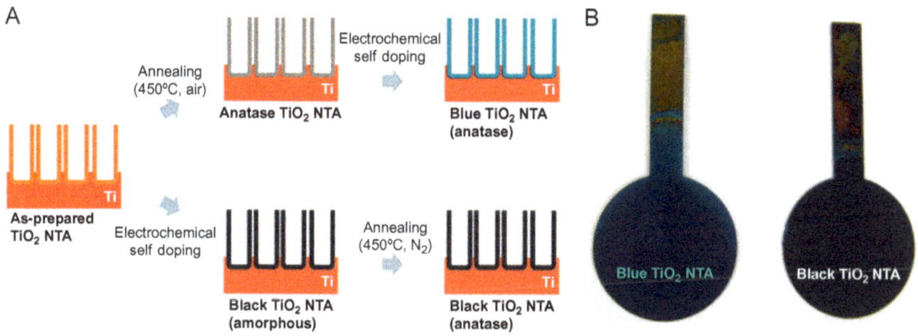

Fig. 12 (A) Schematic diagram for the fabrication process and (B) photographs of blue and black TiO_2 nanotube arrays (NTAs) (Kim et al., 2016). *Reprinted with permission from Kim, C., Kim, S., Hong, S.P., Lee, J., Yoon, J., 2016. Effect of doping level of colored TiO_2 nanotube arrays fabricated by electrochemical self-doping on electrochemical properties. Phys. Chem. Chem. Phys. 18, 14370–14375. Copyright 2016. The Royal Society of Chemistry.*

with ethylene glycol (Kim et al., 2016). Blue TiO_2 nanotube arrays were obtained by annealing at 450°C in air and then electrochemical reduction under constant current (0.017 A cm^{-2}) for 90 s in phosphate buffer solution (Kim et al., 2016). In contrast, black TNTAs were obtained by electrochemical reduction and then annealing at 450°C in nitrogen (Kim et al., 2016).

2.2.2 Electrochemical Oxidation

Dong et al. prepared black TNTAs by a two-step anodization method (Fig. 13A) (Dong et al., 2014). The first anodization was performed at 60 V in ethylene glycol-containing NH_4F (0.25 wt%) and H_2O (2 vol%) for 10 h (Dong et al., 2014). After the oxide layer stripped off by ultrasonication, the Ti foil was subjected to a second stage of anodization with the same conditions (Dong et al., 2014). The obtained TNTAs were washed, dried, and sintered at 450°C for 1 h in ambient atmosphere (Dong et al., 2014). The black TNTAs were obtained after removing anodic TiO_2 layer (Dong et al., 2014). The black TiO_2 showed a dramatically higher visible-light absorbance than the intact TiO_2 (Fig. 13B) (Dong et al., 2014).

2.3 Physical

2.3.1 Ultrasonication

Fan et al. first prepared amorphous black TiO_2 by introducing hydroxyls via ultrasonic irradiation (Fan et al., 2015). The original TiO_2 was prepared by reacting $Ti(SO_4)_2$ solution (12 mL, 8 wt%), ammonia (20 mL, 4 mol L^{-1}), and deionized water (100 mL) at 0°C for 2 h (Fan et al., 2015). The amorphous black TiO_2 was obtained by ultrasonicating the original TiO_2

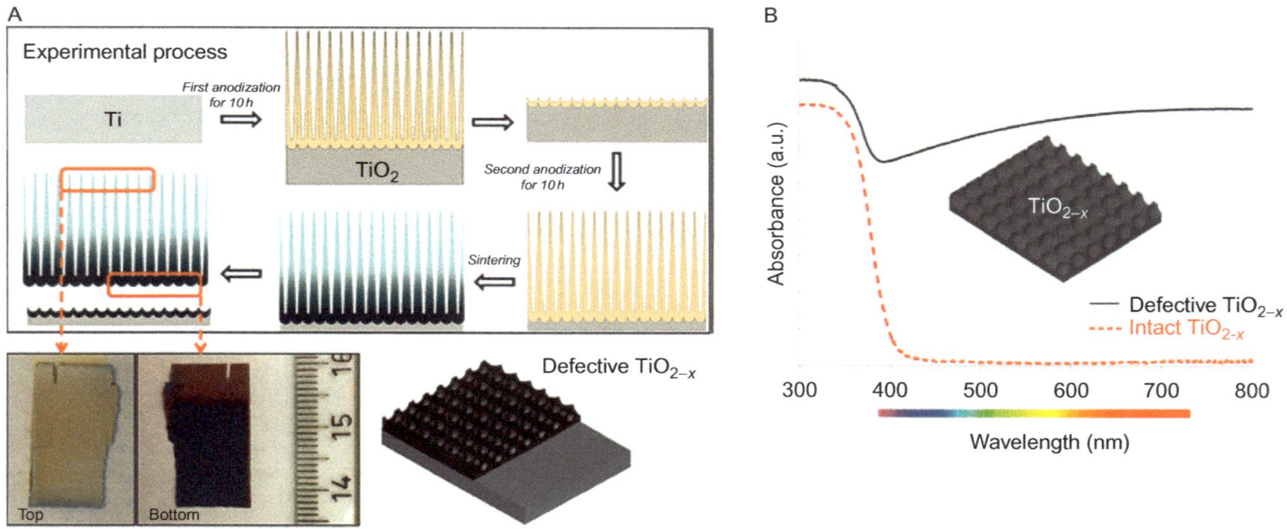

Fig. 13 (A) Experimental process and optical images of the stripped TiO_2 layer. (B) UV–vis absorbance spectra of intact TiO_2 and black TiO_{2-x} (Dong et al., 2014). *Reprinted with permission from Dong, J., Han, J., Liu, Y., Nakajima, A., Matsushita, S., Wei, S., Gao, W., 2014. Defective black TiO_2 synthesized via anodization for visible-light photocatalysis. ACS Appl. Mater. Interfaces 6, 1385–1388. Copyright 2014. American Chemical Society.*

solution at 80°C with an output power density of 1500W/100mL for 0.5–8h (Fan et al., 2015). The amorphous hydroxylated TiO_2 showed a higher absorbance intensity through the whole visible light and near-infrared regions (Fig. 14) (Fan et al., 2015).

2.3.2 Pulsed-Laser Irradiation

Chen et al. prepared black TiO_2 nanospheres with pulsed-laser irradiation (Fig. 15A) (Chen et al., 2015). In a typical experiment, 20mg TiO_2 and 1mL H_2O were transferred into a cuvette, and were irradiated by pulsed laser (λ=355nm, pulse duration=8ns, frequency=10Hz, power=0.35W, instantaneous power=4.4MW) from two sides for 5–120min (Chen et al., 2015). The color of the TiO_2 changed from white to gray (5min), and became darker with the increased time (Fig. 15B) (Chen et al., 2015). Similarly to the optical absorption, the longer laser treated time, the better visible light absorbance (Chen et al., 2015).

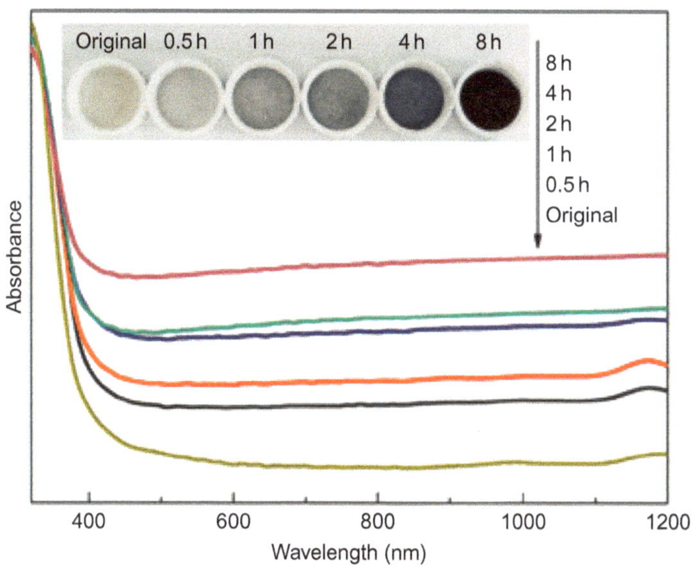

Fig. 14 UV–vis absorbance spectroscopy of original TiO_2 and ultrasonic treated TiO_2 for different hours (*inset* is the photo images) (Fan et al., 2015). *Reprinted with permission from Fan, C., Chen, C., Wang, J., Fu, X., Ren, Z., Qian, G., Wang, Z., 2015. Black hydroxylated titanium dioxide prepared via ultrasonication with enhanced photocatalytic activity. Sci. Rep. 5, 11712. Copyright 2015. Nature.*

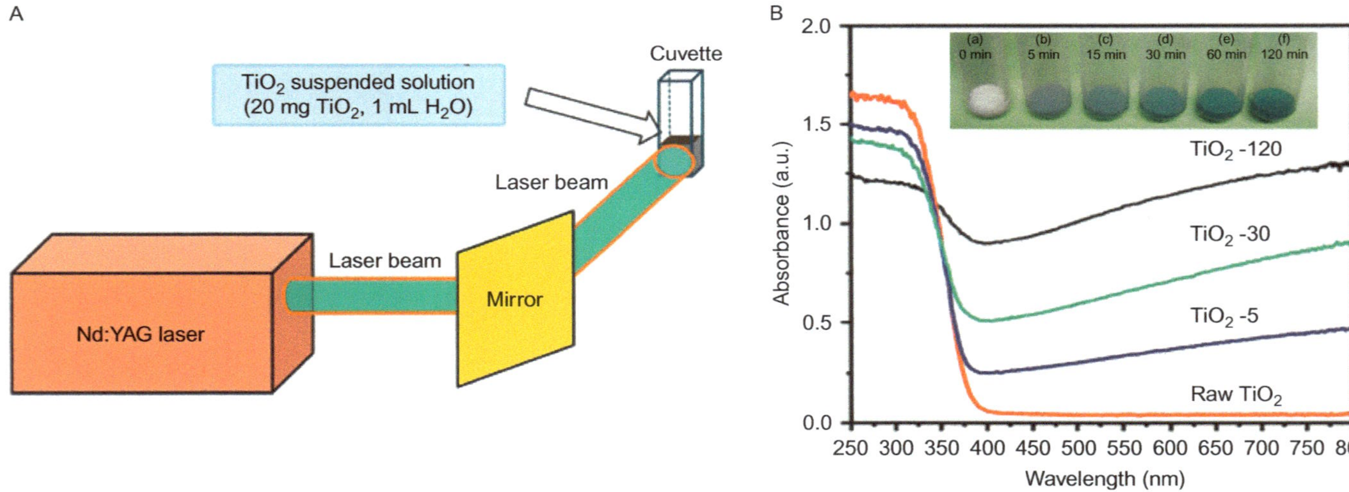

Fig. 15 (A) Schematic of the experimental setup. (B) UV–vis absorbance spectroscopy of TiO_2 before and after laser modification, *inset* is the photographs of suspended TiO_2 solution before and after laser modification (a) raw TiO_2, (b) TiO_2-5, (c) TiO_2-15, (d) TiO_2-30, (e) TiO_2-60, and (f) TiO_2-120 (Chen et al., 2015). *Reprinted with permission from Chen, X., Zhao, D., Liu, K., Wang, C., Liu, L., Li, B., Zhang, Z., Shen, D., 2015. Laser-modified black titanium oxide nanospheres and their photocatalytic activities under visible light. ACS Appl. Mater. Interfaces 7, 16070–16077. Copyright 2015. American Chemical Society.*

3. PROPERTIES

3.1 Structural

The black TiO_2 features with a disordered surface shell surrounding a crystalline core. Chen et al. obtained a disordered layer on the surface of black TiO_2 powder by hydrogenation on TiO_2 powder in a 20.0 bar hydrogen atmosphere at about 200°C for 5 days (Chen et al., 2011; Xia and Chen, 2013). Wang et al. prepared black TiO_2 nanoparticles with a crystalline core–amorphous shell structure by aluminum reduction (Wang et al., 2013b) and hydrogen plasma method (Wang et al., 2013a). Tian et al. obtained black TiO_2 with a nanocrystalline core surrounded by a thin disordered shell by pulsed laser vaporization method (Tian et al., 2015a). High-resolution transmission electron microscopy has been commonly used to observe the disordered phase in the black TiO_2 nanoparticles. Cai et al. also obtained a disordered shell with a thickness of ~2 nm on the surface of TiO_2 after hydrogenation at different temperatures (Cai et al., 2015). X-ray diffraction and Raman also can be used to identify the crystalline phase structures in the black TiO_2 nanoparticles (Cai et al., 2015). Wang et al. found the black $TiO_{2-x}H_x$ nanoparticles prepared by hydrogen plasma had an amorphous shell/crystalline core structure (Wang et al., 2013a). A disordered surface layer with ~2 nm in thickness was coated on a crystalline core (Wang et al., 2013a).

3.2 Chemical

3.2.1 Ti^{3+} Ions and Oxygen Vacancies

Reduction of Ti^{4+} into Ti^{3+} and formation of oxygen vacancies are sometimes observed after pressurized hydrogenation (Naldoni et al., 2012; Shin et al., 2012; Wang et al., 2011), hydrogen plasma treatment (Lepcha et al., 2015), chemical reduction, chemical oxidation, and electrochemical reduction (Yang et al., 2016). Ti^{3+} and oxygen vacancies can be detected by electron spin resonance spectroscopy or electron paramagnetic resonance spectroscopy (EPR) since the paramagnetic Ti^{3+} has a g-value of 1.94–1.99 and the oxygen vacancies appears at $g = 2.004$. For example, after electrochemical reduction, strong signals around at $g = 1.98$ and $g = 2.00$ were corresponded to Ti^{3+} and oxygen vacancy, respectively (Yang et al., 2016). Su et al. certificated the existence of oxygen vacancy in the hydrogenated TiO_2 nanocrystals (Su et al., 2015). Furthermore, the intensity of the signals for these hydrogenated samples increased as the treated temperature rising (Fig. 16) (Su et al., 2015). X-ray photoelectron spectroscopy (XPS) can detect Ti^{3+}. Wang et al. found that the Ti 2p XPS peak of hydrogenated TiO_2 nanosheets had a shift toward lower binding energy with a detectable shoulder (Wei et al., 2012). Jiang and Lu et al. found that there were two extra peaks centered at 457.6–457.9 eV and 463.3–463.5 eV from the Ti 2p 3/2 and Ti 2p 1/2 peaks of Ti^{3+} (Jiang et al., 2012; Lu et al., 2012).

3.2.2 Ti-H and Ti-OH Groups

The appearance of Ti-H and Ti-OH groups was reported in hydrogenated black TiO_2 nanomaterials, and the content changed after different hydrogenation treatment. The Fourier transform infrared (FTIR) spectrum and XPS were often used to detect them. For example, Zheng et al. found one shoulder peak at the lower binding energy side of the Ti 2p in hydrogenated TiO_2 nanowires due to the surface Ti–H bonds (Zheng et al., 2012). Wang et al. found a shoulder peak at around 475.3 eV in the Ti 2p XPS and attributed to the surface Ti–H bonds introduced by hydrogen plasma at 500°C for 4–8 h

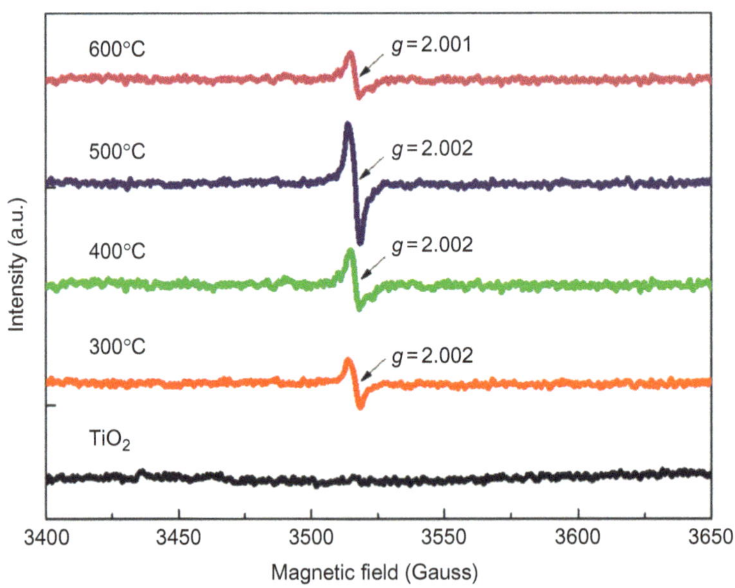

Fig. 16 EPR spectra recorded at 300 K for TiO_2 and hydrogenated TiO_2 samples (Su et al., 2015). *Reprinted with permission from Su, T., Yang, Y., Na, Y., Fan, R., Li, L., Wei, L., Yang, B., Cao, W. 2015. An insight into the role of oxygen vacancy in hydrogenated TiO_2 nanocrystals in the performance of dye-sensitized solar cells. ACS Appl. Mater. Interfaces 7, 3754–3763. Copyright 2015. American Chemical Society.*

(Wang et al., 2013a). A shoulder peak at the higher binding energy side attributed to the Ti-OH was found in the hydrogenated TiO_2 nanowires (Wang et al., 2011). FTIR showed that the intensity of the OH vibration band was different after the hydrogenation treatment. Zhang et al. reported that the intensity of the OH peak for hydrogenated TiO_2 was much lower than pure TiO_2 (Zheng et al., 2012). Xia et al. obtained similar results that no OH absorption bands in hydrogenated TiO_2 nanocrystals, while apparent OH bands existed in the pure TiO_2 (Xia et al., 2013). However, Wang et al. found that more Ti–OH bond on the hydrogenated TiO_2 nanosheet (Wei et al., 2012). Chen et al. reported that both the pure one and the hydrogenated TiO_2 had similar OH bands around 3400 and 3700 cm^{-1} for bridging OH groups and stretching and wagging of O–H vibrations, respectively (Chen et al., 2013).

3.3 Optical

Chen et al. found the hydrogenated black TiO_2 nanoparticle prepared by hydrogen thermal treatment had a significant optical absorption extended to 1200 nm (Fig. 17A) (Chen et al., 2011). Zhou et al. obtained the same optical performance that black TiO_2 has a long-wavelength absorption from UV light to visible and even infrared light regions (Zhou et al., 2014). Liu et al. found the hydrogenation samples showed much stronger light absorption than original TiO_2 nanotubes (Fig. 17B) (Liu et al., 2014).

Su et al. investigated the effect of temperature on the optical absorption of the hydrogenated TiO_2 samples (Su et al., 2015). The TiO_2 nanopowders were prepared via sol–gel hydrothermal method, then heated in tube furnace under a 400 sccm mixture gas flow of 10% H_2 and 90% N_2 for 5 h at 300–600°C (Su et al., 2015). The hydrogenated TiO_2 samples exhibited a larger absorption with the increasing temperature (Fig. 17C) (Su et al., 2015). The remarkable absorption of the H-TiO_2 sample was attributed to the low-energy photon of trapped electron in localized below the conduction band minimum to the conduction band (Su et al., 2015). Yu et al. also found that the absorption of hydrogenated TiO_2 samples depended on hydrogenation temperature and time (Fig. 17D) (Yu et al., 2013).

3.4 Electronic

The electronic structures of TiO_2 nanomaterials are usually modified after hydrogenation treatment, e.g., the valence band shifted. For example, Chen et al. found the hydrogenated TiO_2 nanoparticles treated at 200°C for 4 days

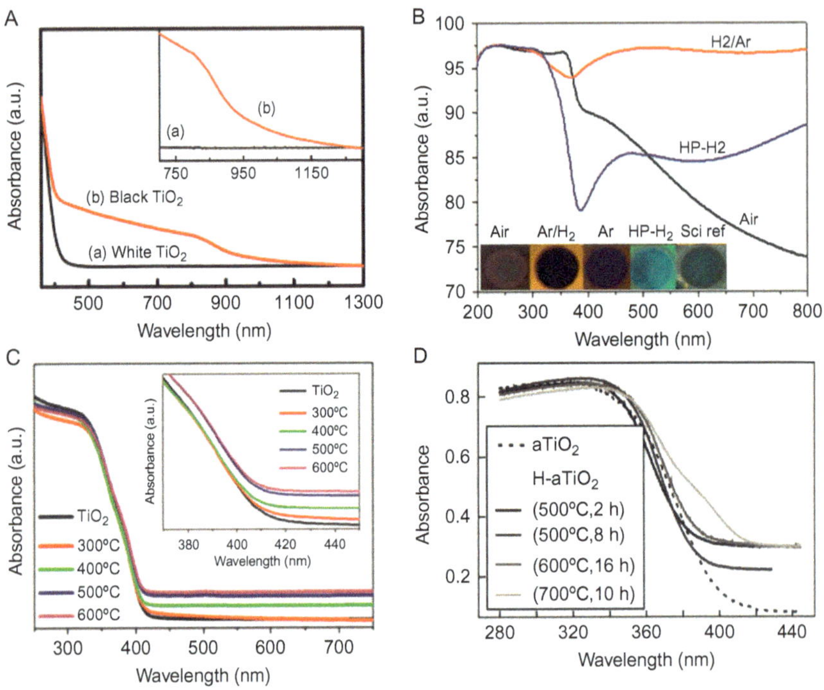

Fig. 17 (A) UV–vis–NIR absorption spectra of (a) white and (b) black TiO_2 nanoparticles (Chen et al., 2011). (B) UV–visible absorption of the TiO_2 nanotubes treated in different atmospheres, *inset* is the optical images for the different treated samples (Liu et al., 2014). (C) UV–vis spectra (*inset* is the partial enlarged drawing) of TiO_2 and H-TiO_2 samples treated at different temperature (Su et al., 2015). (D) UV–visible absorption spectra of anatase TiO_2 nanosheets (aTiO_2) and hydrogenated anatase TiO_2 nanosheets (H-aTiO_2) prepared with a H_2 gas flow at temperatures of 500–700°C (Yu et al., 2013). *Panel (A) Reprinted with permission from Chen, X., Liu, L., Yu, P.Y., Mao, S.S., 2011. Increasing solar absorption for photocatalysis with black hydrogenated titanium dioxide nanocrystals. Science 331, 746–750. Copyright 2011, AAAS. Panel (B) Reprinted with permission from Liu, N., Schneider, C., Freitag, D., Hartmann, M., Venkatesan, U., Müller, J., Spiecker, E., Schmuki, P., 2014. Black TiO_2 nanotubes: cocatalyst-free open-circuit hydrogen generation. Nano Lett. 14, 3309–3313. Copyright 2014. American Chemical Society. Panel (C) Reprinted with permission from Su, T., Yang, Y., Na, Y., Fan, R., Li, L., Wei, L., Yang, B., Cao, W., 2015. An insight into the role of oxygen vacancy in hydrogenated TiO_2 nanocrystals in the performance of dye-sensitized solar cells. ACS Appl. Mater. Interfaces 7, 3754–3763. Copyright 2015, American Chemical Society. Panel (D) Reprinted with permission from Yu, X., Kim, B., Kim, Y.K. 2013. Highly enhanced photoactivity of anatase TiO_2 nanocrystals by controlled hydrogenation-induced surface defects. ACS Catal. 3, 2479–2486. Copyright 2013. American Chemical Society.*

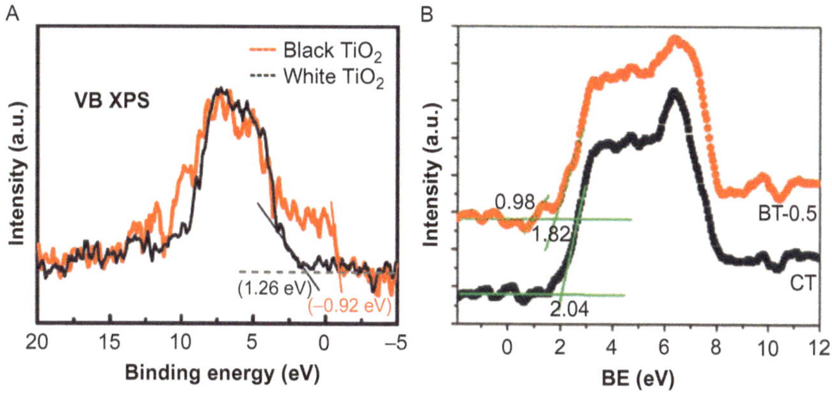

Fig. 18 (A) Valence-band XPS spectra of the white and black TiO_2 nanocrystals (Chen et al., 2011). (B) Valence-band XPS spectra of the commercially nano-anatase TiO_2 (CT) and H-TiO_2 (Sinhamahapatra et al., 2015). *Panel (A) Reprinted with permission from Chen, X., Liu, L., Yu, P.Y., Mao, S.S., 2011. Increasing solar absorption for photocatalysis with black hydrogenated titanium dioxide nanocrystals. Science 331, 746–750. Copyright 2011. Science. Panel (B) Reprinted with permission from Sinhamahapatra, A., Jeon, J.P., Yu, J.S., 2015. A new approach to prepare highly active and stable black titania for visible light-assisted hydrogen production. Energy Environ. Sci. 8, 3539–3544. Copyright 2015. The Royal Society of Chemistry.*

under 20 bar H_2 exhibited a red shift of the valence band from 1.26 to −0.92 eV (Fig. 18A) (Chen et al., 2011). Yu et al. reported that the black TiO_2 nanoparticles obtained by magnesiothermic reduction also had a valence band shift from 2.04 to 1.82 eV and tailed up to 1.06 eV (Fig. 18B) (Sinhamahapatra et al., 2015). However, Wang et al. suggested that hydrogenated TiO_2 nanoparticles and pure TiO_2 had similar valence band position, which was barely influenced by the hydrogen treatment (Wang et al., 2011). Similar conclusions were obtained on hydrogenated TiO_2 nanosheets (Wang et al., 2013a).

4. PHOTOCATALYSIS

4.1 Photocatalytic Degradation of Organic Pollutants

Tian et al. reported that the hydrogenated TiO_2 nanobelts obtained by annealing the TiO_2 nanobelts in hydrogen atmosphere in the temperature range of 600–1000°C exhibited excellent photocatalytic decomposing of methyl orange (Tian et al., 2015b). They attributed the improved photocatalytic property to the Ti^{3+} ions and oxygen vacancies in TiO_2 nanobelts created by hydrogenation (Tian et al., 2015b). Ti^{3+} ions and oxygen vacancies can enhance visible light absorption, promote charge carrier trapping, and hinder the photogenerated electron–hole recombination (Tian et al., 2015b).

An et al. reported that black nanoporous TiO_2 prepared by hydrogen plasma for 30–120 min (H-TiO_2 30, H-TiO_2 120) displayed strong visible absorption and excellent degradation efficiencies for azo dye (An et al., 2016). As shown in Fig. 19, the H-TiO_2 120 photocatalysts almost completely removed the reactive black 5 (RB 5) under UV light or solar light (An et al., 2016). Extra degradation tests of rhodamine B (Rho B) and phenol (Ph) under solar-light irradiation displayed similar results (An et al., 2016). Such superior photocatalytic performance of H-TiO_2 was attributed to its narrowed band gap, the formation of many OH free radicals and large surface area of H-TiO_2 (An et al., 2016).

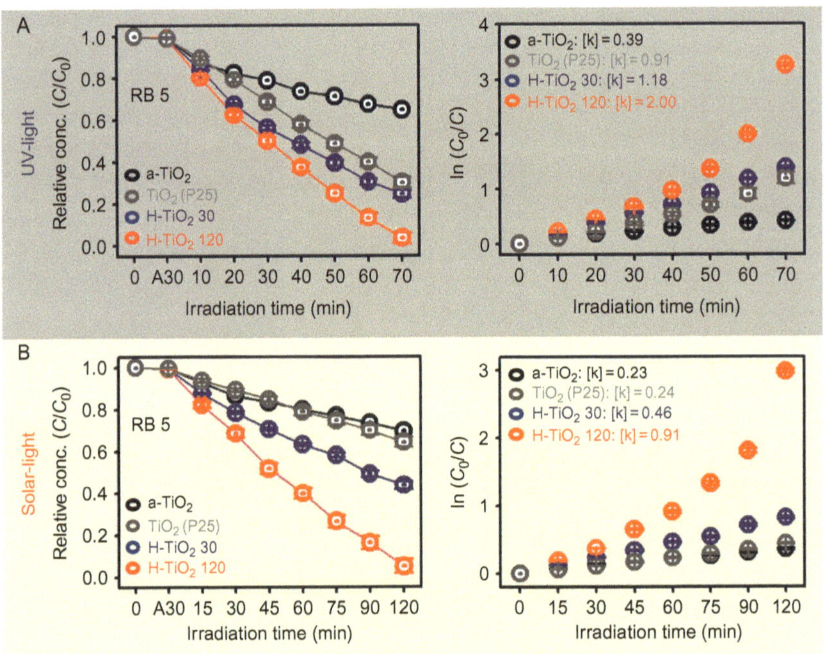

Fig. 19 Removal of RB 5 by commercial TiO_2, as-synthesized TiO_2 (a-TiO_2), H-TiO_2 30, and H-TiO_2 120 under UV- and solar-light irradiation (An et al., 2016). *Reprinted with permission from An, H.R., Park, S.Y., Kim, H., Lee, C.Y., Choi, S., Lee, S.C., Seo, S., Park, E.C., Oh, Y.K., Song, C.G., Won, J., Kim, Y.J., Lee, J., Lee, H.U., Lee, Y.C., 2016. Advanced nanoporous TiO_2 photocatalysts by hydrogen plasma for efficient solar-light photocatalytic application. Sci. Rep. 6, 29683. Copyright 2015. Nature.*

4.2 Photocatalytic Hydrogen Generation

Chen et al. found that the hydrogenated TiO_2 nanoparticles had a broad optical absorption and a much higher photocatalytic activities in generating H_2 from water–methanol solution and decomposing methylene blue and phenol (Chen et al., 2011). Following this report, numerous studies have been carried out to investigate the photocatalytic applications of hydrogenated black TiO_2 nanomaterials (An et al., 2016; Cui et al., 2014; Lepcha et al., 2015; Lin et al., 2014; Liu et al., 2016b; Singh et al., 2016; Sinhamahapatra et al., 2015; Teng et al., 2014; Tian et al., 2015b; Wang et al., 2013a, 2015; Wei et al., 2015). Cai et al. prepared the black TiO_2 (B)-anatase heterophase under different hydrogenation temperature and compared their photocatalytic activities (Cai et al., 2015). The original TiO_2 (B) nanobelts were obtained by hydrothermal method (Cai et al., 2015). Before hydrogenation treatment, 1 wt% Pt was loaded on the dried powder by wetness impregnation method. Then the hydrogenation treatment was performed at 10 vol% H_2/N_2 in a tube furnace at 400, 575, 650, and 750°C for 5 h with the flow rate of 30 mL min^{-1} (labeled as

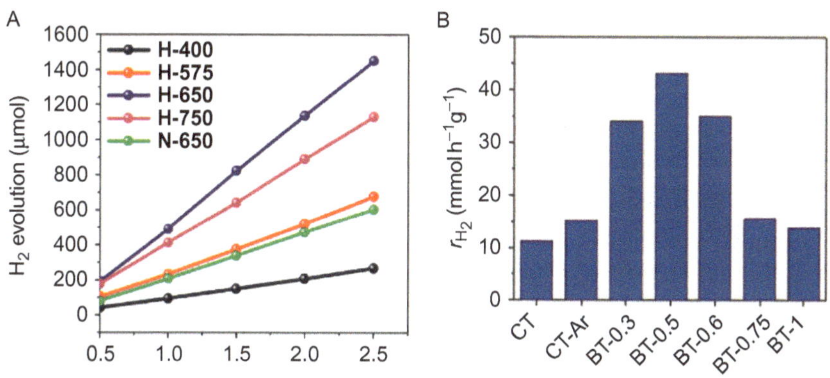

Fig. 20 (A) Photocatalytic hydrogen evolution of the as-prepared different photocatalysts under the simulated solar-light irradiation (Cai et al., 2015). (B) Rate (rH_2) of hydrogen generation for different samples under full solar light (Sinhamahapatra et al., 2015). *Panel (A) Reprinted with permission from Cai, J., Wang, Y., Zhu, Y., Wu, M., Zhang, H., Li, X., Jiang, Z., Meng, M., 2015. In situ formation of disorder-engineered TiO_2(B)-anatase heterophase junction for enhanced photocatalytic hydrogen evolution. ACS Appl. Mater. Interfaces 7, 24987–24992. Copyright 2015. American Chemical Society. Panel (B) Reprinted with permission from Sinhamahapatra, A., Jeon, J.P., Yu, J.S., 2015. A new approach to prepare highly active and stable black titania for visible light-assisted hydrogen production. Energy Environ. Sci. 8, 3539–3544. Copyright 2015. The Royal Society of Chemistry.*

H-400, H-575, H-650, and H-750, respectively) (Cai et al., 2015). As shown in Fig. 20A, the sample H-650 exhibited the highest hydrogen evolution rate of $\sim$580 μmol h^{-1} with 0.02 g of photocatalyst. Both the pure TiO_2 (B) phase (H-400, $\sim$107 μmol h^{-1}) and the anatase phase (H-750, $\sim$452 μmol h^{-1}) show much lower activity than H-650. All the hydrogenated samples had higher photocatalytic activity than that of the sample treated in pure nitrogen atmosphere at 650°C (N-650) (Cai et al., 2015). They suggested that the superior performance of the H-650 was attributed to the synergistic effect of the hydrogenation-induced surface disordered shell, as well as the TiO_2 (B) anatase heterophase junction (Cai et al., 2015).

Yu et al. reported the black TiO_2 prepared by magnesiothermic reduction had an optimum band gap and band position, oxygen vacancies, surface defects, and charge recombination centers, and showed significantly improved optical absorption in the visible and infrared region (Sinhamahapatra et al., 2015). The black TiO_2 material showed remarkable hydrogen production ability in the methanol–water system in the presence of $\sim$1% Pt as a cocatalyst (Sinhamahapatra et al., 2015). The maximum hydrogen production rate was 43 mmol h^{-1} g^{-1} under solar light (Fig. 20B) (Sinhamahapatra et al., 2015).

4.3 Photocatalytic Reduction of CO_2

Yan et al. reported the hydrogenated TiO_2 nanoparticles prepared by hydrogen plasma treatment in several minutes displayed CO_2 reduction activity in aqueous and gaseous media (Yan et al., 2014). As shown in Fig. 21, the slightly hydrogenated TiO_2 (s-H-TiO_2) nanoparticles exhibited enhanced photoactivity than the heavily hydrogenated TiO_2 (h-H-TiO_2) (Yan et al., 2014). They suggested that the higher ratio of trapped holes (O^- centers) and a lower recombination rate induced by the increase of surface defects might be the critical factors for the high activity of s-H-TiO_2; in contrast, h-H-TiO_2 nanoparticles possessed a high concentration of bulk defects, leading to a significantly decreased amount of O^- centers and enhanced nonradioactive recombination, which strongly inhibited their photoactivity (Yan et al., 2014).

5. SUMMARY AND OUTLOOK

Black TiO_2 nanomaterials have been successfully fabricated with many methods for various photocatalytic applications. Up to now, these methods include hydrogenation, chemical reduction/oxidation,

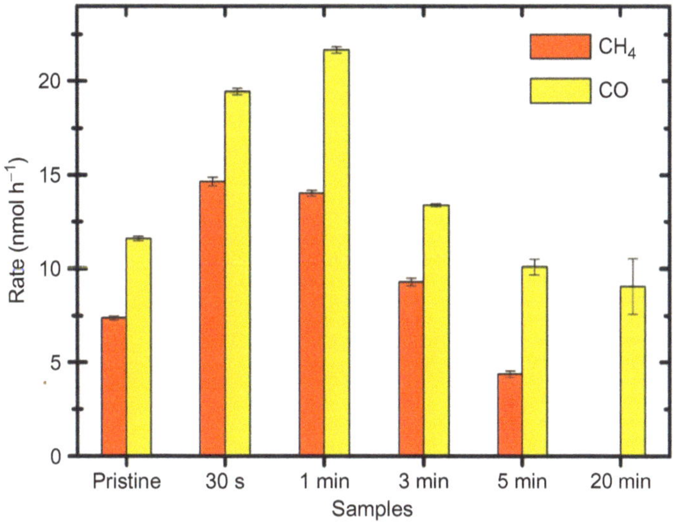

Fig. 21 The formation rates of CH_4 and CO in the photocatalytic reduction of CO_2 with H_2O over pristine-TiO_2 and H-TiO_2 prepared by H_2 plasma treatment after different times of 30 s, 1 min, 3 min, 5 min, and 20 min (Yan et al., 2014). *Reprinted with permission from Yan, Y., Han, M., Konkin, A., Koppe, T., Wang, D., Andreu, T., Chen, G., Vetter, U., Morante, J.R., Schaaf, P., 2014. Slightly hydrogenated TiO_2 with enhanced photocatalytic performance. J. Mater. Chem. A 2, 12708–12716. Copyright 2014. The Royal Society of Chemistry.*

electrochemical reduction/oxidation, ultrasonication, and pulsed-laser irradiation. The hydrogenation includes the treatment under pure, or mixed H_2/inert gas, under high or low pressure, at high or low temperatures for extended time, or the treatment with hydrogen plasma. The chemical reduction methods include the reduction of TiO_2 at high temperatures with Al or Mg powders, or the reduction of TiO_2 in solution with $NaBH_4$ or NaH solution under hydrothermal conditions. Hydrogenation (including hydrogen gas or plasma treatments), chemical reduction with Al and Mg, electrochemical reduction/oxidation, and ultrasonication can effectively lead to the color change of TiO_2 into black color, the reduction with $NaBH_4$ or NaH and pulsed-laser irradiation has so far not yet effectively turn white TiO_2 into black. The color change of TiO_2 depends heavily on the reaction conditions, such as the pressure, temperature, time, and the reducing agent. The variation of the reaction conditions apparently leads to the formation of black TiO_2 featuring different structural (lattice changes or disordering), chemical (formation of Ti^{3+}, oxygen vacancies, Ti-H, Ti-OH), physical properties (such as optical properties), and photocatalytic activities in both hydrogen generation and organic pollutant removal. The variation of the reaction conditions on one hand provides the flexibility of these properties of the black TiO_2, but on the other hand, also increases the complexity of mastering the properties of the black TiO_2 in both fundamental understanding and practical applications. Thus, it is common to see reports from different research groups showing the black TiO_2 possesses somewhat different properties and photocatalytic performances. Regardless of the recent steady progresses in black TiO_2, its photocatalytic activities in the visible-light region are far from satisfactory, although black TiO_2 nanomaterials have shown good capabilities in absorbing a large amount of sunlight and have been sometimes reported with visible-light photocatalytic and photoelectrochemical activities. It is still a good challenge in managing the amount and spatial distributions of the sources of the blackness in the black TiO_2 nanomaterials, such as surface lattice disorder, oxygen vacancies, Ti^{3+} ions, Ti-OH, and Ti-H groups; this means on the other hand opportunities for future improvement. Most focus has so far been on the realization of the black color which only indicates us the optical band gap and the difference between the CBM and VBM, less attention has been paid to the tuning of the electronic structures of the black TiO_2 nanomaterials to match with the requirements of specific photocatalytic reactions yet. Thus adjusting these potential matches from both materials and chemical reactions point of views may provide new opportunities in efficiently utilizing the large amount of light being absorbed by these black TiO_2 nanomaterials, especially in the visible-light region. Apparently, more efforts are needed not only in the synthesis but also in the property and application management, in order to finally improve the efficiency of black TiO_2 nanomaterials for practical photocatalytic applications.

ACKNOWLEDGMENTS

X.C. thanks the support from the U.S. National Science Foundation (DMR-1609061), the College of Arts and Science, University of Missouri—Kansas City and the University of Missouri Research Board. Y.L. appreciates the support from China Scholarship Council for oversee study.

REFERENCES

AlHaddad, A., Wang, Z., Zhou, M., Tarish, S., Vellacheri, R., Lei, Y., 2016. Nanowire arrays: constructing well-ordered CdTe/TiO_2 core/shell nanowire arrays for solar energy conversion. Small 12, 5648.

An, H.R., Park, S.Y., Kim, H., Lee, C.Y., Choi, S., Lee, S.C., Seo, S., Park, E.C., Oh, Y.K., Song, C.G., Won, J., Kim, Y.J., Lee, J., Lee, H.U., Lee, Y.C., 2016. Advanced

nanoporous TiO_2 photocatalysts by hydrogen plasma for efficient solar-light photocatalytic application. Sci. Rep. 6. 29683.
Bai, Y., Yu, H., Li, Z., Amal, R., Lu, G.Q.M., Wang, L., 2012. In situ growth of a ZnO nanowire network within a TiO_2 nanoparticle film for enhanced dye-sensitized solar cell performance. Adv. Mater. 24, 5850–5856.
Bai, Y., Mora-Seró, I., De Angelis, F., Bisquert, J., Wang, P., 2014. Titanium dioxide nanomaterials for photovoltaic applications. Chem. Rev. 114, 10095–10130.
Becker, J.H., Hosler, W.R., 1965. Multiple-band conduction in n-type rutile (TiO_2). Phys. Rev. 137, A1872–A1877.
Borgarello, E., Kiwi, J., Pelizzetti, E., Visca, M., Gratzel, M., 1981. Photochemical cleavage of water by photocatalysis. Nature 289, 158–160.
Breckenridge, R.G., Hosler, W.R., 1953. Electrical properties of titanium dioxide semiconductors. Phys. Rev. 91, 793–802.
Bubenhofer, S.B., Schumacher, C.M., Koehler, F.M., Luechinger, N.A., Grass, R.N., Stark, W.J., 2012. Large-scale synthesis of PbS-TiO_2 heterojunction nanoparticles in a single step for solar cell application. J. Phys. Chem. C 116, 16264–16270.
Cai, J., Wang, Y., Zhu, Y., Wu, M., Zhang, H., Li, X., Jiang, Z., Meng, M., 2015. In situ formation of disorder-engineered TiO_2(B)-anatase heterophase junction for enhanced photocatalytic hydrogen evolution. ACS Appl. Mater. Interfaces 7, 24987–24992.
Cao, F., Xiong, J., Wu, F., Liu, Q., Shi, Z., Yu, Y., Wang, X., Li, L., 2016. Enhanced photoelectrochemical performance from rationally designed anatase/rutile TiO_2 heterostructures. ACS Appl. Mater. Interfaces 8, 12239–12245.
Chambers, S., Thevuthasan, S., Farrow, R., Marks, R., Thiele, J., Folks, L., Samant, M., Kellock, A., Ruzycki, N., Ederer, D., 2001. Epitaxial growth and properties of ferromagnetic co-doped TiO_2 anatase. Appl. Phys. Lett. 79, 3467–3469.
Chen, X., Mao, S.S., 2007. Titanium dioxide nanomaterials: synthesis, properties, modifications, and applications. Chem. Rev. 107, 2891–2959.
Chen, X., Selloni, A., 2014. Introduction: titanium dioxide (TiO_2) nanomaterials. Chem. Rev. 114, 9281–9282.
Chen, X., Liu, L., Yu, P.Y., Mao, S.S., 2011. Increasing solar absorption for photocatalysis with black hydrogenated titanium dioxide nanocrystals. Science 331, 746–750.
Chen, X., Li, C., Grätzel, M., Kostecki, R., Mao, S.S., 2012. Nanomaterials for renewable energy production and storage. Chem. Soc. Rev. 41, 7909–7937.
Chen, X., Liu, L., Liu, Z., Marcus, M.A., Wang, W.C., Oyler, N.A., Grass, M.E., Mao, B., Glans, P.A., Yu, P.Y., Guo, J., Mao, S.S., 2013. Properties of disorder-engineered black titanium dioxide nanoparticles through hydrogenation. Sci. Rep. 3, 1510.
Chen, X., Zhao, D., Liu, K., Wang, C., Liu, L., Li, B., Zhang, Z., Shen, D., 2015. Laser-modified black titanium oxide nanospheres and their photocatalytic activities under visible light. ACS Appl. Mater. Interfaces 7, 16070–16077.
Cheng, C., Amini, A., Zhu, C., Xu, Z., Song, H., Wang, N., 2014. Enhanced photocatalytic performance of TiO_2-ZnO hybrid nanostructures. Sci. Rep. 4, 4181.
Chieh, W.C., Hsun, C.P., 2016. Effects of various hydrogenated treatments on formation and photocatalytic activity of black TiO_2 nanowire arrays. Nanotechnology 27. 325401.
Cong, Y., Tian, B., Zhang, J., 2011. Improving the thermal stability and photocatalytic activity of nanosized titanium dioxide via La^{3+} and N co-doping. Appl. Catal. Environ. 101, 376–381.
Cronemeyer, D.C., 1952. Electrical and optical properties of rutile single crystals. Phys. Rev. 87, 876–886.
Cronemeyer, D.C., 1959. Infrared absorption of reduced rutile TiO_2 single crystals. Phys. Rev. 113, 1222–1226.
Cronemeyer, D.C., Gilleo, M.A., 1951. The optical absorption and photoconductivity of rutile. Phys. Rev. 82, 975–976.

Cui, H., Zhao, W., Yang, C., Yin, H., Lin, T., Shan, Y., Xie, Y., Gu, H., Huang, F., 2014. Black TiO_2 nanotube arrays for high-efficiency photoelectrochemical water-splitting. J. Mater. Chem. A 2, 8612–8616.

Degout, D., Kassabji, F., Fauchais, P., 1984. Titanium dioxide plasma treatment. Plasma Chem. Plasma Proc. 4, 179–198.

Deng, L., Wang, S., Liu, D., Zhu, B., Huang, W., Wu, S., Zhang, S., 2009. Synthesis, characterization of Fe-doped TiO_2 nanotubes with high photocatalytic activity. Catal. Lett. 129, 513–518.

Dette, C., Pérez-Osorio, M.A., Kley, C.S., Punke, P., Patrick, C.E., Jacobson, P., Giustino, F., Jung, S.J., Kern, K., 2014. TiO_2 anatase with a bandgap in the visible region. Nano Lett. 14, 6533–6538.

Dong, F., Zhao, W., Wu, Z., 2008. Characterization and photocatalytic activities of C, N and S co-doped TiO_2 with 1D nanostructure prepared by the nano-confinement effect. Nanotechnology 19. 365607.

Dong, J., Han, J., Liu, Y., Nakajima, A., Matsushita, S., Wei, S., Gao, W., 2014. Defective black TiO_2 synthesized via anodization for visible-light photocatalysis. ACS Appl. Mater. Interfaces 6, 1385–1388.

Eslami, A., Amini, M.M., Yazdanbakhsh, A.R., Bandpei, A.M., Safari, A.A., Asadi, A., 2016. N, S co-doped TiO2 nanoparticles and nanosheets in simulated solar light for photocatalytic degradation of non-steroidal anti-inflammatory drugs in water: a comparative study. J. Chem. Technol. Biotechnol. 91, 2693–2704.

Fan, C., Chen, C., Wang, J., Fu, X., Ren, Z., Qian, G., Wang, Z., 2015. Black hydroxylated titanium dioxide prepared via ultrasonication with enhanced photocatalytic activity. Sci. Rep. 5, 11712.

Feng, H.L., Wu, W.Q., Rao, H.S., Wan, Q., Li, L.B., Kuang, D.B., Su, C.Y., 2015. Three-dimensional TiO_2/ZnO hybrid array as a heterostructured anode for efficient quantum-dot-sensitized solar cells. ACS Appl. Mater. Interfaces 7, 5199–5205.

Ferrighi, L., Fazio, G., Valentin, C.D., 2016. Charge carriers separation at the graphene/(101) anatase TiO_2 interface. Adv. Mater. Interfaces 3, 1500624.

Fujishima, A., Honda, K., 1972. Electrochemical photolysis of water at a semiconductor electrode. Nature 238, 37–38.

Gao, X.F., Li, H.B., Sun, W.T., Chen, Q., Tang, F.Q., Peng, L.M., 2009. CdTe quantum dots-sensitized TiO_2 nanotube array photoelectrodes. J. Phys. Chem. C 113, 7531–7535.

Gao, M., Zhu, L., Ong, W.L., Wang, J., Ho, G.W., 2015. Structural design of TiO_2-based photocatalyst for H_2 production and degradation applications. Catal. Sci. Technol. 5, 4703–4726.

Garcia, J.C., Nolan, M., Deskins, N.A., 2015. The nature of interfaces and charge trapping sites in photocatalytic mixed-phase TiO_2 from first principles modeling. J. Chem. Phys. 142. 024708.

Henrich, V.E., Kurtz, R.L., 1981. Surface electronic structure of TiO_2: atomic geometry, ligand coordination, and the effect of adsorbed hydrogen. Phys. Rev. B 23, 6280–6287.

Hoa, N.T.Q., 2015. Solvothermal synthesis and photocatalytic activity of Co-doped TiO_2 nanowires. Int. J. Nanotechnol. 12, 426–433.

Hoffmann, M.R., Martin, S.T., Choi, W., Bahnemann, D.W., 1995. Environmental applications of semiconductor photocatalysis. Chem. Rev. 95, 69–96.

Hwang, I., Baek, M., Yong, K., 2015. Core/shell structured TiO_2/CdS electrode to enhance the light stability of perovskite solar cells. ACS Appl. Mater. Interfaces 7, 27863–27870.

Jacobs, J.F., van de Poel, I., Osseweijer, P., 2010. Sunscreens with titanium dioxide (TiO_2) nano-particles: a societal experiment. Nanoethics 4, 103–113.

Jiang, X., Zhang, Y., Jiang, J., Rong, Y., Wang, Y., Wu, Y., Pan, C., 2012. Characterization of oxygen vacancy associates within hydrogenated TiO_2: a positron annihilation study. J. Phys. Chem. C 116, 22619–22624.

Jiang, L.L., Lu, X., Xie, C.M., Wan, G.J., Zhang, H.P., Flexible, T.Y.h., 2015. Free-standing TiO_2-graphene-polypyrrole composite films as electrodes for supercapacitors. J. Phys. Chem. C 119, 3903–3910.
Jiao, Z., Zhang, Y., Chen, T., Dong, Q., Lu, G., Bi, Y., 2014. TiO_2 nanotube arrays modified with Cr-doped $SrTiO_3$ nanocubes for highly efficient hydrogen evolution under visible light. Chem.-Eur. J. 20, 2654–2662.
Kang, Q., Liu, S., Yang, L., Cai, Q., Grimes, C.A., 2011. Fabrication of PbS nanoparticle-sensitized TiO_2 nanotube arrays and their photoelectrochemical properties. ACS Appl. Mater. Interfaces 3, 746–749.
Kang, Q., Cao, J., Zhang, Y., Liu, L., Xu, H., Ye, J., 2013. A new approach to prepare highly active and stable black titania for visible light-assisted hydrogen production. J. Mater. Chem. A 1, 5766–5774.
Khan, S.U., Al-Shahry, M., Ingler, W.B., 2002. Efficient photochemical water splitting by a chemically modified n-TiO_2. Science 297, 2243–2245.
Kim, C., Kim, S., Hong, S.P., Lee, J., Yoon, J., 2016. Effect of doping level of colored TiO_2 nanotube arrays fabricated by electrochemical self-doping on electrochemical properties. Phys. Chem. Chem. Phys. 18, 14370–14375.
Kitamura, T., Shibata, K., Takeda, K., 1993. In-flight reduction of Fe_2O_3, Cr_2O_3, TiO_2 and Al_2O_3 by Ar-H_2 and Ar-CH_4 plasma. ISIJ Int. 33, 1150–1158.
Kneiß, M., Jenderka, M., Brachwitz, K., Lorenz, M., Grundmann, M., 2014. Modeling the electrical transport in epitaxial undoped and Ni-, Cr-, and W-doped TiO_2 anatase thin films. Appl. Phys. Lett. 105. 062103.
Kubaschewski, O., Dench, W.A., 1955. The heats of formation in the systems titanium-aluminium and titanium-iron. Acta Met. 3, 339–346.
Kwon, J., Sharma, A.A., Bain, J.A., Picard, Y.N., Skowronski, M., 2015. Oxygen vacancy creation, drift, and aggregation in TiO_2-based resistive switches at low temperature and voltage. Adv. Funct. Mater. 25, 2876–2883.
Lepcha, A., Maccato, C., Mettenbörger, A., Andreu, T., Mayrhofer, L., Walter, M., Olthof, S., Ruoko, T.P., Klein, A., Moseler, M., Meerholz, K., Morante, J.R., Barreca, D., Mathur, S., 2015. Electrospun black titania nanofibers: influence of hydrogen plasma-induced disorder on the electronic structure and photoelectrochemical performance. J. Phys. Chem. C 119, 18835–18842.
Lewicka, Z.A., Yu, W.W., Oliva, B.L., Contreras, E.Q., Colvin, V.L., 2013. Photochemical behavior of nanoscale TiO_2 and ZnO sunscreen ingredients. J. Photochem. Photobiol. A Chem. 263, 24–33.
Li, G., Zhang, Z., Peng, H., Chen, K., 2013. Mesoporous hydrogenated TiO_2 microspheres for high rate capability lithium ion batteries. RSC Adv. 3, 11507–11510.
Li, T., Li, X., Zhao, Q., Shi, Y., Teng, W., 2014a. Fabrication of n-type $CuInS_2$ modified TiO_2 nanotube arrays heterostructure photoelectrode with enhanced photo-electrocatalytic properties. Appl. Catal. Environ. 156, 362–370.
Li, X., Xia, T., Xu, C., Murowchick, J., Chen, X., 2014b. Synthesis and photoactivity of nanostructured CdS-TiO_2 composite catalysts. Catal. Today 225, 64–73.
Li, C., Xi, Z., Fang, W., Xing, M., Zhang, J., 2015a. Enhanced photocatalytic hydrogen evolution activity of $CuInS_2$ loaded TiO_2 under solar light irradiation. J. Solid State Chem. 226, 94–100.
Li, W., Wang, F., Liu, Y., Wang, J., Yang, J., Zhang, L., Elzatahry, A.A., Al-Dahyan, D., Xia, Y., Zhao, D., 2015b. General strategy to synthesize uniform mesoporous TiO_2/graphene/mesoporous TiO_2 sandwich-like nanosheets for highly reversible lithium storage. Nano Lett. 15, 2186–2193.
Li, K., Gao, S., Wang, Q., Xu, H., Wang, Z., Huang, B., Dai, Y., Lu, J., 2015c. In-situ-reduced synthesis of Ti^{3+} self-doped TiO_2/g-C_3N_4 heterojunctions with high photocatalytic performance under LED light irradiation. ACS Appl. Mater. Interfaces 7, 9023–9030.

Li, Y., Cai, Q., Wang, L., Li, Q., Peng, X., Gao, B., Huo, K., Chu, P.K., 2016. Mesoporous TiO_2 nanocrystals/graphene as an efficient sulfur host material for high-performance lithium–sulfur batteries. ACS Appl. Mater. Interfaces 8 (36). 23784.
Lin, T., Yang, C., Wang, Z., Yin, H., Lu, X., Huang, F., Lin, J., Xie, X., Jiang, M., 2014. Effective nonmetal incorporation in black titania with enhanced solar energy utilization. Energy Environ. Sci. 7, 967–972.
Lin, C., Shin, D., Demkov, A.A., 2015. Localized states induced by an oxygen vacancy in rutile TiO_2. J. Appl. Phys. 117. 225703.
Linsebigler, A.L., Lu, G., Yates, J.T., 1995. Photocatalysis on TiO_2 surfaces: principles, mechanisms, and selected results. Chem. Rev. 95, 735–758.
Liu, G., Yang, H.G., Wang, X., Cheng, L., Lu, H., Wang, L., Lu, G.Q., Cheng, H.-M., 2009. Enhanced photoactivity of oxygen-deficient anatase TiO_2 sheets with dominant {001} facets. J. Phys. Chem. C 113, 21784–21788.
Liu, S., Zhang, N., Tang, Z.-R., Xu, Y.-J., 2012. Synthesis of one-dimensional CdS@ TiO_2 core–shell nanocomposites photocatalyst for selective redox: the dual role of TiO_2 shell. ACS Appl. Mater. Interfaces 4, 6378–6385.
Liu, B., Chen, H.M., Liu, C., Andrews, S.C., Hahn, C., Yang, P., 2013a. Large-scale synthesis of transition-metal-doped TiO_2 nanowires with controllable overpotential. J. Am. Chem. Soc. 135, 9995–9998.
Liu, X., Gao, S., Xu, H., Lou, Z., Wang, W., Huang, B., Dai, Y., 2013b. Green synthetic approach for Ti^{3+} self-doped TiO_{2-x} nanoparticles with efficient visible light photocatalytic activity. Nanoscale 5, 1870–1875.
Liu, N., Schneider, C., Freitag, D., Hartmann, M., Venkatesan, U., Müller, J., Spiecker, E., Schmuki, P., 2014. Black TiO_2 nanotubes: cocatalyst-free open-circuit hydrogen generation. Nano Lett. 14, 3309–3313.
Liu, Y., Mu, K., Yang, G., Peng, H., Shen, F., Wang, L., Deng, S., Zhang, X., Zhang, Y., 2015a. Fabrication of a coral/double-wall TiO_2 nanotube array film electrode with higher photoelectrocatalytic activity under sunlight. New J. Chem. 39, 3923–3928.
Liu, Y., Mu, K., Zhong, J., Chen, K., Zhang, Y., Yang, G., Wang, L., Deng, S., Shen, F., Zhang, X., 2015b. Design of a solar-driven TiO_2 nanofilm on Ti foil by self-structure modifications. RSC Adv. 5, 41437–41444.
Liu, Y., Mu, K., Zhang, Y., Wang, L., Yang, G., Shen, F., Deng, S., Zhang, X., Zhang, S., 2016a. Facile synthesis of a narrow-gap titanium dioxide anatase/rutile nanofiber film on titanium foil with high photocatalytic activity under sunlight. Int. J. Hydrogen Energy 41, 10327–10334.
Liu, N., Schneider, C., Freitag, D., Zolnhofer, E.M., Meyer, K., Schmuki, P., 2016b. Noble-metal-free photocatalytic H_2 generation: active and inactive 'black' TiO_2 nanotubes and synergistic effects. Chem. Eur. J. 22, 13810–13814.
Long, G., Mimura, K., Taniuchi, K., 1981. Plasma arc reduction of titanium oxide. J. Jpn. Inst. Light Met. 31, 462–468.
Lu, X., Wang, G., Zhai, T., Yu, M., Gan, J., Tong, Y., Li, Y., 2012. Hydrogenated TiO_2 nanotube arrays for supercapacitors. Nano Lett. 12, 1690–1696.
Lu, H., Zhao, B., Pan, R., Yao, J., Qiu, J., Luo, L., Liu, Y., 2014. Safe and facile hydrogenation of commercial Degussa P25 at room temperature with enhanced photocatalytic activity. RSC Adv. 4, 1128–1132.
Luo, J., Ma, L., He, T., Ng, C.F., Wang, S., Sun, H., Fan, H.J., 2012. TiO_2/(CdS, CdSe, CdSeS) nanorod heterostructures and photoelectrochemical properties. J. Phys. Chem. C 116, 11956–11963.
Ma, Y., Wang, X., Jia, Y., Chen, X., Han, H., Li, C., 2014. Titanium dioxide-based nanomaterials for photocatalytic fuel generations. Chem. Rev. 114, 9987–10043.
Mao, C., Zuo, F., Hou, Y., Bu, X., Feng, P., 2014. In situ preparation of a Ti^{3+} self-doped TiO_2 film with enhanced activity as photoanode by N_2H_4 reduction. Angew. Chem. 126, 10653–10657.

Mathis, J.E., 2015. Visible light absorption characteristics of M-nitrogen (M=Mn, Fe, Co, Ni, Cu) co-doped monodisperse TiO_2 microparticles. MRS Proc. 1806, 19–24.

Matsumoto, Y., Murakami, M., Shono, T., Hasegawa, T., Fukumura, T., Kawasaki, M., Ahmet, P., Chikyow, T., Koshihara, S.-y., Koinuma, H., 2001. Room-temperature ferromagnetism in transparent transition metal-doped titanium dioxide. Science 291, 854–856.

Nakata, K., Fujishima, A., 2012. TiO_2 photocatalysis: design and applications. J. Photochem. Photobiol. C Photochem. Rev. 13, 169–189.

Naldoni, A., Allieta, M., Santangelo, S., Marelli, M., Fabbri, F., Cappelli, S., Bianchi, C.L., Psaro, R., Dal Santo, V., 2012. Effect of nature and location of defects on bandgap narrowing in black TiO_2 nanoparticles. J. Am. Chem. Soc. 134, 7600–7603.

Niu, M., Cheng, D., Cao, D., 2014. Understanding the mechanism of photocatalysis enhancements in the graphene-like semiconductor sheet/TiO_2 composites. Phys. Chem. C 118, 5954–5960.

Nozik, A.J., 1975. Photoelectrolysis of water using semiconducting TiO_2 crystals. Nature 257, 383–386.

Palmer, R.A., Doan, T.M., Lloyd, P.G., Jarvis, B.L., Ahmed, N.U., 2002. Reduction of TiO_2 with hydrogen plasma. Plasma Chem. Plasma Proc. 22, 335–350.

Pan, J.M., Maschhoff, B.L., Diebold, U., Madey, T.E., 1992. Interaction of water, oxygen, and hydrogen with TiO_2(110) surfaces having different defect densities. J. Vac. Sci. Technol. A 10, 2470–2476.

Park, J.H., Kim, S., Bard, A.J., 2006. Novel carbon-doped TiO_2 nanotube arrays with high aspect ratios for efficient solar water splitting. Nano Lett. 6, 24–28.

Pei, Z., Ding, L., Lin, H., Weng, S., Zheng, Z., Hou, Y., Liu, P., 2013. Facile synthesis of defect-mediated TiO_{2-x} with enhanced visible light photocatalytic activity. J. Mater. Chem. A 1, 10099–10102.

Peng, Y.H., Huang, G.F., Huang, W.Q., 2012. Visible-light absorption and photocatalytic activity of Cr-doped TiO_2 nanocrystal films. Adv. Powder Technol. 23, 8–12.

Qin, N., Liu, Y., Wu, W., Shen, L., Chen, X., Li, Z., Wu, L., 2015. One-dimensional CdS/TiO_2 nanofiber composites as efficient visible-light-driven photocatalysts for selective organic transformation: synthesis, characterization, and performance. Langmuir 31, 1203–1209.

Qiu, B., Zhou, Y., Ma, Y., Yang, X., Sheng, W., Xing, M., Zhang, J., 2015. Facile synthesis of the Ti^{3+} self-doped TiO_2-graphene nanosheet composites with enhanced photocatalysis. Sci. Rep. 5, 8591.

Ren, R., Wen, Z., Cui, S., Hou, Y., Guo, X., Chen, J., 2015. Controllable synthesis and tunable photocatalytic properties of Ti^{3+}-doped TiO_2. Sci. Rep. 5. 10714.

Roy, N., Leung, K.T., Pradhan, D., 2015. Nitrogen doped reduced graphene oxide based Pt-TiO_2 nanocomposites for enhanced hydrogen evolution. J. Phys. Chem. C 119, 19117–19125.

Schneider, J., Matsuoka, M., Takeuchi, M., Zhang, J., Horiuchi, Y., Anpo, M., Bahnemann, D.W., 2014. Understanding TiO_2 photocatalysis: mechanisms and materials. Chem. Rev. 114, 9919–9986.

Sekiya, T., Yagisawa, T., Kamiya, N., Das Mulmi, D., Kurita, S., Murakami, Y., Kodaira, T., 2004. Defects in anatase TiO_2 single crystal controlled by heat treatments. J. Phys. Soc. Jpn 73, 703–710.

Shin, J.Y., Joo, J.H., Samuelis, D., Maier, J., 2012. Oxygen-deficient $TiO_{2-\delta}$ nanoparticles via hydrogen reduction for high rate capability lithium batteries. Chem. Mater. 24, 543–551.

Singh, A.P., Kodan, N., Mehta, B.R., Dey, A., Krishnamurthy, S., 2016. In-situ plasma hydrogenated TiO_2 thin films for enhanced photoelectrochemical properties. Mater. Res. Bull. 76, 284–291.

Sinhamahapatra, A., Jeon, J.P., Yu, J.S., 2015. A new approach to prepare highly active and stable black titania for visible light-assisted hydrogen production. Energy Environ. Sci. 8, 3539–3544.

Štengl, V., Bakardjieva, S., Grygar, T.M., Bludská, J., Kormunda, M., 2013. TiO_2-graphene oxide nanocomposite as advanced photocatalytic materials. Chem. Cent. J. 7, 1–12.

Su, T., Yang, Y., Na, Y., Fan, R., Li, L., Wei, L., Yang, B., Cao, W., 2015. An insight into the role of oxygen vacancy in hydrogenated TiO_2 nanocrystals in the performance of dye-sensitized solar cells. ACS Appl. Mater. Interfaces 7, 3754–3763.

Teng, F., Li, M., Gao, C., Zhang, G., Zhang, P., Wang, Y., Chen, L., Xie, E., 2014. Preparation of black TiO_2 by hydrogen plasma assisted chemical vapor deposition and its photocatalytic activity. Appl. Catal. Environ. 148–149, 339–343.

Tian, M., Mahjouri-Samani, M., Eres, G., Sachan, R., Yoon, M., Chisholm, M.F., Wang, K., Puretzky, A.A., Rouleau, C.M., Geohegan, D.B., Duscher, G., 2015a. Structure and formation mechanism of black TiO_2 nanoparticles. ACS Nano 9, 10482–10488.

Tian, J., Leng, Y., Cui, H., Liu, H., 2015b. Hydrogenated TiO_2 nanobelts as highly efficient photocatalytic organic dye degradation and hydrogen evolution photocatalyst. J. Hazard. Mater. 299, 165–173.

Wallace, W.E., Zhong, Q., Genzer, J., Composto, R.J., Bonnell, D.A., 2011. On the use of ion scattering to examine the role of hydrogen in the reduction of TiO_2. J. Mater. Res. 8, 1629–1634.

Wang, L., Sasaki, T., 2014. Titanium oxide nanosheets: graphene analogues with versatile functionalities. Chem. Rev. 114, 9455–9486.

Wang, G., Wang, H., Ling, Y., Tang, Y., Yang, X., Fitzmorris, R.C., Wang, C., Zhang, J.Z., Li, Y., 2011. Hydrogen-treated TiO_2 nanowire arrays for photoelectrochemical water splitting. Nano Lett. 11, 3026–3033.

Wang, Z., Yang, C., Lin, T., Yin, H., Chen, P., Wan, D., Xu, F., Huang, F., Lin, J., Xie, X., Jiang, M., 2013a. H-doped black titania with very high solar absorption and excellent photocatalysis enhanced by localized surface plasmon resonance. Adv. Funct. Mater. 23, 5444–5450.

Wang, Z., Yang, C., Lin, T., Yin, H., Chen, P., Wan, D., Xu, F., Huang, F., Lin, J., Xie, X., Jiang, M., 2013b. Visible-light photocatalytic, solar thermal and photoelectrochemical properties of aluminium-reduced black titania. Energy Environ. Sci. 6, 3007–3014.

Wang, X., Zhang, S., Wang, H., Yu, H., Wang, H., Zhang, S., Peng, F., 2015. Visible light photoelectrochemical properties of a hydrogenated TiO_2 nanorod film and its application in the detection of chemical oxygen demand. RSC Adv. 5, 76315–76320.

Wang, M., Nie, B., Yee, K.K., Bian, H., Lee, C., Lee, H.K., Zheng, B., Lu, J., Luo, L., Li, Y.Y., 2016. Low-temperature fabrication of brown TiO_2 with enhanced photocatalytic activities under visible light. Chem. Commun. 52, 2988–2991.

Wei, W., Yaru, N., Chunhua, L., Zhongzi, X., 2012. Hydrogenation of TiO_2 nanosheets with exposed {001} facets for enhanced photocatalytc activity. RSC Adv. 2, 8286–8288.

Wei, S., Wu, R., Jian, J., Chen, F., Sun, Y., 2015. Black and yellow anatase titania formed by (H,N)-doping: strong visible-light absorption and enhanced visible-light photocatalysis. Dalton Trans. 44, 1534–1538.

Weir, A., Westerhoff, P., Fabricius, L., von Goetz, N., 2012. Titanium dioxide nanoparticles in food and personal care products. Environ. Sci. Technol. 46, 2242–2250.

Williams, G., Seger, B., Kamat, P.V., 2008. TiO_2-graphene nanocomposites: UV-assisted photocatalytic reduction of graphene oxide. ACS Nano. 2, 1487–1491.

Xia, T., Chen, X., 2013. Revealing the structural properties of hydrogenated black TiO_2 nanocrystals. J. Mater. Chem. A 1, 2983–2989.

Xia, T., Zhang, C., Oyler, N.A., Chen, X., 2013. Hydrogenated TiO_2 nanocrystals: a novel microwave absorbing material. Adv. Mater. 25, 6905–6910.

Xin, X., Xu, T., Wang, L., Wang, C., 2016. Ti^{3+}-self doped brookite TiO_2 single-crystalline nanosheets with high solar absorption and excellent photocatalytic CO_2 reduction. Sci. Rep. 6. 23684.

Xing, M., Wu, Y., Zhang, J., Chen, F., 2010. Effect of synergy on the visible light activity of B, N and Fe co-doped TiO_2 for the degradation of MO. Nanoscale 2, 1233–1239.

Xing, M.Y., Qi, D.Y., Zhang, J.L., Chen, F., 2011. One-step hydrothermal method to prepare carbon and lanthanum co-doped TiO_2 nanocrystals with exposed {001} facets and their high UV and visible-light photocatalytic activity. Chem. Eur. J. 17, 11432–11436.

Xing, M., Li, X., Zhang, J., 2014. Synergistic effect on the visible light activity of Ti^{3+} doped TiO_2 nanorods/boron doped graphene composite. Sci. Rep. 4, 5493.

Yan, Y., Hao, B., Wang, D., Chen, G., Markweg, E., Albrecht, A., Schaaf, P., 2013. Understanding the fast lithium storage performance of hydrogenated TiO_2 nanoparticles. J. Mater. Chem. A 1, 14507–14513.

Yan, Y., Han, M., Konkin, A., Koppe, T., Wang, D., Andreu, T., Chen, G., Vetter, U., Morante, J.R., Schaaf, P., 2014. Slightly hydrogenated TiO_2 with enhanced photocatalytic performance. J. Mater. Chem. A 2, 12708–12716.

Yan, X., Tian, L., Tan, X., Zhou, M., Liu, L., Chen, X., 2016. Modifying oxide nanomaterials' properties by hydrogenation. MRS Commun. 6, 192–203.

Yang, Y., Liao, J., Li, Y., Cao, X., Li, N., Wang, C., Lin, S., 2016. Electrochemically self-doped hierarchical TiO_2 nanotube arrays for enhanced visible-light photoelectrochemical performance: an experimental and computational study. RSC Adv. 6, 46871–46878.

Yao, Z., Jia, F., Tian, S., Li, C., Jiang, Z., Bai, X., 2010. Microporous Ni-doped TiO_2 film photocatalyst by plasma electrolytic oxidation. ACS Appl. Mater. Interfaces 2, 2617–2622.

Yu, J., Xiang, Q., Zhou, M., 2009. Preparation, characterization and visible-light-driven photocatalytic activity of Fe-doped titania nanorods and first-principles study for electronic structures. Appl. Catal. Environ. 90, 595–602.

Yu, X., Kim, B., Kim, Y.K., 2013. Highly Enhanced photoactivity of anatase TiO_2 nanocrystals by controlled hydrogenation-induced surface defects. ACS Catal. 3, 2479–2486.

Zhang, J., Pan, C., Fang, P., Wei, J., Xiong, R., 2010. Mo + C codoped TiO_2 using thermal oxidation for enhancing photocatalytic activity. ACS Appl. Mater. Interfaces 2, 1173–1176.

Zhang, Z., Xiao, F., Guo, Y., Wang, S., Liu, Y., 2013. One-pot self-assembled three-dimensional TiO_2-graphene hydrogel with improved adsorption capacities and photocatalytic and electrochemical activities. ACS Appl. Mater. Interfaces 5, 2227–2233.

Zhang, M., Xu, Y., Lv, J., Yang, L., Jiang, X., He, G., Song, X., Sun, Z., 2014. Capability of coupled $CdSe/TiO_2$ heterogeneous structure for photocatalytic degradation and photoconductivity. Nanoscale Res. Lett. 9, 1–7.

Zhang, T., Yu, B., Wang, D., Zhou, F., 2015. Molybdenum-doped and anatase/rutile mixed-phase TiO_2 nanotube photoelectrode for high photoelectrochemical performance. J. Power Sources 281, 411–416.

Zheng, Z., Huang, B., Lu, J., Wang, Z., Qin, X., Zhang, X., Dai, Y., Whangbo, M.H., 2012. Hydrogenated titania: synergy of surface modification and morphology improvement for enhanced photocatalytic activity. Chem. Commun. 48, 5733–5735.

Zhou, Z., Yuan, S., Fan, J., Hou, Z., Zhou, W., Du, Z., Wu, S., 2012. $CuInS_2$ quantum dot-sensitized TiO_2 nanorod array photoelectrodes: synthesis and performance optimization. Nanoscale Res. Lett. 7, 10.1186.

Zhou, W., Li, W., Wang, J.-Q., Qu, Y., Yang, Y., Xie, Y., Zhang, K., Wang, L., Fu, H., Zhao, D., 2014. Ordered mesoporous black TiO_2 as highly efficient hydrogen evolution photocatalyst. J. Am. Chem. Soc. 136, 9280–9283.

Zhou, Y., Chen, C., Wang, N., Li, Y., Ding, H., 2016. Stable Ti^{3+} self-doped anatase-rutile mixed TiO_2 with enhanced visible light utilization and durability. J. Phys. Chem. C 120, 6116–6124.

Zhu, G., Lin, T., Lu, X., Zhao, W., Yang, C., Wang, Z., Yin, H., Liu, Z., Huang, F., Lin, J., 2013. Black brookite titania with high solar absorption and excellent photocatalytic performance. J. Mater. Chem. A 1, 9650–9653.

CHAPTER TWELVE

Effective Charge Carrier Utilization in Visible-Light-Driven CO_2 Conversion

Xiaoxia Chang*,†, Tuo Wang*,†, Jinlong Gong*,†,1

*Key Laboratory for Green Chemical Technology of Ministry of Education, School of Chemical Engineering and Technology, Tianjin University, Tianjin, China

†Collaborative Innovation Center of Chemical Science and Engineering, Tianjin, China

[1]Corresponding author: e-mail address: jlgong@tju.edu.cn

Contents

1. INTRODUCTION

With the fast development of the world economy and the rapid growth of the population, fossil fuels have been exploited as the main energy source for centuries, which are nevertheless nonrenewable and being depleted. The excessive utilization of carbon-based energy resources accelerated the rise in atmospheric CO_2 concentrations, leading to adverse global

Semiconductors and Semimetals, Volume 97
ISSN 0080-8784
http://dx.doi.org/10.1016/bs.semsem.2017.04.004

climate change. Therefore, the energy crisis and global warming are the two main challenges of the 21st century (Dresselhaus and Thomas, 2001; Song, 2006). As a theoretically infinite energy source, solar power can be utilized to reduce CO_2 into chemicals and fuels such as methane or methanol, which has been considered as a promising approach to address both the energy crisis and global warming (Lewis and Nocera, 2006). However, there are many challenges associated with the photoconversion of CO_2. On the one hand, CO_2 is a linear molecule and the double bond between the carbon and oxygen atoms possesses a high dissociation energy of $\sim$750 kJ mol^{-1}, which indicates the requirement of a high energy input for the transformation of CO_2 (Kumar et al., 2012; Xie et al., 2016). On the other hand, the hydrogen evolution reaction (HER) competes with CO_2 reduction in the presence of water, which would lower the overall selectivity for carbonaceous products. However, the utilization of organic solvent to increase the solubility of CO_2 and suppress HER will result in additional costs (Chang et al., 2016b). In addition, the carbon in CO_2 molecule is in its highest oxidation state and the reduction of CO_2 can generate a large variety of products with different carbon oxidation states. The possible gas products range from CO, CH_4 to higher hydrocarbons as well as various oxygenates in the liquid phase such as CH_3OH and HCOOH, which makes the selectivity for our desired products much low and uncontrollable (Hong et al., 2013). Although a number of synthetic photocatalysts have been developed for the reduction of CO_2 with H_2O to organic compounds, most of them still suffer from low energy conversion efficiency, fierce competition with HER, uncontrollable selectivity for desired products, and instability. Therefore, there is an urgent need for the design and fabrication of artificial photosynthesis systems with high efficiency and selectivity for converting CO_2 and H_2O into valuable fuels.

Halmann (1978) reported the seminal investigation of CO_2 photoelectroreduction in 1978 with a p-type GaP photocathode and observed the generation of HCOOH, HCHO, and CH_3OH in the electrolyte (Halmann, 1978). In the following year, Inoue et al. (1979) first reported the CO_2 photoreduction in aqueous suspensions of photocatalysts, such as WO_3, TiO_2, ZnO, CdS, GaP, and SiC, to produce HCOOH, HCHO, and CH_3OH (Inoue et al., 1979). Since then, the photoreduction of CO_2 into fuels using solar energy as the driving force has drawn much attention. There are also several excellent reviews on the design and fabrication of various artificial photosynthesis systems for CO_2 reduction, as well as the description of the reaction mechanism and pathways (Chang et al., 2016a; Habisreutinger et al., 2013; Li et al., 2014b; Ma et al., 2014;

Tu et al., 2014; White et al., 2015; Xie et al., 2016). This chapter describes some basic principles and the critical challenges in both thermodynamics and kinetics for CO_2 photoreduction. The effective utilization of charge carriers is much critical to the efficiency of photoconversion. Therefore, some strategies for this are also presented and discussed. The focus of this chapter will be on the most recent advances in CO_2 photoreduction over visible-light photocatalysts, such as oxide and chalcogenide semiconductors, group IV materials, nitride and III–V semiconductors, perovskites, and hybrid structure photocatalysts. The challenges and perspectives of CO_2 photoreduction over visible-light photocatalysts are presented as well in the last section.

2. FUNDAMENTALS OF CO_2 PHOTOREDUCTION

2.1 Background and Basic Principles

Semiconductor-based photocatalytic (PC) reactions involve complicated processes. It is generally accepted that there are three major crucial steps in photocatalysis: solar light harvesting, charge separation and transportation, and surface reactions (Kudo and Miseki, 2009; Zhang et al., 2016). As shown in Scheme 1, electrons and holes are generated in the bulk of the semiconductor photocatalyst when harvesting the incident photons with energy equal to or greater than the band-gap energy (E_g) of the photocatalyst. The consumed energy excites the electrons from the valence band (VB) to the conduction band (CB), leaving a deficit of negative charge (referred as "hole") in the VB (step (i) in Scheme 1). Then the photogenerated

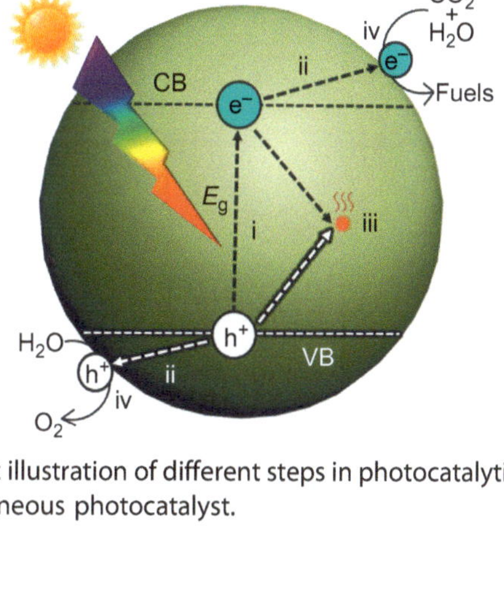

Scheme 1 Schematic illustration of different steps in photocatalytic CO_2 reduction with H_2O over a heterogeneous photocatalyst.

electrons and holes migrate independently to the surface of the photocatalyst in the second step (ii). At the same time, electrons and holes tend to recombine and a large fraction of them may be consumed through recombination and release the energy in the form of light or heat (step (iii) in Scheme 1). In the last step (iv), the photogenerated electrons that reach the surface reduce CO_2 into fuels, while the holes oxidize H_2O into O_2. The above semiconductor-based photocatalysis contains complex photo-, electro-, and chemical processes. Therefore, the effective utilization of charge carriers, such as the promotion of charge generation, migration, and reaction, is highly desirable to improve the efficiency of photoconversion (Zhang et al., 2016).

In general, there are two configurations of reaction systems for CO_2 photoconversion (Chang et al., 2016a; Kudo and Miseki, 2009; White et al., 2015). The first configuration is PC system which utilizes a suspension of photocatalyst particles in a solvent with dissolved CO_2 (Fig. 1A). In this kind of system, the driving force for CO_2 reduction is entirely solar energy. The reaction devices are quite simple and the photocatalysts are accessible, thus making PC system a relatively convenient subject to study. However, due to the limitations of this system, both the reduction and oxidation reactions occur on the same photocatalyst particle, leading to the mixing of all products and the difficulty of product separation. In particular, if there are no hole scavengers, such as H_2O_2, Na_2SO_3, and alcohol, the reduction products

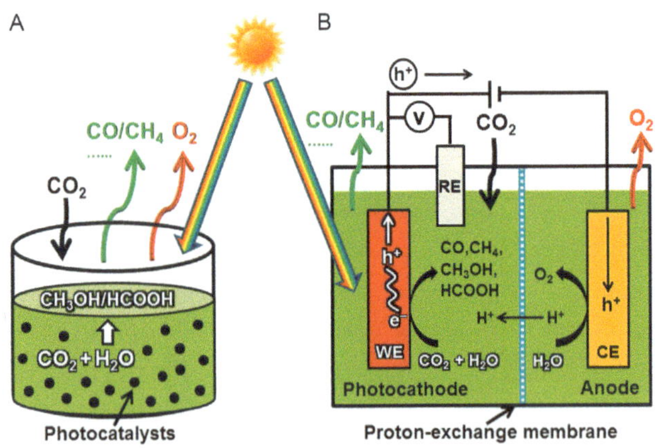

Fig. 1 (A) PC system composed of photocatalyst particles in a CO_2-containing electrolyte to perform both photocatalytic CO_2 reduction and water oxidation. (B) PEC cell system with a photocathode as working electrode (WE) for CO_2 reduction, a counter electrode (CE) for water oxidation and a reference electrode (RE) immersed in a CO_2-containing electrolyte.

from CO_2 could be reoxidized by photogenerated holes or the oxygen produced by water oxidation. However, the use of hole scavengers would result in additional costs. In addition, the used photocatalyst particles in PC system are difficult to recycle and reutilize. The other configuration is known as the photoelectrochemical (PEC) cell system, which is usually composed of a semiconductor photoelectrode (working electrode), a counter electrode, and a reference electrode (Fig. 1B). Under illumination, the photoelectrode harvests photons to generate charge carriers and simultaneously carries out a half-cell reaction, generally on the cathode to reduce CO_2, whereas the other half-cell reaction takes place on the counter electrode. Hence, the reduction and oxidation reactions occur on different electrodes and the products can be spatially separated in different half-cells by a proton exchange membrane, which can suppress the reoxidation of reduction products. The used electrode materials can be easily separated from the reaction mixtures for reuse. In addition to solar energy, an external bias can be applied in PEC system to promote the separation of photogenerated electrons and holes, which usually results in a higher efficiency compared to the PC system (White et al., 2015). Therefore, although the preparation of photoelectrodes and the fabrication of reaction devices are complicated and normally an external bias is needed, it is beneficial to conduct CO_2 photoreduction using a PEC system.

2.2 The Challenges of CO_2 Photoreduction in Thermodynamics and Kinetics

As mentioned earlier, the reduction of CO_2 can generate a large variety of products with different carbon oxidation states. The possible reactions and the related various products in CO_2 reduction as well as the corresponding standard redox potentials obtained from thermodynamic data are shown in Table 1 (Chang et al., 2016a; Hong et al., 2013). Usually, due to the high concentration of protons, HER is prevalent in acidic solutions and CO_2 easily turns into CO_3^{2-} or $HCO_3^{\bullet-}$ in basic solutions, leading to many CO_2 reduction studies being conducted in close neutral electrolyte solutions (Li, 2010; White et al., 2015). Therefore, the potentials in Table 1 are tabulated at pH 7 and referred to a normal hydrogen electrode (NHE) reference in an aqueous solution. CO, CH_4, C_2H_4, C_2H_6, and even higher hydrocarbons are possible products in the gas phase and HCOOH, CH_3OH, C_2H_5OH, and even $H_2C_2O_4$ are the possible products in liquid phase depending on the different reaction mechanisms and pathways. Since

Table 1 The Main CO_2 Reduction Products and the Corresponding Reduction Potentials With Reference to NHE at pH 7 in Aqueous Solution, 25°C, and 1 atm Gas Pressure

Product	**Reaction**	**E^0 (V vs NHE)**	**Equations**
Carbon monoxide	$CO_2+2H^++2e^- \rightarrow CO+H_2O$	−0.51	(1)
Methane	$CO_2+8H^++8e^- \rightarrow CH_4+2H_2O$	−0.24	(2)
Ethene	$2CO_2+12H^++12e^- \rightarrow C_2H_4+4H_2O$	−0.34	(3)
Ethane	$2CO_2+14H^++14e^- \rightarrow C_2H_6+4H_2O$	−0.27	(4)
Formic acid	$CO_2+2H^++2e^- \rightarrow HCOOH$	−0.58	(5)
Methanol	$CO_2+6H^++6e^- \rightarrow CH_3OH+H_2O$	−0.39	(6)
Ethanol	$2CO_2+12H^++12e^- \rightarrow C_2H_5OH +3H_2O$	−0.33	(7)
Oxalate	$2CO_2+2H^++2e^- \rightarrow H_2C_2O_4$	−0.87	(8)
Hydrogen	$2H_2O+2e^- \rightarrow 2OH^-+H_2$	−0.41	(9)

HER is always in fierce competition with CO_2 reduction, its equilibrium potential is also presented in Table 1.

In spite of many years' efforts, the conversion efficiency and product selectivity of CO_2 photoreduction in aqueous solutions are still limited due to several factors.

(i) CO_2 with a linear geometry is one of the most stable and chemically inert molecules. The reduction of CO_2 requires a large energy input to bend the linear molecular structure, break C=O bonds and subsequently form C—H bonds (Freund and Roberts, 1996; Song, 2006). The Gibbs free energy changes for the conversion of CO_2 with H_2O into hydrocarbon fuels such as CH_4 and CH_3OH are 818.3 and 702.2 kJ mol^{-1}, respectively, which are much higher than that of 237.2 kJ mol^{-1} for the conversion of one H_2O molecule into H_2 and $1/2O_2$ (Li et al., 2014b). Therefore, the uphill reaction with highly positive Gibbs-free energy change makes the reduction of CO_2 quite a grand challenge.

(ii) It is generally accepted that CO_2 reduction involves the first step of $CO_2^{\bullet-}$ intermediate formation, which is widely recognized as the first step to activate CO_2 for the subsequent reaction (Rosen et al., 2011;

Surdhar et al., 1989). The addition of a single electron could bend the molecular structure due to the repulsion between the newly acquired electron on the electrophilic carbon atom and the free electron pairs on the oxygen atoms (Freund and Roberts, 1996; Habisreutinger et al., 2013). Due to the large reorganizational energy between the linear molecule of CO_2 and bent radical anion of $CO_2^{\bullet-}$, this step has a very negative equilibrium potential of −1.9 V vs NHE (Eq. 10). This indicates that a high overpotential is required for CO_2 conversion and the real applied potentials are significantly more negative than the equilibrium ones in Table 1.

$$CO_2 + e^- \rightarrow CO_2^{\bullet-} \quad (-1.90\,\text{V vs NHE}) \tag{10}$$

From the band-gap positions of some typical semiconductors and the standard reduction potentials for CO_2 reduction in Fig. 2, it can be seen that the CB edges of most semiconductors are below the equilibrium potential of single electron reduction in Eq. (10). Thermodynamically, photoreduction reactions can occur only when the CB edge of semiconductors is above the standard reduction potential, whereas the VB edge of semiconductors is required to be below the standard oxidation potential for the process of oxidation reactions (White et al., 2015). Therefore, the photogenerated

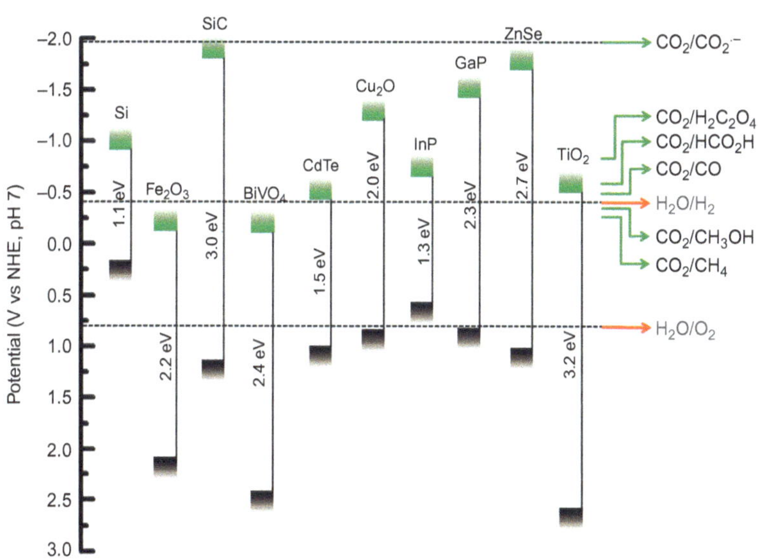

Fig. 2 The conduction band (*green squares*) and valence band (*black squares*) energy levels of some candidate semiconductor photocatalysts and the redox potentials vs NHE of CO_2 reduction and water splitting at pH 7.

electrons in the CB of almost all the candidate semiconductors do not have enough driving force to carry out the single electron reduction of CO_2 in Eq. (10). Partly due to the thermodynamic difficulties in the activation of the CO_2 molecule, the efficiency of CO_2 photoreduction is limited.

(iii) Since it is very thermodynamically unfavorable for the formation of $CO_2^{\bullet-}$ radical, more favorable reaction pathways involving a series of multiple proton-coupled electron transfer (PCET) processes to bypass the single electron reduction of CO_2 have been reported. Through the transfer of protons as well as electrons, the large activation barriers resulted from the significant reorganization energies can be more readily avoided (Huynh and Meyer, 2007). One typical multiple-electron reaction mechanism is the two-electron, two-proton reaction pathway first proposed by Inoue et al. (1979)). The multistep reduction process of CO_2 to CH_4 can be described through the following equations:

$$CO_2 \xrightarrow[+2H^+]{+2e^-} HCOOH \xrightarrow[+2H^+-H_2O]{+2e^-} HCOH \xrightarrow[+2H^+]{+2e^-} CH_3OH \xrightarrow[+2H^+-H_2O]{+2e^-} CH_4 \qquad (11)$$

However, although the large activation barriers could be avoided through PCET, this reaction pathway has a strong dependence on both the concentration of available protons and the partial electron density on the surface of semiconductor photocatalysts. The associated activation energies presenting kinetic barriers in each step of PCET processes must be overcome for the forward reaction (Kumar et al., 2012). If CH_4 and CH_3OH are considered as the main target products, the generation of them needs the transfer of eight and six electrons, respectively, which is more difficult than the two-electron transfer process associated with HER (Table 1). In fact, many great achievements have been made in the reduction of CO_2 into CO and HCOOH. The production of more useful fuels such as CH_4 or CH_3OH, which involves multiple electron and proton transfers, has only been performed with low conversion efficiency and selectivity. Therefore, in addition to the thermodynamic limitations, there are also many kinetic challenges in CO_2 reduction to hydrocarbon fuels.

(iv) Another challenge in photoreduction of CO_2 is the low solubility of CO_2 in aqueous solutions (e.g., 0.033 mol L^{-1} at 25°C under 1 atm), leading to mass transfer limitations for large-scale fuel production (Hara, 1995; Palmer and Van Eldik, 1983). Kinetically, the HER is more favorable and becomes a fierce competing reaction with the reduction of CO_2, resulting in a low selectivity for carbonaceous products. Therefore, the mass transfer limitations of CO_2 should be carefully considered and overcome for high conversion efficiency.

(v) Furthermore, there are many possible products of CO_2 photoreduction presented both in the gaseous and liquid phases (as shown in Table 1) depending on the specific reaction pathway and the number of transferred electrons, which leads to a quite complex process for the separation and detection of products.

Therefore, both in thermodynamics and kinetics, there are many challenges associated with the reduction of CO_2. The promotion of the adsorption/activation of CO_2 molecule and the decrease of the overpotential for $CO_2^{\bullet-}$ intermediate formation are urgently demanded. Meanwhile, the mechanisms and pathways of CO_2 photoreduction on semiconductor photocatalysts are still unclear. The understanding of reaction mechanisms could enable an increase in the conversion efficiency of CO_2 reduction through minimizing the reaction barriers as well as the control of selectivity toward the desired products through guiding the reaction pathways, which is also a grand challenge for the future studies.

2.3 Strategies for the Effective Charge Carrier Utilization in CO_2 Photoreduction

As shown in Scheme 1, there are three major crucial steps in photocatalysis: charge generation, charge separation and transportation, and surface reactions. Maximizing the utilization of photogenerated charge carriers is the key to improving the activity of photocatalysts (Zhang et al., 2016). In recent years, many efforts have been devoted to obtain highly active photocatalysts for effective solar-to-chemical energy conversion, including extending the light absorption range, controlling the morphology, constructing heterojunctions, and promoting surface reaction kinetics.

2.3.1 Promote Charge Carrier Generation

Light-harvesting properties of semiconductors are crucial to the generation of charge carriers. However, many typical semiconductors, such as TiO_2 and ZnO, have wide band gaps and can only harvest the photons in UV region of the solar spectrum, which is only about 4% of the total sunlight energy reaching the earth. The limited utilization of photons in visible region, which is about 48% of the total sunlight energy, leads to the low solar-to-chemical energy conversion efficiency (Schultz and Yoon, 2014). As mentioned earlier, the semiconductor photocatalysts absorb only the incident photons with energy equal to or greater than their E_g, which depends solely on the CB and VB positions, to excite electrons (Marszewski et al., 2015; White et al., 2015). Therefore, the strategies of band engineering are primarily performed to improve light absorption and promote charge carrier generation (Wang et al., 2014a,b).

The most common strategy for band engineering is doping including self-doping (Liu et al., 2012a; Xie et al., 2011) and doping with other elements so as to increase visible-light absorption for CO_2 photoreduction (Richardson et al., 2012; Tsai et al., 2011). Therein, TiO_2 is most extensively studied in this regard with replacing partial O by anion such as N (Ong et al., 2014) and I (Zhang et al., 2011) to raise the VB edge and narrow the band gap of TiO_2 for efficient adsorption of visible region in the solar spectrum. However, the external dopants would also introduce recombination centers and thus might lower the overall utilization efficiency of charge carriers. Self-dopant, such as oxygen vacancy, has recently been demonstrated to extend the light absorption range of photocatalysts with introducing fewer recombination centers (Li et al., 2015a; Liu et al., 2012a; Xie et al., 2011; Yan et al., 2015). When considering the band engineering for enhancing light harvesting, it is worth noticing that the CB and VB edges should be appropriate for the electrons and holes on them to own enough driving force to carry out the redox reactions.

Other strategies for increasing the visible-light excitation include surface plasmon resonance (SPR) effect (Hou et al., 2011; Yan et al., 2013; Zhang et al., 2012a, 2013), solid solution formation (Liu et al., 2011), and dye or quantum dot (QD) sensitization (Cao et al., 2011; Chen et al., 2012; Nguyen et al., 2008), which were discussed in detail in the provided references. In addition, the development of some novel visible-light-responsive semiconductors for CO_2 photoreduction, such as graphitic-C_3N_4 (Lin et al., 2014; Mao et al., 2013), ZnTe (Jang et al., 2014, 2015), Cu_2O (Schreier et al., 2015, 2016), CdS (Wang and Wang, 2015; Yu et al., 2014a), Co_3O_4 (Huang et al., 2013; Shen et al., 2015), $BiVO_4$ (Liu et al., 2009; Mao et al., 2012), $CuFeO_2$ (Gu et al., 2013; Kang et al., 2015), and $InTaO_4$ (Tsai et al., 2011), has become an important and active area of research.

2.3.2 Facilitate Charge Carrier Separation/Transportation

Semiconductor photocatalyst harvests photons and generates electrons and holes as the step (i) in Scheme 1. Then the photogenerated charge carriers migrate independently or recombine on the way to the surface reaction sites (step (ii) and (iii) in Scheme 1). Since the charge separation is crucial to the effective utilization of them, much attention should be paid to increasing the charge separation efficiency. The charge transportation in the bulk of photocatalyst is largely dependent on the crystallinity, crystal structure and particle size of the photocatalyst (Li et al., 2014b). Therefore, a number of strategies associated with improving the above performances have been employed to promote the separation and transportation of charge carriers.

The more time charge carriers spend in the bulk of the photocatalyst, the higher chance they have to recombine. The time spent by a charge carrier in the bulk depends on its mobility and the distance to the surface. Thus, the designed photocatalyst should have a high carrier mobility and a short bulk-to-surface distance for the suppression of bulk recombination. The mobility can be enhanced through the improvement of the crystallinity and purity of the photocatalyst by calcination (Ohtani et al., 1997). Besides, the construction of some special morphologies, such as nanoparticles (0D) (Liu et al., 2012b; Sang et al., 2014), nanorods/nanotubes (1D) (Cao et al., 2016; Zhang et al., 2012b), nanosheets (2D) (Liu et al., 2010; Xu et al., 2013), and ordered porous materials (3D) (Jiao et al., 2014), allows an easy control of the distance for the charge carriers to transport before reaching the surface. Hence, the optimum balance between the maximal path length for photon absorption and the minimal effective charge carriers transport distance can be determined through the nanostructured photocatalysts.

In addition, the construction of different junctions can promote the transfer of photogenerated electrons from the higher CB to the lower CB and the holes in the opposite direction along VB, hence enhancing the separation of photogenerated charge carriers in the interface region (Chang et al., 2015). For example, facets-based homojunctions (Li et al., 2015b; Yu et al., 2014b), crystalline phase-based homojunctions (Li et al., 2008, 2014a), and heterojunctions (Iwase et al., 2016; Shi et al., 2014; Tu et al., 2012) have been extensively applied for CO_2 photoreduction. The internal electric field in the interface region can greatly increase the lifetime of charge carriers and enhance the e^-–h^+ separation and transportation to the surface.

2.3.3 *Enhance Surface Reaction*

When the photogenerated charge carriers reach the surface, electrons reduce CO_2 into fuels and holes oxidize H_2O into O_2 (step (iv) in Scheme 1). However, compared with the generation and transportation of charge carriers in semiconductor photocatalysts (usually within femtoseconds to picoseconds), surface reactions normally take place on a longer time scale (several hundred picoseconds to microseconds) (Kubacka et al., 2012). Many barriers in thermodynamics and kinetics (described in Section 2.3) make surface reaction the rate-determining step of the whole PC process. Therefore, the acceleration of the surface reactions is critical to improve the solar energy conversion efficiency.

Similarly, the more time charge carriers spend on the surface of the photocatalyst, the higher chance they have to recombine and this time depends on how quickly they are consumed in the redox reactions. The adequate supply of reactants, such as CO_2 and H_2O, can increase the consumption rate of charge carriers. Therefore, in aqueous solutions, the enhancement of CO_2 adsorption and activation on the surface of photocatalysts is an effective way to enhance the surface CO_2 reduction. The main strategies for the improvement of CO_2 adsorption and activation include the increase of catalyst surface area (Liang et al., 2015; Mao et al., 2013), surface defects (Lee et al., 2011; Liu et al., 2012a), and basic sites promoted surfaces (Meng et al., 2014; Xie et al., 2013). In addition to the increased CO_2 adsorption, a larger surface area could also lower the surface density of charge carriers and reduce their recombination probability, although this may be disadvantageous to the multielectron reactions.

Another effective strategy to enhance the surface reaction is to accelerate the CO_2 reduction kinetics. Loading cocatalysts has been shown as an effective approach to accelerating the surface reaction kinetics. It is generally accepted that the cocatalysts can provide reaction sites and lower the electrochemical overpotentials associated with the multielectron reduction of CO_2 as well as decrease the activation energy for product evolution (Ran et al., 2014). In addition, the junction formed at the interface of photocatalyst and cocatalyst could promote the transfer of charge carriers and increase the charge separation efficiency (Chang et al., 2015). Lastly, cocatalysts can also have an effect on the selectivity of products in view of the formation and adsorption strength of many different intermediates during the process of CO_2 reduction (Tu et al., 2014). Thereinto, noble-metal (Li et al., 2016; Wang et al., 2012) and molecular (Liu et al., 2016; Schreier et al., 2015, 2016) cocatalysts are the most widely investigated and reported among the CO_2 reduction cocatalysts. As the other half-cell reaction, water oxidation is an important part of the overall process of CO_2 photoreduction. The oxidation of water requires a complicated process involving the four-electron transfer from two water molecules and the removal of four protons (Barber, 2009). Therefore, the enhancement of water oxidation must depend on the efficient separation of the charge carriers and is beneficial to the reduction of CO_2. However, many studies only focus on the detection of products from CO_2 reduction and always ignore the measurement of oxygen from water oxidation. The simultaneous enhancement of surface CO_2 reduction and water oxidation reaction kinetics is becoming a hot research topic.

3. CO_2 REDUCTION ON VISIBLE-LIGHT PHOTOCATALYSTS

3.1 Oxide Semiconductors

3.1.1 Modified TiO_2 Photocatalyst

Over the past few decades, titanium dioxide has been the most extensively researched and reported semiconductor because of its nontoxicity, high stability, and abundance. It is able to perform a variety of PC reactions including organic pollution degradation, water splitting, and CO_2 reduction (Ma et al., 2014). However, TiO_2 is a wide band-gap semiconductor with $E_g \approx 3.0$ eV and can only absorb the photons in UV region of the solar spectrum, which is the biggest drawback of TiO_2. Therefore, much effort has been devoted to extending the light absorption of TiO_2 to the visible region.

As mentioned in Section 2.3.1, band engineering is the primary strategy to narrow the band gap of TiO_2 and make it responsive under visible light. In the band structure of TiO_2, Ti 3d, 4s, 4p orbitals contribute to the unoccupied CB and the filled VB is dominated by O 2p orbitals (Ma et al., 2014). Therefore, the doping of other cations in replacement of Ti could introduce an impurity level in the forbidden band of TiO_2, which could act as either an electron acceptor or donor and make TiO_2 able to absorb visible light (Wang et al., 2014b). Singh et al. (2014) successfully introduced desired states in TiO_2 nanoparticles (NPs) through Pt dopants and the designed catalysts were used for selective reduction of CO_2 into hydrocarbons, alcohols and aldehydes (Singh et al., 2014). Compared with the cation doping, anion doping of TiO_2 is more widely applied for CO_2 photoreduction. Unlike cation doping with the introducing of impurity level, the anion doping could reconstruct and shift the VB of TiO_2 upward to narrow the band gap (Wang et al., 2014b). Several anion doped TiO_2 photocatalysts were found to be efficient in CO_2 photoreduction under visible-light illumination. Ong et al. (2014) synthesized anatase nitrogen-doped TiO_2 NPs with exposed (001) facets deposited on the graphene (GR) sheets. The as-developed N-doped-TiO_2/GR nanocomposites were demonstrated to be photocatalytically active under visible light for the reduction of CO_2 to CH_4 with an activity of 11-fold higher than that of pure TiO_2 (Ong et al., 2014). Truong et al. (2012) synthesized N and C doped TiO_2 NPs with controlled crystalline structure and visible-light harvesting capacity for CO_2 photoreduction to CH_3OH. Thereinto, the bicrystalline anatase-brookite composites with a strong absorption up to 700 nm performed the maximum CH_3OH yield of 0.478 $\mu mol\, g^{-1}\, h^{-1}$ which is 10.6-fold higher than P25 (Truong et al., 2012). However, the introduced external dopants would also become recombination centers, which might lower the overall efficiency of charge carrier utilization. Recently, the oxygen vacancies, acting as self-dopants in TiO_2 have been demonstrated to extend the light absorption range and applied in CO_2 photoreduction. In addition to the extension of light absorption range, oxygen vacancies could also take part in the surface chemical reactions and play important roles in governing the adsorption and activation of CO_2 (Indrakanti et al., 2009). Liu et al. (2012b) first systematically studied the photoreduction of CO_2 with H_2O on oxygen-deficient TiO_2 anatase, rutile, and brookite nanocrystals, respectively. It was found that helium pretreatment of the as-prepared TiO_2 could result in the formation of surface oxygen vacancies and Ti^{3+} sites on anatase and brookite. The defective brookite exhibited the highest activity for CO and CH_4 production with the enhancement factors of 10.3 and 8.2, respectively, as compared to the defect-free surfaces. The simultaneous enhancement of visible-light harvesting and CO_2 activation due to the introducing of oxygen vacancies and Ti^{3+} sites is the main reason for the improvement of CO_2 photoreduction efficiency (Fig. 3A) (Liu et al., 2012b).

In addition to band engineering, another effective strategy to extend the light adsorption range of TiO_2 is the surface sensitization using QDs or dye as light harvester. The QDs or dye molecules can capture the energy of visible light to form excited states and generate electrons. Then the electrons from these excited states can be injected into the CB of TiO_2 and then transfer to CO_2 adsorbed on the surface (Fernando et al., 2015; Woolerton et al., 2011). Wang et al. (2010) synthesized a series of CdSe QDs-sensitized TiO_2 heterostructures for CO_2 photoreduction in the presence of

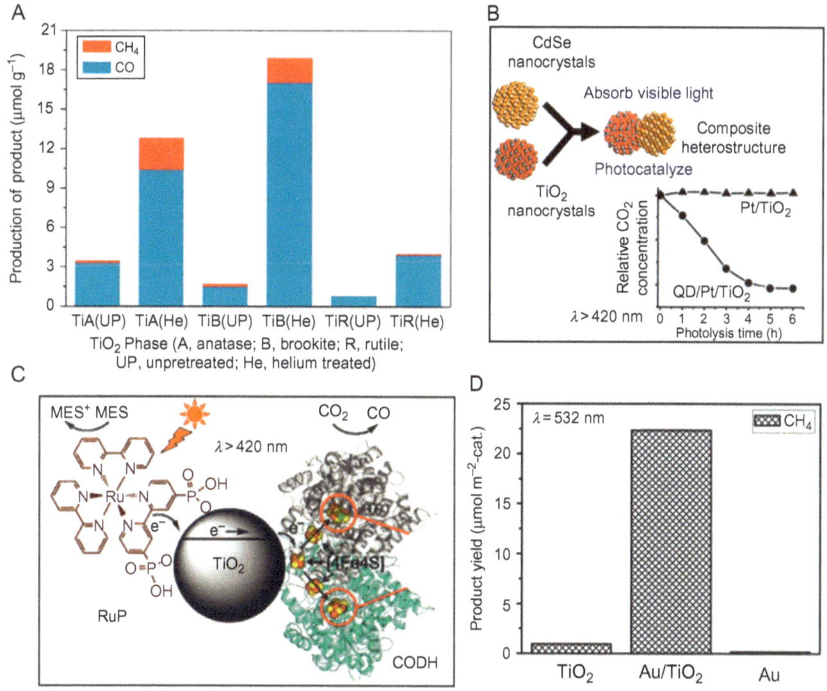

Fig. 3 (A) Production of CO and CH_4 on the three unpretreated and He pretreated TiO_2 polymorphs for 6 h illumination; (B) schematic of $CdSe/TiO_2$ composites and the CO_2 consumption as a function of photolysis time under $\lambda > 420$ nm light irradiation; (C) schematic of the CO_2 photoreduction system using TiO_2 photocatalyst modified by RuP photosensitizer and CO_2 reducing enzyme CODH; and (D) photocatalytic product yields on three different catalytic surfaces after 15 h visible-light irradiation. *(A) Reproduced with permission from Liu, L., et al., 2012b. Photocatalytic CO_2 reduction with H_2O on TiO_2 nanocrystals: comparison of anatase, rutile, and brookite polymorphs and exploration of surface chemistry. ACS Catal. 2, 1817–1828. Copyright 2012 American Chemical Society. (B and C) Reproduced with permission from Wang, C., et al., 2010. Visible light photoreduction of CO_2 using $CdSe/Pt/TiO_2$ heterostructured catalysts. J. Phys. Chem. Lett. 1, 48–53; Woolerton, T.W., et al., 2010. Efficient and clean photoreduction of CO_2 to CO by enzyme-modified TiO_2 nanoparticles using visible light. J. Am. Chem. Soc. 132, 2132–2133. Copyright 2010 American Chemical Society. (D) Reproduced with permission from Hou, W., et al., 2011. Photocatalytic conversion of CO_2 to hydrocarbon fuels via plasmon-enhanced absorption and metallic interband transitions. ACS Catal. 1, 929–936. Copyright 2011 American Chemical Society.*

H_2O. The obtained CdSe QD-sensitized Pt/TiO_2 heterostructures were capable of CO_2 photoreduction under visible-light illumination ($\lambda > 420$ nm) with yields of 48 ppm g^{-1} h^{-1} for CH_4 and 3.3 ppm g^{-1} h^{-1} for CH_3OH (vapor), whereas the Pt/TiO_2 barely exhibited any photoactivity under the same reaction conditions (Fig. 3B) (Wang et al., 2010). Woolerton et al. (2010) reported that the TiO_2 NPs modified with $[Ru^{II}(bipy)_2(4,4'-(PO_3H_2)_2\text{-bipy})]Br_2$ (bipy = 2,2′-bipyridine) photosensitizer and CO_2 reducing enzyme CODH I (carbon monoxide dehydrogenase) exhibited an excellent catalytic activity for CO_2 photoreduction under visible-light illumination ($\lambda > 420$ nm) (Fig. 3C) (Woolerton et al., 2010). Another surface sensitization strategy is to use the localized SPR effect of noble metal NPs (mainly Au and Ag) under visible-light illumination, which would induce the collective oscillation of the conduction electrons (Linic et al., 2011). Therefore, the loaded noble metal NPs would serve as a sensitizer to enhance the visible-light absorption of TiO_2 due to the SPR effects. Hou et al. (2011) reported a 24-fold enhancement in activity for CO_2 photoreduction to CH_4 under visible light due to the intense local electromagnetic fields created by the surface plasmon of Au NPs on TiO_2 (Fig. 3D) (Hou et al., 2011). In addition to the sunlight harvesting, the sensitizers can also inhibit the recombination of e^--h^+ pairs due to the efficient charge separation at the interface of semiconductor/sensitizer (Willkomm et al., 2016). However, the stability of the sensitizers should be improved for the long-term reactions.

3.1.2 Other Oxide Semiconductors

In spite of the great advances made in TiO_2 and the derivative materials for CO_2 photoreduction, the efficiency of this process is still very low due to its adverse band structure. Therefore, although TiO_2 remains a benchmark oxide photocatalyst, there is also a lot of interest in alternative oxide semiconductors.

Cuprous oxide (Cu_2O) with a direct band gap of $\approx$2.0 eV has emerged as a promising material for CO_2 photoreduction both in PC and PEC systems due to its suppression of HER, low toxicity and high abundance (Zhai et al., 2013). However, the main drawback of Cu_2O is its poor stability, which greatly limits its large-scale utilization for CO_2 photoreduction. Therefore, most of the researches on Cu_2O focus on the improvement of its stability. Yu et al. (2016) synthesized carbon layer coated Cu_2O mesoporous nanorods (NRs) on Cu foils for CO_2 photoreduction. This carbon layer could serve as a protective layer to address the common photocorrosion of Cu_2O as well as a conductive layer to facilitate the separation of charge carriers (Fig. 4A). The optimized sample exhibited an apparent quantum efficiency (AQE) of 2.07% for CH_4 and C_2H_4 evolution under visible-light irradiation of 400 nm and an excellent stability after six photoreduction cycles (Yu et al., 2016). Chang

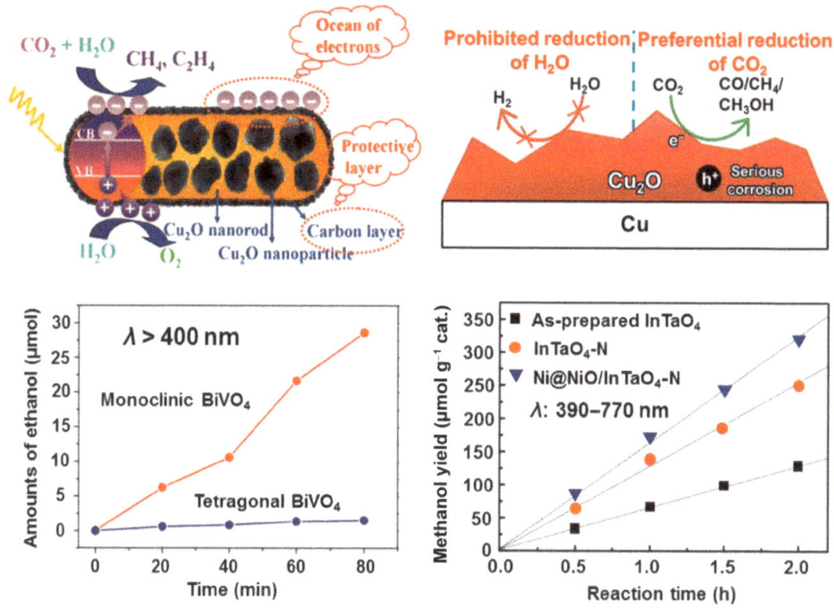

Fig. 4 (A) Schematic of CO_2 reduction and water oxidation over carbon layer coated Cu_2O mesoporous nanorods under visible-light irradiation; (B) schematic of prohibited water reduction and preferential CO_2 reduction on Cu_2O film in aqueous solutions; (C) photocatalytic ethanol evolution over monoclinic and tetragonal $BiVO_4$ under visible-light irradiation ($\lambda \geq 400$ nm); and (D) methanol yield of an NH_3 treatment $InTaO_4$ and Ni@NiO cocatalysts deposited on $InTaO_4$-N under visible-light irradiation (λ: 390–770 nm). *(A and D) Reproduced with permission from Yu, L., et al., 2016. Enhanced activity and stability of carbon-decorated cuprous oxide mesoporous nanorods for CO_2 reduction in artificial photosynthesis. ACS Catal. 6, 6444–6454; Tsai, C.-W., et al., 2011. Ni@NiO core–shell structure-modified nitrogen-doped $InTaO_4$ for solar-driven highly efficient CO_2 reduction to methanol. J. Phys. Chem. C 115, 10180–10186. Copyright 2016 and 2011 American Chemical Society. (B) Reproduced with permission from Chang, X., et al., 2016b. Stable aqueous photoelectrochemical CO_2 reduction by a Cu_2O dark cathode with improved selectivity for carbonaceous products. Angew. Chem. Int. Ed. 55, 8840–8845. Copyright 2016 John Wiley and Sons. (C) Reproduced with permission from Liu, Y., et al., 2009. Selective ethanol formation from photocatalytic reduction of carbon dioxide in water with $BiVO_4$ photocatalyst. Catal. Commun. 11, 210–213. Copyright 2009 Elsevier.*

et al. (2016b) found that the photogenerated holes, instead of electrons, primarily account for the instability of Cu_2O. Therefore, they reported a simple strategy using Cu_2O as a dark cathode to minimize the adverse effects of holes (Fig. 4B). Assembled with TiO_2 NRs as a photoanode, a good stability as well as a high Faradaic efficiency (FE) of 87.45% and selectivity of 92.65% for all the carbonaceous products were achieved during the 3 h CO_2 photoelectroreduction (Chang et al., 2016b).

In addition to the binary oxides, some ternary metal oxides with visible-light responsibility are also utilized for CO_2 photoreduction. Liu et al. (2009) reported the CO_2 photoreduction to ethanol on $BiVO_4$ with two crystal structures of monoclinic and tetragonal zircon. Monoclinic $BiVO_4$ exhibited an ethanol evolution rate of 17 times higher than tetragonal $BiVO_4$ due to the strong adsorption of CO_2 on the former and the larger band gap (2.24 vs 2.56 eV) of the latter (Fig. 4C) (Liu et al., 2009). Tsai et al. (2011) demonstrated the PC activities of $InTaO_4$-based samples, which were synthesized by doping with nitrogen and modified by Ni@NiO core–shell cocatalysts, for CO_2 photoreduction. The Ni@NiO/$InTaO_4$-N exhibited the highest methanol evolution rate of 160 $\mu mol\ g^{-1}\ h^{-1}$ under visible light with wavelengths ranging from 390 to 770 nm (Fig. 4D) (Tsai et al., 2011). Besides, many strategies were conducted to extend the light adsorption range of other nontitania semiconductors. Xie et al. (2011) induced the oxygen defects on $SrTiO_3$ to make it more responsive in visible light. The obtained $SrTiO_{3-x}$ could reduce CO_2 into CH_4 even under the 600 nm light irradiation with a yield of 0.3 $\mu mol\ g^{-1}\ h^{-1}$ (Xie et al., 2011).

3.2 Chalcogenide Semiconductors

In addition to the most extensively researched oxide semiconductors, some chalcogenide semiconductors also emerged as promising photocatalysts for CO_2 reduction. Sulfides initially received a lot of attention because their VB, made of 3p orbitals of the sulfur atom, is shifted upwards compared with those of the oxide analogues (Maeda and Domen, 2007). Thereinto, ZnS and CdS were the most studied sulfides for CO_2 photoreduction. However, ZnS is a direct wide band-gap semiconductor ($E_g = 3.66$ eV in the bulk) and absorbs only in the UV range (Habisreutinger et al., 2013). There are few studies on extending the light adsorption range of ZnS for CO_2 photoreduction under visible-light irradiation. In contrast, CdS with a narrow band gap of 2.4 eV and the absorption onset of 520 nm can perform the visible-light-driven CO_2 reduction. By adding excess Cd^{2+} in *N,N*-dimethylformamide (DMF), Fujiwara et al. (1997) synthesized the CdS-DMF with sulfur vacancies on the surface, which can act as adsorptive sites for CO_2 molecule and increase the PC activity for CO_2 reduction to CO (Fujiwara et al., 1997). Apart from these, other sulfides were used less often for CO_2 photoreduction.

Initially, CdTe with a band gap of 1.5 eV has been the most extensively studied chalcogenide for CO_2 photoreduction due to its p-type conductivity.

Bockris and Wass (1989) conducted the PEC reduction of CO_2 to CO on (100) p-CdTe electrode at a potential of 600–700 mV less negative than an indium cathode with 80%–85% FE and unit quantum efficiency in DMF-5% water under 600 nm light irradiation (Bockris and Wass, 1989). Recently, ZnTe, which was directly formed on a Zn/ZnO nanowire (NW) substrate, has exhibited PC activity toward the reduction of CO_2 to CO under visible light due to its narrow band gap of 2.26 eV. More importantly, ZnTe possesses a high CB edge position of −1.63 V vs reversible hydrogen electrode (RHE) at pH 7.5, which is more negative than the standard reduction potentials for CO_2 reduction (Fig. 2) and can provide large driving force for the conversion. The obtained ZnO/ZnTe core–shell NWs with p-type conductivity exhibited a FE of 22.9% and the maximal photon-to-current conversion efficiency of 85% at −0.7 V vs RHE in 0.5 M $KHCO_3$ (pH 7.5) (Jang et al., 2014).

3.3 Group IV Materials

The most widely investigated photocatalyst in group IV is Si due to its abundance and narrow band gap of 1.1 eV (Fig. 2), which makes it photoresponsive even in the near-IR region (Wang and Gong, 2015). Therefore, many early studies of CO_2 photoreduction have been performed on Si. However, in nearly all of the studies, an external electrical bias was applied on the Si electrode to drive the CO_2 reduction. Taniguchi et al. (1984) reported the seminal investigation of CO_2 photoelectroreduction on p-Si electrode in 1984. CO_2 was reduced to CO in a nonaqueous electrolyte under 600 nm illumination at high potentials of −2.0 V vs saturated calomel electrode (SCE) (Taniguchi et al., 1984). In order to reduce the applied external electrical bias and increase the light energy conversion efficiency, Junfu and Baozhu (1992) synthesized a p^+/p-Si photocathode for the CO_2 photoelectroreduction in an aqueous electrolyte. A potential of −1.2 V vs SCE was found to yield the highest FE for CO_2 reduction to HCOOH on this electrode (Junfu and Baozhu, 1992). The surface modification of p-Si photocathode with either Cu, Au, or Ag is an efficient strategy for converting CO_2 selectively to CO, CH_4, or C_2H_4 in aqueous electrolyte. Kong et al. (2016) fabricated TiO_2-protected n^+p-Si NW arrays in parallel with a Au_3Cu NP catalyst, which exhibited high CO_2-to-CO conversion selectivity close to 80% at −0.2 V vs RHE with suppressed HER and a high stability for 18 h. The overpotential for CO_2 conversion was reduced by 120 mV compared to the planar counterpart due to the optimized spatial arrangement of NP catalysts on the high surface area NW arrays (Kong et al., 2016). In addition to metal NPs, Song et al. (2017) demonstrated that a nanoporous Au thin film, formed by the application of mild electrochemical oxidation and reduction, on n^+p-Si significantly reduced CO_2 reduction overpotential and increased the selectivity of CO generation in aqueous solutions (Fig. 5A). The obtained photocathode exhibited excellent PEC CO_2 reduction with the CO FE of over 90% at a low overpotential of 480 mV (Song et al., 2017).

It is known that the CB edge of SiC is more negative than those of the most other semiconductors leading to a stronger reduction performance of the photoelectrons in SiC. However, a large band gap of ~3.0 eV limits its application under visible-light irradiation. Li et al. (2011) modified SiC NPs with narrow band gap semiconductor Cu_2O and investigated their performance for reducing CO_2 to CH_3OH under visible-light irradiation. The obtained Cu_2O/SiC composite exhibited a CH_3OH generation amount of 191 μmol g^{-1} after CO_2 photoreduction under visible light (200–700 nm) for 5 h (Fig. 5B) (Li et al., 2011).

Photocatalysts based on carbon (such as GR and carbon nanotubes) are another subgroup of group IV materials, which can exhibit a well-developed surface area and unique electronic and optical properties, for CO_2 photoreduction (Marszewski et al., 2015). For example, graphene oxide (GO) can be considered as a wide-band-gap semiconductor and its band gap greatly depends on the carbon-to-oxygen ratio. Therefore, Hsu et al. (2013) investigated this property and synthesized a set of GO materials by varying oxygen content. The obtained samples with band gaps varying from 2.9 to 4.4 eV all exhibited better CH_3OH yield than P25 under visible-light irradiation. The maximal CH_3OH conversion rate can be achieved up to 0.172 μmol g^{-1} h^{-1} on the sample of GO-3, which is sixfold higher than that of pure TiO_2 (Fig. 5C) (Hsu et al., 2013). GR and its derivatives can also act as an electron sink to replace the noble metals hence promoting the charge carrier separation and transportation. An et al. (2014) synthesized the Cu_2O/reduced graphene oxide (RGO) composites through a facile one-step microwave-assisted chemical method. The obtained Cu_2O/RGO samples exhibited an apparent initial quantum yield of approximately 0.344% at 400 nm for CO_2 photoreduction to CO, which was sixfold higher than the bare Cu_2O in the 20 h reaction. Besides, the stability of Cu_2O/RGO was remarkably improved due to the protective function of RGO as evidenced by the linear relationship between the reaction activity and reaction time (Fig. 5D) (An et al., 2014). In addition, due to the well-developed surface

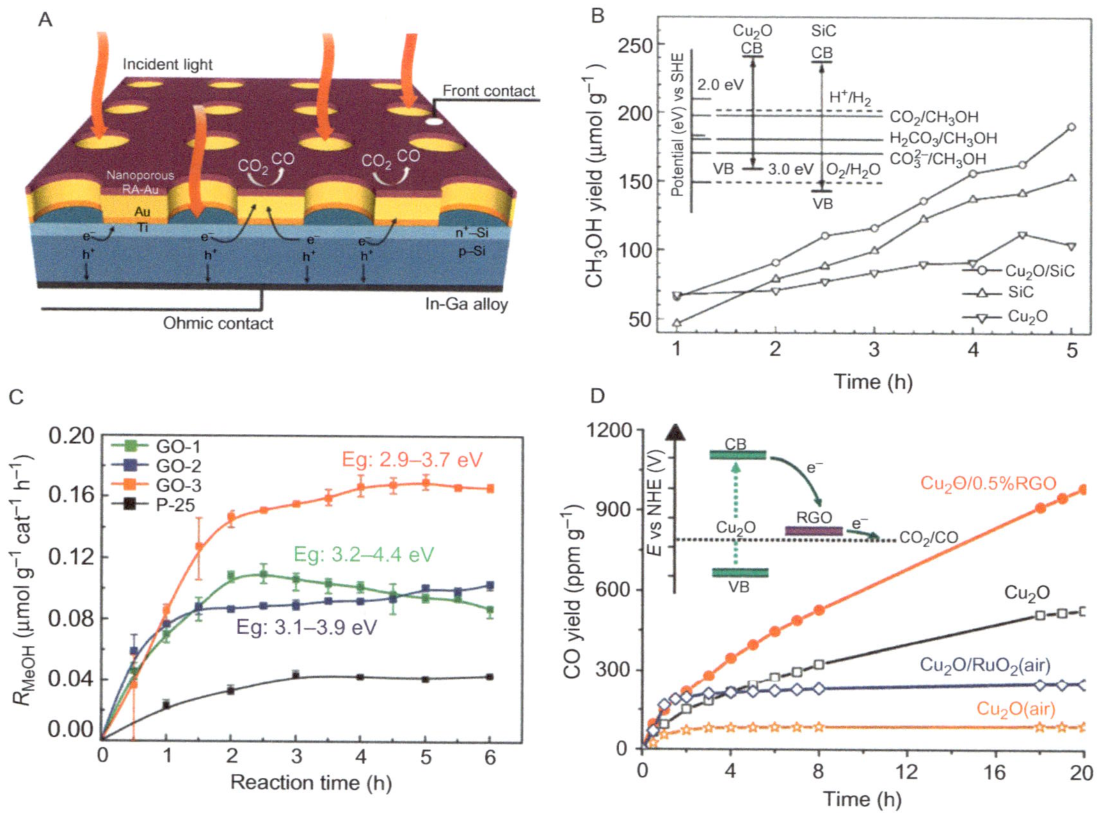

Fig. 5 See figure legend on next page.

area, RGO can enhance the adsorption and activation of CO_2 molecules. Yu et al. (2014a) prepared RGO-CdS NR composites and performed a high activity for CO_2 photoreduction to CH_4. The optimized sample with the RGO content of 0.5 wt% exhibited a high CH_4 production rate of 2.51 μmol g^{-1} h^{-1}, which was over 10 times higher than pure CdS NRs. The adsorption and destabilization of CO_2 molecules by RGO were critical to the enhancement of the CO_2 photoreduction activity (Yu et al., 2014a).

3.4 Nitride Semiconductors

Graphitic carbon nitride (g-C_3N_4) with a narrow band gap of 2.7 eV is yet another promising photocatalyst for CO_2 reduction and water oxidation without the loading of noble metals under visible-light irradiation. It has a GR-like structure consisting of two-dimensional frameworks. The nitrogen-containing groups on the surface make it an excellent photocatalyst to promote CO_2 adsorption and activation (Lin et al., 2014; Yin et al., 2015). Mao et al. (2013) synthesized two kinds of g-C_3N_4 through a pyrolysis process of urea or melamine. The sample derived from urea (u-g-C_3N_4) possessed a mesoporous flake-like structure and a larger surface area than the non-porous flaky sample obtained from melamine (m-g-C_3N_4). The as-prepared u-g-C_3N_4 could produce C_2H_5OH and CH_3OH with amounts of 10.8 and 15.1 μmol in CO_2 photoreduction under 12 h visible-light irradiation, whereas only C_2H_5OH could be obtained over m-g-C_3N_4 with a lower amount of 8.7 μmol (Fig. 6A). The improvement

Fig. 5 (A) Schematic of mesh-type Au cocatalysts form on an Si photoelectrode with a buried p–n junction for CO_2 photoelectroreduction; (B) yields of CH_3OH in the CO_2 photoreduction with H_2O on the three photocatalysts as a function of irradiation time of visible light; (C) photocatalytic CH_3OH formation on different graphene oxide samples with varying E_g and TiO_2 under simulated solar-light source; and (D) time-dependent photocatalytic conversion of CO_2 into CO over Cu_2O and Cu_2O/RGO composites. *(A and D) Reproduced with permission from Song, J.T., et al., 2017. Nanoporous Au thin films on Si photoelectrodes for selective and efficient photoelectrochemical CO_2 reduction. Adv. Energy Mater. 7, 1601103; An, X., et al., 2014. Cu_2O/reduced graphene oxide composites for the photocatalytic conversion of CO_2. ChemSusChem 7, 1086–1093. Copyright 2017 and 2014 John Wiley and Sons. (B) Reproduced with permission from Li, H., et al., 2011. Photocatalytic reduction of carbon dioxide to methanol by Cu_2O/SiC nanocrystallite under visible light irradiation. J. Nat. Gas Chem. 20, 145–150. Copyright 2011 Elsevier. (C) Reproduced with permission from Hsu, H.C., et al., 2013. Graphene oxide as a promising photocatalyst for CO_2 to methanol conversion. Nanoscale 5, 262–268. Copyright 2013 Royal Society of Chemistry.*

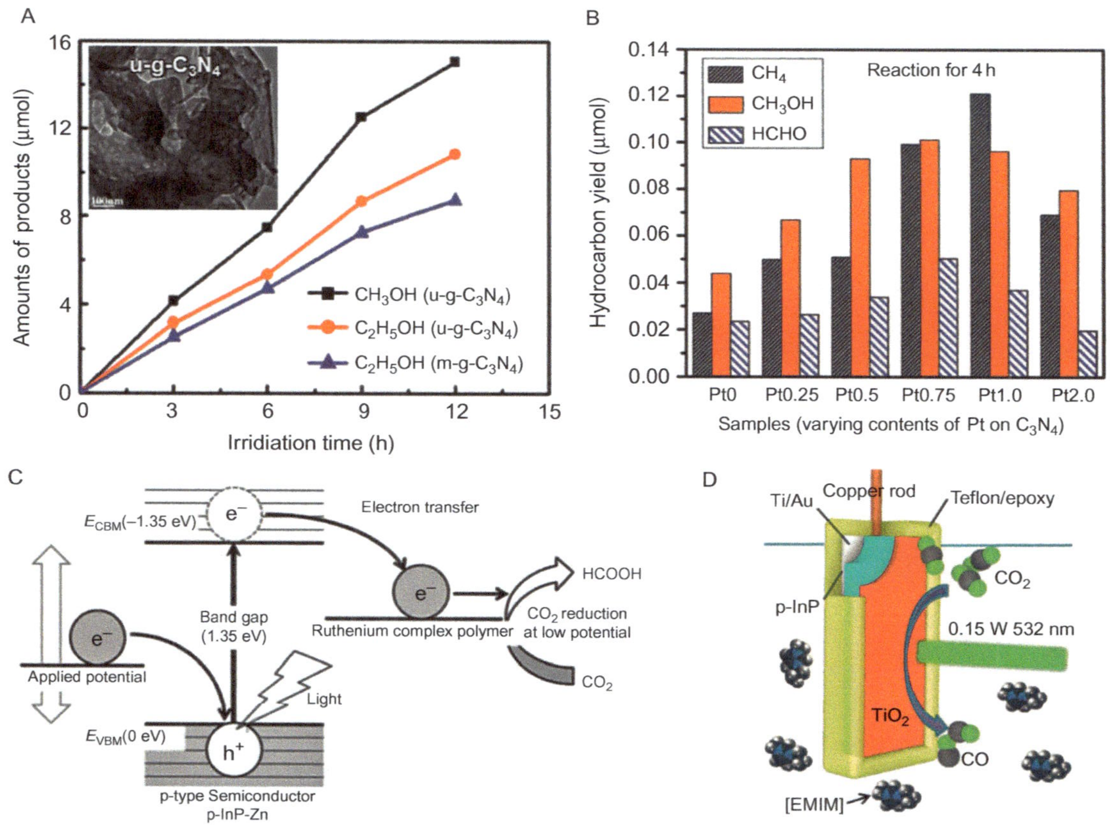

Fig. 6 See figure legend on next page.

of photoactivity was due to the higher surface area of u-g-C_3N_4, which could enhance surface adsorption of reactants and faster photogenerated carrier separation (Mao et al., 2013). Although the loading of noble metals on g-C_3N_4 is not required to facilitate the surface reaction, it can promote the charge separation hence improving the CO_2 photoreduction activity. Yu et al. (2014c) synthesized a series of g-C_3N_4 loaded with various amounts of Pt (0–2 wt%) and investigated their performance for CO_2 photoconversion. The optimal content of Pt was 0.75–1.0 wt% and the corresponding sample exhibited a 4.3-fold increase in CH_4 generation and a 2.2-fold increase in CH_3OH evolution compared to bare g-C_3N_4 (Fig. 6B). The loaded Pt could facilitate electron transfer and enrich more photogenerated electrons on the surface of g-C_3N_4 for CO_2 photoreduction (Yu et al., 2014c).

3.5 III–V Semiconductors

So far, many researches have been focused on the investigation of CO_2 photoreduction over III–V semiconductors, among which GaP, GaAs, and InP (with band gaps of 2.24, 1.4, and 1.34 eV, respectively) are the most commonly studied (Kumar et al., 2012). Halmann (1978) reported the seminal investigation of PEC CO_2 reduction on III–V semiconductor surfaces with a p-GaP photocathode in 1978. GaP is a p-type semiconductor with a

Fig. 6 (A) Generation amounts of CH_3OH and C_2H_5OH over u-g-C_3N_4 or m-g-C_3N_4 as a function of irradiation time of visible light ($\lambda \geq 420$ nm). *Inset*: transmission electron microscopy (TEM) image of the obtained u-g-C_3N_4; (B) comparison of the photocatalytic CH_4, CH_3OH, and HCHO production rate over g-C_3N_4 with the different Pt-loading amounts under simulated solar irradiation for 4 h; and (C) schematic energy diagram of an RCP/p-InP-Zn electrode under visible-light irradiation. (D) Schematic of TiO_2-passivated p-InP photocathode for CO_2 conversion to CO under visible light in a non-aqueous solution consisting of ionic liquid. *(A–C) Reproduced with permission from Mao, J., et al., 2013. Effect of graphitic carbon nitride microstructures on the activity and selectivity of photocatalytic CO_2 reduction under visible light. Catal. Sci. Technol. 3, 1253; Yu, J., et al., 2014c. Photocatalytic reduction of CO_2 into hydrocarbon solar fuels over g-C_3N_4-Pt nanocomposite photocatalysts. Phys. Chem. Chem. Phys. 16, 11492–11501; Arai, T., et al., 2010. Photoelectrochemical reduction of CO_2 in water under visible-light irradiation by a p-type InP photocathode modified with an electropolymerized ruthenium complex. Chem. Commun. 46, 6944–6946. Copyright 2013, 2014 and 2010 Royal Society of Chemistry. (D) Reproduced with permission from Zeng, G., et al., 2015. Enhanced photocatalytic reduction of CO_2 to CO through TiO_2 passivation of InP in ionic liquids. Chem. Eur. J. 21, 13502–13507. Copyright 2015 John Wiley and Sons.*

narrow band gap and a favorable, highly reducing position of the CB. The reported products were HCOOH, HCHO, and CH_3OH for PEC CO_2 reduction in aqueous media. Meanwhile, he found that the optical conversion efficiency was a function of the applied potential. Under 365 nm light irradiation, the maximal optical conversion efficiency of 5.6% was obtained at −0.8 V vs SCE (Halmann, 1978). Barton et al. (2008) have also employed p-GaP as a photocathode for PEC CO_2 reduction to CH_3OH, in which an external quantum efficiency (EQE) of 2.6% under 465 nm light irradiation was obtained at −0.5 V vs SCE. Pyridine was added to the aqueous electrolyte to act as a highly selective electrocatalyst for CO_2 reduction to CH_3OH through intermediate carbamate species (Barton et al., 2008). Later on, they further reported an experimental mechanistic investigation of pyridinium's role as electrocatalyst for CO_2 conversion in 2010 (Cole et al., 2010).

Also by Halmann (1978), the seminal investigation of p-GaAs for PEC CO_2 reduction was conducted in 1978 (Halmann, 1978). After that, extensive researches on p-GaAs ensued in the early 1980s (White et al., 2015). It was found that there existed a similar dependence on CO_2 pressure between p-GaAs and p-GaP during CO_2 photoreduction. The GaAs electrode exhibited higher FE for CO_2 reduction to HCOOH than GaP at −1.0 V vs SCE. However, the stability of GaAs under illumination was much lower than that of GaP, which was ascribed to the reductive decomposition of GaAs into Ga and As (Aurian-Blajeni et al., 1983).

In contrast to the indirect band gap of GaP, InP is a direct band gap III–V semiconductor with a E_g of 1.34 eV and a higher absorption coefficient, which induces extensive investigations on InP. Arai et al. (2010) built a PEC cell with a Zn-doped InP photocathode modified with a ruthenium complex polymer (RCP) $[Ru(L\text{–}L)(CO)_2]_n$, in which L–L is a diimine ligand. This system exhibited a good selectivity for CO_2 photoreduction to HCOOH with a current efficiency of 62.3% in aqueous media under visible-light irradiation (420 nm $< \lambda <$ 800 nm) and −0.6 V vs Ag/AgCl. The loaded RCP could act as an electrocatalyst and enable CO_2 reduction under a lower applied potential than required by conventional electrocatalytic systems (Fig. 6C) (Arai et al., 2010). Nowadays, Zeng et al. (2015) prepared a TiO_2-passivated InP photocathode with a Pt cocatalyst and investigated its performance in PEC CO_2 reduction in a non-aqueous solution consisting of ionic liquid. The reduction product was CO with a FE of 99% under a 532 nm laser and a potential of 0.78 V. The TiO_2 passivation layer deposited by atomic layer deposition (ALD) could form a p–n junction with charge separation effects at the interface and provide surface active sites for CO_2 reduction with low external potential (Fig. 6D) (Zeng et al., 2015).

3.6 Perovskites

Perovskites (AMO_3, e.g., $SrTiO_3$ and $NaNbO_3$) or perovskite-related structures can also be applied for CO_2 photoreduction. A is a cation such as alkalis, alkaline earth or rare earth metal ion and M stands for the transition metals (such as Ga, Ge, In, Sn, or Sb) (Habisreutinger et al., 2013). Perovskite-related structures can be obtained through partially replacing A and M by other metal ions with similar diameters thus maintaining its original crystalline structure. Many of these photocatalysts were initially applied in solar water splitting and they have very recently started to be employed for CO_2 photoreduction.

Niobates with a perovskite structure possess many characteristics such as nontoxicity and good stability. Li et al. (2012) investigated $NaNbO_3$ with two different crystal structures for CO_2 photoreduction. Compared to the orthorhombic $NaNbO_3$ (3.45 eV), cubic $NaNbO_3$ possessed a narrower band gap of 3.29 eV, which was caused by the variant octahedral ligand field. Meanwhile, the high symmetry in cubic $NaNbO_3$ led to its unique electronic structure, which was beneficial for the electron excitation and transportation. Therefore, the CH_4 generation rate in gas phase CO_2 reduction over cubic $NaNbO_3$ (0.486 $\mu mol\ h^{-1}$) was about twice that over orthorhombic $NaNbO_3$ (0.245 $\mu mol\ h^{-1}$) (Fig. 7A) (Li et al., 2012). Later on, they synthesized a $NaNbO_3$ photocatalyst with cubic-orthorhombic surface junctions to improve the charge separation. The obtained mixed-phase $NaNbO_3$ exhibited a 30% and 200% increase in the activity of CO_2 photoreduction into CH_4 than cubic and orthorhombic $NaNbO_3$, respectively (Li et al., 2014a).

Recent work by Iizuka et al. (2011) showed the CO_2 reduction accompanied with water oxidation stoichiometrically. They synthesized Ag cocatalyst-loaded $ALa_4Ti_4O_{15}$ (A = Ca, Sr, and Ba) photocatalysts with layered perovskite structures and investigated their performances for

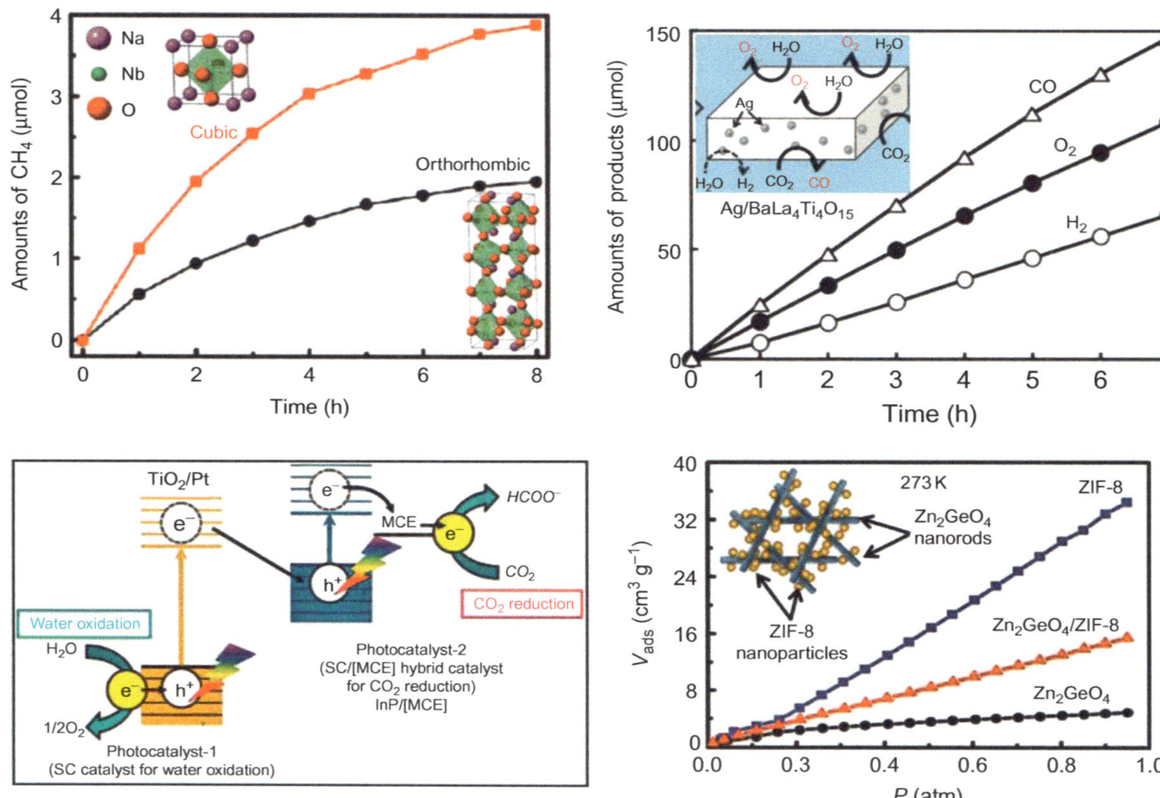

Fig. 7 See figure legend on next page.

CO_2 photoreduction. Ag cocatalyst-loaded $BaLa_4Ti_4O_{15}$ with a liquid-phase chemical reduction loading method was the most active photocatalyst. The optimized Ag/$BaLa_4Ti_4O_{15}$ photocatalyst exhibited the highest evolution rates of 16, 10, and 22 μmol h^{-1} for O_2, H_2, and CO, respectively. The evolution of O_2 in a stoichiometric ratio (H_2+CO:O_2 = 2:1 in a molar ratio) indicated that water was consumed as the only electron donor for the CO_2 reduction (Fig. 7B) (Iizuka et al., 2011).

3.7 Hybrid Structured Photocatalysts

Semiconductor photocatalysts can be coupled with other materials which may improve the light harvesting, facilitate the charge carrier separation or enhance surface reaction of the semiconductor photocatalysts. The obtained hybrid photocatalyst through such combinations could possess superior physicochemical or optical properties, such as a higher dispersion of active centers or enhanced accessibility of catalytic sites (Chen et al., 2015).

Kuriki et al. (2016) synthesized a metal-free organic semiconductor of mesoporous g-C_3N_4 coupled with a Ru(II) binuclear complex containing photosensitizer and catalytic active centers. The obtained hybrid photocatalysts could selectively reduce CO_2 into HCOOH under visible light ($\lambda > 400$ nm) in the presence of a suitable electron donor in aqueous

Fig. 7 (A) CH_4 evolution amounts in gas-phase reaction over cubic $NaNbO_3$ and orthorhombic $NaNbO_3$ samples (with loading 0.5 wt% Pt); (B) CO_2 reduction over $BaLa_4Ti_4O_{15}$ photocatalyst with Ag (2 wt%) cocatalyst loaded by a liquid-phase reduction method in aqueous solution; (C) *Z*-scheme system consisting of TiO_2/Pt and InP/[MCE] photocatalysts for CO_2 reduction and water oxidation; and (D) CO_2 adsorption isotherms (273 K) of the as-prepared samples. *Inset*: schematic of Zn_2GeO_4/ZIF-8 hybrid nanorods. *(A–C) Reproduced with permission from Li, P., et al., 2012. The effects of crystal structure and electronic structure on photocatalytic H_2 evolution and CO_2 reduction over two phases of perovskite-structured $NaNbO_3$. J. Phys. Chem. C 116, 7621–7628; Iizuka, K., et al., 2011. Photocatalytic reduction of carbon dioxide over Ag cocatalyst-loaded $ALa_4Ti_4O_{15}$ (A= Ca, Sr, and Ba) using water as a reducing reagent. J. Am. Chem. Soc. 133, 20863–20868; Sato, S., et al., 2011. Selective CO_2 conversion to formate conjugated with H_2O oxidation utilizing semiconductor/complex hybrid photocatalysts. J. Am. Chem. Soc. 133, 15240–15243. Copyright 2012 and 2011 American Chemical Society. (D) Reproduced with permission from Liu, Q., et al., 2013. ZIF-8/ Zn_2GeO_4 nanorods with an enhanced CO_2 adsorption property in an aqueous medium for photocatalytic synthesis of liquid fuel. J. Mater. Chem. A 1, 11563. Copyright 2013 Royal Society of Chemistry.*

solution. The loading of Ru(II) binuclear complex extended the light absorption edge of g-C_3N_4 from 450 to 460 nm, due to metal-to-ligand charge transfer excitation of the light-harvesting Ru unit (Kuriki et al., 2016). In addition to the light-harvesting enhancement, the modification of semiconductors could also facilitate the charge carrier separation and transportation. Sato et al. (2011) demonstrated that PEC CO_2 reduction to formate over p-type InP/Ru complex polymer hybrid photocatalyst could be highly enhanced by introducing an anchoring complex into the polymer, which greatly enhanced the electron transfer from the CB of InP to the metal-complex electrocatalyst (MCE) (Fig. 7C). When functionally combining the obtained hybrid photocatalyst with TiO_2 for water oxidation, selective CO_2 photoreduction to formate was achieved in aqueous media, where water was used as both an electron donor and a proton source. This so-called *Z*-scheme system could operate without external electrical bias. The selectivity for formate production was >70% and the solar-to-chemical energy conversion efficiency was 0.03%–0.04%, which approached that for photosynthesis in a plant (Sato et al., 2011). Hybrid photocatalysts obtained through the combination of semiconductors and metal–organic frameworks (MOFs) could possess an excellent CO_2 adsorption feature hence enhancing the surface CO_2 reduction reactions. Liu et al. (2013) first directly demonstrated that the adsorption of CO_2 dissolved in water could be enhanced by incorporation of MOFs such as zeolitic imidazolate framework 8 (ZIF-8, composed of $Zn(mIm)_2$ units; mIm = methyl imidazole), hence promoting the PC activity of a semiconductor for CO_2 photoreduction in an aqueous solution. When incorporated with semiconductor Zn_2GeO_4, the as-prepared Zn_2GeO_4/ZIF-8 hybrid NRs inherited both the high CO_2 adsorption capacity of ZIF-8 and the high crystallinity of Zn_2GeO_4. With the optimized ZIF-8 loading amount of 25 wt%, the obtained Zn_2GeO_4/ZIF-8 hybrid NRs exhibited a 3.8 times higher adsorption of dissolved CO_2 and a 62% enhancement in PC CO_2 conversion into CH_3OH than the bare Zn_2GeO_4 NR photocatalysts (Fig. 7D) (Liu et al., 2013).

It is worth noticing that the direct particle-to-particle contact is crucial to efficient operation of such hybrid systems because of the requirement of charge transfer. Some visible-light photocatalysts for CO_2 photoreduction are summarized in Table 2.

Table 2 Some Visible-Light Photocatalysts for CO_2 Photoreduction

Photocatalyst	Light Source	Reaction Conditions	Reaction Activity	Refs.
N-doped-TiO_2/GR nanocomposites	Visible light	Gas-phase reaction at 25 ± 5°C and 1 bar	The highest CH_4 generation rate was 3.7 μmol g^{-1} in 10 h	Ong et al. (2014)
Anatase-brookite TiO_2 NPs	Visible light $\lambda > 400$ nm	50 mg catalyst in 30 mL 0.08 M $NaHCO_3$	The highest CH_3OH yield was 0.478 μmol g^{-1} h^{-1} with QE of 0.0717%	Truong et al. (2012)
Oxygen-deficient TiO_2	$200 < \lambda < 1000$ nm 90 mW cm^{-2}	100 mg catalyst; Gas-phase reaction with continuous-flow rate of 2.0 mL min^{-1}	The highest generation rate of $CO + CH_4$ was 18.9 μmol g^{-1} in 6 h	Li et al. (2012)
CdSe/Pt/TiO_2	Visible light $\lambda > 420$ nm	300 mg catalyst; Gas-phase reaction	The highest yields of 48 ppm g^{-1} h^{-1} for CH_4 and 3.3 ppm g^{-1} h^{-1} for CH_3OH (vapor)	Wang et al. (2010)
Au NPs/TiO_2	532 nm laser 350 mW cm^{-2}	1 m^2 of catalyst surface area; Gas-phase reaction	The highest yield of CH_4 is 22.4 μmol m^{-2}-cat	Hou et al. (2011)
Carbon layer coated Cu_2O mesoporous NRs	Visible light $\lambda > 420$ nm	170 mL 0.1 M $KHCO_3$	The total amount of CH_4 and C_2H_4 reached 0.385 μmol in 12 h	Yu et al. (2016)
Monoclinic $BiVO_4$	Visible light $\lambda \geq 400$ nm	200 mg catalyst in 100 mL H_2O at 0°C with stirring	The highest ethanol evolution amount was ~30 μmol in 80 min	Liu et al. (2009)
Ni@NiO/$InTaO_4$-N	Visible light $390 < \lambda < 770$ nm	100 mg catalyst in 50 mL H_2O at 25°C; The flow rate of CO_2 was 3 kg cm^{-2}	The highest methanol evolution rate was 160 μmol g^{-1} h^{-1}	Tsai et al. (2011)

ZnO/ZnTe core–shell NWs photocathode	Visible light $\lambda > 420$ nm 490 mW cm^{-2}	0.5 M $KHCO_3$ (pH 7.5)	FE of CO was 22.9% at −0.7 V vs RHE	Jang et al. (2014)
p^+/p-Si photocathode	Bromine tungsten lamp; 73 mW cm^{-2}	0.5 M Na_2SO_4	Yield of HCOOH was 32 μmol cm^{-1} h^{-1} at −1.2 V vs SCE	Junfu and Baozhu (1992)
TiO_2-protected n^+p-Si NWs photocathode	Visible light of 740 nm 20 mW $cm^{'2}$	0.1 M $KHCO_3$	CO_2-to-CO conversion selectivity was close to 80% at −0.2 V vs RHE	Kong et al. (2016)
Nanoporous Au thin film on n^+p-Si	100 mW cm^{-2}	20 mL 0.2 M $KHCO_3$ (pH 8.5)	FE of CO was over 90% at a low overpotential of 480 mV	Song et al. (2017)
Cu_2O/SiC composites	Visible light $200 < \lambda < 700$ nm	200 mg catalyst in 0.1 M NaOH +0.1 M Na_2SO_3	CH_3OH generation amount was 191 μmol g^{-1} in 5 h	Li et al. (2011)
GO with different oxygen content.	Visible light	200 mg catalyst Gas-phase reaction at 25 ± 5°C	The highest yield of CH_3OH was 0.172 μmol g^{-1} h^{-1}	Hsu et al. (2013)
Cu_2O/RGO	150 W Xe lamp full spectrum	$Na_2SO_4 + Na_2SO_3$	AQE of CO was 0.344% at 400 nm	An et al. (2014)
RGO-CdS NR composites	Visible light $\lambda \geq 420$ nm	100 mg catalyst; Gas-phase reaction	The highest yield of CH_4 was 2.51 μmol g^{-1} h^{-1}	Yu et al. (2014a)
Pt/g-C_3N_4	300 W simulated solar Xe arc lamp	100 mg catalyst; Gas-phase reaction	The highest generation amount of CH_4 was 0.13 μmol in 4 h	Yu et al. (2014c)

Continued

Table 2 Some Visible-Light Photocatalysts for CO_2 Photoreduction—cont'd

Photocatalyst	Light Source	Reaction Conditions	Reaction Activity	Refs.
p-GaP photocathode	200 W Hg–Xe lamp	10 mM Pyridine	EQE of CH_3OH was 2.6% under 465 nm at −0.5 V vs SCE	Barton et al. (2008)
RCP/p-InP-Zn	Visible light $400 < \lambda < 800$ nm	5 mL H_2O	Current efficiency of HCOOH was 62.3% at −0.6 V vs Ag/AgCl	Arai et al. (2010)
TiO_2/Pt/InP photocathode	532 nm laser	0.02 M [EMIM]BF_4	FE of CO was 99% at a potential of 0.78 V	Zeng et al. (2015)
Pt/$NaNbO_3$	300 W Xe arc lamp	100 mg catalyst in 3 mL H_2O	The yield of CO was 0.486 μmol h^{-1} on cubic $NaNbO_3$ with loading 0.5 wt% Pt	Li et al. (2012)
Ag/$BaLa_4Ti_4O_{15}$	400 W Hg lamp	300 mg catalyst in 360 mL H_2O, CO_2 flow rate of 15 mL min^{-1}	The highest evolution rates of 16, 10, and 22 μmol h^{-1} for O_2, H_2, and CO	Iizuka et al. (2011)
Ru(II) binuclear complex/g-C_3N_4	Visible light $\lambda \geq 400$ nm	4 mg catalyst in 4 mL 10 mM EDTA•2Na	Turnover number of HCOOH was 200 in 24 h	Kuriki et al. (2016)
Zn_2GeO_4/ZIF-8 hybrid NRs	500 W Xe arc lamp	200 mg catalysts in 100 mL 0.1 M Na_2SO_3	Zn_2GeO_4/ZIF-8 nanorods with 25 wt% ZIF-8 exhibited a 62% increase in CH_3OH generation compared with bare Zn_2GeO_4	Liu et al. (2013)

4. SUMMARY AND PERSPECTIVES

Up to now, significant achievements have been made in CO_2 photoreduction with H_2O on visible-light photocatalysts. The insights into the CO_2 photoreduction mechanism have been accumulated for the design of highly efficient and selective photocatalysts. However, the photoconversion efficiency and selectivity for the desired products of CO_2 photoreduction are still very low due to many limiting factors. The effective utilization of charge carriers is much critical to the efficiency of CO_2 photoconversion. Therefore, some strategies to increase light harvesting, facilitate charge carrier separation and enhance surface reaction for CO_2 photoreduction are presented and discussed in this chapter. Recent advances in visible-light photocatalysts applied for CO_2 photoreduction, such as oxide and chalcogenide semiconductors, group IV materials, nitride and III–V semiconductors, perovskites and hybrid structures, are mainly described. The overall efficiency and selectivity for CO_2 photoreduction could be improved through lowering the reaction barriers, tuning the reaction parameters and guiding the reaction pathways based on the knowledge of CO_2 reduction mechanism and pathways.

Therefore, more efforts and deep investigations should be made in this research area in the future. The exploration of the rate-determining steps and reaction barriers in CO_2 photoreduction is important for the further improvement of the conversion efficiency. The in-depth understanding at the molecular level for CO_2 photoreduction through in situ characterization techniques, such as in situ diffuse reflectance infrared Fourier transform spectroscopy or electron paramagnetic resonance spectra, is desirable. Meanwhile, density functional theory calculations for CO_2 photoreduction with the consideration of aqueous environment or negative electrode potential should be conducted. In addition, the other half-reaction of water oxidation, which is considered as the rate-limiting step in water splitting, should be also paid more attention to and improved for the increase of overall efficiency. Due to the organic adsorbates on the surface of as-prepared photocatalysts, which may be formed during fabrication process, it is important to determine whether the products are derived from CO_2 or from carbon impurity intermediates. The utilization of $^{13}CO_2$ isotope labeling as a reactant is essentially important. Because of the relatively negative conduction-band-edge position, which means a large driving force for the reduction reaction, many p-type visible-light semiconductors including

ZnTe, Cu_2O are used in CO_2 photoreduction. However, most of them have poor stability in aqueous electrolytes. Therefore, their stability should be substantially improved for the future utilization. ALD is capable of depositing thin films in a highly controllable manner, which makes it an enabling technique to deposit surface protective layers on the unstable photocatalysts.

ACKNOWLEDGMENTS

We acknowledge the National Key Research and Development Program of China (2016YFB0600901), the National Science Foundation of China (U1463205, 21525626, 51302185), Specialized Research Fund for the Doctoral Program of Higher Education (20120032110024, 20130032120018), the Scientific Research Foundation for the Returned Overseas Chinese Scholars (MoE), and the Program of Introducing Talents of Discipline to Universities (B06006) for financial support.

REFERENCES

An, X., Li, K., et al., 2014. Cu_2O/reduced graphene oxide composites for the photocatalytic conversion of CO_2. ChemSusChem 7, 1086–1093.

Arai, T., Sato, S., et al., 2010. Photoelectrochemical reduction of CO_2 in water under visible-light irradiation by a p-type InP photocathode modified with an electropolymerized ruthenium complex. Chem. Commun. 46, 6944–6946.

Aurian-Blajeni, B., Halmann, M., et al., 1983. Electrochemical measurement on the photoelectrochemical reduction of aqueous carbon dioxide on p-Gallium phosphide and p-Gallium arsenide semiconductor electrodes. Sol. Energy Mater. 8, 425–440.

Barber, J., 2009. Photosynthetic energy conversion: natural and artificial. Chem. Soc. Rev. 38, 185–196.

Barton, E.E., Rampulla, D.M., et al., 2008. Selective solar-driven reduction of CO_2 to methanol using a catalyzed p-GaP based photoelectrochemical cell. J. Am. Chem. Soc. 130, 6342–6344.

Bockris, J., Wass, J., 1989. The photoelectrocatalytic reduction of carbon dioxide. J. Electrochem. Soc. 136, 2521.

Cao, L., Sahu, S., et al., 2011. Carbon nanoparticles as visible-light photocatalysts for efficient CO_2 conversion and beyond. J. Am. Chem. Soc. 133, 4754–4757.

Cao, M., Tang, Z., et al., 2016. The synergy between metal facet and oxide support facet for enhanced catalytic performance: the case of Pd-TiO_2. Nano Lett. 16, 5298–5302.

Chang, X., Wang, T., et al., 2015. Enhanced surface reaction kinetics and charge separation of p-n heterojunction Co_3O_4/$BiVO_4$ photoanodes. J. Am. Chem. Soc. 137, 8356–8359.

Chang, X., Wang, T., et al., 2016a. CO_2 photo-reduction: insights into CO_2 activation and reaction on surfaces of photocatalysts. Energ. Environ. Sci. 9, 2177–2196.

Chang, X., Wang, T., et al., 2016b. Stable aqueous photoelectrochemical CO_2 reduction by a Cu_2O dark cathode with improved selectivity for carbonaceous products. Angew. Chem. Int. Ed. 55, 8840–8845.

Chen, X., Zhou, Y., et al., 2012. Ultrathin, single-crystal WO_3 nanosheets by two-dimensional oriented attachment toward enhanced photocatalystic reduction of CO_2 into hydrocarbon fuels under visible light. ACS Appl. Mater. Interfaces 4, 3372–3377.

Chen, D., Zhang, X., et al., 2015. Synthetic strategies to nanostructured photocatalysts for CO_2 reduction to solar fuels and chemicals. J. Mater. Chem. A 3, 14487–14516.

Cole, E.B., Lakkaraju, P.S., et al., 2010. Using a one-electron shuttle for the multielectron reduction of CO_2 to methanol: kinetic, mechanistic, and structural insights. J. Am. Chem. Soc. 132, 11539–11551.

Dresselhaus, M.S., Thomas, I.L., 2001. Alternative energy technologies. Nature 414, 332–337.

Fernando, K.A., Sahu, S., et al., 2015. Carbon quantum dots and applications in photocatalytic energy conversion. ACS Appl. Mater. Interfaces 7, 8363–8376.

Freund, H.J., Roberts, M.W., 1996. Surface chemistry of carbon dioxide. Surf. Sci. Rep. 25, 225–273.

Fujiwara, H., Hosokawa, H., et al., 1997. Effect of surface structures on photocatalytic CO_2 reduction using quantized CdS nanocrystallites[1]. J. Phys. Chem. B 101, 8270–8278.

Gu, J., Wuttig, A., et al., 2013. Mg-doped $CuFeO_2$ photocathodes for photoelectrochemical reduction of carbon dioxide. J. Phys. Chem. C 117, 12415–12422.

Habisreutinger, S.N., Schmidt-Mende, L., et al., 2013. Photocatalytic reduction of CO_2 on TiO_2 and other semiconductors. Angew. Chem. Int. Ed. 52, 7372–7408.

Halmann, M., 1978. Photoelectrochemical reduction of aqueous carbon dioxide on p-type gallium phosphide in liquid junction solar cells. Nature 275, 115–116.

Hara, K., 1995. High efficiency electrochemical reduction of carbon dioxide under high pressure on a gas diffusion electrode containing Pt catalysts. J. Electrochem. Soc. 142, L57.

Hong, J., Zhang, W., et al., 2013. Photocatalytic reduction of CO_2: a brief review on product analysis and systematic methods. Anal. Methods 5, 1086.

Hou, W., Hung, W.H., et al., 2011. Photocatalytic conversion of CO_2 to hydrocarbon fuels via plasmon-enhanced absorption and metallic interband transitions. ACS Catal. 1, 929–936.

Hsu, H.C., Shown, I., et al., 2013. Graphene oxide as a promising photocatalyst for CO_2 to methanol conversion. Nanoscale 5, 262–268.

Huang, X., Cao, T., et al., 2013. Synergistic photoelectrochemical synthesis of formate from CO_2 on $\{12\bar{1}\}$ hierarchical Co_3O_4. J. Phys. Chem. C 117, 26432–26440.

Huynh, M.H., Meyer, T.J., 2007. Proton-coupled electron transfer. Chem. Rev. 107, 5004–5064.

Iizuka, K., Wato, T., et al., 2011. Photocatalytic reduction of carbon dioxide over Ag cocatalyst-loaded $ALa_4Ti_4O_{15}$ (A = Ca, Sr, and Ba) using water as a reducing reagent. J. Am. Chem. Soc. 133, 20863–20868.

Indrakanti, V.P., Kubicki, J.D., et al., 2009. Photoinduced activation of CO_2 on Ti-based heterogeneous catalysts: current state, chemical physics-based insights and outlook. Energ. Environ. Sci. 2, 745.

Inoue, T., Fujishima, A., et al., 1979. Photoelectrocatalytic reduction of carbon dioxide in aqueous suspensions of semiconductor powders. Nature 277, 637–638.

Iwase, A., Yoshino, S., et al., 2016. Water splitting and CO_2 reduction under visible light irradiation using Z-scheme systems consisting of metal sulfides, CoO_x-loaded $BiVO_4$, and a reduced graphene oxide electron mediator. J. Am. Chem. Soc. 138, 10260–10264.

Jang, J.-W., Cho, S., et al., 2014. Aqueous-solution route to zinc telluride films for application to CO_2 reduction. Angew. Chem. Int. Ed. 53, 5852–5857.

Jang, Y.J., Jang, J.-W., et al., 2015. Selective CO production by Au coupled ZnTe/ZnO in the photoelectrochemical CO_2 reduction system. Energ. Environ. Sci. 8, 3597–3604.

Jiao, J., Wei, Y., et al., 2014. Photocatalysts of 3D ordered macroporous TiO_2-supported CeO_2 nanolayers: design, preparation, and their catalytic performances for the reduction of CO_2 with H_2O under simulated solar irradiation. Ind. Eng. Chem. Res. 53, 17345–17354.

Junfu, L., Baozhu, C., 1992. Photoelectrochemical reduction of carbon dioxide on a p^+/p-Si photocathode in aqueous electrolyte. J. Electroanal. Chem. 324, 191–200.

Kang, U., Choi, S.K., et al., 2015. Photosynthesis of formate from CO_2 and water at 1% energy efficiency via copper iron oxide catalysis. Energ. Environ. Sci. 8, 2638–2643.

Kong, Q., Kim, D., et al., 2016. Directed assembly of nanoparticle catalysts on nanowire photoelectrodes for photoelectrochemical CO_2 reduction. Nano Lett. 16, 5675–5680.

Kubacka, A., Fernandez-Garcia, M., et al., 2012. Advanced nanoarchitectures for solar photocatalytic applications. Chem. Rev. 112, 1555–1614.

Kudo, A., Miseki, Y., 2009. Heterogeneous photocatalyst materials for water splitting. Chem. Soc. Rev. 38, 253–278.

Kumar, B., Llorente, M., et al., 2012. Photochemical and photoelectrochemical reduction of CO_2. Annu. Rev. Phys. Chem. 63, 541–569.

Kuriki, R., Matsunaga, H., et al., 2016. Nature-inspired, highly durable CO_2 reduction system consisting of a binuclear ruthenium(II) complex and an organic semiconductor using visible light. J. Am. Chem. Soc. 138, 5159–5170.

Lee, J., Sorescu, D.C., et al., 2011. Electron-induced dissociation of CO_2 on TiO_2(110). J. Am. Chem. Soc. 133, 10066–10069.

Lewis, N.S., Nocera, D.G., 2006. Powering the planet: chemical challenges in solar energy utilization. Proc. Natl. Acad. Sci. U.S.A. 103, 15729–15735.

Li, W., 2010. Electrocatalytic reduction of CO_2 to small organic molecule fuels on metal catalysts. Advances in CO_2 Conversion and Utilization, vol. 1056. American Chemical Society, Washington, DC, pp. 55–76.

Li, G., Ciston, S., et al., 2008. Synthesizing mixed-phase TiO_2 nanocomposites using a hydrothermal method for photo-oxidation and photoreduction applications. J. Catal. 253, 105–110.

Li, H., Lei, Y., et al., 2011. Photocatalytic reduction of carbon dioxide to methanol by Cu_2O/SiC nanocrystallite under visible light irradiation. J. Nat. Gas Chem. 20, 145–150.

Li, P., Ouyang, S., et al., 2012. The effects of crystal structure and electronic structure on photocatalytic H_2 evolution and CO_2 reduction over two phases of perovskite-structured $NaNbO_3$. J. Phys. Chem. C 116, 7621–7628.

Li, P., Xu, H., et al., 2014a. Constructing cubic-orthorhombic surface-phase junctions of $NaNbO_3$ towards significant enhancement of CO_2 photoreduction. J. Mater. Chem. A 2, 5606–5609.

Li, X., Wen, J., et al., 2014b. Design and fabrication of semiconductor photocatalyst for photocatalytic reduction of CO_2 to solar fuel. Sci. China Mater. 57, 70–100.

Li, L., Yan, J., et al., 2015a. Sub-10 nm rutile titanium dioxide nanoparticles for efficient visible-light-driven photocatalytic hydrogen production. Nat. Commun. 6, 5881.

Li, P., Zhou, Y., et al., 2015b. Hexahedron prism-anchored octahedronal CeO_2: crystal facet-based homojunction promoting efficient solar fuel synthesis. J. Am. Chem. Soc. 137, 9547–9550.

Li, M., Li, P., et al., 2016. Design of a photoelectrochemical device for the selective conversion of aqueous CO_2 to CO: using mesoporous palladium-copper bimetallic cathode and hierarchical ZnO-based nanowire array photoanode. Chem. Commun. 52, 8235–8238.

Liang, L., Lei, F., et al., 2015. Single unit cell bismuth tungstate layers realizing robust solar CO_2 reduction to methanol. Angew. Chem. Int. Ed. 54, 13971–13974.

Lin, J., Pan, Z., et al., 2014. Photochemical reduction of CO_2 by graphitic carbon nitride polymers. ACS Sustainable Chem. Eng. 2, 353–358.

Linic, S., Christopher, P., et al., 2011. Plasmonic-metal nanostructures for efficient conversion of solar to chemical energy. Nat. Mater. 10, 911–921.

Liu, Y., Huang, B., et al., 2009. Selective ethanol formation from photocatalytic reduction of carbon dioxide in water with $BiVO_4$ photocatalyst. Catal. Commun. 11, 210–213.

Liu, Q., Zhou, Y., et al., 2010. High-yield synthesis of ultralong and ultrathin Zn_2GeO_4 nanoribbons toward improved photocatalytic reduction of CO_2 into renewable hydrocarbon fuel. J. Am. Chem. Soc. 132, 14385–14387.

Liu, J.-Y., Garg, B., et al., 2011. CuxAgyInzZnkSm solid solutions customized with RuO_2 or $Rh_{1.32}Cr_{0.66}O_3$ co-catalyst display visible light-driven catalytic activity for CO_2 reduction to CH_3OH. Green Chem. 13, 2029.
Liu, L., Zhao, C., et al., 2012a. Spontaneous dissociation of CO_2 to CO on defective surface of Cu(I)/TiO_{2-x} nanoparticles at room temperature. J. Phys. Chem. C 116, 7904–7912.
Liu, L., Zhao, H., et al., 2012b. Photocatalytic CO_2 reduction with H_2O on TiO_2 nanocrystals: comparison of anatase, rutile, and brookite polymorphs and exploration of surface chemistry. ACS Catal. 2, 1817–1828.
Liu, Q., Low, Z.-X., et al., 2013. ZIF-8/Zn_2GeO_4 nanorods with an enhanced CO_2 adsorption property in an aqueous medium for photocatalytic synthesis of liquid fuel. J. Mater. Chem. A 1, 11563.
Liu, X., Inagaki, S., et al., 2016. Heterogeneous molecular systems for photocatalytic CO_2 reduction with water oxidation. Angew. Chem. Int. Ed. 55, 14924–14950.
Ma, Y., Wang, X., et al., 2014. Titanium dioxide-based nanomaterials for photocatalytic fuel generations. Chem. Rev. 114, 9987–10043.
Maeda, K., Domen, K., 2007. New non-oxide photocatalysts designed for overall water splitting under visible light. J. Phys. Chem. C 111, 7851–7861.
Mao, J., Peng, T., et al., 2012. Selective methanol production from photocatalytic reduction of CO_2 on $BiVO_4$ under visible light irradiation. Catal. Commun. 28, 38–41.
Mao, J., Peng, T., et al., 2013. Effect of graphitic carbon nitride microstructures on the activity and selectivity of photocatalytic CO_2 reduction under visible light. Catal. Sci. Technol. 3, 1253.
Marszewski, M., Cao, S., et al., 2015. Semiconductor-based photocatalytic CO_2 conversion. Mater. Horiz. 2, 261–278.
Meng, X., Ouyang, S., et al., 2014. Photocatalytic CO_2 conversion over alkali modified TiO_2 without loading noble metal cocatalyst. Chem. Commun. 50, 11517–11519.
Nguyen, T.-V., Wu, J.C.S., et al., 2008. Photoreduction of CO_2 over ruthenium dye-sensitized TiO_2-based catalysts under concentrated natural sunlight. Catal. Commun. 9, 2073–2076.
Ohtani, B., Ogawa, Y., et al., 1997. Photocatalytic activity of amorphous-anatase mixture of titanium(IV) oxide particles suspended in aqueous solutions. J. Phys. Chem. B 101, 3746.
Ong, W.-J., Tan, L.-L., et al., 2014. Self-assembly of nitrogen-doped TiO_2 with exposed {001} facets on a graphene scaffold as photo-active hybrid nanostructures for reduction of carbon dioxide to methane. Nano Res. 7, 1528–1547.
Palmer, D.A., Van Eldik, R., 1983. The chemistry of metal carbonato and carbon dioxide complexes. Chem. Rev. 83, 651–731.
Ran, J., Zhang, J., et al., 2014. Earth-abundant cocatalysts for semiconductor-based photocatalytic water splitting. Chem. Soc. Rev. 43, 7787–7812.
Richardson, P.L., Perdigoto, M.L.N., et al., 2012. Manganese- and copper-doped titania nanocomposites for the photocatalytic reduction of carbon dioxide into methanol. Appl. Catal. Environ. 126, 200–207.
Rosen, B.A., Salehi-Khojin, A., et al., 2011. Ionic liquid-mediated selective conversion of CO_2 to CO at low overpotentials. Science 334, 643–644.
Sang, L., Zhao, Y., et al., 2014. TiO_2 nanoparticles as functional building blocks. Chem. Rev. 114, 9283–9318.
Sato, S., Arai, T., et al., 2011. Selective CO_2 conversion to formate conjugated with H_2O oxidation utilizing semiconductor/complex hybrid photocatalysts. J. Am. Chem. Soc. 133, 15240–15243.
Schreier, M., Gao, P., et al., 2015. Efficient and selective carbon dioxide reduction on low cost protected Cu_2O photocathodes using a molecular catalyst. Energ. Environ. Sci. 8, 855–861.

Schreier, M., Luo, J., et al., 2016. Covalent immobilization of a molecular catalyst on Cu_2O photocathodes for CO_2 reduction. J. Am. Chem. Soc. 138, 1938–1946.
Schultz, D.M., Yoon, T.P., 2014. Solar synthesis: prospects in visible light photocatalysis. Science 343, 1239176.
Shen, Q., Chen, Z., et al., 2015. High-yield and selective photoelectrocatalytic reduction of CO_2 to formate by metallic copper decorated Co_3O_4 nanotube arrays. Environ. Sci. Technol. 49, 5828–5835.
Shi, H., Chen, G., et al., 2014. Polymeric g-C_3N_4 coupled with $NaNbO_3$ nanowires toward enhanced photocatalytic reduction of CO_2 into renewable fuel. ACS Catal. 4, 3637–3643.
Singh, V., Beltran, I.J., et al., 2014. Photocatalysis deconstructed: design of a new selective catalyst for artificial photosynthesis. Nano Lett. 14, 597–603.
Song, C., 2006. Global challenges and strategies for control, conversion and utilization of CO_2 for sustainable development involving energy, catalysis, adsorption and chemical processing. Catal. Today 115, 2–32.
Song, J.T., Ryoo, H., et al., 2017. Nanoporous Au thin films on Si photoelectrodes for selective and efficient photoelectrochemical CO_2 reduction. Adv. Energy Mater. 7, 1601103.
Surdhar, P.S., Mezyk, S.P., et al., 1989. Reduction potential of the carboxyl radical anion in aqueous solutions. J. Phys. Chem. 93, 3360–3363.
Taniguchi, I., Aurian-Blajeni, B., et al., 1984. The reduction of carbon dioxide at illuminated p-type semiconductor electrodes in nonaqueous media. Electrochim. Acta 29, 923–932.
Truong, Q.D., Le, T.H., et al., 2012. Synthesis of TiO_2 nanoparticles using novel titanium oxalate complex towards visible light-driven photocatalytic reduction of CO_2 to CH_3OH. Appl. Catal. Gen. 437-438, 28–35.
Tsai, C.-W., Chen, H.M., et al., 2011. Ni@NiO core–shell structure-modified nitrogen-doped inTaO_4 for solar-driven highly efficient CO_2 reduction to methanol. J. Phys. Chem. C 115, 10180–10186.
Tu, W., Zhou, Y., et al., 2012. Robust hollow spheres consisting of alternating titania nanosheets and graphene nanosheets with high photocatalytic activity for CO_2 conversion into renewable fuels. Adv. Funct. Mater. 22, 1215–1221.
Tu, W., Zhou, Y., et al., 2014. Photocatalytic conversion of CO_2 into renewable hydrocarbon fuels: state-of-the-art accomplishment, challenges, and prospects. Adv. Mater. 26, 4607–4626.
Wang, T., Gong, J., 2015. Single-crystal semiconductors with narrow band gaps for solar water splitting. Angew. Chem. Int. Ed. 54, 10718–10732.
Wang, S., Wang, X., 2015. Photocatalytic CO_2 reduction by CdS promoted with a zeolitic imidazolate framework. Appl. Catal. Environ. 162, 494–500.
Wang, C., Thompson, R.L., et al., 2010. Visible light photoreduction of CO_2 using CdSe/Pt/TiO_2 heterostructured catalysts. J. Phys. Chem. Lett. 1, 48–53.
Wang, W.N., An, W.J., et al., 2012. Size and structure matter: enhanced CO_2 photoreduction efficiency by size-resolved ultrafine Pt nanoparticles on TiO_2 single crystals. J. Am. Chem. Soc. 134, 11276–11281.
Wang, T., Luo, Z., et al., 2014a. Controllable fabrication of nanostructured materials for photoelectrochemical water splitting via atomic layer deposition. Chem. Soc. Rev. 43, 7469–7484.
Wang, Z., Liu, Y., et al., 2014b. Progress on extending the light absorption spectra of photocatalysts. Phys. Chem. Chem. Phys. 16, 2758–2774.
White, J.L., Baruch, M.F., et al., 2015. Light-driven heterogeneous reduction of carbon dioxide: photocatalysts and photoelectrodes. Chem. Rev. 115, 12888–12935.
Willkomm, J., Orchard, K.L., et al., 2016. Dye-sensitised semiconductors modified with molecular catalysts for light-driven H_2 production. Chem. Soc. Rev. 45, 9–23.
Woolerton, T.W., Sheard, S., et al., 2010. Efficient and clean photoreduction of CO_2 to CO by enzyme-modified TiO_2 nanoparticles using visible light. J. Am. Chem. Soc. 132, 2132–2133.

Woolerton, T.W., Sheard, S., et al., 2011. CO_2 photoreduction at enzyme-modified metal oxide nanoparticles. Energ. Environ. Sci. 4, 2393.

Xie, K., Umezawa, N., et al., 2011. Self-doped $SrTiO_{3-\delta}$ photocatalyst with enhanced activity for artificial photosynthesis under visible light. Energ. Environ. Sci. 4, 4211.

Xie, S., Wang, Y., et al., 2013. Photocatalytic reduction of CO_2 with H_2O: significant enhancement of the activity of Pt-TiO_2 in CH_4 formation by addition of MgO. Chem. Commun. 49, 2451–2453.

Xie, S., Zhang, Q., et al., 2016. Photocatalytic and photoelectrocatalytic reduction of CO_2 using heterogeneous catalysts with controlled nanostructures. Chem. Commun. 52, 35–59.

Xu, H., Ouyang, S., et al., 2013. High-active anatase TiO_2 nanosheets exposed with 95% {100} facets toward efficient H_2 evolution and CO_2 photoreduction. ACS Appl. Mater. Interfaces 5, 1348–1354.

Yan, S., Wang, J., et al., 2013. Zinc gallogermanate solid solution: a novel photocatalyst for efficiently converting CO_2 into solar fuels. Adv. Funct. Mater. 23, 1839–1845.

Yan, J., Wang, T., et al., 2015. Tungsten oxide single crystal nanosheets for enhanced multichannel solar light harvesting. Adv. Mater. 27, 1580–1586.

Yin, S., Han, J., et al., 2015. Recent progress in g-C_3N_4 based low cost photocatalytic system: activity enhancement and emerging applications. Cat. Sci. Technol. 5, 5048–5061.

Yu, J., Jin, J., et al., 2014a. A noble metal-free reduced graphene oxide–CdS nanorod composite for the enhanced visible-light photocatalytic reduction of CO_2 to solar fuel. J. Mater. Chem. A 2, 3407.

Yu, J., Low, J., et al., 2014b. Enhanced photocatalytic CO_2-reduction activity of anatase TiO_2 by coexposed {001} and {101} facets. J. Am. Chem. Soc. 136, 8839–8842.

Yu, J., Wang, K., et al., 2014c. Photocatalytic reduction of CO_2 into hydrocarbon solar fuels over g-C_3N_4-Pt nanocomposite photocatalysts. Phys. Chem. Chem. Phys. 16, 11492–11501.

Yu, L., Li, G., et al., 2016. Enhanced activity and stability of carbon-decorated cuprous oxide mesoporous nanorods for CO_2 reduction in artificial photosynthesis. ACS Catal. 6, 6444–6454.

Zeng, G., Qiu, J., et al., 2015. Enhanced photocatalytic reduction of CO_2 to CO through TiO_2 passivation of InP in ionic liquids. Chem. A Eur. J. 21, 13502–13507.

Zhai, Q., Xie, S., et al., 2013. Photocatalytic conversion of carbon dioxide with water into methane: platinum and copper(I) oxide co-catalysts with a core-shell structure. Angew. Chem. Int. Ed. 52, 5776–5779.

Zhang, Q., Li, Y., et al., 2011. Visible light responsive iodine-doped TiO_2 for photocatalytic reduction of CO_2 to fuels. Appl. Catal. Gen. 400, 195–202.

Zhang, N., Ouyang, S., et al., 2012a. Mesoporous zinc germanium oxynitride for CO_2 photoreduction under visible light. Chem. Commun. 48, 1269–1271.

Zhang, X., Han, F., et al., 2012b. Photocatalytic conversion of diluted CO_2 into light hydrocarbons using periodically modulated multiwalled nanotube arrays. Angew. Chem. Int. Ed. 51, 12732–12735.

Zhang, Z., Wang, Z., et al., 2013. Au/Pt nanoparticle-decorated TiO_2 nanofibers with plasmon-enhanced photocatalytic activities for solar-to-fuel conversion. J. Phys. Chem. C 117, 25939–25947.

Zhang, P., Wang, T., et al., 2016. Effective charge carrier utilization in photocatalytic conversions. Acc. Chem. Res. 49, 911–921.

INDEX

Note: Page numbers followed by "*f*" indicate figures, "*t*" indicate tables, and "*s*" indicate schemes.

D

E

F

G

H

Q

R

S

T

U

V

W

Z

MIX
Papier | Fördert gute Waldnutzung
FSC® C083411

Kolibri 360 GmbH
Ferdinand-Jühlke-Straße 7
99095 Erfurt, Deutschland
produktsicherheit@kolibri360.de